THE ROUTLEDGE COMPANION TO MEDIA TECHNOLOGY AND OBSOLESCENCE

Edited by
Mark J. P. Wolf

Routledge
Taylor & Francis Group
NEW YORK AND LONDON

First published 2019
by Routledge
52 Vanderbilt Avenue, New York, NY 10017

and by Routledge
2 Park Square, Milton Park, Abingdon, Oxon OX14 4RN

Routledge is an imprint of the Taylor & Francis Group, an informa business

Library of Congress Cataloging in Publication Data
Names: Wolf, Mark J. P., editor.
Title: The Routledge companion to media technology and obsolescence / edited by Mark J.P. Wolf.
Description: New York : Routledge/Taylor & Francis Group, 2019. | Includes bibliographical references and index.
Identifiers: LCCN 2018032532 |
Subjects: LCSH: Analog electronic systems—History. | Mathematical instruments—History. | Information retrieval—Equipment and supplies—History. | Product obsolescence.
Classification: LCC TK7809 .R68 2019 | DDC 609—dc23
LC record available at https://lccn.loc.gov/2018032532

ISBN: 978-1-138-21626-6 (hbk)
ISBN: 978-1-315-44268-6 (ebk)

Typeset in Bembo
by Swales & Willis Ltd, Exeter, Devon, UK

Printed and bound in Great Britain by
TJ International Ltd, Padstow, Cornwall

CONTENTS

ABOUT THE CONTRIBUTORS

Meredith A. Bak is an Assistant Professor in the Department of Childhood Studies at Rutgers University-Camden. She holds a Ph.D. in Film and Media Studies from the University of California, Santa Barbara. Her research concerns the relationship between children and new media from the 19th century to the present. She writes on historical and contemporary children's media, toys, visual, and material culture, and her work appears in *Early Popular Visual Culture*, *Film History*, *The Moving Image*, and *The Velvet Light Trap*. She is working on a book manuscript about the role of pre-cinematic visual media from optical toys to early pop-up books in shaping children as media spectators. A second project in development considers the history and theory of animate toys from talking dolls to augmented reality apps. [meredithabak@gmail.com]

Paul Benzon is a Visiting Assistant Professor in the Department of English at Skidmore College. His work has appeared in electronic book review, *Media-N*, *PMLA* (where it won the William Riley Parker Prize for an Outstanding Article in 2010), and *Narrative* (where it received the James Phelan Prize for the Best Contribution to *Narrative* in 2013), and is forthcoming in *The Routledge Companion to Media Studies and Digital Humanities*. He is currently at work on a project entitled *A Partial History of Deletion: Absence, Obsolescence, and the Ends of Media*. [pbenzon@skidmore.edu]

Peer D. Bode is an internationally exhibiting new media artist. He is Professor of Video Arts, co-founder and co-director of the Institute for Electronic Arts (IEA), School of Art and Design, NYSCC at Alfred University. As a new media arts practitioner and educator, Bode harnesses historical and emerging new media technologies in a reflexive investigation that explores and pushes their historical, technical, and semiotic conditions. Bode's work has been shown and collected worldwide, including the Whitney Biennial, the Museum of Modern Art, and the Video Data Bank's video anthology *Surveying the First Decade: Video Art and Alternative Media in the U.S. 1968–1980* (2008), curated by Chris Hill. Peer Bode has collaborated on numerous electronic tool building projects with video systems designer David Jones and artist Ralph Hocking. Bode is a member of the Carrier Band with Andrew Deutsch, Steven Vitiello, and Rebekkah Palov. [peerbode@hotmail.com]

M. M. Chandler holds a Ph.D. in Visual Studies from the University of California, Irvine. Her dissertation offers a critical reconsideration of contemporary digital film preservation through a material history of celluloid acetate plastics and film stock. Her work has appeared in the journals *Visual Culture & Gender* and *Porn Studies*, as well as the anthologies *Theorizing Visual Studies: Writing Through the Discipline* (2012) and *Dolls Studies: The Many Meanings of Girls' Toys and Play* (2015). Currently, she is an Adjunct Professor in Art History at Santa Monica College and Film Studies at Irvine Valley College. [meghanmchandler@gmail.com]

Jason Curtis is a medical librarian at Shrewsbury and Telford Hospital NHS Trust, UK. He began collecting examples of physical media formats in 2006, and subsequently created the Museum of Obsolete Media website (www.obsoletemedia.org) in 2013, that now documents over 525 collected formats spanning audio, video, film, and data storage. [webmaster@obsoletemedia.org]

Anne C. Deger is an instructor in the Communications and Media Studies Department at Fordham University. Her research interests include digital cultural studies, Internet history, and on-line community formation. She is currently completing her Ph.D. dissertation, a cultural history of AutoCorrect, at Stony Brook University. [acdeger@gmail.com]

Joshua Greenberg directs the Digital Information Technology program at the Alfred P. Sloan Foundation, and is the author of *From Betamax to Blockbuster* (2008). He holds a Ph.D. in Science and Technology Studies from Cornell University. [epistemographer@gmail.com]

David Hochfelder is Associate Professor of History at the University at Albany, SUNY. He is the author of *The Telegraph in America: 1832–1920* (2012). He is researching a book on the history of thrift, tentatively titled *Thrift in America: From Franklin to the Great Recession*. He is also collaborating with two other historians on a digital history project about urban renewal in Albany, New York. Preliminary results can be found at https://98acresinalbany.wordpress.com/. [dhochfelder@albany.edu]

Peter M. Hopp C. Eng. MBCS CITP, is a retired Chartered Electronic Engineer by training, Computer Systems Engineer by conversion, and former Business Development Manager by preference. He has been collecting, researching, and writing about slide rules and their industrial archaeology for about 30 years. His first book: *Slide Rules, Their History, Models and Makers* was published in 1999 and subsequently he has had published two further books on specialist rules: *2-foot, 2-fold Slide Rules* (2009) and *Pocket-Watch Slide Rules* (2011). He regularly writes for the various specialist slide rule and scientific instrument collectors' organizations and has lectured on slide rules and other scientific instruments and their history. Being one of the dwindling band of engineers who has used a slide rule in anger during his career in computers and computer systems gives him the ability to look at these delightful devices with a unique perspective when their demise is viewed due to the advent of the Digital Revolution. [peterhopp678@btinternet.com]

Matthew Kirschenbaum is Professor of English and Director of the Graduate Certificate in Digital Studies at the University of Maryland. His most recent book is *Track Changes: A Literary History of Word Processing* (2016). He still keeps his floppies in a shoebox. [mkirschenbaum@gmail.com]

Peter Krapp is Professor and Chair of Film & Media Studies at the University of California, Irvine, where he is also a member of the English Department and of the Department of Informatics. He is the author of *Déjà Vu: Aberrations of Cultural Memory* (2004) and of *Noise Channels: Glitch and Error in Digital Culture* (2011), as well as an editor of the *Handbook Language–Culture–Communication* (2016), and of *Medium Cool* (2002). [krapp@uci.edu]

Lori Landay is Professor of Cultural Studies at Berklee College of Music, and an interdisciplinary scholar and new media artist exploring the making of visual meaning in 20th- and 21st-century culture. She is the author of two books, *I Love Lucy* (2010) and *Madcaps, Screwballs, and Con Women: The Female Trickster in American Culture* (1998), and articles on virtual worlds, digital narrative, silent film, television culture, and other topics. Her creative work includes animation, graphic design, creative documentary, machinima, interactive virtual art installations, and music video. Her current project combines critical and creative work to explore subjectivity, presence, and the "virtual kino-eye" in interactive media, continuing the inquiry begun during her NEH Enduring Questions Grant for "What is Being?" in 2010–2012. [llanday@berklee.edu]

Gary Locklair is Professor and Chair of Computer Science at Concordia University Wisconsin, where he has served since 1986. He worked in the computer industry for a number of years prior to teaching, most notably at Hewlett-Packard where he was a program manager for LaserJet printers. He earned his B.S., M.S., and Ph.D. degrees in Computer Science. Prof. Locklair teaches both undergraduate and graduate classes in Computer Science and Information Technology at Concordia University Wisconsin. He has developed programs and courses outside of his major field and has taught for a variety of schools at CUW (Business, Education, Human Services, Arts and Sciences). His professional interests include software engineering, artificial intelligence, and theoretical foundations of computing. Prof. Locklair is listed in several editions of *Who's Who* and is a member of Upsilon Pi Epsilon, the computer science honor society, and the Alpha Chi honor society. At CUW he has been honored as the Advisor of the Year, the outstanding graduate faculty member, and the Faculty Laureate. He and his wife, Karen, have five children and seven grandchildren. They enjoy living on a small hobby farm with rescue animals. Occasionally they drive their ZR-1 Corvette for pleasure. [gary.locklair@cuw.edu]

Ken S. McAllister is Professor of Rhetoric and the College of Humanities Associate Dean of Research and Program Innovation at the University of Arizona. [mesmer@email.arizona.edu]

Amanda McQueen completed her Ph.D. in Film Studies at the University of Wisconsin-Madison in 2016. Her primary areas of research include the history of media industries, particularly Hollywood; sound studies; and the musical genre on stage and screen. Between 2012 and 2016, she assisted on the Wisconsin Nitrate Film Project, an interdisciplinary grant project funded by the National Endowment for the Humanities. She currently works as a Faculty Assistant at UW-Madison. [almcqueen@wisc.edu]

Stephen Mamber is a Professor in the UCLA Department of Film, Television, and Digital Media. He is the author of *Cinema Verite in America: Studies in Uncontrolled Documentary* (1974) and numerous articles about film and new media. He is also the author of three iPad apps: "Clipnotes", an annotation and presentation tool, "Who Shot Liberty Valance?", "7 Thursdays: Looking at Brief Encounter", and "The Seventh Race: Kubrick at the Starting Gate". All are available for free at the iOS app store. [smamber@ucla.edu]

Sheila C. Murphy is the author of *How Television Invented New Media* (2011). She teaches Digital Studies and Media Studies as an Associate Professor in Screen Arts & Cultures and Digital Studies at the University of Michigan. [scmurphy@umich.edu]

Richard Osborne is Senior Lecturer in Popular Music at Middlesex University. Prior to becoming a lecturer he worked in record shops, held various posts at PRS for Music and co-managed a pub. His blog on popular music is available at: http://richardosbornevinyl.blogspot.co.uk. His book, *Vinyl: A History of the Analogue Record*, was published by Ashgate in 2012. [r.osborne@mdx.ac.uk]

John Reid Perkins-Buzo is interested in the areas where technologies of the past and present meet the arts of the moving image (animation, cinema, and gaming). He has advanced degrees in both the fine arts and applied mathematics/computer science. He continues to develop software for moving image projects, including for a variety of wearable devices. He is Assistant Professor of Digital Media in the Radio-TV-Digital Media Department at Southern Illinois University, Carbondale. [reidop@me.com]

Richard Polt is Professor of Philosophy at Xavier University in Cincinnati. His publications include *Heidegger: An Introduction* (1999) and *The Typewriter Revolution: A Typist's Companion for the 21st Century* (2015). He is the editor of *ETCetera*, the magazine of the Early Typewriter Collectors' Association. [polt@xavier.edu]

Judd Ethan Ruggill is an Associate Professor of Computational Media at the University of Arizona and an inveterate (albeit accidental) media archivist. [jruggill@email.arizona.edu]

Paige Sarlin is a filmmaker, scholar, and political activist. Her first film, *The Last Slide Projector*, premiered at the Rotterdam International Film Festival in 2007. Her writing has appeared in *October*, *Re-Thinking Marxism*, *Afterimage*, *Reviews in Cultural Theory*, and *Framework: A Journal of Film and Culture*. She is an Assistant Professor in the Department of Media Study at the University at Buffalo, SUNY. [p.sarlin@gmail.com]

Matt Soar is an intermedia artist and filmmaker teaching in the Department of Communication Studies at Concordia University, Montreal. He has written extensively about graphic design, visual communication, and cultural production for scholarly and professional audiences. Soar's project *Lost Leaders* (2011 onwards) is an artistic, archival, and scholarly exploration of the histories of U.S. film leader standards. Creative outcomes have screened at ATA/Other Cinema (2015), the Poetics & Politics documentary symposium (UCSC 2015), the Montreal Underground Film Festival (2014), the Orphan Film Symposium (Amsterdam 2014, Culpeper VA 2016), and FUSE #2: Mobile Interactive Microcinema (Ann Arbor, 2014). His article "The Beginnings and the Ends of Film: Leader standardization in the US and Canada (1930–1999)" appears in the Fall 2016 issue of *The Moving Image*. Soar is co-founder and director of the Montreal Signs Project, a growing collection of commercial and civic signs from around the city. In 2017, he co-curated, with Danica Evering, *YMX: Migration, Land, and Loss after Mirabel*, artist Cheryl Sim's media installation exploring the legacy of the Montréal-Mirabel International Airport. [mattsoar@gmail.com]

Braxton Soderman is an Assistant Professor in the Department of Film & Media Studies at the University of California, Irvine. He researches digital media, video games, new media aesthetics, the history of technology, and critical theory. He has published articles in *Journal of Visual Culture, Space and Culture, differences: A Journal of Feminist Cultural Studies, Dichtung Digital: A Journal of Art and Culture in Digital Media, Transformative Works and Cultures*, and elsewhere. [asoderma@uci.edu]

Brent Strang is a Lecturer at Fordham University and a doctoral candidate with a Women's and Gender Studies certificate in Stony Brook University's Cultural Analysis and Theory Program. His dissertation is a media history of the remote control device. He has published in *Journal of Visual Culture, Cinephile*, and has two chapters in edited collections: *The Western in the Global South* (2015) and *Race, Gender and Sexuality in Post-Apocalyptic Film* (2015). [tardycannon@gmail.com]

Michael Thomasson is one of the most widely respected video game historians in the field today. He currently teaches college-level video game history, design, and graphics courses and, in 2014, *The Guinness Book of World Records* declared that Thomasson had "The Largest Video Game Collection" in the world. For television, Michael conducted research for MTV's video game-related program *Video MODS*. In print, Michael authored *Downright Bizarre Games: Videogames that Crossed the Line* (2016), and has contributed to nearly a dozen texts including the publication by the inventor of home video games, *Videogames: In the Beginning* (2005) by Ralph H. Baer; *Bill Kunkel's Confessions of the Game Doctor* (2005); both volumes of the *Encyclopedia of Video Games* (2012), and more. His historical columns have been distributed worldwide in newspapers and magazines. Michael has written business plans for several video game vendors and managed almost a dozen game-related retail stores spanning three decades. He has also contributed to or published dozens of games for several consoles, such as the SEGA CD, Colecovision, 3DO, Intellivision, Vectrex, CD-i, and several Atari platforms. Michael's classic gaming business also sponsors retro-gaming tradeshows and expos across the United States and Canada. Mr. Thomasson, his wife JoAnn, and daughter Anna reside in New York. His website is www.GoodDealGames.com. [service@gooddealgames.com]

Robert S. Wahl is an Associate Professor of Computer Science at Concordia University Wisconsin. He has an M.S. in Computer Science from North Central College and a Ph.D. in Information Technology from Capella University specializing in information assurance and security. In addition to computer security, he is interested in computer animation and databases. He started his journey with computers using punched cards to write his first programs. [robert.wahl@cuw.edu]

Mark J. P. Wolf is a Professor in the Communication Department at Concordia University Wisconsin. He has a B.A. (1990) in Film Production and an M.A. (1992) and Ph.D. (1995) in Critical Studies from the School of Cinema/Television (now renamed the School of Cinematic Arts) at the University of Southern California. His books include

Abstracting Reality: Art, Communication, and Cognition in the Digital Age (2000), *The Medium of the Video Game* (2001), *Virtual Morality: Morals, Ethics, and New Media* (2003), *The Video Game Theory Reader* (2003), *The Video Game Explosion: A History from PONG to PlayStation and Beyond* (2007), *The Video Game Theory Reader 2* (2008), *Myst and Riven: The World of the D'ni* (2011), *Before the Crash: Early Video Game History* (2012), the two-volume *Encyclopedia of Video Games: The Culture, Technology, and Art of Gaming* (2012), *Building Imaginary Worlds: The Theory and History of Subcreation* (2012), *The Routledge Companion to Video Game Studies* (2014), *LEGO Studies: Examining the Building Blocks of a Transmedial Phenomenon* (2014), *Video Games Around the World* (2015), the four-volume *Video Games and Gaming Cultures* (2016), *Revisiting Imaginary Worlds: A Subcreation Studies Anthology* (2017), *Video Games FAQ* (2017), *The World of Mister Rogers' Neighborhood* (2017), *The Routledge Companion to Imaginary Worlds* (2017), and two novels for which his agent is looking for a publisher. He is also founder and coeditor of the Landmark Video Game book series from University of Michigan Press, the founder and editor of the Imaginary Worlds book series from Routledge, and the founder of the Video Game Studies Scholarly Interest Group and the Transmedia Studies Special Interest Group within the Society of Cinema and Media Studies. He has been invited to speak in North America, South America, Europe, Asia, and Second Life; has had work published in journals including *Compar(a)ison*, *Convergence*, *Film Quarterly*, *Games and Culture*, *New Review of Film and Television Studies*, *Projections*, and *The Velvet Light Trap*; is on the advisory boards of Videotopia, the International Arcade Museum Library, and the *International Journal of Gaming and Computer-Mediated Simulations*; and is on several editorial boards including those of *Games and Culture* and *The Journal of E-media Studies*. He lives in Wisconsin with his wife Diane and his sons Michael, Christian, and Francis. [mark.wolf@cuw.edu]

Josh Zimmerman is a new media scholar specializing in technology-enhanced community relations and a Learning Consultant with TTG, Inc. [joshuazim@gmail.com]

PREFACE

Snow, on the television screen, after the end of the broadcast day. The radiant glow of projected Kodachrome imagery on a silver lenticular screen. The high-speed *clickety-clack* of a 16mm film projector, or the *tack-tack-tack* of typewriter keys hitting the page. The smell of warm Bakelite and the humming fan of an Airequipt slide projector operating. The looping, scratchy crackle of a record player needle after the record has ended. The fresh scent of newly printed purple mimeograph ink. The whirring crunch of a disk drive searching for data. The cacophonic ambience of a video game arcade. The buzzing, beeping, and dinging tones of a modem receiving data. All of these sensations, and many more, are now largely things of the past; memories and nostalgic recollections that my generation might be the last to have experienced as a part of normal daily life. Digital technology, for all its advantages, will never be able to accurately reproduce the sights, sounds, smells, and feel of yesterday's analog technologies.

The idea for this book began in 2015. For Valentine's Day that year, my wife bought us an HDTV television set, to replace the NTSC cathode-ray tube television set that we had had since 2001, which had been a wedding gift. But having the HDTV set sitting in its box in the family room, waiting to replace the CRT, somehow made me realize the nostalgia I had for the old television technology, which was of course the kind I had when I was growing up. I realized it would probably be the last CRT we would ever own, and I became a bit loath to give it up. Ever-thankful for my wife's patience with me, it wasn't until October that we finally switched over to the HDTV, and of course I enjoy the high-resolution imagery and flat screen technology as much as anyone else. The old CRT now sits in the basement, my wife once again patiently waiting for me to get rid of it entirely, as its bulky shape takes up space in the playroom.

My nostalgia for the CRT led me to consider writing about it, along with the disappearance of so many different technologies that I had known in my childhood, an idea that grew into a proposal for an anthology I planned to call *Analog Sunset: On the Impending Demise of Vanishing Media*. Not only did Routledge like the idea for the anthology, they asked me to expand it and turn into *The Routledge Companion to Media Technology and Obsolescence*. This allowed me to add essays on some obsolete digital technologies, a reminder that obsolescence is not limited to analog media; in fact, anyone who fondly remembers using home computers of the 1970s and 1980s will note that digital media artifacts often become obsolete even faster than analog media ones. All these media, however, comprise a technological milieu that has largely passed away.

Analog Sunset

The rapid advancements in technology mean that our children's childhoods and lives will be very different from our own, widening the technological generation gap; but there is even more at stake here. The technological developments going on all add up to a much larger-scale changeover, which began slowly in the latter half of the 20th century, gradually picking up speed; the change from analog technology to digital technology. This involves not just new technologies but a whole new lifestyle, with a different set of metaphors ordering society, a paradigm shift as great as that separating the industrial age from the agricultural age that preceded it.

As well as evoking a sheen of nostalgia, the term "Analog Sunset" describes the end of the production of technology with analog video inputs and comes from section 2.2.2.1 of the "Advanced Access Content System ("AACS") Adopter Agreement"[1] published in June 2009 by the Advance Access Content System Licensing Administrator LLC (AACS LA). The AACS LA is a consortium organized by IBM, Intel, Microsoft, Panasonic, Sony, Toshiba, The Walt Disney Company, and Warner Bros. to develop the AACS, which they describe as "a specification for managing content stored on the next generation of prerecorded and recorded optical media for consumer use with PCs and CE devices".[2] The AACS was designed for the protection of copyrighted material against illegal copying and distribution, and the consortium determines not only new systems to be used, but also the ending, or "sunset" of older systems and technologies. Thus, while some technologies decline and fall out of use simply because newer or (supposedly) better technologies become available, and especially ones that are cheaper than the technologies they replace, other technologies are deliberately designated as obsolete by the companies and institutions that used to produce and support them.

Of course, the idea of declaring something obsolete for the purpose of selling newer things to consumers goes back to 1924 and the "dynamic obsolescence" of Alfred P. Sloan Jr., head of General Motors (his critics referred to it as "planned obsolescence", and so it has been remembered to this day). Likewise, companies can also retain a tight grip on older technologies that are well-entrenched, rejecting a move to something provably better; consider RCA's ruthless rejection and suppression of Howard Armstrong's FM radio, which RCA saw as a threat to its radio networks and production of AM radios. And sometimes in the marketplace, for a variety of reasons, the best technology does not win out over its inferior competitors; consider the battle between VHS and Beta videotapes, which VHS won. The rise and fall of technologies are complex and overdetermined things, difficult to analyze, and require a great deal of historical context. The essays in this volume admirably take on this challenge, giving an overview of their chosen technologies, and for the sake of some consistency across the essays of the book, contributors were urged to include the following: a description of the technology and its purpose; a brief history of the technology leading up to its peak; an analysis of its decline and descent into obsolescence, and the reasons for its decline or demise; its influence on the technology that is replacing it; discussion of the trade-offs involved in replacing it, what was gained, and more importantly, what might be lost due to the transition; and finally, the transformation of the dying technology into a luxury item, cult object, collectible, object of nostalgia, or curiosity; the continuing legacy that the technology leaves behind.

Types of Obsolescence

The essays are organized chronologically, according to their subject's time of appearance. This was done not only because earlier technologies often influence the form of later ones, but also because many technologies never completely die out; for example, a quick search on the Internet reveals that people are still making Daguerreotypes today, even though the technology nearly died out after 1860, when better photographic techniques became available. Thus, we can demarcate several different types or stages of obsolescence. The first is when a technology is no longer the dominant one of its kind, when it has functionally been replaced by something which is hailed as better (whether or not it might actually be so); for example, when LCD and LED screen technologies brought an end to the dominance of the cathode-ray tube, or when Windows 95 replaced MS-DOS-based Windows 3.1, or when the confusingly-named Xbox One replaced the Xbox 360. Such shifts are usually the result of new or improved technologies, coupled with corporate production and marketing, requiring consumers to buy in and adopt to a great enough degree that the dominant technology changes.

The second type of obsolescence is when a technology is no longer mass-produced for everyday consumers; it is no longer just a less-popular choice in the marketplace, but something that is considered unprofitable to continue to produce or for stores to carry at all. Some technologies have enough of a niche that they might not reach the second stage for some time, even after reaching the first; AM radio, for example, lost its dominance to FM, but still remains a viable medium. But many technologies lose their viability along with their dominance, find themselves unable to compete, and thus go out of production. At this point, a technology becomes difficult to find for those still using it; one might have to resort to buying used items, or find unsold inventories still lingering in back storerooms. Such a stage of obsolescence might come as a surprise, especially if it does so rapidly, turning what was once a commonly found object into a rare item.

This can lead to the third stage, when a technology's rarity, or even nostalgia for the technology, make it collectible, and people begin competing for the acquisition of what remains of it. Occasionally, if there is interest in the technology beyond just being a collectible, it might even spark a niche market, becoming a specialty or even luxury item; no longer mass-produced, it will likely be more expensive than when it was dominant, but there will be a small audience still willing to pay for it. Such resurgences can be minimal (as in the Daguerreotype example given earlier), or might be surprising widespread, like the comeback made by vinyl records in recent years.

Finally, the last stage of obsolescence is when a technology almost completely disappears, leaving only stories about it, and perhaps a few extant artifacts (naturally, if a technology *completely* vanished without any trace, we would no longer even know about it). At this point, such a technology is truly a collectible or museum piece, valued only for its rarity, and perhaps used very little in order to keep its condition intact. In the most extreme cases, a lost technology might only have left us a few puzzling remains, allowing speculation as to what it might have been, like the gears of the mysterious Antikythera mechanism of which we have only a few remains from the ancient world. Despite greater interest in preservation and museums, some technologies, even more recent ones, still have managed to slip into extreme obscurity and rarity.

But there is still much technological history to be brought to light, and I have had the pleasure of working with many of the finest scholars writing about declining and obsolete technologies, whose research and writings have certainly broadened and enriched my own knowledge of the field. Together, the essays in this volume reveal a number of themes and issues regarding technology, such as how technologies become obsolete and the many factors contributing to their demise; shifts in the way technologies are used and how this affects their longevity; how technology, culture, and society mutually shape each other; how the status of vanishing technology changes what we think of it; and why certain technologies become objects of nostalgia and enjoy comebacks, while other technologies do not.

As mentioned above, the essays are arranged chronologically according to each technology's appearance, and each tracks a technology's growth and decline, and its cultural legacy which continues today.

Following the essays is a Timeline of Obsolescence, which charts the disappearance of a number of technologies over the last hundred years or so. Seeing all these dates, many of which mark the discontinuations and departures of technologies, makes one realize just how quickly the world is changing, and also that such changes are accelerating, a feeling that many of the individual essays also convey.

Naturally, a single volume can give only a sampling of the many topics and ideas through which media technology, even obsolete media, can be considered and studied.[3] Hopefully readers will find the *Companion* useful and interesting, as well as a point of departure for further research, and perhaps even an influence on how we think of new and emerging technologies, and the place they should have in society and culture.

Notes

1 See AACS LA, "Advanced Access Content System ('AACS') Adopter Agreement", June 2009, available at www.aacsla.com/license/AACS_Adopter_Agrmt_090605.pdf.
2 According to the AACS website overview, available at www.aacsla.com/what/overview.
3 For example, I would have loved to include essays on hand-cranked motion picture cameras, cassette tapes, dial telephones, 8mm and super-8 film formats, specific brands of early home computers, and other obsolete media technologies.

ACKNOWLEDGMENTS

A *Companion* like this is possible with the contributions of readers and scholars whose interest and desire bring about the study of media technology and obsolescence. Therefore, great thanks go out to all of the contributors to this volume, who graciously agreed to join in this endeavor: Meredith A. Bak, Paul Benzon, Peer D. Bode, M. M. Chandler, Jason Curtis, Anne C. Deger, Joshua Greenberg, David Hochfelder, Peter M. Hopp, Matthew Kirschenbaum, Peter Krapp, Lori Landay, Gary Locklair, Stephen Mamber, Ken S. McAllister, Amanda McQueen, Sheila C. Murphy, Richard Osborne, John Reid Perkins-Buzo, Richard Polt, Judd Ethan Ruggill, Paige Sarlin, Matt Soar, Braxton Soderman, Brent Strang, Michael Thomasson, Robert S. Wahl, and Josh Zimmerman. Thanks also to all those who gave suggestions for this volume; to Erica Wetter and Routledge Press for accepting the book, asking to turn it into a *Routledge Companion* (the third one that I have edited), and supporting it along the way; and to all those who will use it in the classroom and elsewhere. Finally, I also must thank my wife Diane Wolf and sons Michael, Christian, and Francis who were patient with the time taken to work on this book. And, as always, thanks be to God.

1

PAPER SLIPS

The Long Reign of the Index Card and Card Catalog

Peter Krapp

Before fading toward obsolescence, the index card and card catalog had developed into an influential technology of knowledge management and discovery: a mere clutch of paper scraps deployed to great effect not only in libraries but in academic research and in offices, for business and creative pursuits alike, permitting storage, processing, and transmission of data in discrete, mobile, uniform chunks that can be rearranged according to various principles.

Yet, is this range of applications for index cards completely obsolete? Certainly the index card as an informative object has faded in importance, and while you can still find purveyors of normed index cards among stationery or school and business supplies, it is a safe assumption that librarians, office managers, and writers no longer rely much on index cards, despite the fact that the card catalog long reigned supreme in those information environments. Few students today cram vocabulary, for instance, or formulae with index cards, yet a certain type of hipster will proudly own a piece of furniture originally designed for a library card catalog. However, while the object as such might have faded, arguably the affordances of a card index have not. Few among us maintain our own system of cross-references among browser bookmarks, recipe collections, metadata for CDs ripped to our gadgets, or any other sort of data collection, yet most of us have grown accustomed to associative indexing, from Amazon's reading suggestions based on your past browsing to streaming music service recommendations.

Certainly under the conditions of hypertext, as manifested across networked computers, the storing, processing, and transmitting of data (business data, library data, audio recording metadata, etc.) allows for a kind of serendipitous discovery of correlations and cross-references that were one strength of index cards, as valued by generations of writers, artists, and academics. One might say the card index lives on in a number of related formats: from hypercard stacks as introduced by Apple—maintained from 1987 to 2004 as a multimedia programming environment, for CD-ROM interactive content and games like *Myst* (1993)—to the generalized footnote we now call hypertext, and even to the ubiquitous slide decks, be they collated in PowerPoint or Keynote or Prezi. Each of these media exhibits features of what made index cards a success for centuries.

From Library Catalogs to Accounting and Business

A scholar is only a librarian's way of creating another scholar.

—Daniel Dennett[1]

Establishing origins is, so often, hazardous terrain. A British historian of science, Staffan Mueller-Wille at the Centre for Medical History at the University of Exeter, recently claimed that Swedish natural scientist Carl Linnaeus (1707–1778), the father of modern taxonomy, had "invented" the card index to manage his information storage and retrieval. Working with paper slips that could be shuffled, updated, and sorted according to different criteria, Linnaeus certainly helped change the understanding of the natural world, away from linear filiation models and toward networks of characteristics that could be mapped.[2] Despite such claims, one can find index card systems that predate Linnaeus.

At the end of the 17th century, a comparison of techniques for excerpting led the German lawyer and librarian Vincent Placcius (1642–1699) to develop a "learned box" to enable the relational manipulation of notes.[3] German polymath Gottfried Wilhelm Leibniz (1646–1716) was able to buy such a piece of furniture to accommodate his paper slips in 1676.[4] And in the 16th century, Swiss doctor Conrad Gessner (1516–1565) reflected openly on how to generate and copy excerpts for a register, although then paper slips were usually threaded together.[5] For rhetorical memory, it was preferable not to work with loose sheets, as this could imperil the entire project if their positions were variable.[6] The ability to sort and shift entries in varying correlations was long perceived not as a valued feature of knowledge management, but as a dangerous weakness of excerpting, copying, and note-taking. Although secretaries in 17th-century France or Italy were forbidden to speak of their work in public, their confiscated speech never dampened their drive to express the master-medium dialectic of their employment. As Foucault demonstrates, doctors, like confessors, figured as stenographer of a client's secrets, until the birth of the clinic forced them out of their secretarial role. Discussing the documentary system of surveillance, Foucault points to a "partly official, partly secret hierarchy" in Paris that had been using a card index to manage data on suspects and criminals at least since 1833. In a note, he dryly remarks: "Appearance of the card index and constitution of the human sciences: another invention the historians have celebrated little".[7] Soon, card catalogs were used not just in a learned scholar's study but in libraries and in business.

Upon taking office, librarians often complained about the lack of order in the stacks and catalogs, and went about reorganizing shelves and finding aids. Document mobility requires addressing and recombination both of what is cataloged and of catalogs themselves. The Viennese Imperial Library established a card catalog (around 300,000 paper slips in 205 boxes) of its holdings in 1780, featuring instructions for the cataloger, along with a flowchart for dividing indexing labors. As Krajewski tells it, however, it was an accidental reinvention at the Harvard College Library in 1817 that brought the card catalog to the New World. Instead of tackling the overwhelming task of cataloging all stock, William Croswell cut up the partially bound catalogs compiled by his predecessors, allowing him to prepare a complete card index for over 20,000 volumes in less than six months.[8] But before the card index could also reign in office management, technical questions had to be settled.

In many places, the search for a normed paper slip size was conveniently settled: playing cards were in use for indexing at least since the French Revolution. On May 15, 1791, the French government decreed that a list of nationalized holdings was needed to make them accessible to the public. Librarians working for aristocrats and clergy resisted, since they had reason to fear that after an index went to Paris, the items themselves would soon follow. Thus, new instructions were issued to aides who would take stock where intractable librarians procrastinated. Regardless of local cataloging, they were to copy each item's identifying information on a numbered playing card. The operation netted the commission 1.2 million cards, soon used to add 300,000 volumes to the national library.[9]

By the time energetic reformer Melvil Dewey returned from Europe to his roots in the United States (having played a lot of cards on the transatlantic voyage), the country was ready for the standardization by Dewey's business, Library Bureau. Patenting the card index and furnishing drawers that held 1,000 slips in two rows, he succeeded in getting the American Library Association to bless his index card format in 1877. Within a few years, the business found more demand from offices rather than libraries.[10] By 1896, Library Bureau supported census data in several countries, in major contracts with the Hollerith Tabulating Machine Company (renamed IBM as of 1924). Before punched cards took over, the humble paper slip economy made inroads in government and business offices around the globe.

Elsewhere, this method for a flexible knowledge repository was soon adapted and adopted by historians, writers, lawyers, and philosophers. And while the memory crutch and administrative kludge long goes unacknowledged, soon one sees card index techniques openly credited: while John Locke had published a description of his card index in 1686 anonymously, by 1796 Jean Paul could publish a novel called *The Life of Quintus Fixlein, pulled from 15 card indexes*. Whatever occurred to Leibniz while reading or even on his walks, he scribbled onto slips for which he had a special cabinet constructed.[11] As contemporaries of Hegel describe in detail, he systematically hoarded ideas and excerpts on note cards, and carried them with himself from his school days, when he started at age 15, to his death.[12] A similar system was described by Charles Darwin:

> I keep from thirty to forty large portfolios, in cabinets with labeled shelves, into which I can at once put a detached reference or memorandum. I have bought many books and at their ends I make an index of all the facts that concern my work. Before beginning on any subject I look to all the short indexes and make a general and classified index, and by taking the one or more proper portfolios I have all the information collected during my life ready for use.[13]

One can find index cards at play all the way into the 20th century, for instance in Walter Benjamin's unfinished *Arcades Project* (1983/2002). Pioneering social scientist Beatrice Webb reported in her autobiography, *My Apprenticeship* (1980), of her attempts to persuade Oxbridge graduates that her index cards were "an indispensable instrument in the technique of sociological enquiry", and C. Wright Mills notes that what he called cross classification was crucial in keeping index cards.[14] And indeed, all the way into the 20th century, the playing card remains one model for how to interact with paper slips to generate new knowledge.

From the Scholarly to the Literary Card Index

> Only a historian of playing cards might find this relevant.
>
> —Jean-Baptiste Labiche[15]

Despite a respectable lineage, the card catalog mostly remained an anonymous, furtive factor in text generation, acknowledged merely as a memory crutch. Since the enlightened scholar is expected not just to reproduce knowledge but to produce innovative thought (not just as a recombination of good quotations, but opening new arguments and lines of investigation), knowledge management is a private matter, with rare exceptions. The question remains whether there is indeed a departure from the "neolithic mind" anthropologist Claude Lévi-Strauss glosses over in an interview, when he admits that his own memory "is a self-destructive thief" counter-balanced only by his extensive use of a card index:

> I get by when I work by accumulating notes—a bit about everything, ideas captured on the fly, summaries of what I have read, references, quotations . . . And when I want to start a project, I pull a packet of notes out of their pigeonhole and deal them out like a deck of cards. This kind of operation, where chance plays a role, helps me revive my failing memory.[16]

In his subversion of the rigorous constraints of memorial order by dint of chance and play, Lévi-Strauss seems to allow that his notes might either restore memory, or else restore the possibility of contingency which gives thinking a chance under the conditions of modernity. That hypertext may instantiate such an epistemology of chance and play on-screen is therefore no innovation; the encoding and deciphering practices of computer-linked textuality merely recapture what had been possible already with the means of note cards or playing cards.

Ludwig Wittgenstein's papers, dispersed between Britain, Norway, Austria, and elsewhere, presented the executors of his estate with a conundrum when they found a box labeled ZETTEL ("paper slips"), containing 717 loose fragments, the earliest dating from 1929, the latest from 1948 (the bulk was dictated between 1945 and 1948). Were they excess material, occasional ideas, sources and excerpts? Should the typescripts and hand-written notes be published, destroyed, classified? Posthumous version control proved to be arduous. Not presuming to reconstruct what Wittgenstein had "meant" to say in unfinished notes, the editors ordered and published what they deemed significant from this card index. A typescript of 768 pages (labeled simply *The Big Typescript*) dated from 1933 had been in the estate since 1951, but only in 1967 were the "Zettel" recognized from which it was compiled. Cut-and-paste was integral: "Usually he continued to work with the typescripts. A method which he often used was to cut up the typed text into fragments ('Zettel') and to rearrange the order of the remarks".[17]

Another important 20th-century thinker to rely on index cards was pioneering media theorist Harold Innis.[18] The executors of his estate published a tome called *The Idea File* (1980), composed of 18 inches of index cards, plus five inches of reference cards. Innis had a selection of hand-written index cards typed up and numbered, 1 through 339. It is unclear if these ruminations on television and art, communication and trade, secrecy and money, literature and the oral tradition, archives and history were intended to constitute a book project; the decision to publish the cards balances the putative will to posterity of an author, and the potential embarrassment of incomplete work. Clearly Innis intended to work synchronically rather than diachronically, to focus less on logical connections than on analogies, to practice pattern recognition—and the associative links of a card index lend themselves perfectly to this kind of project.

Similar features can be discerned in the silicon sociology of Niklas Luhmann's recombinant excerpts.[19] His card index cost him more time, he claimed, than writing his many books: little surprise that they demonstrate systematic redundancy.[20] Shortly after Luhmann's death in 1998, a dictionary and a glossary facilitated access to his thought, and an interactive database, marketed as "Luhmann on your computer", was offered on disk. A provocative question is whether from the depths of such a memory bank, further texts could be generated. Users of the Luhmann CD-ROM might try their hand at emulating his arguments within the recursive parameters of his systems theory.[21] A different approach to Luhmann's associative indexing is explored in another collaborative database tool, called nic-las in homage to the late sociologist ("nowledge integrating communication-based labeling and access system"), and billed as a "software prototype of an *autopoietic* knowledge landscape for social systems".[22] Intriguingly, deleted elements end up, for a while, in a digital unconscious: they remain searchable, and can return in unforeseen ways. The system distinguishes between a Freudian and a Deleuzian unconscious; while the former pushes some deleted objects back onto the documentation surface, the latter generates a random selection of deleted and undeleted objects in the form of new virtual index cards.

With this transition to multimedia software imitating the card index, we arrive at the surmise that hypercard systems and hypertext online obey the index card logic of associative links. George Landow and other adopters of this convergence hypothesis claim that French cultural theorist Roland Barthes anticipated this.[23] Be it Proust, the daily newspaper, or the television screen—to Barthes, it was all text, so in the age of the Internet, it was going to be Barthes who always already anticipated its structures and strictures.[24] Barthes' writing lends itself to this, because he often read in a manner that generated, despite all categorical, classificatory zest, a déjà vu effect.[25] In *S/Z* (1970), Barthes goes so far as to claim that, faced with the impure communication or "intentional cacophony" that is literature, one must accept "the freedom of reading the text as if it had already been read"—and asserts that faced with the plural text, there is no such thing as forgetting its meaning: one truly reads only in such quasi-forgetting.[26] No surprise that distinctions Barthes made in 1960 between writerly and readerly texts return in 1968, and his semiological definition of text crops up in publications from 1963 through 1976. "Though most of Barthes' now 'canonical' formulations on textuality occur in the period from 1968 to 1975, the issues that pushed him toward it were organizing his writing much earlier," observed John Mowitt, "in essence adumbrating the move that directed his attention to the work's status".[27] Mowitt notices how "articulation", Barthes' term in "The Structuralist Activity" of 1963, "reappears eight years later in the Preface to *Sade/Fourier/Loyola*"—and such continuities abound:

> Though I might be accused of stretching the point, it is also worth noting that in order to exemplify the procedural category of "dissection" (articulation's twin) Barthes has recourse in this essay to the sonoric distinction between s and z—precisely the distinction that Barthes later exploited in his most ambitious demonstration of how one might read "textually", namely, *S/Z*.[28]

Faced with such textual echo, Mowitt concludes "it becomes difficult to dismiss this tangle of associations as merely fortuitous." The reason became widely evident when the *Centre Pompidou* mounted a big exhibition on Barthes: he had worked, daily throughout his intellectual life, with an extensive card index. In an interview, Barthes described his method:

> I'm content to read the text in question, in a rather fetishistic way writing down certain passages, moments, even words which have the power to move me. As I go along, I use my cards to write down quotations, or ideas which come to me, and they do, curiously, already in the rhythm of a sentence, so that from that moment on, things are already taking on an existence as writing.[29]

From 1942 to his death, Barthes amassed 12,250 index cards, constantly rewritten and reordered. "There is a kind of censorship," he said, "which considers this topic taboo, under the pretext that it would be futile for a writer to talk about his writing, his daily schedule, or his desk". But as Barthes confessed:

> I have my index-card system, and the slips have an equally strict format: one quarter the size of my usual sheet of paper. At least that's how they were until the day standards were readjusted within the framework of European unification.

But Barthes found solace about his mental health in this unwelcome change: "Luckily, I'm not completely obsessive. Otherwise, I would have had to redo all my cards from the time I first started writing."[30] Once his papers became accessible to manuscript researchers, the scope of his card index could be studied. Written in pencil or blue ink, cards show quotes, observations, or diagrams; words or phrases are underlined, crossed out, or corrected. In the left or right top corner, he would note the date and page numbers of publications where he used the information on the card (e.g., a fiche on "acting out" refers to *S/Z* pages 71–72). Many cards show more than one use—including the passages noted by Mowitt.[31] Underlining or circling a word indicates it is taken up on another card (some cards list up to three such links). Outing his card catalog as co-author of his texts was "an anti-mythological action", he said: "it contributes to the overturning of that old myth which continues to present language as an instant of thought, inwardness, passion, or whatever." The editors of the exhibition catalog concluded that Barthes' *fiches* are not the carcass of an unfinished project, despite his sudden death in 1980.[32] The last course Barthes taught, however, was called *La préparation du roman*, preparing the novel. Spread over two years, it simulates exercises leading up to a novel; soon after the last class, Barthes died from injuries sustained in a traffic accident. On the one hand, his death might have prevented him from actually writing his novel; on the other hand, the entire seminar, now published as a notebook, marks the novel as a lost object from the start. A postscript to his *Lover's Discourse: Fragments* (1977) was going to discuss his card index and method of writing, as found only later among his papers.[33]

Tension between academic and literary production also propels a Swiss novel published posthumously in 2016, in a hybrid edition (in print and online) by the Swiss Literary Archives, presenting the textual genesis of a complex project. Hermann Burger's *Lokalbericht* is a playful book written between 1970 and 1972.[34] The typescript of 177 pages had rested in the archives in part because of its provocative format—it is construed as the mutual contamination of two expansive decks of index cards, one working toward an academic dissertation and the other toward a quasi-autobiographical novel by a doctoral candidate. Their mixing up and cross-fertilization (page 45f.) is owed to a purported challenge tossed off by the protagonist's thesis advisor, who joked about the career potential of an interpretation of a novel that does not yet exist—an invented time, place, and plot, an unknown author, an extended index card catalog on some 600 fragmentary pages somewhere between impressionism and expressionism, and *voilà*—the makings of a chair in new discoveries in literature. But realizing that creative ideas of this sort are all too rare in academia, the protagonist decides to explore this fantasy, and sets out to construe such a house of index cards, without completely abandoning his expected thesis on street names and places in the works of Günter Grass. Thus this card index novel starts with an imaginary letter to the advisor, along with the inevitable response the protagonist expects he would receive. In a historical context that sees Swiss literary figures and critics debate whether regional focus in writing is a limitation or a strength, a weakness or an intentional fountain of creative inspiration, the ridicule heaped on a dry-as-dust dissertation about place names and streets is only one elaboration of this debate, as Burger, throughout his career as a writer, emphasized the poetic potential of the local.[35]

The archival publication of the novel (in around 550 pieces) documents not only how Burger developed his verbal acrobatics, but also how, having been an advanced graduate student of literature at the University of Zurich for quite a while, he parodied and criticized academic prose in his work. The framing meta-fiction of a researcher struggling with two writing projects is an aspect of the novel that lends itself particularly well to a hyper-textual presentation.[36] That digital edition, prepared by the Swiss Literary Archive in collaboration with Cologne Center for eHumanities, not only presents high-resolution images of the fragmentary typescript, but also documents the text-genesis with a range of variants and corrections Burger made, as well as an edited digital text version without any micro-genetic variances. Commingling traits of various genres, including but not limited to the *nouveau roman*, campus novel, ironic *Bildungsroman*, city novel, *roman a clef*, and picaresque novel, *Lokalbericht* is, above all, a meta-novel: a novel about writing and about the stakes of the 20th-century novel, with detailed reflections about production processes and conditions for crafting the narrative. When the protagonist interrupts his "local report" to intersperse letters to the reader or to characters in the book, he provides details not only about locations, place names, views, and other circumstances, but also, in one memorable passage, about the two typewriters he uses: ostensibly one for the dissertation and one for the novel, but soon they enter into other levels of competition. Describing them as a sporty red convertible and a classy grey-green sedan, a stylish Ferrari and a comfortable cruiser, he speculates about the best use of their different typefaces, and begins to worry about their rivalry. Soon he feels he needs to write about one on the other and vice versa—the well-damped luxury of the Hermes Media describes the thrill of the Olivetti Valentino (*sic*), the white letters on black keys here, black letters on off-white keys there, and so on (pages 21–26). The same recursive structure is observed in the two growing card indices that mutually contaminate each other, one aiming at a novel, one at a dissertation:

> Who pulls the hollow tooth within which the paper scrap with the story of the hollow tooth is hidden? Once there was an old man who had a hollow tooth. In that tooth there was a box, and in the box, a piece of paper that said: once there was an old man.
>
> (page 207)

Much the same mockery is directed at the academic and critical figures that are part of the framing narrative; Kleinert the professor and Neidthammer the literary agent are the beginning and the end of the literary frame, and both figures are barely veiled representations of real people (the Zurich academic Emil Staiger, who was in fact Hermann Burger's doctoral advisor, and the local literary critic Anton Krättli whom Burger had known since 1963). Indeed, the critic has the last word, advising the protagonist not to write the projected novel but to let the manuscript age a year, two years, ten years—the book closes with the critic's advice not to finish and publish that very book (page 228). And the more the protagonist accidentally mixes his notes for the novel into the notes for the dissertation and vice versa, the more obvious it becomes that the incompletion of the novel is a mere simulation, while the completion of a dissertation recedes into the distance with the increasing poetic use of the academic ideas about contemporary novels (page 101). Beyond this rivalry, however, the project becomes legible as an archival fiction, and archive novel, which the reader puzzles together from the index card notations that form a montage of varied textual and fictional or metafictional levels. Unsurprisingly, one finds references to other novels that rely explicitly on index cards, for instance Arno Schmidt's notorious *Zettel's Traum* (1970).

Voraciously citing, inveterately punning, Schmidt, like Burger, distilled his card index into literary texts, published as complex typescripts, photo-mechanically reproducing his montages without editing. Between 1963 and 1969, Schmidt worked on his 130,000 cards for up to 16 hours per day, producing a text of 1,130 pages, 13 by 17.5-inches in size, and managed to publish it as *Zettel's Traum* the following year. But he sought recognition not only as a creative writer, but also as a theorist of linguistic and stylistic elements of modern prose. According to Schmidt, only diaries constitute a serious attempt at dealing with internal human processes—they help recollect, just as a photo album does, and Schmidt calculated the graphic dimensions of his textual arrangements so as to assist you in following certain associations and connections. Critics even speak of Schmidt's guidance "luring the reader into identification, into the *déjà vu* conviction that these recollections are his own".[37] Joining impulses from Joyce and Freud, among others, Schmidt documents how literature springs from less than divine sources. *Zettel's Traum* is an extended essay on E. A. Poe; over the course of 24 hours, the four protagonists discuss Poe's works, and Schmidt arranged his text in three parallel columns: the center column contains the action, the left one the Poe discussion, and the right column is made up of comments, footnotes, and auctorial opinions. Page (or card) 914 of this proto-hypertext contains the passage most critics view as the key to this gigantic structure.[38] Each of the four characters in this card index fiction is spaced out on Schmidt's pages in a collective score, and here, the book is allegorized as a quartet of voices—the voluptuous unconscious, the mean super-ego, the observant ego, and a fourth instance—something which, according to Schmidt, happens to men in their fifties, when the sex drive wanes and gives way to what the detached, smiling alter-ego of the author represents. Such unrelenting artifice stands in the way of naive investments in make-believe, auctorial inspiration, or genius.[39]

These textual devices have a long literary history, although it is relatively rare that creative writers make them known. Gerhart Hauptmann "wrote his nocturnal ideas on the wallpaper near his bed", then cut it up to paste it into his daily output.[40] Similar textures are also evident in Michel Butor's *Mobile* (1962), or in Vladimir Nabokov's *Pale Fire* (1962), a self-declared novel that falls into four parts: a preface, a poem, a lengthy annotation, and an index focusing almost exclusively on the notes.[41] In the preface, Nabokov recommends that readers start with the annotations, then return to them after cursorily picking the poem apart; he even goes so far as to suggest taking the book apart in order to cut-and-paste pages together at will, or at least buying a second copy to read them side by side. The poem itself is said to be written on 80 index cards of 14 lines each, as the preface dryly describes.[42] Similar concerns accompanied the posthumous publication of another Nabokov novel, or scraps for one, which is extant on index cards; indeed Nabokov wrote most of his novels, including *Lolita* (1955) and *Pale Fire*, on index cards. His novel *Ada or Ardor: A Family Chronicle* (1969) takes up over 2,000 cards, *The Original of Laura* (2009) consists of 139 transcribed cards.[43] Jules Verne's writing is equally illuminated by the reflective fire of a card index, since the source code for his science fiction was a box of some 20,000 excerpts and notes on scientific journals and books.[44] Raymond Carver taped citations and fragments on three- by five-inch cards to the wall beside his desk; Georges Perec, who had worked as an archivist in a scientific laboratory, likewise yielded to the "temptation towards an individual bureaucracy" and developed a complex filing system, using his index cards for most of his literary publications.[45]

From Individual Collections to Art Installations

> The card index marks the conquest of three-dimensional writing, and so presents an astonishing counterpoint to the three-dimensionality of script in its original form as rune or knot notation.
>
> —Walter Benjamin[46]

By 1969, it had become possible for Lucy Lippard to curate an art exhibit in Seattle titled *557,087* with index cards she had solicited, including from notables such as Eva Hesse and Robert Smithson, arranging black and white photographs and the index cards in glass cases.

Taking its title from the 1960 census figure for Seattle, the show was archived as revolutionary, despite and because of the fact that it did not leave behind paintings and sculptures, but a stack of 4- by 6-inch cards from around 60 artists, among them many names now famous for conceptual art or minimalism. The concept also traveled to Vancouver (where its title became *955,000*) and Buenos Aires (as *2,972,453*) before returning to the Seattle Art Museum.[47]

What art historian Aby Warburg laid out in his *Mnemosyne Atlas*, namely pattern recognition that operates by analogy and associative linking rather than diachronic filiations, finds its purest expression in art installations pivoting on index cards. But is notation on mobile paper slips outdated in the computer age, and reduced to ad-hoc jottings on sticky notes? Arguably, the card index influenced not only knowledge management, but interface design and creative processes.[48] A late example for the former: in 1981, when the Internet consisted of just 256 computers, Bob Kahn—co-designer of the TCP/IP networking protocol—was in charge of issuing Internet addresses, and carried around index cards in his shirt pocket to keep track of newly issued addresses.[49] As for the creative potential: it would appear to reside in part in material resistance on the one hand, and in harnessing chance on the other—as when Brian Eno designed a deck of inspirational cards titled "oblique strategies" (1975), or when Marshall McLuhan sold a deck of playing cards with provocative quotes as a management game called "Distant Early Warning" (1969). One wonders whether despite all the continuities in card index use over the centuries, there are not aspects of the index card catalog that are in peril of disappearing in the transition of valuable traits and affordances of index cards into other formats. Can everything be transcoded? This question motivated the artist David Bunn, who found pencil marks, hand-written corrections, drawings, finger prints, chocolate smears, and other manifestations of what he calls "subliminal messages" in the discarded card catalog of the Los Angeles Central Library. Focusing on these aesthetic communications that the electronic catalog did not preserve, Bunn developed art installations in dogged pursuit of contingent traces.[50] As if offering to make a connection between the aforementioned Roland Barthes exhibit at the Pompidou and David Bunn's art installations a continent away, Christian Marclay also mounted index cards so as to fill the walls of an art gallery, calling it "White Noise".[51]

A famously more conspiratorial example in the art world of the use of index cards involves Mark Lombardi. His drawings, based on his own index card catalog of public sources, trace relationships between powerful financial and political figures, such as oil companies, the Bush family, the Bin Laden family, and various banks. A few weeks after the September 11, 2001 terrorist attacks, an FBI agent called the Whitney Museum of American Art and asked to see a drawing on exhibit there.[52] Lombardi allegedly committed suicide the year before. Using just a pencil and a huge sheet of paper, Lombardi had created an intricate pattern of curves and arcs to illustrate the links between global finance and international terrorism. Meanwhile, a collector made a substantial offer to the show's curator, Robert Hobbs, a professor of art history at Virginia Commonwealth University, for the purchase not of any drawings, but of Lombardi's extensive index card collection.[53] Thus it appears that a poetics of intellectual capital can be embodied in the card index.

Other artists noted that an "index" can also denote repression and censorship. *The File Room* (1994) by Antoni Muntadas is one of the first widely recognized art works on the World Wide Web—a pioneering work of net art inviting online collaboration to document censorship (thefileroom.org). On display at the Randolph Street Gallery of the Chicago Cultural Center as well as online, *The File Room* started in May 1994 with 450 entries on censorship, from Athens in the fifth century BC to Salman Rushdie's *The Satanic Verses* (1988); viewers could ponder Diego Rivera's dispute with the Rockefeller Center over his depiction of Lenin, or TV moderator Ed Sullivan's request to The Doors to change one line of their lyrics in "Light My Fire". Moreover, the installation invited members of the public not only to browse the card index or website, but also to add entries about current or historical bias regarding religion, ideology, or sexual orientation. Visitors in Chicago and online were able to interact and contribute, emphasizing that an archive of censorship can never be closed or complete. The installation featured a computer on a desk, surrounded by 138 black metal filing cabinets of four drawers each; seven of the 552 cabinet drawers were taken up by computer monitors. *The File Room* offers definitions of censorship, an archive of cases, an interface used to submit additional cases, a bibliography, and a search tool—by date, subject, location, and medium. Today, the National Coalition Against Censorship maintains *Censorpedia* (wiki.ncac.org) as a participatory wiki of censorship from antiquity to the present, building on Muntadas's *File Room*.

Censorship is a thorny topic, as it seeks not only to suppress images, sounds, and words, but also to hide the means of suppressing them. Muntadas called himself an "information analyst".[54] As Edward Shanken writes, the creators of *The File Room* were "concerned about the potential of technology both to support and resist censorship".[55] As with his pioneering contributions to CD-ROM art in the 1990s, Muntadas put some thought into affording interactivity without yielding control over the installation to viewers, balancing access with maintenance, both in the card index and online. Announcing *The File Room* during a residency in September 1993 at the University of Illinois, Muntadas worked with gallery director, Paul Brenner, as project manager and Maria Roussos as hypertext developer for over two years. Drawing on the capabilities of the NCSA Mosaic browser (1993–1997) and starting with definitions before branching out into cases, *The File Room* comprises examples from visual art, music, dance, and literature. Curator Steve Dietz associates Muntadas's art with the "dream of the open work" as inspired by Umberto Eco: "one of the strongest shifts of emphasis in the digital age has been on the production side and on the movement from creating finished works of art to creating systems for the production of art".[56] As Muntadas moved beyond the gallery's index cards onto the Internet, he described the project as "a social sculpture *à la* Joseph Beuys which gains its meaning through a group effort".[57] Institutions taking on net art and web art (such as the ZKM in Karlsruhe, the Walker Art Center in Minneapolis, the Whitney Museum in New York, and the San Francisco Museum of Modern Art) emphasize that this is not merely a different exhibition space, but a different modality for aesthetic communication.[58] Muntadas's *The File Room* is indebted to conceptual works of the Art & Language collective—card stacks such as *Index 01* (1972), eight cabinets of variable dimensions (like columns topped with drawers) and photostats; *Index 2* (1972), consisting of a similar installation and surrounded by a wallpaper of index cards, plus file boxes on a table; and *Index 5* (1973), offering "instructions for reading the index". These installations, pillars of database art, illustrate how information lies dormant until it is accessed through an interface, but also how that same interface might distort information. They illustrate the perennial tension between attempts to erase, suppress, or hide information, and efforts to document historical, geographical, and topical dimensions of creation and censorship. This tension motivates art projects with index cards in the computer age, counting on the material resistance of analog remainders.

Notes

1 Daniel Dennett, *Darwin's Dangerous Idea* (London: Penguin, 1995), 202, alluding to Samuel Butler (who wrote in his *Life and Habit*, 1877, that "a hen is only an egg's way of making another egg").

2 British Society for the History of Science, "Carl Linnaeus Invented the Index Card," *Science Daily* (June 16, 2009), www.sciencedaily.com/releases/2009/06/090616080137.htm. Compare Jonathan

Schiffman, "How the Humble Index Card Foresaw the Internet," Popular Mechanics (February 11, 2016) www.popularmechanics.com/culture/a19379/a-short-history-of-the-index-card/

3 Vincent Placcius, "De scrinio litterato," *De arte excerpendi* (Stockholm and Hamburg, 1689), 121–159.

4 Christoph Gottlieb von Murr, "Von Leibnizens Exzerpirschrank," *Journal zur Kunstgeschichte und allgemeinen Litteratur* (1779), #7, 210–212.

5 Conrad Gessner, *Pandectarum sive partitionum universalium libri XXI* (Zurich 1548); see H. Wellisch, "How to Make an Index—16th Century Style: Conrad Gessner on Indexes and Catalogs," *International Classification* 8 (1981), 10–15.

6 Christoph Meinel, "Enzyklopädie der Welt und Verzettelung des Wissens: Aporien der Empirie bei Joachim Jungius," in Franz Eybl, Wolfgang Harms, Hans-Henrik Krummacher, and Werner Welzig eds. *Enzyklopädien der frühen Neuzeit. Beiträge zu ihrer Erforschung* (Tübingen: Niemeyer, 1995), 162–187.

7 "Apparition de la fiche et constitution des sciences humaines: encore une invention que les historiens célèbrent peu." Michel Foucault, *Surveillir et punir. Naissance de la prison* (Paris: Gallimard, 1975), 287, referring to A. Bonneville, *De la recidive* (Paris, 1844), 92–93.

8 https://hollisarchives.lib.harvard.edu/repositories/4/resources/4004. Compare Markus Krajewski, *Paper Machines: About Cards & Catalogs, 1548–1929* (Cambridge: MIT Press, 2011).

9 Hans Petschar, "Einige Bemerkungen, die sorgfältige Verfertigung eines Bibliothekskatalogs für das allgemeine Lesepublikum betreffend." In Hans Petschar, Ernst Strouhal, and Heimo Zobernig eds., *Der Zettelkatalog. Ein historisches System geistiger Ordnung* (Vienna: Springer, 1999), 17. Compare Heike Gfereis and Ellen Strittmatter, eds., *Zettelkästen. Maschinen der Phantasie* (Marbach: Deutsche Schillergesellschaft, 2013).

10 Wayne Wiegand, *Irrepressible Reformer: A Biography of Melvyl Dewey.* Chicago: American Library Association, 1996.

11 John Locke, "Méthode nouvelle de dresser des Recueils communiquée par l'Auteur," *Bibliothèque Universelle et Historique* (Amsterdam, 1668), vol. 2, 315–340; Jean Paul, *Das Leben des Quintus Fixlein* (Stuttgart: Reclam, 1987) and Jean Paul, "Die Taschenbibliothek," in *Sämtliche Werke* II:3 (Frankfurt: Zweitauseneins, 1996), 772; Ch. G. von Murr, "Von Leibnizens Excerpirschrank," *Journal zur Kunstgeschichte und allgemeinen Litteratur* VII (1779), 211; Markus Krajewski, "Zitatzuträger. Aus der Geschichte der Zettel/Daten/Bank." *Anführen—Vorführen Aufführen. Das Zitat in Literatur und Theorie*, eds. Nils Plath and Volker Pantenburg (Bielefeld: Aisthesis, 2002), 177–195.

12 Johann Jacob Moser, "Einige Vortheile für Cantzley-Verwandte und Gelehrte in Absicht auf Acten-Verzeichnisse, Auszüge und Register," *Lebensgeschichte, von ihm selbst geschrieben* (Frankfurt and Leipzig, 1777), vol. 3; Karl Rosenkranz, *Georg Friedrich Wilhelm Hegels Leben* (Berlin, 1844), 12; Hermann Schmitz, "Hegels Begriff der Erinnerung," *Archiv für Begriffsgeschichte* 9 (1964), 37–44; Friedrich Kittler, *Die Nacht der Substanz* (Bern: Benteli, 1989), 18.

13 Nora Barlow ed., *The Autobiography of Charles Darwin 1809–1882* vol. 1 (London: Collins, 1958), 137.

14 Beatrice Webb, *My Apprenticeship* (Cambridge: Cambridge University Press, 1926), 426–433; C. Wright Mills, *The Sociological Imagination* (London: Penguin, 1970), 217–245.

15 Jean-Baptiste Labiche, *Notices sur les depots littéraires et la révolution bibliographique* (Paris: Parent, 1880), 64; Hellmut Lehmann-Haupt, *Gutenberg and the Master of the Playing Cards* (New Haven, CT: Yale University Press, 1966). There is at least one book structured as a card game: Marc Saporta, *Composition numéro 1*: Roman (Paris: Seuil, 1962); see Reinhold Grimm, "Marc Saporta oder der Roman als Kartenspiel," *Sprache im Technischen Zeitalter* 14 (1965): 1172–1184.

16 Didier Eribon, *Conversations with Claude Lévi-Strauss* (Chicago: University of Chicago Press, 1991), vii–viii; Claude Lévi-Strauss, *Structural Anthropology* (New York: Basic Books, 1963), 129f.

17 Georg Henrik von Wright, "The Wittgenstein Papers," *The Philosophical Review* 78:4 (1969), 483–563, here: 487.

18 Innis Papers, Archives of the University of Toronto, Thomas Fisher Library, Box 8. The cards themselves appear lost, but a typescript based on them was published posthumously: William Christian, *The Idea File of Harold Adams Innis* (Toronto: University of Toronto Press, 1980).

19 Niklas Luhmann, "Kommunikation mit Zettelkästen. Ein Erfahrungsbericht," *Universität als Milieu*, ed. André Kieserling (Bielefeld: Haux, 1993), 53–61. Compare *Evernote* (http://evernote.com) and *Zettelkasten* (www.verzetteln.de/synapsen).

20 Niklas Luhmann, *Archimedes und wir. Interviews.* (Berlin: Merve, 1987), 142–149; William Rasch, "Theory of a Different Order: A Conversation with Katherine Hayles and Niklas Luhmann," *Cultural Critique* 31:2 (autumn 1995), 7–36.

21 Detlev Krause, *Luhmann-Lexikon* (Stuttgart: UTB, 2001); Claudio Baraldi, Giancarlo Corsi, Elena Esposito, *GLU. Glossar zu Niklas Luhmanns Theorie sozialer Systeme* (Frankfurt: Suhrkamp,

1997); Theodor M. Bardmann and Alexander Lambrecht, *Systemtheorie verstehen: Eine multimediale Einführung in systemisches Denken* (Wiesbaden: Westdeutscher Verlag, 1999).

22 Nic-las, 1999–2005, www.nic-las.com (30.11.2005); compare www.iasl.uni-muenchen.de/links/GCA-VI.2e.html

23 George P. Landow, "Hypertext, Metatext, and the Electronic Canon," in Myron C. Tuman ed., *Literacy Online: The Promise (and Peril) of Reading and Writing with Computers* (Pittsburgh: University of Pennsylvania Press, 1992), 67–94.

24 Katherine Hayles, "Information or Noise? Competing Economies in Barthes's S/Z and Shannon's Information Theory," in George Levine ed., *One Culture: Essays in Literature and Science* (Madison: University of Wisconsin Press, 1987), 119–142.

25 This was Paul de Man's attack on Barthes' literary-historical assumptions: "You distort history because you need a historical myth to justify a method which is not yet able to justify itself by its results," in Richard Macksey and Eugenio Donato eds., *The Structuralist Controversy: The Languages of Criticism and the Sciences of Man* (Baltimore: Johns Hopkins University Press, 1972), 150.

26 Roland Barthes, *S/Z* (Paris: Plon, 1970), 9–28, esp. iv, v, ix.

27 John Mowitt, *Text: The Genealogy of an Antidisciplinary Object* (Durham: Duke University Press, 1992), 117.

28 Mowitt, *Text*, 118. See Mowitt, "What is a Text Today?" *PMLA* 117:5 (2002), 1217–1221.

29 "An almost obsessive relation to writing instruments" (interview with Jean-Louis de Rambures of *Le Monde*, September 27, 1973), in Roland Barthes, *The Grain of the Voice* (Berkeley: University of California Press, 1985), 177–182.

30 Barthes, *The Grain of the Voice*, 182. My reading of Barthes' *fichier* has been indexed, as it were, by Rowan Wilken, "The Card Index as Creativity Machine," *Culture Machine* 11 (2010), 7–30.

31 Barthes' note card titled "fiches" reads: "D'origine érudite, la fiche devient le coin vengeur que le désir insère dans la loi compacte du travail. Principe poétique: ce carré savant ira dans le tableau de l'écriture, non dans celui du savoir."

32 "Le fichier n'est pas le livre à venir: il n'y a pas d'oeuvre manquante que quelques milliers de fiches inédites viendraient constituer. Barthes a écrit tout ce qu'il avait à écrire." Nathalie Leger, "Immensément et en detail," *R/B* (Paris: Centre Pompidou/Seuil/IMEC, 2002), 94. Co-editor Marianne Alphant thinks the notes for his last course limn the ichnographic *moi-poisson* book he was working toward: Marianne Alphant, "Presque un roman," *R/B*, 125–128. The executor of Barthes' unpublished papers also believes "these courses revolve around the idea of a possible novel, a novel that death prevented him from writing." Eric Marty, "Interview with Jacques Henric," *Art Press* 285 (December 2002), 51.

33 Roland Barthes, "Comment est fait ce livre," *Art Press* 285 (December 2002), 55; Daniel Ferrer, "Genetic Criticism in the Wake of Barthes," in Jean-Michel Rabaté ed., *Writing the Image: After Roland Barthes* (Philadelphia: University of Pennsylvania Press, 1997), 217–227. See Denis Hollier, "Notes (on the Index Card)," *October* 112 (spring 2005), 35–44. *Roland Barthes par Roland Barthes* "manifests the pleasure of auto-commentary and of reflexivity which includes the relation of the author to his manuscript," asserts Anne Herschberg Pierrot, "Les manuscripts de *Roland Barthes par Roland Barthes.* Style et genèse," *Genesis* 19 (2002), 195.

34 Hermann Burger, *Lokalbericht* (Zurich: De Gruyter, 2016).

35 Magnus Wieland and Simon Zumstieg, "Hermann Burgers Lokabericht: Von der Archivfiktion zur Archivedition," *Germanistik in der Schweiz* 9 (2012), 91–109.

36 www.lokalbericht.ch

37 F. Peter Ott, "Tradition and Innovation: An Introduction to the Prose Theory and Practice of Arno Schmidt," *German Quarterly* 51:1 (1978), 26.

38 Siegbert Prawer, "Bless Thee Bottom! Thou Art Translated," in WD Scott-Robson ed., *Essays in German and Dutch Literature* (London: Institute of Germanic Studies, 1973), 156–191.

39 Arno Schmidt, "Der Platz, an dem ich schreibe," *Essays und Aufsätze* vol. 2 (Zurich: Haffmanns Verlag, 1995), 28–31.

40 Günter Kunert, "Zettel," *Akzente* 33:5 (1986), 391–394. Also Francesco Sacchini, *Über die Lektüre, ihren Nutzen und die Vortheile sie gehörig anzuwenden* (Karlsruhe, 1832), 101–102.

41 Vladimir Nabokov, *Pale Fire* (New York: Putnam, 1962).

42 Brian Boyd, *Nabokov's Pale Fire: The Magic of Artistic Discovery* (Princeton: Princeton University Press, 1999); Markus Krajewski, "Ver(b)rannt im Fahlen Feuer. Ein Karteikartenkommentar," *Kunstforum International* 155 (June–July 2001), 288–292.

43 Vladimir Nabokov, *The Original of Laura* (London: Knopf, 2009). See also Richard Sieburth, "Leiris/Nerval: A Few File Cards," *October* 112 (spring 2005), 51–62.

44 Vladimir Stibic, *Tools of the Mind* (Amsterdam: Elsevier, 1982), 77; Stibic also mentions Jack London's index cards.

45 Raymond Carver, "On Writing," *Fires: Essays, Poems, Stories* (New York: Vintage, 1968), 22–27. Georges Perec, "Notes Concerning the Objects That Are on My Work-Table," *Species of Places and Other Pieces* (New York: Penguin, 1999), 145 and 152. Perec's novel *Life: A User's Manual* (London: Harvill, 1987) features characters who share his obsession with indexing; see also David Bellos, *Georges Perec: A Life in Words* (London: Harvill, 1999), 207.

46 Walter Benjamin, "Vereidigter Bücherrevisor," *Gesammelte Schrift en* vol. IV.1 (Frankfurt: Suhrkamp, 1991), 102–104.

47 Jen Graves, "Dematerialized: A 1969 Exhibition on Index Cards," *The Stranger* (May 3, 2013), and Lucy Lippard, "Curating by Numbers", *Tate Papers No. 12*, www.tate.org.uk/research/publications/tate-papers.

48 A pair of journalistic articles in the same business magazine explores the half-life of the sticky note: just four years after running a piece declaring the sticky note obsolete, *Fast Company* speculates it could indeed become the latest innovation technology. James Hunt, "Why Designers Should Declare Death to the Post-It" (May 20, 2010), and David Lavender, "How the Post-It Note Could Become the Latest Innovation Technology" (March 26, 2014).

49 Katie Hafner and Matthew Lyon, *Where Wizards Stay Up Late: The Origins of the Internet* (New York: Simon and Schuster, 1996); Janet Abbate, *Inventing the Internet* (Cambridge, MA: MIT Press, 1999); Michael Hauben, "Behind the Net: The Untold History of the ARPANET and Computer Science," www.columbia.edu/~rh120/ch106.x07

50 David Bunn, *Subliminal Messages* (Cologne: Walter König, 2004). Compare David Bunn, "A Place for Everything and Everything in its Place," *Discourse* 20:3 (fall 1998), 175–178; and David Bunn, "Bodysnatching," *Discourse* 24:1 (winter 2002), 120–148.

51 Christian Marclay, "White Noise," Kunsthalle Bern, Switzerland, 1998; Fawbush Gallery New York, 1994; and daadgallerie Berlin, 1994. See Russell Ferguson, *Christian Marclay* (UCLA Hammer Museum 2003), 184–187.

52 NPR Weekend Edition, Saturday, November 1, 2003. Compare Patricia Goldstone, *Interlock: Art, Conspiracy, and the Shadow Worlds of Mark Lombardi* (London: Counterpoint, 2015).

53 See Frances Richard, "Obsessive – Generous: Toward a Diagram of Mark Lombardi," in *Mark Lombardi: Global Networks* (New York: Independent Curators Inc., 2003), 115–118. A photo of Lombardi's pink and green index cards appears there.

54 Slavko Kacunko, *Closed Circuit* (Berlin: Logos, 2004), 305/372; see Antoni Muntadas & Anne-Marie Duguet, *Muntadas: Media Architecture Installations* (Paris: Centre Georges Pompidou, 1999).

55 Edward Shanken, *Art and Electronic Media* (London: Phaidon, 2009), 35.

56 Steve Dietz, "Ten Dreams of Technology," *Leonardo* 35:5 (2002), 509–522, here: 512.

57 Margot Lovejoy, *Digital Currents: Art in the Electronic Age* (London: Routledge, 2004), 248.

58 Steven Wilson, *Information Arts* (Cambridge: MIT, 2002), 563.

2

FROM HERO TO ZERO

The Rise and Fall of the Slide Rule as the Calculating Tool of Choice

Peter M. Hopp

The slide rule, the tool for all mathematical occasions. So it was for some 300 years. Those of us who used one for real at school, college, and at work in pre-calculator days, be it as engineer or academic, automatically reached for one when calculation of any sort requiring multiplication and division and more complex maths such as trigonometrical and exponential functions, was required. Most of us owned at least two slide rules, one 5-inch pocket device and a 10-inch desk model—though our transatlantic cousins were rumored to carry a 10-inch one in a scabbard hooked to their belts! Nothing as "flash" for us stiff-upper-lipped British. We had one in our brief cases. I cannot say I mourned my slide rules' passing when I bought my first calculator, a Sinclair Cambridge in kit form, in the early 1970s. That calculator, despite its idiosyncratic implementation, really was much quicker and more accurate. But was that extra accuracy really necessary? We had certainly got used to "good enough" answers with numbers of significant figures on our trusty slide rules. In any event, the slide rule's demise was not quite as quick as it could have been—I well remember my boss who had given me some horrible mathematical task to do being amazed when I completed the task well ahead of schedule. "How the blazes have you done that?" he asked. I proudly took out my shiny new Sinclair calculator and showed him. "Great heavens!" said he. "Go away and do it properly with a slide rule." Over the next few years the electronic pocket calculator became much more common and financially available and the slide rule was gradually relegated to a desk drawer. Like all new technology we had to learn the foibles of calculators. I can remember an inordinate amount of time being spent at work on an "invention" that was to revolutionize our workplace and make all of our fortunes. This turned into a chimera based on an apparent improvement in performance created by the fact that one divided by three and then multiplied by three again gave an answer 0.9999 on most early calculators! It was only after realizing that the next engineering generation was not as mathematically adept as we were that the loss of the slide rule and its attendant numerical methods became something to be valued. There was further evidence of this when recently a maths teacher relation confessed that she had no idea what a slide rule was or indeed what were log-tables. And what was more to the point, they did not teach such things anymore! Realization of the almost complete disappearance of a legendary and ubiquitous tool is now complete.

Genesis

The slide rule was invented at a period in history when mathematical calculation was becoming more complex, just as the technical world of that time such as astronomy and other philosophical arts required the ability to perform reliable calculation much more quickly and accurately. The speed of the seminal inventions, their evolution and the further discoveries that allowed a device that we would now recognize to be developed, all followed each other in short order to create the slide rule.

The Reverend William Oughtred (1574–1660) is now universally recognized as the inventor of the slide rule sometime about 1622. Even so, at that time there was a heated argument about who claimed to have invented the device. Richard Delamain(e) (*c*.1600–*c*.1644), another mathematician also involved in the invention of a type of sundial, published his *Grammelogia or The Mathematical Ring* in about 1630. He was declared to be a charlatan and plagiarist, and his claim found to have actually depended on Oughtred's invention. But he was the first to write about slide rules. His claim was probably due to their common mathematical instrument maker Elias Allen. The date of the initial publications relating to Oughtred's slide rules is about 1631, although this date is by no means certain. Oughtred was an old-fashioned believer in understanding the craft of calculating at a time when few could even add reliably, let alone multiply or divide. Users of mathematical "trickery" such as a slide rule to perform calculations were branded "Jongleurs" (jugglers or tricksters, who did not necessarily understand the basics, but worked by rote). Thus it was Oughtred's pupil, William Forster, who persuaded Oughtred to allow Forster to edit on his behalf the first treatise on his slide rules, the *Circles of Proportion and the Horizontal Instrument* (1632), thus recording that Oughtred had invented a slide rule some years previously. This book described a circular slide rule initially made by Elias Allen (1588–1653), a leading mathematical instrument maker of his time, and then by others. At about the same time—and hence the priority debate—Forster published the information on Oughtred's slide rule with an Appendix: *"To the English Gentrie . . . the just Apologie of W. Oughtred against . . . the . . . insinuations of R. Delamain in a pamphlet called Grammelogia or Mathematical Ring, . . .".* The contents of "*Grammelogia*" were described as a "*patchery and confusion of disjointed stuffe.*" Fascinating stuff indeed.

The slide rule is an analog logarithmic device. It takes advantage of the mathematical feature whereby the addition of two logarithms is the equivalent of multiplying the numbers whose logarithm they are. The inverse is also true for division. It is thus also true that multiplying the logarithms by two is equivalent to squaring the numbers and so on. It therefore follows that there had to be further inventions, developments, and transpositions to logarithms before a slide rule as we know it could come into existence.

The seminal invention had to be logarithms themselves. Various efforts at producing a tool that we would now recognize as logarithms were attempted throughout Europe by Burghi and others. However, it is now recognized that it was John Napier, Baron of Murchiston near Edinburgh in Scotland, a true polymath of his period (1550–1617) who in 1614 actually was the inventor of logarithms. We have just recently celebrated, rather quietly, the Quadricentenary of this momentous work. This follows the much more extensive Tercentenary celebrations in Scotland in 1914, illustrating to some extent the demise in importance of logarithms. Napier's initial design of logarithms was not directly usable for a slide rule, nor indeed particularly convenient to use by the average "calculator". The design required translating into a more usable form—the now well-known logarithms to the base 10—by Henry Briggs (1561–1630), Professor of Mathematics at Oxford in about 1617, although most of his work was done at Gresham College. These are the logarithms that generations of school children and mathematicians alike used for calculation through to the present day. Many hundreds of different forms of logarithmic tables, with different systems of presentation and accuracy, were calculated and published in the intervening period. However, it was the four-figure log table which was the most popular and generally used format. Seven-figure and higher precision tables were available for those requiring greater accuracy, and most laboratories and other major users would have had a set.

The next requirement on the path to the slide rule was for these logarithms to be transposed into a physical form. This form would be either a circular scale for a circular slide rule, or more commonly into a rectilinear form, onto a ruler. The best known of these is from Edmund Gunter (1581–1626), Professor of Astronomy at Gresham College in London, another polymath and prolific inventor who possibly as early as 1621, invented the Gunter Scale. This enabled a precursor to the slide rule, indeed an alternative form of logarithmic calculator—Gunter's rule—to be invented and come into use in its own right before the slide rule itself appeared.

Let us consider these inventions in greater detail. Without doubt, the momentous invention was Napier's logarithms. Interestingly and probably unsurprisingly, in parallel events on the continent we have attempts to develop similar calculating tools, but without the same success. These attempts attested to the recognized need for an improved method of performing sophisticated calculation at this time. We have observed that Napier was a polymath with many different ideas under his consideration. Not least of these was his work on theology, necromancy, and alchemy. He had a particular antipathy to the papacy, evidenced in what many still consider his most important work—ahead of his work on logarithms—his book published in 1593, *A Plaine Discovery of the Whole Revelation of St John* where he predicted the end of the world in 1688 or 1700. His *Mirifici Logarithmorum Canonis Descriptio*, which contained 57 pages of explanatory matter and 90 pages of tables of numbers related to natural logarithms, was published in 1614. This description of his great new invention, logarithms, started the exciting new world of accurate and speedy calculation. Sadly, Napier was to die shortly after his invention of logarithms, and even though the mathematical world largely recognized what a seminal point this was, Napier himself might not have been quite so convinced. In his *Rabdologie*, published posthumously in 1617, he proposed yet another mechanical method of multiplication, this time using a set of rods based on Arabic lattice multiplication—Napier's Bones. It is thus not unreasonable to speculate that perhaps he was not aware of how vital his previous invention of logarithms was, or indeed how useful it would prove to be for future generations of mathematical practitioners. He also produced other calculating tools such as the little known "Promptuary", an extension to his Bones, which was described later in the second edition of his *Rabdologie*.

The conversion of Napier's "natural" logarithms to the more useful logarithms to base 10 was something Napier himself had considered but had not attempted due to his ill health. Henry Briggs, who taught Geometry, Astronomy, and Navigation at Oxford, might have been prompted by Gunter to make the first of several four-day journeys to visit Napier in Scotland in 1616 after he had obtained a copy of Napier's *Mirafici*. He went to "congratulate" Napier and to "marvel" at this important invention. What a marvellous picture is painted! Briggs was ultimately charged by Napier with doing the conversion to base 10, for which he is now well known. To some extent, Briggs had a vested interest, as it was astronomical calculations among many others that would initially really benefit from this improved and simplified method of reliable calculation.

Edmund Gunter was another famous English clergyman, mathematician, geometer, and astronomer. Gunter is particularly remembered for his several inventions of Gunter's Chain, Gunter's Quadrant—yet another calculating device—and finally in 1620, the Gunter's Scale. This was another analog device initially invented to calculate logarithmic tangents. Gunter was mentored in mathematics by Briggs and he, too, eventually became Gresham Professor of Astronomy in 1619, a post he held until his death.

William Oughtred was the rector of Albury Church in Surrey, a mathematician and teacher, and the inventor of several important mathematical constructs including the decimal point and "X" for the multiplication sign, as well as inventing the slide rule. It is notable that Oughtred did not take the existing Gunter's scale, but instead used a circular form of the logarithmic scale to invent what was the first slide rule described in his *Circles*, though it is rumored that he had in fact taken two of Gunter's scales and run them adjacent to each other as the first rectilinear slide rule. It is also notable that the first circular slide rule used a single circular logarithmic scale and a pair of opening cursors to traverse the scale and perform the calculations. This is exactly the method many other later designs of circular and pocket-watch slide rules would use some 250 years later. In all probability, this was because it was much simpler to engrave only one set of logarithmic scales rather than the two that would have been necessary if he was to have logarithmic scales moving adjacent to each

Figure 2.1 William Oughtred by Wenceslas Hollar, unknown date, circa 1640 (public domain)

other. Oughtred, our hero as inventor of the slide rule, was a man of some character. He was colorfully described by John Aubrey (1626–1697), a biographer of the day:

> He was a little man, had black haire, and blacke eies (with a great deal of spirit). His head was always working. He would draw lines and diagrams on the dust. . . . he used to lye a bed till eleaven or twelve a clock, with his doublet on . . . studyed late at night, went not to bed till 11 a clock, had his tinder box by him, and on top of his bed-staffe, he had his inke-horne fixed. He slept but little. Sometimes he went not to bed in two or three nights, and would not come downe to meales till he had found out the quaesitum.
>
> (Aubrey, 1693)

He was both a famous teacher whose pupils included Delamain, Christopher Wren, and John Wallis and also a prolific writer. Among many other mathematical works, his *Clavis Mathematicae* (The Key to Mathematics) was an important mathematical work of the day, published in 1631. Just before he died he produced a book on watch-making for his son.

Figure 2.2 Elias Allen, also by Wenceslas Hollar, 1653 (public domain)

He was a truly great mathematician and a real character. Oughtred died after he had been rector of Albury Church for over 50 years and shortly after the restoration of Charles II to the throne.

Elias Allen was the pre-eminent London mathematical instrument maker of the time. He made the first example of "Circles of Proportion" for Oughtred, and subsequently made others. While he was making Oughtred's "Circles . . ." it was likely that Delamain saw them either at Allen's workshop or when he was with Oughtred and rushed to produce his own "Mathematical Ring". A set of these was presented to King Charles I with a claim for a patent on the device, which was never awarded. Florin Cajori (1920) provides a fascinating analysis of the whole saga. We are very fortunate that several examples of Elias Allen and other makers' Circles of Proportion are preserved in museums both in the UK and the USA and thus we are able to appreciate this leap in technology and the beauty and functionality of the instrument makers' art in these devices. No example of the Mathematical Ring is known.

This period of history, the beginning of the "Age of Enlightenment", is interesting in that many new and exciting methods of simplifying calculations were invented and coming into use. We see many earlier designs of calculating equipment such as the several different designs of the Sector already in existence, Gunter's Quadrant and Gunter's scales as well as "logarithms" from Burgi; Prostapharesis, the new logarithms from Napier and so on, all jostling for position and in many cases being further developed for a particular specialization. For example, Gunter's scales were used for navigating, both in the Merchant Navy as well as the Royal Navy, for many years through to the late 19th century. However, despite these minority methods of calculation, the slide rule gradually achieved ascendance in nearly all fields of calculation, both general and specialist. Other tools and devices fell by the wayside.

Evolution

Figure 2.3 An Elias Allen-made Circles of Proportion (circa 1648) from the National Museums of Scotland collection

We are able to follow the initial evolution of the slide rule via many general as well as specialist designs. The very first specialist design was produced by Oughtred himself in about 1633 for the Company of Vintners for gauging. This was a traditional rectilinear slide rule. Another very recognizable rectilinear slide rule due to Robert Bissaker is in the Science Museum and was made in about 1654, showing that interest was widening relatively swiftly. The earliest specialized slide rule to use a format that was subtly different was Henry Coggeshall's (1623–1690) timber contenting slide rules developed from about 1677 through to 1767, almost a century after its invention. This was then followed by Thomas Everard's (about whom very little is known other than he rose from Officer in the Excise in Southampton to Commissioner for Gauging in the Excise) gauging and alcohol and tax calculating designs in about 1683 which became established and continued in use through to the 20th century. Coggeshall's designs were embodied as a slide and logarithmic scales in one arm of a 2-foot, 2-fold measuring rule as already invented some 150 years previously and which would have been used by most artisans and craftsmen working in timber at that time. Everard's designs returned to an ingenious rectilinear design which were built into a different square-form rectilinear rule with slides initially on two sides, but soon expanding onto a third and finally onto all four sides of the square format. These were available in all lengths from 6-inch to 36-inch, allowing considerable accuracy in calculation at the longer lengths. It will be noted that these two major evolutionary designs were remarkably long-lived. All were available probably through to the beginning of World War I.

As other scientific development continued, we find many other special designs produced to achieve a particular result for specific fields. Many special scales came and went, however, some such as the log-log scales invented by Peter Mark Roget (1779–1869) in 1815 were used for different new developments in thermodynamics and other engineering fields. Roget is probably better known for his development of the Thesaurus bearing his name. Log-log scales achieved worldwide acceptance and many variants became standard on many designs of slide rule for general usage. This is yet another example of an idea that was re-invented on more than one occasion—by Thomson in 1881, and yet again by Perry (who patented his version) in 1901, and finally by Boardman in 1933. All these had some "improvements" on Roget's original 1815 idea, over a century earlier. The evolutionary period of the slide rule probably ended with the 18th century, and the final development period began.

Development

The slide rule had arrived and by the mid- to late-18th century had become generally accepted and used across the technological world. Development of the basic ideas continued on several fronts. The rectilinear slide rule was produced in several lengths to give increased accuracy, the most common standard size being the 10-inch desk model. However, the 5-inch pocket model was also produced in huge quantities. There were longer devices, the 20-inch being reasonably common, but somewhat unwieldy. We can identify several specific developments that characterized the advancement of the slide rule into a ubiquitous general calculating tool. We will look at two of these in greater detail. One particular development is the Soho engineering slide rule developed by James Watt (1736–1819) with the assistance of Mathew Bolton in his Soho "manufactory" in Birmingham in about 1790. Soho slide rules were especially recognized for their scale layout, and for their accuracy of scale design and production. The layout with two-cycle logarithmic scales on the top of the stock and on both edges of the slide continued through to the 20th century. This ensured that squares, required in many engineering calculations, were easily calculated by one two-cycle scale

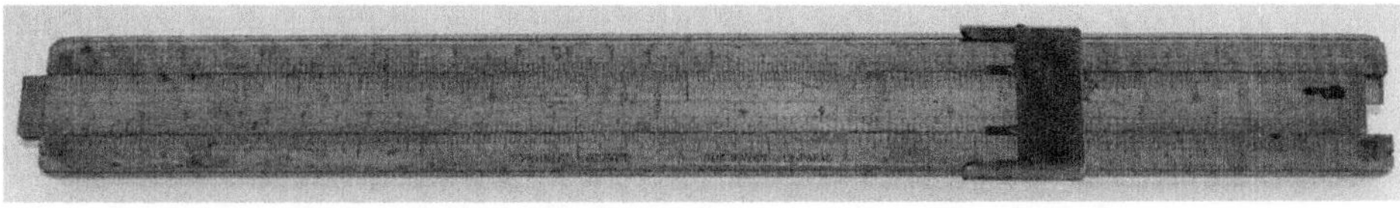

Figure 2.4 Mannheim slide rule made by Taverier-Gravet of Paris, the original manufacturers of such slide rules; circa 1900 (from the author's collection of images)

being adjacent to the single-cycle logarithmic scale on the second edge of the stock. This layout obviated the need for a cursor as the results could be read off directly. Soho rules also featured much sharper scale production with more and finer markings, taking advantage of the recently developed dividing engines and so ensuring increased accuracy.

Our second example is the cursor. This is also the final item of slide rule equipment needed to make the 20th-century slide rule ubiquitous. The cursor was the device that simplified slide rule use by enabling results found on one set of calculating scales to be transferred to another scale, or different type of scale, and if necessary back again. These pairs of scales could be either on the front and/or the back of the slide rule, and intermediate results did not have to be memorized. It is said that Newton (1643–1727) used the first "cursor"—a hair—to enable squares to be calculated. However, the cursor as we know it with a fine hairline etched on a glass screen, in its movable frame, became the final part of the slide rule design we all know and recognize and was to be omnipresent until the end of its life. The cursor plus the Mannheim scale layout, named after Amédée Manheim, a French Artillery officer (1831–1906), who developed and gave his name to this collection of features, was the final improvement in what became a standard slide rule layout. It appeared in about 1851. There were many further detail changes, but the slide rule was now set. Among other important details we can include the reciprocal scale, now positioned generally in the middle of the slide, a change attributed to Max Rietz, (1872–1956) a German steam engineer who gave his name to the Rietz (or Reitz depending which side of the English Channel was writing) pattern of slide rules which existed from about 1910. We now have the final physical design of slide rule and layout of scales that was instantly recognizable across the technological world and continued to be manufactured worldwide until the demise of the slide rule.

We can identify further design features that contributed to the total *oeuvre* of slide rules. One type that became very popular was the true duplex slide rule with scales on both sides of the stock and/or stator, which were synchronized—that is, the index marks all coincided. There had been duplex and indeed multi-scale rules such as the many designs of alcohol and revenue calculating rules, with two, three, and four slides and sets of scales, but perhaps surprisingly these were never synchronized (the index points of the scales did not all coincide). The first true duplex slide rule was invented in the United States and patented by William Cox in about 1891, and sold by instrument makers Keuffel & Esser in New York. Extreme examples of duplex slide rules with over 30 different scales were developed and manufactured enabling a mighty suite of calculating tools to be accessed by the user. There are also very specialized designs for many and varied uses, as wide ranging as esoteric black body radiation calculations and simple speed–time–distance calculation; with many other even more arcane specializations along the way.

It is also educational to look at the original circular designs and spiral scale designs which started with John Brown and Henry Sutton's designs of 1660 and 1663 and continued in various shapes and sizes through the whole lifespan of the slide rule. There were also several helical scale designs which had their champions. These all found their adherents and produced very interesting designs which were all manufactured throughout the existence of the slide rule.

Demise

The demise of the slide rule started in about 1970, contemporaneously with the appearance of the first affordable electronic calculators. The demise occurred amazingly quickly. It took place over very few years, with most of the big manufacturers ceasing manufacture of all slide rules after about 1980. The industrial archaeologist would have great difficulty in finding any other device of such importance which disappeared from general use as quickly as the slide rule. The generally accepted reason for this swift demise was the advent of the electronic calculator, followed swiftly by the affordable pocket electronic calculator. Definitions of "affordable" and "pocket" notwithstanding, the appearance of the HP-35 (35 from the number of keys it had) electronic slide rule calculator in 1972 probably signaled the real beginning of the end for slide rules. The HP-35, at \$365 in the USA and £365 in the UK, was many times more expensive than even the finest quality slide rule. However, the sheer convenience of being able to press buttons to perform sophisticated calculations—including addition and subtraction—with results to many significant figures, was very appealing. Other designs of small electronic calculator had preceded the HP-35. Bowmar in the USA is generally credited with the very first design, and many other well-known and less well-known makers all produced electronic calculators in an attempt to get a foothold in this new and exciting market. It is interesting and notable that many early designs were marketed as "slide rule calculators" because initially the electronic calculator struggled to find its niche in the calculating firmament, and this marketing trick served to slot the new calculators into a place in the calculating cosmos. As a young electrical engineer at that time there was no way that I could have afforded an HP-35. Their price was equivalent to many weeks' wages and there was not the same appreciation of gadgets or high technology that is a feature of modern day marketing and life. I believe I was typical of my peers at that time and it was the advent of Sir Clive Sinclair's much more affordable—and attainable—calculator technology a very few years later, such as the Sinclair Cambridge in 1973 in its many forms, that finally convinced me to move to a calculator from my trusty slide rule. These Sinclair calculators were also available in even cheaper kit form, and most engineers, electrical as well as mechanical and civil, at that time were well able to put together electronic kits, whether they were for high-fi, amateur radio, calculating, and later for computing. Sinclair's marketing strategy was also brilliant; if it did not work, return it to the company and a working calculator would be returned! This was a quite incredible offer, and was just one of the elements that made them so attractive to buyers. Even so, it was still several years after I bought my first calculator that I finally consigned my slide rule to a peaceful end in my desk drawer—although it was still somehow easier to use a slide rule for quick and dirty calculation, and I still have one on my desk as I write this. There was a down-side; early calculators used batteries at a phenomenal rate. This was a cost not found with a slide rule!

There were numerous ill-fated attempts at extending the life of the slide rule. Perhaps the saddest was the attempt by Faber-Castell in Germany (one of the largest manufacturers of slide rules in the world) of adding an electronic calculator to the reverse of a slide rule. Several such models, with increasingly sophisticated calculators, were produced. How many owners of what was a very expensive slide rule ever used the slide rule in preference to the calculator is not known! It was an understandable development when it is realized that

Faber-Castell, together with other manufacturers, had previously manufactured both 5-inch and 10-inch slide rules with a Pescaline/Troncett adder on the back, called an "Addiator", and these had successfully sold. While this form of adder had been a useful development allowing addition and subtraction, it was definitely redundant to have a sophisticated electronic calculator performing every function that was available on the slide rule—effectively rendering the slide rule obsolete!

The slide rule ethos means that there is now a whole generation, which is fast disappearing, who can remember what a special occasion the purchase of our first slide rule was. This would have followed an extensive and careful selection process with much discussion and deliberation. This culminated in the actual day of purchase—usually preceding another special day or rite of passage—the first day in 6th form, start of National Certificate or Diploma courses, technical college, university, and so on, complete with that shiny new slide rule. The slide rule was not something that was changed for a newer and better example on an almost annual basis as is so much of today's technology. As the saying goes: "A slide rule was for life, not just Xmas!" It was possible that one would buy a better slide rule, better in either or both quality of manufacture or scale selection if it was found that one's first bought was lacking in specific capability or falling to pieces. In general though, the selected slide rule was a one-off purchase for life, an attitude that is totally at variance with today's technological lifecycle. While I now have a multitude of calculators of varying cost and capability, I only ever had one pocket and one desk slide rule. As a collector, I now have over one thousand, but that is as a collector and never as the real-life user that I used to be.

Slide Rule Mathematics

The ubiquity of the slide rule is not immediately obvious until one studies the history of technology across the developing world. When this is done, one finds that every one of the modern wonders of the technological world were developed using a slide rule. A few moments' thought will show that this has to be so—there was no other form of calculator available! Thus all technologists who had to calculate would have had a slide rule, and undeniably they were so ubiquitous that this fact of life is almost never mentioned in autobiographies of the famous designers and engineers. Just a few examples will show what I mean. Sir Frank Whittle inventor of the jet engine was a slide rule user, and James Watt, inventor of the steam engine was a slide rule inventor and user. Many of the Manhattan Project members have been proudly pictured with their slide rules. Authors Robert Heinlein and Arthur C. Clarke were slide rule users, Clarke being christened "Fastest slide rule in Whitehall". Aircraft designers Frederick Handley Page of the UK, Sergei Korolev, "father of the Soviet Space Race", as well as Wernher von Braun his opposite number in the USA, were slide rule users. Architect Frank Lloyd Wright and David Packard of Hewlett Packard fame were users. This is a wide variety of technologists, and there are many others we could also name. How do we know this? They were all proudly photographed with their slide rules. It is an enduring legacy of which we can all be proud. The engineer and technologist were synonymous with a slide rule!

The slide rule by its very design required a familiarity with the order of magnitude of the numbers involved in a calculation. Virtually all books on the slide rule and any training course concentrated for a proportion of their time on the fact that, for example, the "2" on the "C" scale could mean 0.002, 0.02, 0.2, 2, 20, 200, 2,000, etc. Performing the calculation then took place with the final decimal point being placed after an "order of" calculation had been performed. Most of my generation of technologists and mathematicians will look at a sum written down and mentally work out its order of magnitude. This sum could involve many multiplications and divisions as well as trigonometrical or other functions. Quick mental arithmetic together with an essential knowledge of some common trigonometrical and square function values would enable one to come to this order of magnitude value fairly rapidly and almost automatically. The design of the slide rule with its single-cycle "C" or "D" logarithmic scale used in the majority of calculations was numbered from "1" at the left-hand end of the scale to "10" at the right-hand end. However, depending on the calculation at hand, those numbers could mean any decade one wished, provided one was consistent throughout the calculation and, most importantly, kept track of the decimal point. As mentioned, the slide rule user had already instinctively looked at a calculation and came up with an "order of answer" before carrying out the calculation in detail. The slide rule itself, by the nature of its markings, was only capable of giving two or three significant figures (depending where on the logarithmic scale the answer was found) which meant that users developed a "good enough" approach to the accuracy of that answer. Where greater accuracy was essential, one used either a longer slide rule—20-inch was not unusual, and there were some special monsters designed—or else some alternative extended format such as helical or spiral scales to give a much greater scale length within a workable slide rule size. If all else failed, then one had to resort to 7-figure logarithms or alternatively some form of mechanical or early electronic calculator. I can well remember that while at college we were recommended to attend a course on the use of the Brunsviga mechanical calculator (and indeed, the incredible tearing noise it made when subtracting 1 from 0.9999). Later, I can remember having to "book out" and use the one and only early Facit electronic desk calculator we had on site, complete with its row of glowing "Nixie" number tubes, capable of working to 13 digits. This was required to perform electronic equipment reliability calculations which were impossible on a slide rule and really beyond 7-figure logarithms as well.

The slide rule's inability to be used for addition and subtraction meant that many people who had to work with such numbers—for example, monetary calculations in £. s. d., with its 12 pennies to the shilling, and 20 shillings to the pound—developed a special capability in addition and subtraction. There were many people who were capable of adding a column of figures as fast as the pencil could be run down the column. This was most impressive to witness in action. There were also many mental tricks that were used in simplifying multiplication and division. Certainly all of my peers would have known to the penny what change they expected from a tendered note for any shopping—without having to resort to looking at the till! It is sad that there are films (such as *Apollo 13* (1995)) which show a slide rule apparently being used for addition and subtraction. When ignorance is bliss it is truly folly to be wise!

The modern generation's reliance on the electronic calculator has resulted in a loss of numeracy and facility or comfort with numbers and an over-reliance on spurious accuracy. An example is the asinine statement that one hears where in any conversion, say, from miles to kilometers, the result is regularly stated to three significant figures (at least!). All this is now even further exaggerated when Microsoft's on-screen Personal Computer calculator in the latest versions of Windows can present results to no fewer than 32 significant figures and with a heady four-digit exponent in scientific mode. This inevitably means that conversions are presented with totally unrealistic and unnecessary accuracy, something that the slide rule user would find most unsatisfactory. This is but a part of our loss of understanding of the significance of the numbers being quoted.

The Digital Revolution is no longer part of the "burning heat of technology" and there is a certain irony that the very calculating device that will have produced that revolution is a major victim of it. Purely "bashing in" numbers with a decimal point into an electronic calculator, and not understanding the value of the figures that are to hand, results in a lack of understanding of the result, and hence not recognizing mistakes when they occur. Here in the United Kingdom, arguments are presented for and against the learning of multiplication tables by rote; the one side claims that it removes all spontaneity of the calculation process, the other noting that it adds fluency to the understanding of numbers. In the UK, our reliance on a very rigidly defined National Curriculum which does not give any scope for maneuver by the maths teacher means that the use of the slide rule by the modern school pupil is the rare exception. In general, only outside-school activities allow for alternative methods of calculation to be studied—for example, the Bill Tutte (he of Enigma and Bletchley Park fame) Maths Club in Newmarket. It would appear that on the other side of the Atlantic, American scholars are more welcoming of such "retro" activities. The American Oughtred Society which looks after mainly US-based slide rule collectors in conjunction with the web-based International Slide Rule Museum (ISRM) has a flourishing "loan" program to schools and colleges, which we do not mirror here in the UK. It means that future generations of trans-atlantic scholars will probably learn to be more fluent and numerate in their number work, to their everlasting benefit.

The minor counterpoint is that there still are a very few areas of learning where the use of a slide rule features. Fledgling pilots must still demonstrate their capability to use their "Whiz Wheel"—the long-established ASA E6B Flight Computer which includes several logarithmic scales—before they will be awarded their first Private Pilot's License. This supposedly covers the possibility where their modern navigational device fails due to lack of power or malign influence. A second area where the convenience of a simple slide rule still features is in the calculation of gestation period for a new mother; this is still the most expedient way for an obstetrician to calculate the birth date. It is also interesting to note that the UK Open University still quotes a set of log-tables and a slide rule as a pre-requisite for their maths and science courses—are they dinosaur or forward-looking geniuses? Only time will tell.

Inevitably, collecting slide rules is a popular specialist hobby. There are as many reasons given for collecting these devices as there are collectors. There is some feature of a slide rule available to interest all types of collector. At one end, we have the nerd who might enjoy some of the esoteric scales or else the specialist types and formats. For the technologically aware collector, we have slide rules dealing with specializations in technology from nautics to nuclear. For all collectors, we have different complexity, size and value, maker, specific technical specializations, and many more personal reasons. Indeed, there is something to cover the full strata of collectors' ambitions (Hopp, 1999, 2009, and 2011). Many countries have their own collecting organization. The United Kingdom Slide Rule Circle (UKSRC), the Oughtred Society in the USA, the Kring Historische Rekeninstrumenten (Kring) in Holland, and the Rechenschieber Treff (RST) in Germany are examples. In some cases, these organizations have a wider remit such as mathematical instruments and calculators, both mechanical and electrical. Links between the organizations are surprisingly strong. There has been a long-established "round" of annual International Meetings of slide rule collectors which circulate between different countries, so far with meetings in six different countries. Unusually—and unexplained—most collectors started collecting some 20 years after the slide rule fell out of favor, and sadly, as with many collectors, it is an "old persons" game. Much of the youth of today do not have the time or the inclination for collecting. What will happen to these collections? Probably nothing. They will molder in a loft somewhere until a future generation throws them away in complete ignorance of the inherent beauty and value, real and historic, of the collection.

Conclusion

It is quite remarkable, and possibly unique, that an item of analog technology can have gone from ubiquity to virtually forgotten in less than 400 years, where 350 of those years included an inexorable and exciting development path which mirrored the new technologies that required calculation, producing ever more sophisticated calculating equipment. An item of essential equipment for the technologist in all guises has now become a collector's item. Fortunately there is much for the industrial archaeologist to study. The methods of manufacture and marketing exhibit all the characteristics of the appropriate period in history, and the devices themselves are quite incredibly beautiful with plenty of different attributes to tempt the specialist within the collecting field. Indeed from "Hero to Zero" is a pithy way to characterize this delightful piece of industrial, social, and technological history!

References

Aubrey, John (1693), "Schediasmata: Brief Lives". John Aubrey stated collecting his biographical sketches which he called "Schediasmata: Brief Lives" in about 1680 and continued till about 1693 when the collection was presented to the Ashmolean Museum in Oxford. These have subsequently been edited into a number of versions of "Brief Lives . . ." by several editors. They present a fascinating contemporary view of many famous figures of that time including Oughtred.

Cajori, Florin (1920), "On the History of Gunter's Scale and the Slide Rule during the Seventeenth Century", *University of California Publications in Mathematics*, Vol. 1, No. 9, February 17, 1920, pages 187–209.

Hopp, Peter M. (1999), "Slide Rules, Their History Models and Makers", Apple Valley, Minnesota: Astragal Press, 1999.

Hopp, Peter M. (2009), "Joint Slide Rules, Sectors, 2-foot, 2-fold and similar slide rules", San Francisco, California: Hexagon Press, 2009.

Hopp, Peter M. (2011), "Pocket-Watch Slide Rules", Apple Valley, Minnesota: Astragal Press, 2011.

3

THE HISTORY OF PUNCHED CARDS

Using Paper to Store Information

Robert S. Wahl

My great-grandfather was in the textile industry in Switzerland. Based on his work creating a portrait of then-president William McKinley using a loom and punched cards, he was recruited by a major American textile firm as a result of the complexity demonstrated in weaving the portrait, and immigrated to the United States. Many years later, my introduction to computers involved the use of punched cards. This personal anecdote illustrates the longevity of punched cards as a storage medium.

Punched cards are small pieces of heavy paper stock that were used in a variety of industries for the control of machinery. This was achieved by punching a series of holes in the cards along a series of rows and columns. When these cards were run through a reading machine, a set of feeler arms (small lightweight pieces of metal) would detect a hole and its location on the card, and make associated changes to the machinery. When a hole was encountered, the feeler arm dropped through the opening and a mechanical or electronic relay on the lever arm closed indicating to the machine that a hole was detected.

One of the first industries to use punched cards was the textile industry starting with the Jacquard loom, first demonstrated in 1801. Looms for producing textiles used punched cards to control both the weft threads and the warp threads. If you are unfamiliar with looms, the warp threads are the threads that are wound back and forth in rows and the weft threads are sent through separations in the warp threads. By using this technique and by frequently changing the color of the thread, intricate patterns could be woven into materials. The use of punched cards was an improvement over the earlier use of paper tape to control the weaving process. Paper tape was first developed by Sir Charles Wheatstone in 1857, and was used to control computers from 1944 to the mid-1970s.[1]

Over the period of time punched cards were used, they came in a variety of sizes and layouts and their use became more sophisticated and the patterns that were woven became more intricate. Today, many modern fabrics have the patterns printed on them, instead of woven into them, and an easy way to tell the difference is that woven fabrics appear the same on both sides as the color threads are woven around each other. In printed fabrics, the "inside" of the material usually has a washed-out appearance where only some of the printed color has bled through the material.

Punched cards were used for many years in both the textile industry (1801–present)[2] and for election voting (1890–2012)[3] before they were used for the control of computers starting in the 1940s. Punched cards were also used for timekeeping in businesses. Employees would write their hours for the week onto a punched card and then a keypunch operator, working on a keypunch machine, would later key the information in by physically punching the appropriate holes in the card. These cards were then used as input to a mechanical tabulating machine or to a computer program running on a mainframe computer. The technique of using punched holes in paper was also utilized in controlling player pianos and fairground organs. Changing the punched roll of paper changed the melody. The use of punched paper was, in itself, an improvement and cheaper than using metal discs to change melodies as found on earlier music boxes and fairground organs. These early metal discs used a series of holes in the metal with raised nibs to pluck a comb with little metal prongs of varying lengths, to produce the musical notes. While this is a different method than was used by punched cards on computers, the basic concept of using holes in a material to control devices was similar.

As mentioned previously, punched cards were used long before the invention of computers. The textile industry was an earlier adopter of punched-card technology and their use was credited to Joseph Marie Jacquard starting around 1810. Jacquard developed a system to control textile machines and to automate the process of switching between different colors of thread by changing the card input. Using this technique, intricate textile patterns could be achieved.

Figure 3.1 A Jacquard loom (photograph taken by Edal Anton Lefterov).

Heading toward Computers

In the early 1820s, a mathematician and polymath named Charles Babbage used punched cards to control a machine known as the difference engine. The work of Charles Babbage was influenced by the earlier work of Jacquard.[4] This was an early predecessor of today's computers and it contained many of the concepts that are incorporated into today's machines such as internal memory and an ALU (arithmetic logic unit). Babbage developed this idea but did not actually build a working machine largely due to funding issues and machinery limitations of the day. The difference engine was followed by several analytical engine and difference engine designs, which progressed in abilities and in simplified designs. Babbage's son Henry completed the first working Babbage analytical engine after Charles Babbage's death. Babbage's first analytical engine design had the ability to produce output on punched cards. Babbage's designs included several other features that were incorporated into computer programs used on future computers including sequential control, looping, and branching between instructions.

If you were alive and working with computers in the 1960s, 1970s, or 1980s, there is a very good chance that you were exposed to the use of punched cards, which were one of the main ways of inputting data and information into large-scale computer systems of the day. During this era of early computers, the personal computer (PC) was just starting to emerge with revolutionary machines such as the TRS-80, Commodore PET, and Apple 1. With the introduction of PCs, a gradual transition from using punched cards as the primary means for entering data, to the use of keyboards, began. Thus, for the vast majority of people, using mainframe computers was their only exposure to working on a computer and punched cards were the way to enter input and to program the computer. The IBM punched cards were initially designed in 1928 and were used by computers from their inception until the late 1960s. IBM had the largest market share for punched-card sales followed by other vendors such as Remington Rand.[5]

The 1960s and 1970s, then, were the heyday of using punched cards as the primary means of entering data into computers. A less common but still important use was for computer output; the cards used for output were then used for input into another computer program. In this way, computers could produce card output, which could then be used as input for a different computer job; punched cards occupied a unique position in the history of computing as they could be used for input as well as output. They could also be used for storage as punched-card decks could be stored and reused over and over again as long as they did not become damaged. For some applications, the cards were created for a single use only and the cards were destroyed after the processing was completed. Single-use cards were more prevalent in the early use of cards and the use of punched cards for storage became more common later in their history.[6]

Punched Paper Tape

Punched tape was also used during the early history of computing (1940s–1960s) but its use never obtained the volume of punched cards likely as a result of the increased storage and functionality of the punched card.[7] The first use of perforated paper tapes was to control textile looms in 1725 by Basile Bouchon. Early paper tapes were made from a series of punched cards joined together and fed through a machine one after another. The chain of cards was easier to deal with than individual cards for the textile industry. Historically, punched paper tapes were used for telegrams, textiles, stock market "ticker" machines, teletypewriters, and minicomputers (see Braxton Soderman's essay on ticker tape in Chapter 6).

Figure 3.2 Paper tape (from the author's collection).

Other uses for punched tape included newspapers (used until the mid-1970s), early cash registers (1970s), and even reporting baseball scores between stadiums. Over the time it was used, punched tape changed from being a series of connected punched cards to being one long strip of paper. While various standards existed, punched tape typically consisted of either five or eight rows of holes to maintain consistency with Telex machines.[8] The punched paper strips varied in width from three-quarters of an inch to one-inch and were limited to storing only a few characters of data per inch of paper tape.[9]

The paper material that was punched from cards was collected in a bin under the card-punch machines so it could easily be disposed of. The small bits of paper were typically called chads. An interesting reuse of this material was for "ticker tape" parades where the punched out material was tossed from high-rise buildings onto parade routes.

Punched cards had been in use since the 1890s for the U.S. census,[10] but this technology was in use even earlier. For the 1890 United States census, a 24-column card was used. This same format was used up until the 1940 census where a 45-column card was used.[11] Following their use in the 1890 census, punched cards were used by the United States Army during World War I to keep track of medical records and supplies. Insurance companies also used punched cards to track customer data beginning as early as 1896.[12] Large retail chains used punched cards for inventory control and railroads were adopters of punched cards starting in 1896.[13] Railroads used punched cards to track all of their rolling stock and to track merchandise that was being shipped by rail.

The turnpike systems in the eastern United States used punched cards as the primary means to track toll amounts 1940–1983.[14] I remember as a child riding through Pennsylvania with my family and stopping at the first toll plaza to pick up a punched card. This smaller sized card had all of the tollgate information and costs printed on it so that drivers would know how much it would cost to exit the turnpike down the road. These cards were mechanically read at the exit to determine the toll cost. On these family vacations, getting to hold the tollbooth ticket was a close second to the excitement of riding through the Pennsylvania Turnpike tunnels.

What Need Did They Fill?

Although using punched cards for computer input seems awkward and slow by today's standards, it was a giant leap forward over hardwired computer programming such as was done on the ENIAC (Electronic Numerical Integrator And Computer), considered to be the first general-purpose electronic computer. Before punched cards were adopted, the method used for entering information into computers was via hardwiring. Entering instructions into computers was even more difficult using a hardwired setup:

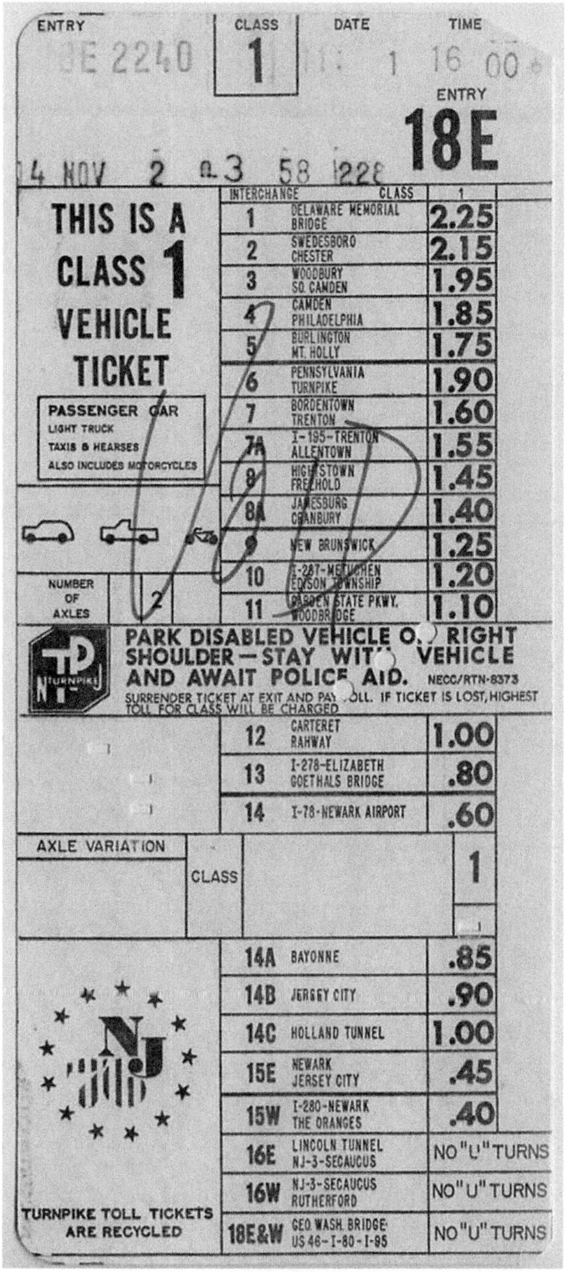

Figure 3.3 A punched card used for the New Jersey Turnpike (photo used with permission from Douglas W. Jones, University of Iowa).

One of the peculiarities that distinguished ENIAC from all later computers was the way in which instructions were set up on the machine. It was similar to the plug boards of small punched-card machines, but here we had about 40 plug boards, each several feet in size. A number of wires had to be plugged for each single

> instruction of a problem, thousands of them each time a problem was to begin a run; and this took several days to do and many more days to check out. When that was finally accomplished, we would run the problem as long as possible, i.e. as long as we had input data, before changing over to another problem. Typically, changeovers occurred only once every few weeks.[15]

For this generation of computer operators, users, and programmers, the development of punched cards was as exciting as the release of any new technology today. Looking back at this technology, it appears to be highly outdated and inefficient, but at the time it was heralded with announcements such as "The wondrous magic of punch card maintenance of records".[16] Punched cards, while obsolete today, "came to symbolize all that was up to date and businesslike".

Importance

Punched cards played an important role in the history of computers and accomplished a feat that will likely never be surpassed. Punched cards were used to control computers for almost 50 years (1950s–2000s).[17] If you take into consideration the use of punched cards for non-computer use, they have been used for more than 200 years. Pre-dating Jacquard's loom was an invention by Basile Bouchon in 1725, which used perforated paper rolls.[18] This means that we are approaching 300 years of use. None of today's popular technologies are likely to achieve this length of use, as technology changes seem to be happening more rapidly in recent years. Floppy disks were used for approximately 40 years until they were replaced by other, smaller, more durable storage media such as USB sticks.[19] CD-ROMs, used for storing computer data and applications, came into commercial use in 1985 and remained popular until DVDs emerged in 1995, which were in turn followed by Blu-ray optical disks in 2003. CD-ROMs, DVDs, and Blu-ray disks are collectively known as optical media, the use of which is dwindling, as many new laptops are being shipped without any optical drives. While it is too early to write off this technology, its use in the home is shrinking and being replaced with streaming technologies.

The ubiquitous USB flash drive became popular around the year 2000. Since this time, storage capacity has increased and cost per megabyte has fallen. As of the time of this writing (mid-2017), 2 Terabyte USB flash drives are available. Even with this growth, USB flash drives will be replaced by a new technology down the road such as cloud storage or streaming data. Given the rapid changes in technology and in customer demands, the nearly 50-year reign of punched cards is truly remarkable.

Punched cards were so ubiquitous during this period of time that their look became popular in other areas. The Engineering Research Building on the campus of the University of Wisconsin–Madison was built in 1966 and the window design and the overall shape of the building resembles a punched card standing on end, an interesting interior design in which some offices benefit from having many windows while others have only a few or none.

Different Types of Cards

Throughout the years and across various industries, numerous sizes, types, and layouts of cards have been used, with no universal card design. The earliest cards and paper tape used to control textile looms, went through many iterations. While Joseph Marie Jacquard is frequently mentioned as the inventor of this system, his work was an advancement on the

Figure 3.4 The UW-Madison Engineering Research Building, which resembles a punched card standing on end (photo from the author's collection).

punched-card designs of Jean-Baptiste Falcon, Jacques Vaucanson, and Basile Bouchon.[20] Basile Bouchon is credited as the inventor of paper tape to control textile looms but his design was limited by using paper tape and had a limited number of needles. Jean-Baptiste Falcon modified the initial design and substituted a series of paper cards, which improved the durability of the system. Jacques Vaucanson attempted to have the looms run unattended and, later, Jacquard was successful in achieving full automation.

Herman Hollerith is credited with developing the use of punched cards for controlling computers, and his work was based on the earlier work of Jacquard and Babbage, the latter of which is credited with the origination of the concept of a programmable computer. Herman Hollerith, working in the 1880s, used punched cards and developed a system that detected punches when the cards were passed over a set of electrical contacts using the previously described feeler arms. While Hollerith is fairly well known in the computer industry, credit for the development of a punched-card system for the United States census also belongs to medical doctor John Shaw Billings.

Figure 3.5 Herman Hollerith and his glorious mustache.

John Shaw Billings developed the initial concept of the punched-card tabulator for the census, while Herman Hollerith is credited as the inventor and builder of the system following discussions with John Shaw Billings. John Shaw Billings was well known as a lover of reading, and had amassed a large personal collection of books (that became the National Medical Library) for which he had developed a tracking system.[21] Billings might have read about the early attempts at replacing hand calculations with mechanical means such as those undertaken by Charles Babbage, Blaise Pascal, and Gottfried Wilhelm Leibniz (who developed a mechanical calculating machine in 1673).[22] According to Chapman:

> If there is a single unifying theme in Billings' illustrious career it is information storage and retrieval; and although he may never have heard the word computer, he was brought to consider ways and means to manage great masses of numerical data by his involvement with the census and with vital statistics.[23]

Billings recalled information that he read on mechanical calculators by Blaise Pascal and Charles Babbage and this information was passed on to Herman Hollerith. In fact, he is quoted as saying to Hollerith "there ought to be some mechanical way of doing this job, something on the principle of the Jacquard loom, whereby holes in a card regulate the pattern to be woven."

Hollerith took Billings' suggestion and developed a punched card and an electromechanical system that read the cards and sorted them based on the location of the hole punches. While Billings was concerned with public service and his duties, Hollerith realized the commercial potential and developed the Tabulating Machine Company in 1896.[24] The paths of Billings and Hollerith were not at odds with each other, and it is clear that Billings was well aware of Hollerith's commercial pursuits. So Billings came up with the initial idea to use punch cards based on earlier work on Jacquard and others, and Hollerith was responsible for the implementation and development, which led to a successful 1890 census and future commercial success. Hollerith's Tabulating Machine Company merged with three other companies and became the International Business Machines Corporation (IBM).

There have been numerous designs, formats, and uses for punched cards over their 50-year history as a computer input device. While some of these designs have been lost to history, several were popular and widely used in their day. The original IBM or Hollerith cards used for the 1890 and 1910 censuses were 3¼-inch high and 6⅝-inch wide (matching the size of US paper currency) and contained 24 columns of holes. The first four columns on the left side of this design were used to punch in a unique card identifier. The remaining columns could be used for other information. It is interesting to note that these first Hollerith cards did not have anything written on them, unlike the typical computer punched cards from the 1970s and 1980s. Yet even though the cards had no written information, people working with the cards became adept at reading the cards visually.[25]

In a typical card deck of the 1970s and 1980s, the first part of the deck contained control information on what was called control cards. These were used to identify what compiler to use (e.g., Fortran or PL/1), how much disk space to allocate, how much of the core processor to use, and any accounting information.[26] This information on IBM machines was typically written in Job Control Language (JCL). The control cards were followed by the program cards, which were typed by either the programmer or a keypunch operator. A stack of data cards followed the program cards. All of these cards needed to be in their exact order for everything to run correctly.

An interesting variant of punched cards was developed by IBM and was known as the Port-a-Punch. This was a small handheld device that was developed to allow users to enter information onto punched cards manually without the need for a large cardpunch machine. The idea was that manually punched information could be used for recording product sales, taking a physical store inventory, tracking shipment information, or for surveys. This product was introduced in 1958 and was also used by computer operators and by programmers to manually reproduce a card if a card deck failed in the middle of processing,[27] eliminating the need to return to a cardpunch machine to punch a replacement card. The idea was that if a card deck had one thousand or more cards and one was bent, torn, or incorrectly punched, the replacement card could be quickly produced and the processing resumed. Punching was accomplished by placing a new punched card into the small, portable device and then using a stylus that resembled a pen to punch out the holes. The stylus had a small metal tip to help it pop out the chad. The Port-a-Punch device was only slightly larger than a standard Hollerith card, and the cards were pre-scored to facilitate the punching.

Another application of punched-card technology that shows the wide variety of disciplines in which they were used was in the field of corrosion resistance where a system was developed for tracking and quickly locating technical articles. This system used unique punched cards that contained two rows of holes around all four edges of the cards. A classification system was developed, and by punching the correct holes on the cards, technical papers could be classified and quickly located, like an early form of Google Scholar.

This system was known as the McBee system and was used from its inception in the early 1950s until the end of the decade.[28] When a mechanical card reader processed the cards, cards that met the search criteria were physically dropped from the deck, while articles that did not match the search criteria were not dropped.

As mentioned earlier, punched cards found uses in a variety of industries and applications. A common use for punched cards was for bookkeeping. Businesses used punched cards to track corporate data and for input to simple systems that could add, subtract, multiply, and divide.[29] An interesting application that combined technologies of the day was the use of punched cards for tracking magazine subscriptions. The system's punched cards were of a different format and layout than the ubiquitous IBM format. The cards were based on the Remington Rand format with three of the four corners cut off and areas on the card that contained no punches which were used for mailing labels and text. These punched cards could be read by machines and by magazine subscribers, and contained information such as the shipping address, the amount due, expiration date, and type of subscription. When the card was returned from a subscriber, a computer could read it and sort the records by city, state, and subscriber last name. Using this system, records could be filed and sorted much faster and more accurately than by hand-processing. An interesting side benefit, used by the American Nurses' Association, was that the information on the punch cards could be used for additional research. As nurses subscribed to the journal, information was included on what their areas of specialization were. By gathering this new, additional information, research and statistical analysis could be performed to determine statistics such as the percentage of nurses specializing in one area within a given geographic region.

IBM was the leading proponent of punched cards for controlling computers, and during the 1950s, sales of punched cards were responsible for 20% of the company's revenue. While IBM dominated the early computer industry and required the use of IBM-sourced cards for use in their machines, they were not the only player in the punched-card game. Remington Rand Corporation had its own punched-card format, different from the standard IBM format, and the company acquired the Powers Accounting Machine Company,[30] founded by James Powers, Hollerith's co-worker at the Census Bureau. Hollerith and Powers set the stage for the two large companies that would compete from the late 1920s to the mid-1950s. One major difference between the IBM system and the Remington Rand system was that the Hollerith IBM system used electronics to determine the location of punched holes in the cards, while the Remington Rand system was purely mechanical. In some cases, the Remington Rand system might have been superior to the IBM system, in that it was faster and more reliable.

The most commonly used punched-card format was the 80-column IBM card, which were 80-columns wide and 10–12 rows high. Small numbers labeled 1 through 80 were printed on the cards to visually indicate to the card puncher which column was being punched. The first row contained a string of zeros, the second row contained a string of ones, and so on. Representing numbers from 0 through 9 was very simple; if you wanted to start a line with the number 2, the corresponding 2 was punched out. When the feeler arm ran across the opening, an electrical connection was made and the number 2 was entered into the computer.

A–Z characters were slightly more complicated to represent. The top two rows of the IBM card were not labeled, but could be used for control. These top two rows were known as zone rows, and the holes were known as zone punches. The topmost row was known as the Y-row or the 12-zone, and the second row was known as the X-row or the 11-zone. Depending on the card format being used and the character representation, a single character could require

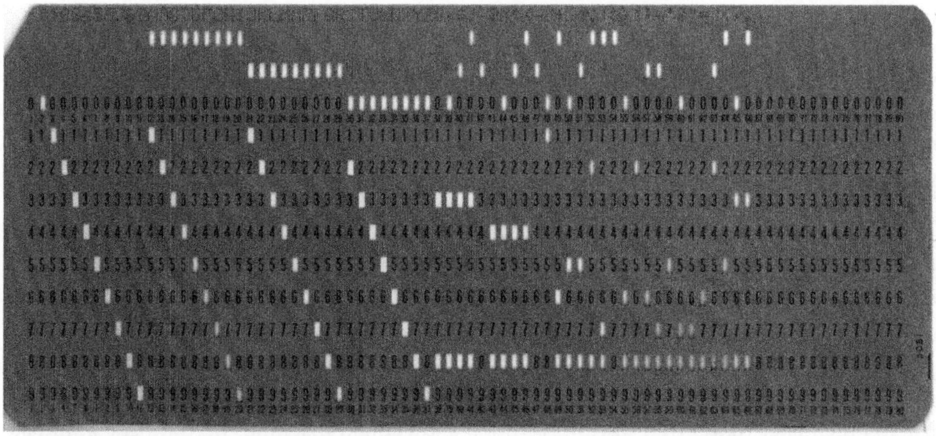

Figure 3.6 An IBM punched card.

the use of up to six punches within a single column. These characters were based on IBM's EBCDIC encoding, rather than today's more common ASCII or Unicode encodings.

The IBM system, including the System 360 mainframe and all of the associated peripherals, was so popular that in the early days of computing, it seemed like there were only two types of machines that people worked on: "a 360" or "something else". Whatever the "something else" was did not truly matter, as IBM had such a large dominance in the industry. IBM's 360 held a 65.3% share of the market in 1965, and the closest competitor was Sperry-Rand with 12.1%.[31]

Two Parts Needed

Two devices were needed in a system, the keypunch machine and the card reader. The keypunch machine was used to punch the rectangular (or round) holes into the punched cards. Keypunch machines contained a hopper that held a stack of unpunched cards, which were then fed individually into an area called a punch station where the keypunch operator would type on a keyboard and the corresponding holes would be punched into the cards. The challenge was to get a completely clean punch and to separate the punched-out material cleanly from the card (the problem of incomplete punches will be described later in this essay). The second piece of equipment needed, the card reader, was the machine that was able to detect the holes in the punched card and to identify the columns and rows of the punches.

Often, computer rooms had a third piece of equipment; a card sorter was used to physically sort the cards into order. By off-loading this process from the mainframe, expensive processing time was saved. A card sorter would take multiple passes to completely sort a deck, based on one field of information on a card. Cards were sorted in order based on the lowest column, then by the second column, etc., in a technique known as a *radix sort.*[32] One of the most popular card sorters of the day was the IBM Type 80 sorter, which had the ability to sort up to 450 cards per minute, and later models were capable of sorting up to 2,000 cards per minute.

The concept that seems most foreign to users of today's systems is that rarely were any of these pieces of equipment connected, either by wires, or wirelessly; each of the machines was essentially standalone. The programmer could write code on paper at a desk and then pass it on to a keypunch operator or punch the information personally. The keypunch machine produced a stack of punched cards that were physically carried to the next device, so there was simply no need to have the machines interconnected. Next, a card reader could be used to check or verify the deck; this machine was also standalone and not connected to the mainframe. Eventually, when everything was checked out, the cards were physically walked over to the mainframe where an attached card reader loaded the information into the computer. Output was either sent to another cardpunch machine or printed on green-bar paper.

My first experience with punched cards occurred while I was pursuing an Associate's degree in "Business Data Processing" during the early 1980s. I vividly remember typing PL/1, Fortran, and RPG programming language assignments onto punched cards and submitting a deck of cards to the computer operator so they could be run. This process is, of course, very different from writing computer code today, but in some ways it forced programmers to be more thoughtful and precise. Computer code was typically written on paper first and then later transferred to cards through a keypunch machine. If you made a mistake and typed information into the wrong column on the card, the card needed to be tossed out and a new card typed. Often information typed into a wrong column was only discovered after the computer program was run. If a card got torn or bent, it needed to be tossed out and a new card typed. Once your card deck was correct, you carefully walked it over to the operator and tried hard to avoid dropping the deck. By default, the cards were not numbered, so a drop meant that you might have needed to re-punch the entire deck since there was no easy way to see what order the cards should be in. When creating a card deck, you had the option of adding a card number to each card but I remember that this was not commonly done as it was extra work. One trick that was used by some clever programmers of the day was to draw a diagonal line across the top of the card deck. With this line in place, if a deck was dropped, it was fairly easy to put everything back in the correct order, and/or identify where cards were possibly missing.

The computer operator then ran your program in the order it was received by loading your cards into a card reader. Once the job was finished, you would pick up your deck of cards and a printout of the results on green-bar paper. As a student, further problems arose from transporting these cards to and from school in the bottom of a backpack. There were many opportunities for a card to become bent and unusable. One of the fond memories I have from this time was the rather satisfying sound of the card readers. As the cards were mechanically separated and fed into the machine, there was a sound produced that was not unlike the sound of a small engine running. Apparently, I am not alone in the enjoyment of the sounds from punched-card machines; a book review written in 1942 contains the following quote:

> Some day, perhaps, a study will be made on the psychological effects of punched-card machines. The aesthetics of the machine, the whirling noises and fluttering cards, with the neatly ordered columns of figures as the end product, seem to have a fatal attraction for the human mind.[33]

This complicated and cumbersome process meant that a programmer could only run a program typically once or twice in an afternoon. This is very different from today's rapid-fire process of changing a line of code, running it, changing something else, running it, etc. Because of the amount of time it took to complete this cycle, programmers tended to be very careful about the code submitted and more time was spent on bench-checking code before even attempting to run it.

This process, while inefficient, had one advantage. The punching of cards could be done on a separate machine without the need to use the mainframe computer. These keypunch machines offloaded the card work and allowed the mainframe to be used solely for processing a computer program. This was a large advantage over the use of hardwiring programs, which effectively kept the mainframe as a single-user machine during both the "coding" and "running" phases. Keypunch machines were relatively inexpensive when compared to the cost of mainframe computers, so offices and schools typically had many rows of keypunch machines available.

Why Did Punched Cards Fall Out of Use?

As mentioned above, there were limitations to punched cards, which led to a search for faster and more efficient ways to enter, store, process, and output computer information. Over time, the use of punched-card technology decreased, but replacing this technology was a slow process. The United States Social Security Administration used punch cards up until 1995.[34] Contributing to the decline of punched-card use were problems of storage, cost, and speed.

Storage

Individually, a single punched card took up only a small amount of space, but collectively the number of punched cards needed to run a program took up quite a bit of space, and the total number of punched cards needed to run a business took up a large amount of space. By 1937, IBM was producing 10 million punched cards each day and this continued until the 1950s when other storage media started being used.[35] Determining the total storage space used by computers today is difficult and estimates vary. In 2017, estimates of the total size of storage needed for all material on the Internet only (not counting personal and business storage not on the Internet) are between 10 and 11 zettabytes and predicted to grow to 20 zettabytes by 2020.[36] Based on back-of-the-envelope calculations made by the author, a 2-inch stack of IBM punched cards contained 508 cards. Each card contains a maximum of 80 characters.[37] (With one byte needed for each character, a punched card could hold, at best, 80 characters.) My 2-inch stack of cards shown in Figure 3.7(b) could contain 40,640 bytes. If we

(a) (b)

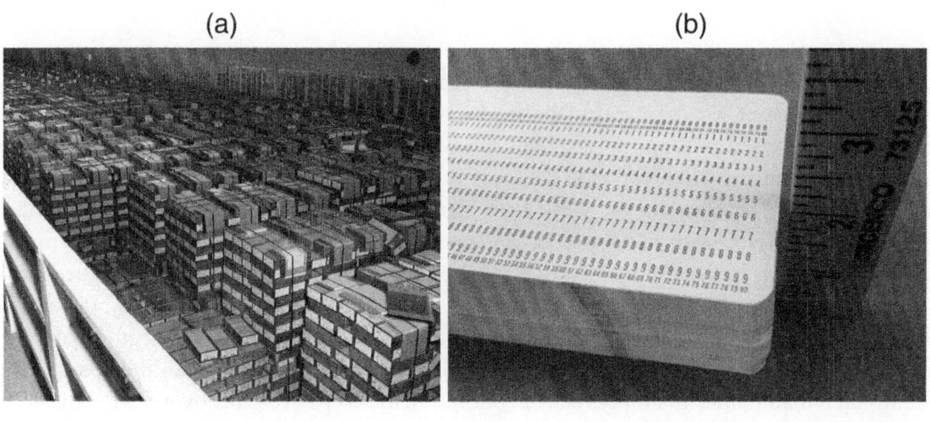

Figure 3.7 (a) Boxes of punched cards in storage; each carton held 2,000 cards. (b) 508 punched cards form a 2-inch stack.

use 10 zettabytes as the size of the Internet storage in 2017, we would produce a card stack 4,101,049,868,766,405 feet or about 776,713,990,296 miles high!

In addition to the space needed, punched cards needed to be stored correctly. The bottom of a student's backpack was not safe and businesses had similar concerns with the proper storage of cards. Cards needed to be protected from moisture to avoid curling and damage caused by water and mildew. Punched cards also needed to be protected from physically getting bent. A very common phrase that was printed on punched cards, as well as on signs hanging in data centers was "Do not fold, spindle, or mutilate". While fold and mutilate are rather obvious terms, the term "spindle" has itself become largely obsolete; it referred to a slender metal spike that was used in offices to collect messages and notes by pressing the paper down on the spike. Obviously, punching an extra hole on a punched card was a bad idea; hence the warning. This phrase became so popular that it was used as the name of a movie (an ABC Movie of the Week in 1971), some popular books, and even printed on T-shirts back when punched cards were popular. Another term that was popular in the day was "face down, 9 edge first". "Face down" meant that the side of the card with the writing on it should be loaded into the card hopper pointing downwards, and "9 edge first" meant that the long edge along the bottom of the card was loaded first. Keep in mind that a single card generally represented a single line of programming code, so if a program contained 2,000 lines of code, 2,000 cards needed to be punched.

Cost

While the cost of punched cards was not excessive, they were a business expense and when lower-cost alternatives such as magnetic disks became available, the use of punched cards began to decrease. Some companies had specially designed and printed cardstock, which added to the overall cost. In addition to the cost of originally purchasing cardstock, there was the cost of storage and the cost of replacing bent or damaged cards. In 1995, when the United States Social Security Administration decided to replace their existing punched-card system, it was estimated that they would save $10 million per year. With this kind of potential savings, businesses and governments had to examine alternatives.

Speed

Because punched cards used an extra step in the process when compared to today's systems, there were delays inherent in the system. Even a great typist working rapidly and accurately still needed to wait for the cards to be read into the computer system. Early in the use of punched cards to control computers, the cards were frequently typed twice, by two different keypunch operators. Then the two decks of cards could be fed to a verifier machine that identified cards that were not the same, thus identifying errors before they were even sent to the mainframe. Even with the crosschecking process in place, this only eliminated keypunch errors and not syntax errors in the code. Those error types would only be identified once the program was loaded into the mainframe and run. While, by today's standards, the cost of having two clerks type identical information and double-check each other seems excessive, in the era when this was done, the cost of clerical help was vastly lower than the cost of a job failing on a mainframe. It simply made economic sense to pay for this overhead. This delay contributed to the demise of punched cards and they were ultimately replaced by direct entry via terminals. No more bent, incorrect, or dropped cards, as data were entered directly into the computer.

Advantages

While both punched cards and punched paper tapes became obsolete for controlling computers, they did offer some advantages over their successors, including longevity. If printed on good paper stock, punched cards and punched paper tape could outlast magnetic tapes. There were several other subtle advantages that punched cards and punched tape provided over modern media used for storage and to control computers. First, was durability; while paper products can be damaged with improper storage and handling, if preserved correctly, punched cards and tape from 50 years ago could be read by machines today. Over time, punched paper strips were replaced with Mylar plastic, which proved to be tougher and more durable than paper strips. Most modern storage uses magnetic media and this is actually less durable over time than punched cards. Magnetic storage can be damaged and become unreadable if the device is dropped and/or if the mechanical arm used to read the disk comes in contact with the storage platter (an accident known as a head crash). Extremes in temperature can also damage magnetic media, as well as strong magnetic fields.

A second advantage of punched cards and punched tape over magnetic storage is that they are visually readable. In the event that a mechanical reader was not available, it was possible for a person to determine what was on a card visually by examining the rows and columns of the punches. This same technique is, of course, not possible with magnetic media.

A last advantage of punched cards and punched tape is that they are easily destroyed. While this might seem contrary to the advantage of durability mentioned above, it actually is a positive. If information needs to be destroyed for security reasons, paper-based storage such as punched cards and punched tape can easily be shredded or burned. Destroying magnetic media is more difficult.

Disadvantages

Punched cards and punched paper tape had several disadvantages when compared to other computer input/output devices. Perhaps the largest limitation is that punched material was single-use; once a card or tape was punched, it could not be altered and needed to be replaced if it was typed incorrectly. Many cards needed to be redone, not necessarily because of incorrect information but simply because the holes were placed in the wrong row or column. Newly emerging magnetic media had a large advantage in that they could be reused multiple times.

Punched cards and punched paper tape had a second large limitation that contributed to their decline as information technology improved; relative to other emerging storage technologies, they had a low information density. In other words, it took a significant amount of paper to store a fairly small amount of information.

Several key problems might have contributed to the decline in the use of punched cards. One of the main problems was discussed earlier. If a card was typed incorrectly or damaged, folded, or bent, it needed to be replaced; no corrections could be made once a character was typed and a hole was punched. If a wrong character was typed, or the right character was typed in the wrong column, the card was useless and a new card was needed. Although probably more legend than fact, stories about programmers dropping a deck of cards abounded at the height of punched-card use in computers. The small paper disks or squares ejected when a hole was punched in a punched card were known as chads, chaff, or sometimes chips. These waste fragments in computer rooms were collected in a bin underneath the keypunch machine. Each chad was roughly ⅛-inch long and when a handful of chads were dropped into your shirt, they became rather itchy (a story for another time).

Partial Punches

While the use of punched cards diminished in IT departments, they were still used in election systems as recently as 2000. Throughout the history of elections in the United States, there has been a level of confidence in the voting process and, with some small exceptions, voters were comfortable that the voting system worked adequately.[38] A punched-card system was behind the controversy in the 2000 United States presidential election. Florida, and other states, used an election ballot commonly known as a Butterfly ballot, the name referring to the ballot's shape, with the two sides resembling the wings of a butterfly (see Figure 3.8). The idea was to vote for your choice of candidates by punching out a marked area on the card.

During the balloting, pointing devices, about the shape of a lead pencil, were used to punch out the voter's preferences and these ballots were then read by a card reader machine. In the case of the 2000 presidential election, two problems occurred. First, during the presidential election, some of the chads did not punch out cleanly but were left partially attached. In the popular press at the time, these became known as "hanging chads". These partial punches were euphemistically called "swinging doors" as their shape resembled a door and they often appear to hang from one hinge. In some cases, the chads were not punched out at all, but the card paper was merely indented. These mistakes were called pregnant chads, dimpled chads, or fat chads. Adding to the confusion at the time, another problem is depicted in Figure 3.8. Because of the confusing ballot design, it was difficult to know which punch-out was meant for which candidate.

Incidentally, the problems that occurred in the 2000 election were not new and the potential issues were known prior to the election.[39] A similar problem occurred in the 1996 election in the state of Massachusetts.[40] Voting systems using these types of punched cards are known as Votomatic systems and had higher error rates than other systems that were available at the time, and they also had a higher impact when it came to voters with lower levels

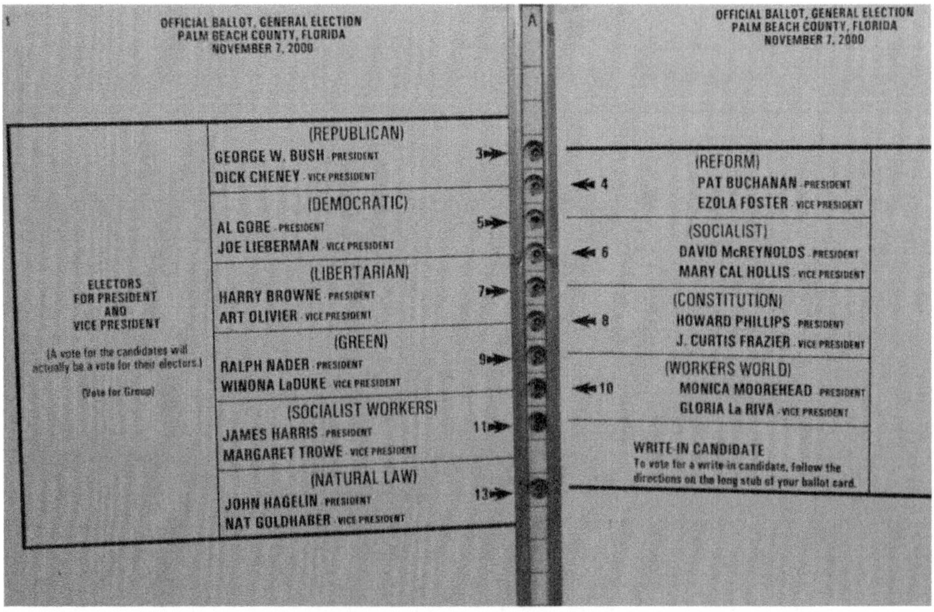

Figure 3.8 Hanging chads – the 2000 election Butterfly ballot.

of education and among poorer people. The irony of this entire 2000 presidential election problem is that while the punched cards contributed to the problem, at the same time, they provided a perfect audit trail in that the original cards could be reexamined by hand. This is not the case with forms of electronic voting.

A common problem for punched-card programmers was when a card was nearly perfect with the exception of one extraneous punch. Typically, this extra punch would make a card effectively useless and the official solution was to completely re-punch the card. Since re-punching a card was annoying and time-consuming, some enterprising programmers discovered that a previously punched chad could be placed into the extra hole. By using a soft lead pencil and lightly scribbling over this chad and the edges of the hole, the replacement chad would stay in place and make it through the card reader. This was only a temporary fix but it demonstrated the inventiveness of the programmers at the time.[41]

Current and Future Uses

As surprising as it might be, punched cards have not totally disappeared from the computing landscape, as their use in voting machines demonstrates. Punched-card systems were initially used for voting during the 1964 elections, and some are still in use. As recently as 1996, 37.3% of United States voters voted using a punched-card system. By 2012, only .02% of the United States voted using punched-card systems, which was limited to four counties in the state of Idaho. This conversion was largely the result of the aforementioned debacle during the 2000 U.S. presidential election. This decrease to .02% is a dramatic change from the 2000 presidential election, where 32.1% of the voters in the United States used a form of punched cards for voting, making this technology the most widely used one in the election. Punched-card ballot systems are being replaced with newer methods including mark-sense ballots and systems where the vote is directly recorded into a computer system.

The current use of punched cards was recently detailed in an article on a small manufacturing company in Texas.[42] Sparkler Filters of Conroe, Texas currently (as of early 2017) uses an IBM 402 system that uses punched cards. Several reasons are given for the continued use of punched cards in the business, but when a large computer system is in place and works effectively, it can be expensive and troublesome to switch to a new system. Basically, "if it works don't fix it" applies.

What's old is what's new; after more than 50 years, the basic concept behind punched cards could be used once again in computers. An emerging technology being developed by IBM creates holes in polymer material at a density of a trillion bits per square inch.[43] The end result of this technology is that information can be stored in an extremely small space but the system is basically the same as a punched card; a hole in the polymer represents a zero in the data and the absence of a hole is used to represent a 1. This project is known as "IBM Millipede", and while it might never see the light of day as a commercially available product, it does leave open the door for the possibility of future "punched" technology. Potential advantages of this technology are very fast read and write rates compared to flash memory and very high data density rates.[44]

Punched cards are still used today on knitting machines such as those manufactured by Brother, Elna, and PASSAP to control patterns in the stitches using methods similar to the original Joseph Marie Jacquard system devised more than 200 years ago. Punched-card sets that produce woven patterns are readily available and some new machines from manufacturers such as PASSAP, are still commercially sold.

Punched cards are also still a popular option for retail businesses to track customer loyalty. Small business-card sized pieces of paper are handed out and every time an eligible purchase is made, a uniquely shaped punch is entered on the card. When the punches are complete, the customer receives a discount or a free item. One drawback of this approach is that the paper punched cards can easily be lost or misplaced which results in a less than satisfying experience for the customer and potential lost sales for the business.[45] This use of punched cards is slowly dwindling with bar-coded tags for key rings, which in turn is slowly being replaced by the use of smartphone apps.

Punched-card use had a long life during the history of computing and could have new uses in the future. Just as computer scientists at the beginning of the computer age repurposed existing punched-card technology from other industries, punched cards might not be done yet. For computer scientists and business users, punched cards played a large and vital role in getting technology to where it is today. Knowledge of this era in computer history is beneficial as we look forward to future advances in technology.

Notes

1 See www.eetimes.com/author.asp?doc_id=1285484 and www.pcworld.com/article/188661/evolution_of_removable_storage.html#slide2
2 www.computerhistory.org/revolution/punched-cards/2/4
3 www.computerhistory.org/tdih/April/7/
4 https://en.wikipedia.org/wiki/Jacquard_loom
5 www.museumwaalsdorp.nl/computer/en/punchcards.html
6 Mounier-Kuhn, Pierre-E. "Book review of *Punched-Card Systems and the Early Information Explosion, 1880–1945*. By Lars Heide." *Business History Review*, Vol. 85, No. 1 (2011), pp. 233–236.
7 https://en.wikipedia.org/wiki/Ticker_tape
8 https://en.wikipedia.org/wiki/Punched_tape
9 www.pcmag.com/encyclopedia/term/48807/paper-tape
10 http://whatis.techtarget.com/reference/History-of-the-punch-card
11 Truesdell, Leon E. "The Development of Punch Card Tabulation in the Bureau of the Census, 1890–1940; with Outlines of Actual Tabulation Programs." Washington, DC, U. S. Govt. Print. Off., 1965.
12 Lubar, Steven. "'Do Not Fold, Spindle or Mutilate': A Cultural History of the Punch Card." *The Journal of American Culture*, Vol. 15, No. 4 (1992).
13 www-03.ibm.com/ibm/history/ibm100/us/en/icons/tabulator/
14 http://statemuseumpa.org/pennsylvania-turnpike-irwin-interchange-tollbooth/
15 Alt, Franz, "Archaeology of Computers – Reminiscences, 1945–47." *Communications of the ACM*, July 1972.
16 Anonymous, "You're on a punch card now!" *The American Journal of Nursing*, Vol. 56, No. 3, March, 1956.
17 www-03.ibm.com/ibm/history/ibm100/us/en/icons/punchcard/
18 http://history-computer.com/Dreamers/Bouchon.html
19 https://en.wikipedia.org/wiki/Floppy_disk
20 http://history-computer.com/Dreamers/Bouchon.html
21 Chapman, Carleton B. "John Shaw Billings, 1838–1913: Nineteenth Century Giant." *Bulletin of the New York Academy of Medicine*, Vol. 63, No. 4 (May 1987), pp. 386–409.
22 www.britannica.com/biography/Gottfried-Wilhelm-Leibniz
23 Chapman, Carleton B. "John Shaw Billings, 1838–1913."
24 www-03.ibm.com/ibm/history/history/decade_1890.html
25 Lubar, Steven. "'Do Not Fold, Spindle or Mutilate': A Cultural History of the Punch Card."
26 http://wiki.c2.com/?HollerithPunchCard
27 https://calculating.wordpress.com/2014/01/27/ibm-port-a-punch/
28 http://ethw.org/Early_Punched_Card_Equipment,_1880_-_1951
29 Mounier-Kuhn. "Book review of *Punched-Card Systems and the Early Information Explosion, 1880–1945*."
30 http://ethw.org/Early_Punched_Card_Equipment,_1880_-_1951
31 www.computerhistory.org/tdih/April/7/
32 www.righto.com/2016/05/inside-card-sorters-1920s-data.html

33 Cooper, William, "Reviewed Work: *Principles of Punch-Card Machine Operation* by Harry Pelle Hartkemeier." *The Accounting Review*, Vol. 17, No. 4 (Oct., 1942), pp. 417–418.

34 Jackson, William. "SSA punch card readers unplugged." *Government Computer News*, 3 July 1995.

35 www.extremetech.com/computing/90156-the-history-of-computer-storage-slideshow/2

36 www.northeastern.edu/levelblog/2016/05/13/how-much-data-produced-every-day/

37 http://gizmodo.com/if-data-was-stored-on-punch-cards-how-much-space-would-1354520013

38 Bullock III, Charles, Hood III, M.V., and Clark, Richard (2005) "Punch Cards, Jim Crow, and Al Gore: Explaining Voter Trust in the Electoral System in Georgia, 2000." *State Politics & Policy Quarterly*, Vol. 5, No. 3 (Fall 2005), pp. 283–294.

39 Buchler, J., Jarvis, M. and McNulty, J.E. "Punch Card Technology and the Racial Gap in Residual Votes." *Perspectives on Politics*, Vol. 2, No. 3 (Sep., 2004), pp. 517–524.

40 Analysis: Massachusetts election battle in 1996 that was caused by punch-card ballot controversy. *Weekend All Things Considered*: 1. Washington, D.C.: NPR. (Nov. 18, 2000).

41 "'9 Edge First': An ode to the punch card." *National Review*, December 31, 2000.

42 www.pcworld.com/article/249951/computers/if-it-aint-broke-dont-fix-it-ancient-computers-in-use-today.html

43 "News Briefs", *Computer*, Vol. 35, No. , pp. 22–24, September 2002, doi:10.1109/MC.2002.10078

44 Walsh, Conor, "IBM's Millipede." *Friction and Wear of Materials*, RPI Hartford, December 13, 2012.

45 Google Inc., "Patent Application for Digital Punch Card for Mobile Device." April 30, 2014.

4

A HISTORY OF THE ELECTRICAL SIGNAL

From the Atlantic Telegraph Cable to the Quest for Artificial Intelligence

David Hochfelder

I admit it—I'm a nerd. I indulge in a hobby known as DXing, the reception of long-distance radio broadcasts ("DX" is the radiotelegraph abbreviation for distance). In effect, I collect AM radio stations. This is a hobby that originated with the dawn of radio broadcasting nearly 100 years ago, and has a close relationship with the community of licensed amateur radio operators.[1]

By definition, every hobby is a waste of time and money. DXing is no exception—I probably own 20 radios of various sizes, capabilities, and costs. The main reason for my radio hoarding is that over time the technology has improved dramatically, allowing for greater performance in ever-smaller packages.

Before the turn of the 21st century, top-grade radios used by DXers took up a couple square feet of desk space and weighed about 15 pounds. Today, radios offering similar performance are about the size of a pack of cigarettes and connect to laptop computers or even tablets. This vast reduction in size and improvement in capability is due to an accelerating shift from analog to digital signal processing, a shift that is now largely complete. It is therefore a testament to the ongoing validity of Moore's Law, the celebrated 1965 observation by integrated circuit pioneer Gordon Moore that the performance of digital electronics doubles roughly every 18 months. Like other media described in this volume, the analog radio receiver—and the analog signal, more generally—is in its twilight, thanks to ever-cheaper and more powerful digital circuits.[2]

I offer these admittedly personal observations about radio gear to prompt some questions fundamental to the theme of this volume. What exactly does "analog" mean? Is it defined only in relation to, or opposition to, "digital?" Analog or digital *what*? Quite simply, the terms "analog" and "digital" refer to electrical impulses and how they are converted into sight and sound. Historian Fredrik Nebeker defines a signal as "a time-variant quantity that conveys information", and signal processing as "the changes made to signals so as to improve transmission or use of the signals". The *Comprehensive Dictionary of Electrical Engineering* defines "analog signal" as "a signal represented in a continuous form with respect to continuous time, as contrasted with digital signal represented in a discrete (discontinuous) form in a sequence of time instant". The *Oxford English Dictionary* is more straightforward, defining "analogue" as "the manipulation of continuously variable physical quantities (as voltage, spatial position, or time) which are analogues of the quantities being computed". Like the more technical dictionary, *OED* goes on to note that "analogue" is "typically contrasted with *digital*".[3]

My argument here is that the electrical signal has served as a powerful metaphor for concepts of self and social interaction, especially since World War II. Thanks to the evangelizing work of engineers Claude Shannon, Warren Weaver, and Norbert Wiener, engineering concepts helped to shape the research of social scientists after about 1950. To give brief examples, the work of role sociologist Erving Goffman is suffused throughout with concepts about information exchange and feedback. At the same time, the ongoing quest for artificial intelligence sought to blur the boundary between human and machine by searching for mathematical or functional equivalencies. This quest relied on modeling both human and machine as signal-processing entities.

Before exploring these concepts further, I offer a technical history of signal processing in three overlapping phases. Signal processing arose with the design of sending and receiving equipment for long undersea telegraph cables in the last third of the 19th century. These efforts were non-electronic—that is, they used only passive components that could not amplify signals. Starting in the early 20th century, telephone companies, especially the Bell System in the United States, turned to early electronics in the form of vacuum tubes (then later discrete transistors) to improve the range, reliability, and bandwidth of telephone circuits. Finally, the introduction of the integrated circuit in 1958 led to a shift from analog signal processing methods to digital. I will conclude by posing some conjectures on the propagation of concepts from the engineering discipline of signal processing into social-scientific thought regarding selfhood and social interactions, as well as the quest for artificial intelligence.

Submarine Telegraph Cables and the Dawn of Signal Processing

The distinction between analog and digital signals was, of course, moot before the harnessing of electricity to transmit information. The electric telegraph, developed simultaneously by several researchers around 1840, was the first such device. Early telegraphs, like those developed by William Fothergill Cooke and Charles Wheatstone in Great Britain and Samuel F. B. Morse in the United States, used the recently discovered phenomenon of electromagnetism to send ephemeral signals that indicated letters of the alphabet visually or aurally. These systems were technologically straightforward, consisting essentially of a battery, length of wire, switch to open and close a circuit, and an electromagnet.[4]

This changed during the 1850s. British and American engineers and entrepreneurs began to consider the construction of an underwater telegraph cable spanning the Atlantic Ocean. This undertaking presented major technical challenges. Unlike landline telegraphs, signals sent through long underwater cables suffered high levels of attenuation (loss of strength) and distortion (loss of definition). This is because oceanic salt water is an electrical conductor. No matter how well insulated a telegraph cable might be, its immersion in a conducting medium changes the physics of signal transmission. The longer the cable, the worse the attenuation and distortion. Because of this, the success of an Atlantic telegraph cable was in doubt. Indeed, an 1858 cable worked for only a few weeks at low speeds before failing. After failed attempts to lay working cables in 1857, 1858, 1865, and 1866, two cables were installed in August 1866 between Ireland and Newfoundland, placing the continents in permanent electrical communication.

The Atlantic cable project was a massive engineering and financial undertaking that involved major advances in marine engineering, electrical engineering, and even theoretical physics. The problem occupied much of the attention and career of Scottish physicist and engineer William Thomson (later Sir William and then Lord Kelvin).[5] In the 1850s, Thomson derived the basic equations governing the transmission of electrical signals through long ocean cables. (Interestingly, neuroscientists today also use these equations to model the behavior of neurons in the central nervous system.)[6] With these equations, Thomson was able to modify both medium (the dimensions and structure of the cable itself) and signal to maximize transmission efficiency. He also developed sending and receiving equipment, including the siphon recorder, an instrument that permanently recorded incoming messages in ink. Armed with these recorded signal traces, cable engineers refined their sending and receiving methods and improved the cables themselves. Thus, the siphon recorder ushered in the modern electrical engineering discipline of signal processing.

Thomson first turned his attention to the problem of transmitting signals through long ocean cables in 1854 and 1855. His papers on the subject combined scientific research with hard-nosed engineering. While scientific research is concerned with identifying and understanding natural phenomena, engineering is the application of these phenomena in a way that balances cost, schedule, and desired performance. The problem Thomson set for himself was "the practicability of sending distinct signals" through a very long submarine cable. Using the mathematics of differential equations, he set out to "show exactly how much the sharpness of the signals will be worn down" and to calculate the strength of a received signal for a given level of transmitter current. Thomson also sought to understand the tradeoff between cost and performance. His equations could "easily . . . determine the dimensions of wire and covering which, with stated prices of copper, gutta-percha, and iron, will give a stated rapidity of action with the smallest initial expense." In a later article, he wrote, "Capitalists ought to require a very 'matter-of-fact' proof of the attainability of a sufficient rapidity in the communication of actual messages . . . before sinking so large an amount of property in the Atlantic."[7] In modern language, Thomson developed methods that related the material cost of a cable to its bandwidth, the amount of information that could be transmitted through the cable in a given time.

Thomson not only developed the mathematics that described signaling through long underwater cables, but he also designed transmitting and receiving equipment that embodied his theories. A submarine cable stores electric charge, which slows down transmission speed. To overcome this problem, Thomson designed a transmission method that sent positive and negative pulses of electricity (Morse code dots being positive pulses, dashes negative pulses),

Figure 4.1 Sir William Thomson's mirror galvanometer. (From Silvanus P. Thompson, *Elementary Lessons in Electricity and Magnetism* (London: Macmillian & Co., 1881), page 169.)

thus clearing the stored electric charge when signal polarity reversed in the course of transmitting a message. More importantly, he constructed receivers that could detect signals at very low signal strengths. His first receiver was the mirror galvanometer, a very light silvered mirror (weighing 32 milligrams) housed within an electromagnet. When the galvanometer received a telegraphed signal, it deflected a beam of light to the right or left, depending on whether a dot or dash was transmitted. A pair of receiving operators, one to call out dots and dashes, and one to write them down, decoded messages from Morse code to English text.

Signals received by the mirror galvanometer were ephemeral—receiving operators and engineers could deduce little about the sharpness and strength of the incoming signals. This changed when Thomson invented the siphon recorder, which was first put into service around 1870. This instrument made a permanent visual record of received signals, allowing engineers to determine with accuracy their strength and sharpness. The use of the siphon recorder on long submarine telegraph cables marks the dawn of the engineering discipline of signal processing. Armed with these signal traces, cable engineers set about to optimize the strength and sharpness of received signals.

One of the major engineering challenges in the early history of signal processing on cables was the design and operation of duplex equipment. The duplex was an instrument that allowed for the simultaneous transmission of two messages, one in each direction. The heart of the duplex system was the "artificial line", a rack of equipment that matched the resistance (opposition to the flow of current, and the reason for the diminution of signals over distance) and capacitance (the storage of charge in the cable, and the reason for the distortion of signals over distance) of the actual telegraph cable. The siphon recorder that received incoming signals was placed in an electrical balance between the cable and artificial line, thus rendering it insensitive to outgoing transmissions. The cable's resistance and capacitance changed with varying weather and geomagnetic conditions, and the artificial line had to be constantly

Figure 4.2 Thomson's siphon recorder. (From Sir William Thomson, "On Signaling through Telegraph Cables", *Mathematical and Physical Papers, Vol. II* (Cambridge: Cambridge University Press, 1884), page 170.)

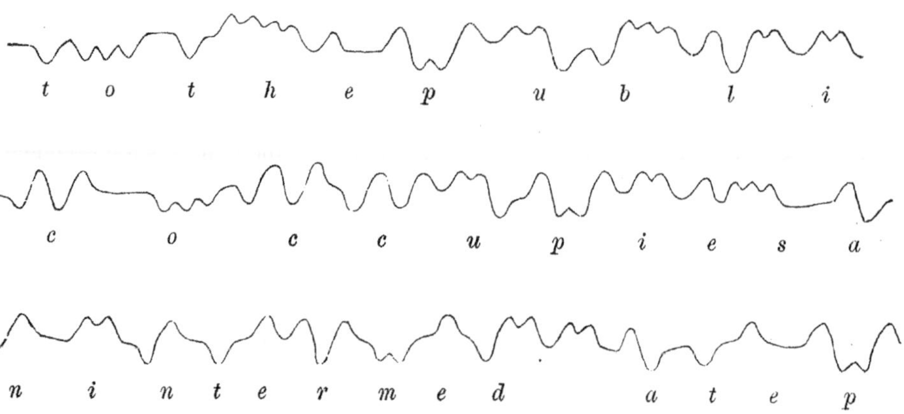

Figure 4.3 Siphon recorder signal trace, using Morse code, with decoded text below. Note that dashes are negative pulses, and dots are positive pulses. (From Sir William Thomson, "On Signaling through Telegraph Cables", *Mathematical and Physical Papers, Vol. II* (Cambridge: Cambridge University Press, 1884), page 172.)

adjusted to match the cable's parameters. In addition, operating a duplex required real-time coordination between the stations at opposite ends of a cable. Thus, the successful installation and operation of duplex equipment needed cooperation between equipment manufacturers and cable engineers, and between operators and engineers at the cable stations.

By the turn of the 20th century, submarine cable telegraphy had become a mature technology. Although some improvements in transmission speed and bandwidth occurred afterward, the locus of engineering innovation moved from the cables to the telephone industry and, later, the radio broadcasting industry.

Telephone Network Engineering, Early Electronics, and Information Theory

Around the turn of the 20th century, telephone engineers confronted a similar set of challenges that cable telegraphers had encountered—improving the distance, reliability, and bandwidth of their communication networks. This was especially true for Bell System engineers and managers who were building a national long-distance telephone network. As M. D. Fagen, official historian of Bell Telephone Laboratories, phrased it, the challenge was "to transmit the speech waves to a distance with adequate magnitude to be heard by the listener, sufficiently free from distortion to be understandable, and with unobjectionable amounts of interference from extraneous sources of sound."[8]

Bell's search for solutions to the problems of building and operating a nationwide telephone network led to the establishment of one of the most successful industrial research laboratories in the world, Bell Telephone Laboratories. Bell Labs published several highly regarded technical journals, conducted basic research in physics and electrical engineering, developed key advances in electronics, and garnered several Nobel Prizes along the way. The developments important to the purposes of this chapter are the development of electronic amplifiers and oscillators using vacuum tubes and Claude Shannon's work on communication and information theory.

The first major technological challenge for the Bell System was to increase the range of telephone calls. Until about 1890, nearly all informed observers thought that the telephone would remain a short-haul communication network, suitable for calls within a city and its environs. The telegraph and postal system were likely to remain the only communication networks capable of long-distance transmission of messages. However, from about 1890 onward, Bell System managers deliberately sought to compete with telegraphy and began to extend the range of their telephone circuits. By the turn of the century, Bell offered reliable long-distance service up to a few hundred miles. These long-distance circuits used a technique called inductive loading that reduced the distortion of signals.

Under its ambitious president Theodore N. Vail, Bell sought to do better—to create a nationwide long-distance network. Placing a call from New York to San Francisco required a new engineering innovation, a means to amplify telephone signals. In landline telegraphy, this amplification was straightforward: one simply placed repeaters at key points in a telegraph circuit. A repeater was essentially a relay that kicked in a fresh battery. But a telephone signal carried information about the human voice, a much more complicated signal than the dots and dashes of a telegram. A telephone amplifier needed to boost the strength of a telephone signal while preserving the frequency characteristics of the human voice.

The solution lay in the triode vacuum tube, developed during the first decade of the 20th century by Lee DeForest and others. Its principle of operation was, in the crudest sense, similar to that of an electromechanical telegraph relay—it took a weak signal as its input and delivered a strong signal as its output. But the triode allowed for precise control of a signal's amplification, making possible long-distance transmission of telephone calls. In a celebrated public-relations event, the Bell System showcased its pioneering use of electronic amplification when Alexander Graham Bell, in New York City, called his erstwhile assistant Thomas A. Watson in San Francisco on January 25, 1915. Repeaters using triode amplifiers lay at the heart of the Bell System's long-distance network until the mid-twentieth century.[9]

Figure 4.4 A Bell System telephone repeater using vacuum tube amplifier circuits. (Bancroft Gherardi and Frank B. Jewett, "Telephone Repeaters", *Transactions of the American Institute of Electrical Engineers* 38:2 (1919): Plate LII following page 1342.)

Vacuum tubes also offered a way to increase the bandwidth of a telephone line, to allow it to carry multiple simultaneous conversations. When positive feedback (feeding a portion of the output of an amplifier circuit into the input, with suitable choice of auxiliary components) is used, a triode circuit can act as an oscillator. That is, it can generate an output of precise and controllable frequency. After about 1920, the oscillator circuit allowed the Bell System to develop carrier-current multiplexing, a technique frequently called "wired wireless" that used several circuits oscillating at different frequencies to permit the simultaneous transmission of several telephone conversations over a single wire.

Vacuum tube electronics was also fundamental to the rise of the modern broadcast media. Commercial radio broadcasting, which began in 1920, relied on vacuum tubes as amplifiers and oscillators used in transmitting stations and receiving sets. Many Americans became familiar with vacuum tube electronics through their radio sets. Radio was a major growth industry of the 1920s, and by 1930, 39% of the nation's households had a radio receiver, and by 1940, 73% did.[10]

The second major development to arise out of Bell Labs was information theory. The important figure in the articulation of a mathematical theory of communication and information was Claude Shannon (1916–2001). His thought on this subject grew out of his World War II work on cryptography. Using concepts borrowed from the statistical thermodynamics of gases, Shannon developed a theory that elegantly captured all of the trade-offs inherent in a communication system—bandwidth, transmission speed, and accuracy of transmission. Fellow Bell Labs engineer and popular writer Robert Lucky wrote, "I know of no greater work of genius in the annals of technological thought." Prominent signal-processing engineer Sergio Verdu has aptly called Shannon's 1948 papers "the Magna Carta of the information age, . . . a unifying theory" that "continues to set the stage for the development of communications, data storage and processing, and other information technologies."[11]

Most major engineering advances do not come to widespread public notice. Shannon's theory would likely not have either, except for the evangelism of fellow engineer Warren Weaver (1894–1978). In 1949, the pair published Shannon's 1948 papers in book form with a long introductory essay by Weaver that explained Shannon's ideas—and their cultural and philosophical implications—to nontechnical audiences. Indeed, it is because of Weaver's lay translation that Shannon's sophisticated mathematical theory began to influence the social sciences and general cultural ideas about the nature of selfhood and interpersonal communication.[12]

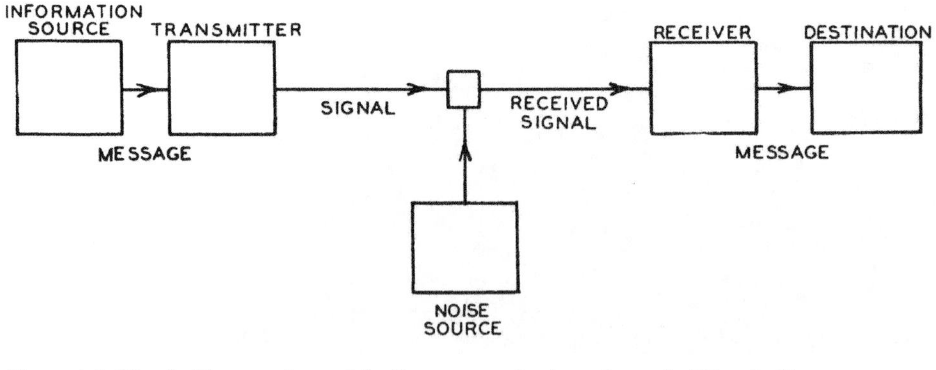

Figure 4.5 Claude Shannon's model of a communication channel. (Claude Shannon, "A Mathematical Theory of Communication", *The Bell System Technical Journal* 27:3 (July 1948), page 381.)

Weaver began by identifying three levels to the science of communication:

> Level a. How accurately can the symbols of communication be transmitted? (The technical problem.)
>
> Level b. How precisely do the transmitted symbols convey the desired meaning? (The semantic problem.)
>
> Level c. How effectively does the received meaning affect conduct in the desired way? (The effectiveness problem.)

Weaver noted that Level A referred to "the engineering details of good design of a communication system; while B and C seem to contain most if not all of the philosophical content of the general problem of communication". Yet the reader would be mistaken in thinking "that Level A is a relatively superficial one". Instead, "the analysis at Level A . . . overlaps the other levels more than one could possibly naïvely suspect". Shannon wrote that the "semantic aspects of communication are irrelevant to the engineering problem. The significant aspect is that the actual message is one *selected from a set* of possible messages", each having equal probability of being chosen.[13] However, Weaver argued that the reverse was not true: the engineering work in building a communication channel is highly relevant to the semantic meaning of a received message.

To make his case, Weaver first warned readers that "*information* must not be confused with meaning". In Shannon's theory, "this word information" refers not "to what you *do* say" but "to what you *could* say. That is, information is a measure of one's freedom of choice when one selects a message." Mathematically, "the amount of information is defined" by the base-2 logarithm "of the number of available choices". This use of the base-2 logarithm transforms the number of informational choices available to the creator of a message into digital bits. The simplest illustration of this is a message with only two choices, say, "yes" or "no". This can be stored digitally with only one bit of information ($\log_2 2 = 1$). Or take a situation in which a message can have eight possible choices, say the numbers zero through seven. This can be represented in binary notation as 000, 001, 010, 011, 100, 101, 110, 111 and therefore requires only three bits of digital storage ($\log_2 8 = 3$). (The implications of Shannon's theory for digital signal processing and data storage are, of course, clear.) In the case where a sender of information chooses to send a string of numbers between zero and seven, all of which are equally likely to be sent, the choice of preceding numbers has no effect on the likelihood of a particular number being sent afterward.

Instead of the simple case of merely choosing from a set of equally likely numbers, consider a message in English or another language. In this case, the characters or words sent previously have great influence on the likelihood of what subsequent characters or words will be sent. This probabilistic nature of natural human language means that there is a great deal of redundancy in the information conveyed. Weaver estimated that English was about 50% redundant, so that "about half of the letters or words we choose . . . are really controlled by the statistical structure of the language." This statistical structure and redundancy permits data compression—not all symbols need be transmitted for the recipient of a message to understand it. This linguistic compression has been a feature of communication technology since the very early telegraph. Telegraph pioneer Alfred Vail likewise estimated that telegraphers only sent about half of the characters that customers wrote down on telegram forms. This aspect of Shannon's theory dealing with the statistical nature of natural language has informed so-called "machine learning", efforts to make human language intelligible to computers. This is, in effect, how autocorrection works on today's smartphones.[14]

The Digital Turn

Historian Frederik Nebeker has called 1948 "the *annus mirabilis* in the emergence of the discipline of signal processing". Although I have argued that the discipline of signal processing originated with the efforts of engineers in the late 19th century to optimize the transmission capabilities of long ocean telegraph cables, Nebeker is correct to note the importance of 1948. In addition to the publication of Shannon's papers and Norbert Wiener's book on cybernetics, that year witnessed two key technological developments: the demonstration of the first digital computer using a stored program and Bell Labs' announcement of the invention of the transistor (it had been successfully demonstrated on December 23, 1947). Both developments would be of great importance to the shift from analog to digital signal processing.[15]

The transistor was a major improvement over the vacuum tube with respect to size, power consumption, and reliability. But engineers initially used it in much the same way as the tube, as a device that amplified a signal or switched a signal on and off. The full significance of the transistor awaited the development of integrated circuits (ICs) after 1958. The early digital computer, likewise, promised to be a major technological breakthrough, but because of size and power limitations, its initial applications were limited to brute-force uses such as scientific calculation or simple data processing. Several engineers recalled that the professionalization of signal processing did not truly arise until the mid-1960s, after the IC made possible the construction of more powerful computers. James Kaiser of Bell Labs recalled that before that time, engineers thought of signal processing as analog "signal shaping or that kind of thing", similar to what telegraph cable engineers and telephone engineers had been doing since the late 19th century. Most of the work in analog signal processing during the 1950s and early 1960s took place in the fields of telephone engineering and audio engineering.[16]

The advent of a self-defined profession of signal processing occurred only during the mid-1960s after sufficient computing power became available. Kaiser recalled that "the development of computer processor power is what prompted the realization" among engineers that robust signal processing could be done by converting analog signals to digital information. Quite simply, signal processing required the IC and the resulting miniaturization of electronics.

The IC is the building block of all modern electronics. It is at the heart of the computer, mobile telephone, and all other electronic devices, and entered commercial service in 1960. The IC arose out of the increasing demand for smaller and more reliable circuits, both by the U. S. military and by the commercial sector. Unlike the transistor, which was a major scientific and engineering breakthrough, the IC was more of a "manufacturing innovation", according to historians David Morton and Joseph Gabriel. Of course, the magnitude of this manufacturing innovation is immense: new IC fabrication facilities can cost up to $10 billion to build today.[17]

As the term "integrated circuit" implies, an IC contains many electronic components on a single silicon substrate. The technology advanced rapidly. In 1965, IC pioneer Gordon Moore predicted that "integration will bring about a proliferation of electronics, pushing this science into many new areas", including "such wonders as home computers[,] automatic controls for automobiles and personal portable communications equipment". Moore also predicted correctly that the number of components on ICs would double every 18 months or so. This exponential rise in complexity meant that by the mid-1970s, microprocessor chips contained tens of thousands of transistors, by 1989 over a million, and by 2005, over a billion.[18]

The IC transformed the engineering discipline of signal processing because it brought large amounts of computing power to bear. Consider that engineers model the signal-processing behavior of a circuit by referring to its "transfer function", the mathematical relationship between input signal and output signal. In an analog circuit, only certain transfer functions can be implemented. In digital signal processing using a modern computer and appropriate mathematical techniques, virtually any transfer function can be implemented. (This is how today's software-defined radios used by DXers can best the performance of older tabletop radios.)[19]

In 1990, in language reminiscent of the theme of this volume, engineer John G. Truxal argued that "the advantages of digital signals are so overwhelming that essentially all communication technologies are changing to this form." Today, thanks to cheap and powerful ICs, digital signal processing is ubiquitous, as any purchaser of music on CDs or online can attest. While Richard Osborne and others in this volume have argued for the persistence of analog media like the vinyl LP record, from an engineering standpoint, the sun has already set on the analog signal.[20]

Conclusion

In 1950, mathematician Alan Turing published one of the most famous papers in the history of engineering—an eclectic mix of mathematics, electronics, and philosophy of mind. At the very beginning, he posed the provocative question, "Can machines think?" To test this, he proposed what he called "the imitation game", what computer scientists now call the Turing Test. A machine passes this test by fooling a human interlocutor into thinking that it is also a human being. Turing went on to note that the digital computer, only recently invented, stood the best chance of passing his test. He concluded his paper with the "hope that machines will eventually compete with men in all purely intellectual fields", and suggested that language acquisition or chess would be promising places to start.[21]

Within a few years, other researchers began to investigate whether machines could mimic other human cognitive skills such as pattern recognition and generalized learning. In 1957, Frank Rosenblatt, psychologist at the Cornell Aeronautical Laboratory, proposed a machine he called a "perceptron" that would have "such human-like functions as perception, recognition, concept formation, and the ability to generalize from experience". Such a machine "would be capable of conceptualizing inputs impinging directly from the physical environment … rather than requiring the intervention of a human agent to digest and code the necessary information." Such a device would be "closely analogous to the perceptual processes of a biological brain". Until the mid-1960s, Rosenblatt's suggested line of investigation held sway among artificial-intelligence researchers. Although computer scientists Marvin Minsky and Seymour Papert later claimed that Rosenblatt's approach was of limited utility, it has continued to inform work in neural networks. Many computer scientists regard neural networks as a promising avenue to deep learning, a computer's ability to process images, natural language, and even handwriting.[22]

What Turing and Rosenblatt proposed would have been impossible without two developments: ever more powerful digital computers (thanks to Moore's Law) and Claude Shannon's theory of information and communication. To generalize further, their ideas depended on a longer historical dynamic, resting on the engineering discipline of signal processing that began on the long ocean telegraph cables of the late 19th century. After all, learning to play chess or to speak a language is a constant process of information exchange, feedback, and error correction.

The ongoing 70-year search for artificial intelligence is the logical culmination of the discipline of signal processing. Engineers, of course, have long tried to automate what they have perceived as manual drudgery and rote mental work. What is equally interesting is

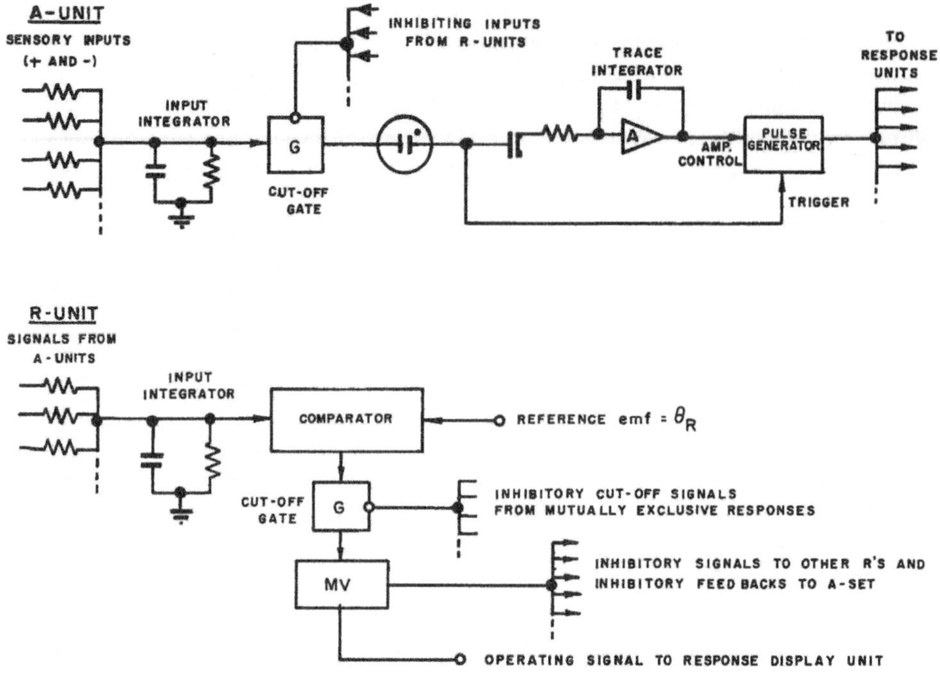

Figure 4.6 Circuit block diagram for Frank Rosenblatt's perceptron. (Rosenblatt, *The Perceptron: A Perceiving and Recognizing Automaton* (Buffalo: Cornell Aeronautical Laboratory, 1957), page 10.)

how engineering concepts influenced the thought of social scientists, and how they also blurred the boundary between human and machine. The foundational work in this effort was Norbert Wiener's 1948 book on cybernetics, which he described as the theory of "control and communication in the animal and the machine". As Warren Weaver did when he translated Shannon's theory of information and communication for a lay audience, Wiener sought to persuade psychologists, sociologists, anthropologists, and economists to incorporate signal-processing concepts into their analysis.

However, Wiener warned that the new science of cybernetics "has unbounded possibilities for good and evil". These possibilities arose because "the modern ultra-rapid computing machine", coupled with appropriate sensory input devices and output "motor organs", could form the basis for "artificial machines of almost any degree of elaborateness of performance". The future automated world could be a technological paradise of leisure or could create new forms of oppression. The shadow of World War II hung over the new science, and "those of us who have contributed" to it "thus stand in a moral position which is, to say the least, not very comfortable . . . . We can only hand it over into the world that exists about us, and this is the world of Belsen and Hiroshima."[23]

While Wiener's most extreme hopes or fears have not come to pass, his ideas significantly influenced the work of social scientists. Psychologist George A. Miller, a pioneering researcher working at the intersection of linguistics and machine learning, drew heavily on Shannon's and Wiener's work. Miller began his landmark 1951 book *Language and Communication* with a description of "the idealized communication system", which closely matched Shannon's model. Miller wrote that "five components—source, transmitter, channel, receiver, and destination—comprise the idealized communication system. In one form or another these five components are present in every kind of communication." Some of his articles give insight into how Shannon's and Wiener's ideas shaped his research agenda. One line of research used the redundancy of the English language to investigate the best methods to reconstruct the meaning of a mutilated message. Miller regarded this as interesting from a psycholinguistic perspective, but also noted that this research had implications for the electronic transmission and processing of messages. In another article, Miller and colleagues investigated the extent to which English text could be predicted from preceding words and letters. They used a UNIVAC computer to analyze the rank and frequency of over 36,000 words. When they combined these statistics with "a two-state Markov process that will generate strings of letters interrupted by spaces", they obtained a passable imitation of English text. Miller's 1960 book, *Plans and the Structure of Behavior* relied heavily on Wiener's cybernetics. In his preface, Miller noted that "we reviewed once more the cybernetic literature on the analogies between brains and computers, between minds and programs." Much of the research grew out of a 1958 RAND Corporation seminar on the Simulation of Cognitive Processes. Miller gently rebuked cyberneticists for having been "remarkably innocent about psychology—the creatures whose behavior they want to simulate often seem more like a mathematician's dream than like living animals." His book grew out of the search for "a favorable intersection" between psychology and cybernetics.[24]

Similarly, pioneering role sociologist Erving Goffman borrowed engineering concepts to explain the dynamics of social interaction. While Goffman characterized his approach as "dramaturgical", he used information theory to describe the "arts of impression management". He began his analysis by describing a feedback process:

> When an individual enters the presence of others, they commonly seek to acquire information about him or to bring into play information about him already possessed . . . . Information about the individual helps to define the situation, enabling others to know in advance what he will expect of them, and what they may expect of him. Informed in these ways, the others will know how best to act in order to call forth a desired response from him.[25]

For Goffman, public behavior combines theatrical performance with a perpetual feedback-and-control loop. In language remarkably similar to Turing's "imitation game", Goffman described social interaction as "a kind of information game—a potentially infinite cycle of concealment, discovery, false revelation, and rediscovery". The goal of this cycle is "to control the conduct of the others, especially their responsive treatment of him".[26]

Machines and engineering concepts have long served as metaphors for selfhood and social organization.[27] But what I consider unique to the postwar intersections between communication and information theory and the social sciences is the extent to which researchers have sought to blur the boundary between human and machine. This has occurred from both directions of the boundary. Social scientists and engineers have tried to reduce human cognition to rules and systems that can be implemented technologically, in order therefore to imbue machines with human-like attributes.

Researchers have identified four capabilities required for true artificial intelligence: perception, action, cognition, and language. Some technological optimists—Ray Kurzweil especially comes to mind—have predicted that artificial intelligence will soon exceed biological intelligence, thanks to Moore's Law and the exponential growth of computing power. Others have warned that the so-called "technological singularity" could result in intelligent machines that will not only exceed but will supersede biological intelligence, to the detriment of humanity. At the present time, these hopes and fears are far-fetched. Researchers have shown that it is possible to automate some aspects of human cognition—computers can now beat human experts at games such as chess or *Jeopardy!* and possess rudimentary linguistic abilities. But the full machine integration of these four attributes is still a long way off. If, as I have argued, the analog signal is in its sunset, it is unlikely that the sun will ever set on human cognition.[28]

Notes

1 Susan J. Douglas, *Listening In: Radio and the American Imagination from Amos 'n' Andy and Edward R. Murrow to Wolfman Jack and Howard Stern* (New York: Times Books, 1999): 55–82.

2 Gordon E. Moore, "Cramming More Components onto Integrated Circuits", *Electronics* 38:8 (19 April 1965): 114–17; Moore, "Moore's Law at 40", in David Brock, ed., *Understanding Moore's Law: Four Decades of Innovation* (Philadelphia: Chemical Heritage Foundation, 2006): 67–84.

Historians of technology debate whether Moore's Law represents the internal logic of electronics technology, or is merely an industry standard to which integrated-circuit manufacturers adhere by choice. Paul Ceruzzi, "Moore's Law and Technological Determinism: Reflections on the History of Technology", *Technology and Culture* 46:3 (July 2005): 584–93; Cyrus C.M. Mody, *The Long Arm of Moore's Law: Microelectronics and American Science* (Cambridge: MIT Press, 2016).

3 Frederik Nebeker, *Signal Processing: The Emergence of a Discipline, 1948 to 1998* (New Brunswick, NJ: The IEEE History Center, 1998): 7, 10; "Analog signal", Phillip A. Laplante, Editor-in-Chief, *Comprehensive Dictionary of Electrical Engineering* (Boca Raton: CRC Press, 1999); "analogue, n. and adj." *OED* Online. December 2016. Oxford University Press. www.oed.com/view/Entry/7029?redirectedFrom=analog (accessed February 5, 2017).

4 David Hochfelder, *The Telegraph in America: 1832–1920* (Baltimore: Johns Hopkins University Press, 2012): 1–5; Hochfelder, "Joseph Henry: Inventor of the Telegraph?" Smithsonian Institution Archives website, available at http://siarchives.si.edu/oldsite/siarchives-old/history/jhp/joseph20.htm (accessed February 5, 2017).

5 Crosbie Smith and M. Norton Wise, *Energy and Empire: A Biographical Study of Lord Kelvin* (Cambridge and New York: Cambridge University Press, 1989): 446–58, 660–84.

6 Sebastian Seung, *Connectome: How the Brain's Wiring Makes Us Who We Are* (Boston and New York: Houghton Mifflin Harcourt, 2012): 47.

7 Sir William Thomson, "Letters on 'Telegraphs to America'", *Mathematical and Physical Papers*, Vol. II, (Cambridge: Cambridge University Press, 1884), 93.

8 M.D. Fagen, ed., *A History of Engineering and Science in the Bell System: The Early Years (1875–1925)*, (New York: Bell Telephone Laboratories, 1975): 196.

9 Fagen, *History of Engineering and Science in the Bell System*, 256–77.

I need to distinguish between the terms "electrical" and "electronic". Although more technical definitions are available, for purposes of this essay, the distinction is between passive and active means to act upon signals inputted into a circuit. The former generally refers to techniques and components that are passive; that is, they cannot amplify a signal, and are linear in their effects upon an incoming signal. The latter generally denotes techniques and components that are active—they can amplify a signal and act upon it in a nonlinear fashion. An analog electronic circuit typically affects an inputted signal through amplification (gain) or oscillation (the generation of a time-varying output signal of precise frequency). A digital electronic circuit typically affects an inputted signal by creating an output that is either on or off, depending on the desired Boolean mathematical operation performed by the circuit.

10 U.S. Census Bureau, *Statistical Abstract of the United States* (Washington, DC: Government Printing Office, 1999): 885.

11 Claude E. Shannon, "A Mathematical Theory of Communication", *The Bell System Technical Journal* 27:3 & 4 (July & Oct. 1948): 379–423, 623–56. Robert Lucky, *Silicon Dreams: Information, Man, and Machine* (New York: St. Martin's Press, 1989): 37. Sergio Verdu, "Fifty Years of Shannon Theory", *IEEE Transactions on Information Theory* 44:6 (Oct. 1998): 2057.

12 My discussion here closely follows Weaver's introduction in Claude E. Shannon and Warren Weaver, *The Mathematical Theory of Communication* (Urbana: University of Illinois Press, 1949): 3–28.

13 Shannon, "A Mathematical Theory of Communication", 379.
14 Hochfelder, *Telegraph in America*, 74–82.
15 Nebeker also includes two other developments in 1948: Pulse Code Modulation (PCM), a technique that converts the amplitude (strength) of an analog signal into a digital binary value, represented as a series of pulses; and Richard W. Hamming's digital error-correction code. Nebeker, *Signal Processing*, 14–21.
16 James Kaiser: An Interview Conducted by Andrew Goldstein and Janet Abbate, Center for the History of Electrical Engineering, 11 February 1997. http://ethw.org/Oral-History:James_Kaiser (accessed March 3, 2017).
17 David L. Morton, Jr. and Joseph Gabriel, *Electronics: The Life Story of a Technology* (Baltimore: Johns Hopkins University Press, 2004): 75–82; Jim Handy, "Why Are Computer Chips So Expensive?" *Forbes*, April 30, 2014. www.forbes.com/sites/jimhandy/2014/04/30/why-are-chips-so-expensive/; and Ed Sperling, "How Much Will That Chip Cost?" *Semiconductor Engineering* March 27, 2014. http://semiengineering.com/how-much-will-that-chip-cost/ (both accessed March 3, 2017).
18 Moore, "Cramming More Components onto Integrated Circuits", 114; Peter Clarke, "Intel Enters Billion-Transistor Processor Era", *EE Times*, October 14, 2005. www.eetimes.com/document.asp?doc_id=1297860 (accessed March 4, 2017).
19 Nebeker, *Signal Processing*, 66–74.
20 John G. Truxal, *The Age of Electronic Messages* (Cambridge: MIT Press, 1990): 227. Truxal's book is a good lay introduction to the disciplines of communication engineering and signal processing.
21 A.M. Turing, "Computing Machinery and Intelligence", *Mind* 59 (1950): 433–60. For a 50-year retrospective, see Ayse Pinar Saygin, Ilyas Cicekli, and Varol Akman, "Turing Test: 50 Years Later", *Minds and Machines* 10:4 (Nov. 2000): 463–518.
22 Frank Rosenblatt, *The Perceptron: A Perceiving and Recognizing Automaton* (Buffalo: Cornell Aeronautical Laboratory, 1957) https://blogs.umass.edu/brain-wars/files/2016/03/rosenblatt-1957.pdf (accessed March 7, 2017); Mikel Olazaran, "A Sociological Study of the Official History of the Perceptrons Controversy", *Social Studies of Science* 26:3 (Aug. 1996): 611–59.
23 Norbert Wiener, *Cybernetics: Or Control and Communication in the Animal and the Machine* (Cambridge: MIT Press, 1948): 27–28.
24 George A. Miller, *Language and Communication* (New York: McGraw-Hill, 1963 [1951]): 6–8; Miller et al., "Length-Frequency Statistics for Written English", *Information and Control* 1(1958): 370–89; Miller and Elizabeth A. Friedman, "The Reconstruction of Mutilated English Texts", *Information and Control* 1 (1957): 38–55; Miller et al., *Plans and the Structure of Behavior* (New York: Holt, Rinehart and Winston, 1960): 2–4.
25 Erving Goffman, *The Presentation of Self in Everyday Life* (New York: Doubleday, 1959): 1.
26 Erving Goffman, *The Presentation of Self in Everyday Life*, 1–16.
27 See, for example, Otto Mayr, *The Origins of Feedback and Control* (Cambridge: MIT Press, 1971); and Mayr, *Authority, Liberty, and Automatic Machinery in Early Modern Europe* (Baltimore: Johns Hopkins University Press, 1986).
28 For contemporary reviews, see "Am I Human?" *Scientific American* 316:3 (March 2017): 58–63; and John Pavlus, "The New Turing Tests", *Scientific American* 316:3 (March 2017): 61–62.

5

THE LIFE, DEATH, AND REBIRTH OF THE TYPEWRITER

Richard Polt

Our word-processing applications have accustomed us to searching, modifying, and rearranging text that is held in flexible digital memory. The essence of a word processor is the freedom it gives the writer during the period between initial typing and the finalization of the text. In contrast, text printed by a typewriter onto unforgiving paper cannot easily be altered. Despite this obvious advantage of digital writing, typewriters have been remarkably persistent. Although they have long been displaced from their traditional roles, they have not disappeared, and in recent years they have taken on a fresh cultural significance.

A typewriter is a "writing machine", as the device is called in most languages. It combines the standardized nature of type, as in printing, with the sequential writing action of a user, as in handwriting. The United States Patent and Trademark Office provides a suitably general definition:

> an apparatus wherein a user of the apparatus causes an intelligible character to be imprinted on a record-medium by a type-member that is impressed on the record-medium, said type-member being selected from a plurality of different type-members, and the selected type-members being impressed serially to form a sequence of characters that record intelligible information.[1]

The large majority of typewriters have been keyboard devices used to print one letter at a time onto paper, in immediate response to the touch of a user. However, the basic function of a typewriter can be achieved by a wide variety of mechanisms and user interfaces. Some features of typical typewriters that we take for granted, such as the use of a keyboard and separate sheets of paper, are not in fact universal: index typewriters do without a keyboard, and book typewriters type into bound volumes. Although most typewriters are intended to produce linguistic text, some, like music typewriters, are designed to print other symbols. While most typewriters are intended to print a character immediately, some incorporate a delay between input and output, and some have been adapted to type prepared text that is stored on punched paper or a magnetic medium. Some devices that are not usually called typewriters are closely akin to them, such as stenotypes (which print several characters simultaneously) and teleprinters (which print text transmitted from a remote source).

Compared to handwriting, typewriters have the advantage of speed, legibility, and standardization. In their heyday—roughly, 1880 to 1980—they were basic equipment in businesses, government offices, newspapers, schools, and many homes. Typewriters were used for correspondence, record keeping, and the composition and communication of fiction and nonfiction. Today, even though typewriters have left the technological mainstream, they have found new life as countercultural, individualistic tools.

A Brief History of Typewriters

The utility of a typewriter did not become obvious until the late 19th century, but the mechanical principles necessary to create one were known long before. Fourteenth-century clocks used escapements, and movable type was used by Gutenberg around 1439. Typewriters typically use an escapement to achieve the step-by-step motion of a carriage that holds the paper, on which characters are printed using type slugs similar to traditional printer's type. Other simple mechanical parts, such as levers, springs, gears, bearings, and screws, provide the elements of a manual typewriter. However, before the Industrial Revolution, such a device would have been quite expensive to produce and would not have been generally recognized as useful.

In 1714 a certain Henry Mill received English patent No. 395 for what sounds like a typewriter:

> an artificial machine or method for the impressing or transcribing of letters, singly or progressively, one after another, as in writing, whereby all writings whatsoever may be engrossed in paper or parchment so neat and exact as not to be distinguished from print.[2]

We do not know what Mill's invention looked like or what, exactly, it did. The earliest typewriter that we know existed was constructed by Pellegrino Turri in 1808 for his blind friend Countess Carolina Fantoni da Fivizzano. The device does not survive today, but numerous letters written by the Countess, typed in regularly spaced capital letters, testify to the fact that it worked.[3]

Later in the 19th century, various pioneers invented writing machines from the simple to the extremely complex, from tiny to huge, and from practical to absurd. Some significant early developers include William Austin Burt (whose invention dates from 1829), Charles Thurber (1843), Alfred Ely Beach (1847–1856), and Samuel W. Francis (1857). Other notable inventors include the Italian Giuseppe Ravizza, the Frenchman Xavier Progin, and the Tyrolean Peter Mitterhofer—who carved his typewriter frames out of wood and trundled a prototype to Vienna by wheelbarrow to appeal for funds from Emperor Franz Joseph. These three inventors all came up with the concept of placing types (or in Mitterhofer's case, arrangements of needle points) on the ends of long, thin levers, known as typebars. Their machines, probably invented without mutual influence, arranged the typebars in a circle and swung them upwards to reach the paper.

Mass production began when an American, John Jones, established a factory in 1852 to produce his "Mechanical Typographer", a straightforward device that placed the types on a large, horizontal wheel. Unfortunately, the building burned down and most of Jones's 130 machines were destroyed.

Finally, two inventors created machines that were successfully manufactured in a series. The first was Pastor Rasmus Malling-Hansen of Denmark, whose *skrivekugle* or Writing Ball resembled a pincushion. The machine was a finely constructed piece of engineering that was patented in 1870. Its best-known user was Friedrich Nietzsche. Unfortunately, Nietzsche's machine was damaged before he ever got it, and he gave it up after using it for just six weeks in 1882. Malling-Hansen designed several models of the Writing Ball, including an electric version, but production was very limited. Today, the Writing Ball is the most valuable typewriter, worth as much as $100,000.

The next factory-produced invention was far more influential: it launched an enduring industry, and even introduced the term for the device that we use in English today. The Sholes & Glidden Type Writer was born in Kleinsteuber's Machine Shop, a post-Civil War gathering place for manufacturers and tinkerers in Milwaukee, Wisconsin. Its primary inventor was Christopher Latham Sholes, who was also a newspaperman, printer, and politician. Sholes's machine, with its typebars hanging in a circle, resembled the inventions of Ravizza, Progin, and Mitterhofer. Sholes also created the QWERTY keyboard, which is still very much with us today despite numerous attempts to overthrow it. His friend Carlos Glidden got credit in the name of the machine, but more credit is due to financier James Densmore, who urged Sholes to perfect his initially clumsy contraption.[4] The Type Writer was commercially manufactured starting in September 1873, by the Remington Arms Company of Ilion, New York, and reached the market in 1874. It was a fancy creation, often sporting colorful decals and hand-painted decorations that were unique to a particular machine. Like the original Writing Ball, it wrote in capitals only. Its most famous early adopter was Mark Twain, who proudly sent his publishers a typescript of *Life on the Mississippi* (1883).

The Sholes & Glidden was an expensive curiosity that attracted few buyers, but the novel idea of a keyboard for writing did make people pay attention. As far away as Australia, newsman J. H. Clark wrote in 1875:

> Who is the ingenious inventor of the type-writing machine that opens such a wide field of hope for the cautious caligraphist while it takes from the feeble spellist his only safeguard, illegibility? My *Register* describes it as an instrument somewhat resembling a family sewing-machine, with small keys by which you manipulate your correspondence. It also says that with practice a person can play upon it with as much ease and rapidity as a skilful pianist. . . . You can lounge over this silent but powerful agent of love and business, and as the seething thoughts crowd through your burning brain they will glide out at your finger tips and impress themselves on the rose-tinted paper ere you can say Jack Robinson! . . . If they could afford it, would not the poor compositors present each and every one of the *Register* staff . . . with such an instrument, and no more have to wade wearily through crabbed characters like dissipated Assyrian hieroglyphics![5]

Clark's fantasy came true; in the 1880s, the typewriter became accepted as a helpful and even essential tool for a modern newspaper, business, or government office, a tool that could produce multiple copies of easily legible documents in a short time. A typewriter industry gradually developed: not just factories but distributors, shops, salesmen, and manufacturers of accessories such as typing paper and ribbons.

With the mechanical typewriters came human "typewriters" or "operators" who had been trained to work the devices. These were usually unmarried young women, who had generally been banned from the business world before then, but who were considered to have the manual dexterity appropriate for the job. In a time when few women could work outside the home, the typewriter created new economic opportunities for them. Of course, these opportunities were limited: Women were unapologetically paid less than traditional male clerks, who resented being undercut by the new cheap labor. A firm and low glass ceiling was in place, and the division of labor by gender—male boss, female secretary—persisted for nearly a century.[6]

The machines of the 1880s included the Remington No. 2 (the successor to the Sholes & Glidden, which was retroactively called the Remington No. 1 after the typewriter company spun off from the arms manufacturer); the Model 2 introduced a shift key, to switch between uppercase and lowercase letters. Some of its competitors, including the Caligraph, Yost, and Smith Premier, spurned the shift and developed a "full" keyboard that assigned a separate key to each lowercase and uppercase character. The Hammond and the Crandall featured interchangeable type elements (called a shuttle and type cylinder, respectively) and two-row curved keyboards with double shifts—one shift for capitals, another for numerals and punctuation. With its inlaid mother-of-pearl, the New Model Crandall is often considered the most beautiful typewriter ever made.[7]

The decade of the 1890s sported the greatest biodiversity in the typewriter ecosystem. Hundreds of patents were filed and dozens of companies, mostly American, competed for a share of the growing market. To research these inventions is to discover a period of creative experimentation that pushed the boundaries of how letters could be inked onto paper.

Inventors might put the type on typebars, swinging sectors, or typewheels. The greatest of the typewheel devices was the ingenious little Blickensderfer, made in Stamford, Connecticut. Dozens of different typewheels were available for "Blicks", which were made into the 1910s.

Typewriters might have full keyboards, four-bank (four-row) keyboards, three-bank keyboards, or even no keyboard at all—as in "index" typewriters. These gadgets, such as the Hall, People's, American, or Odell, required the user to select a character on an index and then perform a separate action to print it. This concept survived into the twentieth century in cheap German machines such as the Mignon and Gundka, as well as in label makers and toy typewriters such as the Simplex.

Machines might use Sholes's QWERTY keyboard or another supposedly more efficient arrangement, such as Blickensderfer's DHIATENSOR. They might use ribbons, ink pads, or ink rollers. There were expensive hundred-dollar machines (comparable to the cost of a high-end laptop today) and trinkets that sold for a dollar. Most of these models were failures at the time, and they are rare treasures for collectors today.

The typewriter factories of the 1890s experimented with modern production processes, sometimes employing hundreds of men and women on assembly lines. They also used innovative materials such as aluminum and ebonite, an early plastic based on hardened rubber.

The greatest technical challenges for the inventors of the time were durable alignment and immediate visibility for the typed text. The Sholes & Glidden, the Caligraph, and many other early typewriters were "understroke" designs, or "blind writers": they typed on the bottom of the platen (the rubber-covered cylinder), so that the user had to raise the carriage to see the work. Some argued that visible writing was unnecessary, since a skilled typist would have her eyes on her shorthand copy rather than the typewriter itself (this fact also explains the need for a bell that signals the end of the line). But the advantages of immediate visibility for all but nearly infallible typists were obvious. Some inventors tried typebars that stood upright in front of the platen (Bar-Lock, Franklin) or behind it (Waverley, Fitch, Brooks, North's) and swung down onto the paper. On the Williams, the typebars rested horizontally on ink pads and jumped like grasshoppers onto the paper from both front and back; unfortunately, only one or two lines of text were visible at a time. The Oliver found great success with its typebars shaped like inverted U's that swung down from the sides; this design survived until the 1920s in the US, and into World War II in the UK. The Wellington used a "thrust" design that shoved typebars horizontally against the front of the platen; this approach was eventually adopted by Adler in Germany, where the thrust mechanism was popular for decades.

But the best solution to visibility was the "frontstroke" design pioneered by the Daugherty (1891) and perfected by the Underwood (designed by Franz X. Wagner and introduced in 1896). In this arrangement, typebars are arranged in an arc above the keyboard, and swing up against the front of the platen. The speedy, well-engineered Underwood was eventually so successful that within a couple of decades, the vast majority of typewriters followed its model: They were single-shift, frontstroke, typebar typewriters with four-bank QWERTY keyboards, inked by a ribbon.

The creative chaos of the 1890s inevitably waned in the face of competition and consolidation. The Union Typewriter Company, also known as the typewriter trust, was controlled by Remington and included several other major producers of blind writers. The companies banded together to keep the price of their typewriters at a hundred dollars, and they succeeded in dominating the market for some time, despite various attempts by smaller enterprises to undercut them. By the early 20th century, however, independent makers of visible writers—including Underwood, Royal, and L. C. Smith (led by the same brothers who formerly made the blind Smith Premier)—were giving the trust a run for its money.

After World War I, American typewriter manufacturing was dominated by a few large companies that had survived the challenging market of earlier years and the stresses of the war: Remington, Underwood, Royal, and Smith-Corona (the result of a 1926 merger between the L. C. Smith office typewriter and the Corona portable). Underwood and Royal

Figure 5.1 Fox No. 3 understroke typewriter, circa 1904

built gigantic factories in Hartford, Connecticut, that employed thousands and produced millions of machines. There were a few smaller players, such as the elegant Fox and the unique Hammond, which survived under the name Varityper and found a niche market as a "cold typesetting" machine into the 1970s.

Meanwhile, typewriter manufacturing had spread to Europe, especially Germany, where dozens of factories were built in the first few decades of the century. Major German companies included Adler, Triumph, Continental, Torpedo, Seidel & Naumann, and AEG/Olympia. The largest British company, Imperial, exported many machines to other countries in the Commonwealth. Other European nations also became typewriter exporters, including France, Switzerland, and Italy (with the great Olivetti). Labor conditions at typewriter factories varied wildly; Remington was a ruthless strikebreaker, while Olivetti provided a spectrum of services to its workers and built innovative spaces in the various cities where it established factories.[8]

The early 20th century saw the rise of portable typewriters, which could serve as ready tools for students, housewives, journalists, or traveling businessmen. Building on the success of the Standard Folding, an aluminum portable with a carriage that folded over onto the keyboard (1907), the Corona 3 was introduced shortly before World War I and went on to sell by the hundreds of thousands. It was succeeded by many other machines designed for personal or travel use. While the Corona and other early portables had a three-bank keyboard with double shift, the Remington portable of 1920 introduced a four-bank keyboard and was a big success. By the early thirties, the three-bank design was confined to a few offbeat machines and toys—that is, until its revival on smartphones and tablets of the 21st century.

In the mid-20th century, portables were further differentiated into machines that were as compact as possible (such as the Hermes Baby) and midsize typewriters that incorporated some of the sophisticated features of office machines. Styling evolved with the times. Olivetti was a style leader, employing internationally recognized designers such as Marcello Nizzoli and Ettore Sottsass and introducing iconic machines such as the Valentine—a bright red, informal portable marketed to countercultural youth.

In addition to developing portables, manufacturers sought advantages over their competitors by introducing inventions such as the "noiseless" typewriter, which reduced the shock of the type's impact, or Royal's "magic margin" device, which set the margin at the touch of a lever. Some companies developed specialized applications such as machines for stenography, musical notation, writing on flat surfaces, or typing Chinese or Japanese (using large trays or cylinders of characters).

Another important development was the rise of electric typewriters. Malling-Hansen had already designed an electric *skrivekugle*, as well as another electric machine, the Takygraf, which was capable of 1,200 strokes a minute. The Cahill Electric (1900) and Blickensderfer Electric (1902) were produced in series, but failed to gain many users. Electric typewriters reached wide acceptance only in the 1930s, led by IBM's massive Electromatic. The IBM Selectric of 1961 followed the Blick Electric in providing interchangeable elements that could type in different styles or alphabets. The Selectric eliminated a moving carriage; instead, its "golfball" type element traveled across the page. Electronic typewriters, which use integrated circuits to control the behavior of the machine, arrived in 1978 with the Olivetti ET 101 and Exxon Qyx, both of which employed rotating daisy wheel type elements. (The daisy wheel concept dates back to the 1889 Victor index typewriter.)

By the 1970s, typewriter production was no longer centered in the United States and Europe. The machines were produced in countries including Japan, India, Brazil, Mexico, and South Africa. They were mass-produced, affordable, and usually less durable than their predecessors.

How were all these machines used? The romantic picture we have of the novelist or newsman smoking at the typewriter is not wrong, but the majority of typewriters, especially large standards, were used by secretaries and clerical workers, who were not expected to do any creative writing. Often the boss would dictate a letter while his secretary took shorthand notes, or he would speak into a recording device such as a Dictaphone; then the secretary would expertly turn those words into neatly typed sheets, making a carbon copy or two, and bring the letter to the boss for his signature. In the accounting department, workers would type up spreadsheets with long tables of figures, sometimes using wide-carriage typewriters to handle big forms, or decimal tabulators that helped the user line up columns of figures on the decimal point. Some workers would manipulate massive devices that grafted typewriters onto complex adding machines. Copies were made directly with a typewriter, using carbon paper, or a typewriter was used to create a master document such as a mimeograph stencil that could then be reproduced in large numbers. The basis of all these uses was paper, which was taken for granted as a universal medium for conveying and storing symbols.

Of course, not all uses of typewriters were prosaic. During the Cold War, behind the Iron Curtain, the machines were tightly monitored. When combined with mimeographs or photocopiers, typewriters could be used to produce *samizdat*, or illicit publications.[9] In the West, typewriters could create publications that were legal but offbeat: everything from rough, photocopied, fly-by-night zines to more sophisticated journals promoting alternative points of view.[10]

The typewriter accumulated a host of cultural meanings during its heyday.[11] As a tool used by typically female secretaries, who took dictation from typically male superiors, the typewriter was associated with drudgery, but also with romantic and sexual possibilities. Early on, this situation generated a genre of erotic and comic material picturing businessmen's fantasies and fears; on postcard after postcard, the boss's vengeful wife catches him with his typist on his lap.[12] Machines belonging to creative writers or journalists were perceived as absorbing some of the romance and adventure of those professions. Typewriters could also be imagined as haunted objects that channeled users of the past or spirits on another plane.[13] But predominantly, the typewriter was a symbol of standardization, mechanization, efficiency, and bureaucratic order.

Meanwhile, a few electronics hobbyists with names like Jobs, Wozniak, and Gates were tinkering in their garages. Their work would spell the beginning of the end of the typewriter's popularity. With the rise of affordable electronic information devices, which allow faster and more flexible manipulation of symbols, the obsolescence of the typewriter began.

It had long been recognized that computers could be used to produce text, but the size and cost of early computers made such an application seem frivolous. Dedicated word-processing units began to appear in the late 1970s; they were often expensive and involved a steep learning curve, but many creative writers reacted enthusiastically to the new freedom word processing gave them to edit text, move it, and copy it without depending on ink and paper. The word processor's apparent capacity to produce "perfect" text appealed to its early adopters, who could become entranced by the technicalities of perfecting the output, so that the labor of typing was replaced by the not inconsiderable labor of mastering a complex system.[14]

A landmark memo issued by Apple CEO Mike Scott on February 1, 1980 decreed that no typewriters should be in use at the company by the end of the year: "If word processing is so neat, then let's all use it! . . . We believe the typewriter is obsolete. Let's prove it inside before we try and convince our customers."[15] Over the following decade, affordable personal computers swiftly supplanted many traditional functions of typewriters, first serving as means to the end of putting ink on paper, and then bypassing paper altogether.

However, typewriters did not disappear, and manufacturers such as Smith-Corona, IBM, Canon, and Brother attempted to adapt to the new conditions. Like the typewriter explosion of the 1880s and 1890s, the typewriter obsolescence of the 1980s and 1990s was a period of innovation and experimentation, which is currently understudied.[16] The incorporation of electronics into typewriters gave them a variety of new functions, and manufacturers experimented with new approaches such as dot-matrix printing using impact, inkjet, and thermal technologies. One electronic device of the mid-1980s, the Brother Type-O-Graph, even used pens to draw individual letters.[17] (This invention challenges the USPTO definition of a typewriter.) Often, typewriters made since 1980 have digital features or can play a part in the digital environment. Some include minimal word-processing capabilities that allow the user to correct errors on a small screen before printing; others can function either as word processors or as typewriters (that is, text can be either printed immediately or manipulated on screen and printed later). Other typewriters of this period can be used as input devices or printers in conjunction with computers; intermediary devices were also available that fit over typewriter keyboards in order to convert them to printers. The dominant form of typewriter that survived this period of experimentation and culling was the electronic daisy wheel system.

In 2012, the last typewriter made in the United Kingdom, an electronic Brother, rolled off the assembly line in Wrexham, but production of electronic typewriters continues in Asia. The major brands include Swintec, Nakajima, TA Adler-Royal, and Brother.

Purely mechanical typewriters have also persisted into the 21st century. In 2009, manufacture of standard manual office typewriters finally ended when the Indian company Godrej & Boyce announced it would no longer produce them. However, portable manual typewriters based on several Japanese designs were still being made in China as of 2018.

Throughout the history of typewriters, manufacturers acquired examples of their competitors' machines for research purposes. Museums and individual collectors have also been attracted by rare and unusual typewriters, particularly those that illustrate the creative experimentation of the late 19th century. The Early Typewriter Collectors' Association has published a quarterly magazine, *ETCetera*, since 1987. The German-language Internationales Forum Historische Bürowelt, which comprises collectors of typewriters, mechanical calculators, and other antique office machines, has published *Historische Bürowelt* since 1982. Today, typewriter collecting is an unusual but increasingly popular hobby that is no longer limited to antique and peculiar-looking inventions; now that typewriters in general have become somewhat exotic devices, many collectors are also interested in 20th-century machines.

The Technological Legacy of the Typewriter

Even as typewriters have receded into the fringes of contemporary culture, our most advanced devices continue to show their influence. The very concept of using a keyboard to create a text was, of course, promoted by the typewriter, and this system remains ubiquitous despite the rise of voice recognition technology. In particular, the Sholes & Glidden Type Writer continues to influence the daily hand movements of billions of people. Christopher Latham Sholes' QWERTY keyboard layout, along with its closely related variants (such as the German QWERTZ and the French AZERTY), is still by far the most popular system for inputting Roman-alphabet text on computers, smartphones, and other devices.

The drawbacks of QWERTY as a system for typing English text have been decried since the 19th century. A ten-fingered typist must often use the left hand and weaker fingers to type common letters, and at least in theory, a faster system would allow more frequent alternation between left and right hands. Many alternatives to QWERTY have been proposed. For instance, Blickensderfer's "Scientific" keyboard placed the most frequent English letters on the bottom row of the typewriter, in the sequence DHIATENSOR; in this system, the most common letters are typed by the strongest fingers and require the least movement of hands and mechanism. Blickensderfer also developed independent systems for different languages, again based on studies of letter frequency. In the 20th century, the best-known alternative keyboard was developed by Dr. August Dvorak in 1936. The Dvorak layout was studied by the U. S. Navy, was available on some typewriters, and has its advocates today.

However, no non-QWERTY system has achieved popularity beyond a cadre of devotees, even now, when digital technology makes it relatively easy to implement such systems. The simplest explanation for QWERTY's success is "path dependence"—that is, the inertia created by generations of trained typists and their accompanying QWERTY typewriters. Muscle memory and the user interface have historically reinforced each other, and few people are willing to face the inconvenience of learning a different system.[18]

What was the origin of QWERTY? We do not know the answer with certainty. It does seem evident that Sholes began with an alphabetical order, as we can see from the DFGHJKL sequence. It is also clear that the typebars of the Sholes & Glidden, which rely on gravity to return to their resting position, can easily jam if adjacent typebars are activated in quick succession. This might be the origin of the belief that Sholes deliberately created an inefficient system that would slow typists down.[19] However, ten-fingered typing was not invented until the 1880s, so Sholes could not have been trying to slow down such a typist. A more plausible theory is that he was trying to separate frequently used pairs of typebars so that the machine would jam less often—resulting in greater, not lesser speed. This project would have required a study of the frequency of letter combinations in English. It has been demonstrated that on the Sholes & Glidden, commonly used pairs of typebars are in fact separated.[20] It seems, then, that a system that was intelligently designed to be efficient given the particular construction of the first Type Writer was transferred to typewriters that used different mechanical systems, and then to digital devices that do not print at all.

The "shift" key is another legacy of the typewriter. The Sholes & Glidden wrote only in capital letters, but a carriage that could be shifted forwards and backwards was introduced on the Remington No. 2; this mechanism allowed every printing key to produce two characters, such as a lowercase and an uppercase letter. Left and right shift keys are still found on modern keyboards, along with other modifier keys. The three-row keyboards used on tablets and smartphones have revived the layout of many small typewriters that used one shift for capital letters and another for numerals and punctuation.

The influence of the typewriter is also evident in today's dominant word-processing application, Microsoft Word. Particularly in its Print Layout View, Word is a WYSIWYG system (what you see is what you get), where the assumption is that the user wants to "get" a stable text on paper as the finished product. The default format for a Word document for North American users is thus an 8.5- by 11-inch page, the dimensions of US letter paper, with margins and line spacing that can be adjusted, as on a typewriter. In an age when most text is stored and transmitted without the intervention of paper, these formatting conventions seem increasingly quaint.

Typewriting in the 21st Century

Contemporary typewriters are used for niche applications in business and government. For instance, the New York Police Department and 17 other New York City agencies use over a thousand typewriters, to the amazement and scorn of many. "They still have a function and your belief that typewriters have gone away is just erroneous," said pugnacious Mayor Michael Bloomberg in 2012. "It's like books. Some people, believe it or not, still read books in paper."[21] Prison inmates, forbidden to use computers, use typewriters with transparent shells that cannot hide contraband. Libraries type labels. Funeral homes type official death certificates. Typewriters may find employment at medical, financial, or legal offices.

Some older writers, such as historians David McCullough and Robert Caro and novelists Larry McMurtry and Paul Auster, have continued to rely on typewriters. They are joined by an increasing number of younger writers, including children and teenagers, who grew up without touching a typewriter but now want to write poetry on one, or even use one to participate in National Novel Writing Month, typing at least 50,000 words in 30 days.

Why would digital natives turn to typewriters? Recent decades have brought a flowering of the possibilities of electronic text—notably, with the rise of the Internet, the ability for anyone to publish words immediately to a worldwide audience, at little or no cost. However, users have also grown warier of some risks and disadvantages of digital information technology from which non-digital devices, such as typewriters, are relatively free.

The escalating complexity of digital technology and its interconnected nature imply an alienation of the individual user from the tool. The typical user does not and cannot understand the technical and legal dimensions of the devices that he or she uses every day. For most users, computers typify the "device paradigm" described by Albert Borgmann, in which the user becomes a consumer of commodities rather than developing a physical and intellectual engagement with the tool that makes those commodities available.[22] In contrast, a mechanical typewriter operates with macroscopic pieces whose function the user can understand through common sense, observation, and logic. The typical typist is able to comprehend the device and can learn to repair it, thus creating greater self-reliance and a closer connection to the machine.

The privacy of all digital data is uncertain, despite the availability of sophisticated encryption algorithms. In contrast, a typewritten text cannot be hacked as long as it exists purely as ink on paper. For this reason, intelligence agencies and other sensitive governmental organizations use typewriters today for top-secret documents.[23]

Whereas most word processors are embodied in networked computers that offer countless other activities and distractions, a typewriter is a single-purpose machine. Many writers find that using a typewriter focuses them on writing and helps them avoid procrastination. As Michael Heim noted during the rise of word processors, the transition from the typewritten environment to the digital "information-rich world" quickly plunges us into "an information-polluted world".[24] Novelist Will Self writes, "People are deceived into believing that writing on a computer is faster, but it's not. Using a typewriter is more disciplined."[25]

The life cycle of digital devices is typically very short; they are designed to be part of a process of upgrades and replacements that will make them obsolete within a few years. In contrast, most typewriters were made to be durable, repairable tools, and a typewriter made 50 years ago could easily survive for another 50 years. It is one of the few useful items that can serve as a lifetime companion, and thus it represents stability and personal meaning in a time of flux and disposable objects.

The relative difficulty of correcting typewritten text can be an advantage if one wants to avoid the habit of constantly second-guessing and over-editing one's writing. The typewriter can be an excellent tool for brainstorming or creating a first draft.

Finally, an annotated or edited typescript gives us tangible evidence of the writer's process, whereas a digital document typically does not. (Of course, some writers save older versions of their text files.) Digital writing thus tends to obscure its own history and the creative personality of the author.

Such contrasts to the dominant digital technology have reversed the traditional meaning of typewriters. What was once primarily a tool for efficient, impersonal communication has become an individual, quirky, and artistic device that is cherished for its own sake. For instance, in 2016, Michaels, a US-based chain of over a thousand arts and craft stores, began carrying the We R Memory Keepers Typecast typewriter, a new, Chinese-made manual portable clad in a mid-20th-century-style body, available in several colors. The typewriter was conceived as the center of hobby activities such as scrapbooking and artistic correspondence, and was offered with several accessories such as colored ribbons, labels, and stationery. The Michaels venture illustrates the new meaning of typewriters in the 21st century; it marks the first time that a new typewriter was deliberately marketed as an artistic tool, not only in its ability to record language but by virtue of the mechanical technology that it uses. Of course, older typewriters, which are often more durable and attractive than recently made machines, can also be appropriated with new significance and for new purposes; in fact, 21st-century typists generally prefer typewriters made between 1920 and 1970.

To use a typewriter today is to signal one's individuality and one's resistance to the digital norm. That spirit of rebellion is most evident in practices such as digital Sabbath (including organized events such as the National Day of Unplugging), digital detox parties, and digital-free retreats (such as Camp Grounded, a weekend camp for adults). All these happenings require participants to set aside their electronic devices, and typewriters are often on offer as alternative technology and a reminder of physical forces and tools. A similar message of rebellion is conveyed by the literary journal *Harlequin Creature*, which is produced without any digital technology; each individual copy is created on typewriters by volunteers mobilized in "typing bees".

However, most contemporary typists do not avoid electronic technology permanently, nor do they wish to. Instead, they combine typewriters with digital technology in creative ways. The practice of publishing digital photos or scans of typewritten text, sometimes called "typecasting", is popular on social media such as Instagram, where typewritten poetry is particularly prevalent and the most successful typecasting poets have hundreds of thousands of followers.[26] Typecasting sends the message that the writer cares about the process and craft of writing; it reveals the fallibility and materiality of the writer and the text; and in an atmosphere dominated by the slickness of digital imagery, typecasting comes across as raw and genuine. Other forums for typecasting include the "typosphere", a typewritten and typewriter-loving subset of the blogosphere.

Typewriters can also be connected functionally to modern digital devices. For example, the USB Typewriter, designed by Jack Zylkin, is a kit that allows many manual typewriters to gain extra functionality as digital keyboards. Zylkin is one of many "makers" who are interested in understanding, controlling, and altering technology. Typewriters have attracted interest among this population, as well as within the steampunk subculture, which explores the potential of 19th-century technology.

Typewriters can be used to create new human connections, as well. The first "type-in" was held in December 2010 in Philadelphia, and has been followed by similar events in cities on several continents. Typewriter lovers gather in a coffeehouse, pub, or other venue to admire and trade machines, to talk, and to type.

The multiform "slow movement" takes the form of slow communication in contemporary letter writing, which often enhances correspondence by using typewriters. Typewritten

Figure 5.2 Street typists in Cincinnati, Ohio, 2015. Photo taken by the author.

"snail mail" creates personal connections between sender and recipient, but it can also be incorporated in social events such as letter-writing parties.

Street typists, who write poems, letters, or short stories at the request of passersby, use typewriters as a way to focus attention, stand out from the norm, and create drama.[27]

A more ambitious project was initiated by St. Louis poet Henry Goldkamp, who in 2013 placed over 70 typewriters in public places around the city and invited the public to type anonymous reflections. Goldkamp then selected the most striking and telling texts to publish in a volume titled *What the Hell is St. Louis Thinking?*[28]

Typewriters have also been repurposed to create music by artists such as the Boston Typewriter Orchestra, and have been incorporated into other forms of art. Artistic uses of typewriters include representations of the machines; art created using typewriters as tools; typewriters that have been repainted or modified; sculptures formed from typewriter parts; and installations and performance art that include typewriters. For instance, artist Tim Youd retypes a novel by one of his favorite writers, reusing a single sheet of paper with one backing sheet, employing the same kind of typewriter that the author originally used. Youd types in places that are relevant to the story or the author, and may create sculptures or paintings that commemorate the event. His performance art produces a diptych—a ragged, ink-soaked rectangle and a battered backing sheet, mounted side by side.

Perhaps the most dramatic symbol of the typewriter revival is the Blunderwood Portable, an installation at the 2015 Burning Man festival. The Blunderwood was a giant typewriter, with keys made of barstools, on which people could climb, dance, and type messages that were projected on a giant sheet through a digital connection.

Nearly four decades after the Apple memo declared that "the typewriter is obsolete", typewriters continue to be produced and to engage the 21st-century imagination, appearing in everything from fashion to fonts. Like vinyl records and Polaroid cameras, they are associated with "hipsters" but exceed the boundaries of any subculture, appealing to a diverse group of enthusiasts who are finding new possibilities in old technology. It remains to be seen whether the rebirth of typewriters will prove to be a passing trend or an enduring source of refreshment and perspective in an increasingly digital world.[29]

Notes

1 See www.uspto.gov/web/patents/classification/uspc400/defs400.htm, accessed December 27, 2016. I have omitted asterisks that refer to further definitions of some terms used in this definition.

2 Bennet Woodcroft, *Appendix to Reference Index of Patents of Invention, from March 2, 1617 to October 1, 1852* (London: Great Seal Patent Office, 1855), page 49.

3 This first documented typist is the heroine of Carey Wallace's novel *The Blind Contessa's New Machine* (New York: Viking, 2010). For an account of Turri's achievement and a survey of other pioneering inventions, see Michael H. Adler, *The Writing Machine* (London: George Allen & Unwin, 1973).

4 Richard Nelson Current, *The Typewriter and the Men Who Made It* (Urbana: University of Illinois Press, 1954).

5 Geoffrey Crabthorn (pseudonym of John Howard Clark), "Echoes from the Bush", *South Australian Register*, September 14, 1875. I thank Robert Messenger for this text and for his research on early typewriter inventors, available on his blog ozTypewriter (http://oztypewriter.blogspot.com).

6 Margery W. Davies, *Woman's Place is at the Typewriter: Office Work and Office Workers, 1870–1930* (Philadelphia: Temple University Press, 1982).

7 Illustrated overviews of early typewriters are provided by Darryl Rehr, *Antique Typewriters and Office Collectibles* (Paducah, KY: Collector Books, 1997); Michael H. Adler, *Antique Typewriters from Creed to QWERTY* (Atglen, PA: Schiffer, 1997); and Paul Robert and Peter Weil, *Typewriter: A Celebration of the Ultimate Writing Machine* (New York: Sterling, 2016).

8 Patrizia Bonifazio and Paolo Scrivano, *Olivetti Builds: Modern Architecture in Ivrea: Guide to the Open Air Museum* (Milan: Skira, 2001).

9 In *Burying the Typewriter: A Memoir* (Minneapolis: Graywolf Press, 2012), Carmen Bugan recalls how her father kept a typewriter buried in the backyard; he would dig it out at night to write pamphlets against Romanian dictator Nicolae Ceauşescu. The Stasi's obsession with controlling typewriters in East Germany is dramatized in the 2006 film *The Lives of Others*, dir. Florian Henckel von Donnersmarck.

10 An example is *The Match!*, an anarchist periodical published by Fred Woodworth since 1969, and still produced at present using a Varityper typewriter.

11 See Friedrich A. Kittler, *Gramophone, Film, Typewriter*, translated by Geoffrey Winthrop-Young and Michael Wutz (Stanford: Stanford University Press, 1999); and Darren Wershler-Henry, *The Iron Whim: A Fragmented History of Typewriting* (Ithaca: Cornell University Press, 2007).

12 Paul Robert, *Sexy Legs and Typewriters: Women in Office-Related Advertising, Humor, Glamour, and Erotica* (Netherlands: Virtual Typewriter Museum, 2003).

13 See John Kendrick Bangs's novel *The Enchanted Type-Writer* (New York and London: Harper & Brothers, 1899).

14 For a wide-ranging account of creative writers' use of early word processors, see Matthew G. Kirschenbaum, *Track Changes: A Literary History of Word Processing* (Cambridge, MA: Belknap Press of Harvard University Press, 2016); on the quest for perfection and the technical challenge, see Chapters 2 and 3. Michael Heim's *Electric Language: A Philosophical Study of Word Processing* (New Haven: Yale University Press, 1987) is a valuable reflection on the experience of the computerization of text.

15 See https://archive.org/stream/Apple_Memo_No_Typewriters, accessed December 27, 2016.

16 The best source for researching this period is currently typewriterdatabase.com, a growing website that provides images, data, and patents for typewriters of all times.

17 This concept was pioneered in 1845 by Charles Thurber's "mechanical chirographer" (U.S. patent 4,271).

18 Paul A. David, "Clio and the Economics of QWERTY", *American Economic Review* 75:2 (1985): 332–7. David's account, along with the conventional wisdom that the Dvorak system is far superior, are challenged in Stan J. Liebowitz and Stephen E. Margolis, *The Economics of QWERTY: History, Theory, and Policy*, ed. Peter Lewin (New York: New York University Press, 2002).

19 The story is found, for instance, in Stephen Jay Gould, "The Panda's Thumb of Technology", *Natural History* 96:1 (Jan. 1987): 14–23.

20 The E and R typebars have only one typebar between them, but even this degree of separation significantly reduces the likelihood of jams. All other common pairs are farther apart, and no two adjacent typebars are normally paired in English. See Richard E. Dickerson, "Did Sholes and Densmore Know What They Were Doing When They Designed Their Keyboard?" *ETCetera: Newsletter of the Early Typewriter Collectors Association* 6 (Feb. 1989): 6–9; Neil M. Kay, "Rerun the

Tape of History and QWERTY Always Wins", *Research Policy* 42 (2013): 1175–85. Kay also makes a mathematical argument (1177–78) that the fact that the word "typewriter" can be typed on the top row of the keyboard is unlikely to be an accident, lending credence to an old story that this feature is deliberate.

21 See http://newyork.cbslocal.com/2012/01/30/bloomberg-defends-city-spending-on-typewriters/, accessed December 27, 2016.

22 Albert Borgmann, *Technology and the Character of Contemporary Life: A Philosophical Inquiry* (Chicago: University of Chicago Press, 1984), pages 40–8.

23 "Kremlin Security Agency to Buy Typewriters 'to Avoid Leaks'", *BBC News*, July 12, 2013; Rahul Bedi, "Indian High Commission Returns to Typewriters", *The Telegraph*, September 27, 2013; and Philip Oltermann, "Germany 'may revert to typewriters' to counter hi-tech espionage", *The Guardian*, July 15, 2014.

24 Michael Heim, *The Metaphysics of Virtual Reality* (New York: Oxford University Press, 1993), page 12.

25 See www.shortlist.com/news/will-selfs-long-term-relationship, accessed December 29, 2016.

26 Poet Tyler Knott Gregson made a successful transition from typecasts posted on social media to print publication with his *Chasers of the Light: Poems from the Typewriter Series* (New York: Perigee, 2015).

27 Poet Abigail Mott and other street typists are documented in the film *A Place of Truth*, dir. Barrett Rudich (2013). The experiences of a band of public typists are collected in Jodi Egerton, David Fruchter, Sean Petrie and Kari Anne Holt, *Typewriter Rodeo: Real People, Real Stories, Custom Poems* (Kansas City, MO: Andrews McMeel, 2018).

28 For more examples of how typewriters invite anonymous confessions, see Michael Gustafson and Oliver Uberti, *Notes from a Public Typewriter* (New York: Grand Central Publishing, 2018).

29 For more information on the contemporary uses of typewriters discussed here, see Richard Polt, *The Typewriter Revolution: A Typist's Companion for the 21st Century* (Woodstock, VT: Countryman Press, 2015). The contemporary typewriting revival is also explored in the documentaries *The Typewriter (In the 21st Century)*, dir. Christopher Lockett (2012) and *California Typewriter*, dir. Doug Nichol (2016).

6

THE LURE OF THE TICKER

Braxton Soderman

> Every day, when down town, I talk to scores of men in offices, banking-rooms, and hotels. Every now and then they walk quickly away from me to the little machine known as the stock "ticker". Picking up the tape, they glance anxiously over it, come back in a half-forgetful way, and resume the conversation, although I see plainly that they are thinking more of the tape than of the talk. And the machine goes on with its rattling, whirring, clicking, nervous "tick!" "tick!" "tick!" What is this "ticker?" The machine is of little consequence, but the tally it keeps is a wonderful one. In the past few years it has ticked off every minute, every hour, every day, the ebb and flow of the prosperity of the country.
>
> —Rufus Hatch, 1882

While Rufus Hatch explains that the ticker "machine is of little consequence", nothing could be further from the truth. The stock ticker was a key technology of modernity, mobilizing increasingly "real-time" networks of information gathering and dissemination, linking financial markets and raising public awareness of these markets. While the massive reorganization of space and time brought about by the introduction of the railroads and telegraphy is well known, the stock ticker has not received the same attention. This essay explores the history of the stock ticker as a key technology of modernity while also arguing that an analysis of the ticker allows for a deeper understanding of the present moment as the world disembarks analog shores for the oceans of digital data that glimmer on the horizon. First, I offer a brief history of the stock ticker, its function, and decline. Second, I analyze the ticker in relation to theories of modernity, arguing that discourses surrounding the stock ticker and its visual and auditory aspects train attention within states of distraction and produce modern, speculative subjects calibrated to financial markets. Third, I discuss the ticker's legacy after it was engulfed by processes of computerization. Throughout the essay I explore how the ticker illuminates a variety of discourses surrounding topics such as storage and memory, the materiality of information, and the production of real-time networks. Indeed, the ticker was a consequential machine that prefigured future developments, and as such, offers a largely untapped archive through which to historicize contemporary issues that populate the digital world.

A Brief History of the Stock Ticker

The ticker was invented in 1867 by Edward Calahan. The original device employed two wheels which printed characters on a single strip of paper. According to Alex Preda,

"The first wheel had the letters of the alphabet on it; the second wheel had figures, fractions, and some letters. The inked wheels printed on a paper tape divided into two strips: the security's name was printed on the upper strip and the price quote on the lower one, beneath the name. In the 1870s, the ticker also began to record the traded volume, printing it on the tape immediately before the price" (Preda, 2006: 754). Instead of two human telegraph operators—one to send out price information in Morse code and one to decode at the destination—the ticker allowed price data to be automatically recorded in letters and numbers at the destination (Knorr Cetina and Preda, 2007: 119). Thomas Edison introduced his Universal Stock Ticker in 1869, the first mass-produced ticker. These tickers were rented out to various establishments throughout New York City and soon beyond. Employees of the New York Stock Exchange (NYSE) called "reporters" listened for completed securities transactions (occurring on the floor of the exchange), jotted down the sell price and volume and quickly delivered the information to telegraphers stationed on the floor itself. These "quotations" were transmitted upstairs to a central distribution office and then marshaled over the wires and through various relays until they soon appeared on distant, unfurling ticker tapes (Anonymous, 1908). Thus, the tickers reported the data of completed stock trades in near "real-time", and the dissemination of these data represented the spread of the market. The noise emitted from the ticker gives it its name, and the word "tick" was also used to indicate the smallest increment of change that a stock can increase or decrease—upticks and downticks. These "ticks" were measured in various fractions, historically becoming more refined, moving from one-eighths to one-sixteenths, etc.

An article from the *New York Times* in 1902 describes this burgeoning market which extended not one but multiple nerves from its origin in Manhattan:

> Wall Street has reached almost perfection in its system of telling the world what it is doing. Within forty seconds from the consummation of a sale on the Stock Exchange the news of it may be in the hands of every member of the Exchange, and within sixty seconds it may be read on the tape of thousands of 'tickers' from the Battery to New Haven.
>
> (Anonymous, 1902)

The author mentions over nine different tickers operating within this network: stock tickers transmitting price of stocks, produce tickers spreading the current value of wheat, financial news tickers, and general news tickers transmitting "flash" bulletins of worthy events. At the turn of the century the NYSE created a vast network of input (agents gathering information from exchanges) and output (a number of tickers dedicated to the spread of distinct content). While Preda argues that "The ticker was not wanted for efficient, accurate and broad diffusion of price data", but "because it helped reinforce social status and a monopoly over authoritative price data", the diffusion of a network and price data inevitably followed the expansion of the ticker.

The stock ticker was not an isolated technology seen only in various exchanges, specialized brokerage houses, and within the confines of selected financial businesses. Just 15 years after the introduction of the ticker, in the article from which the epigraph for this paper was drawn, Hatch wrote, "The 'tick!' 'tick!' 'tick!' goes on in every city of our land", and further, "The 'ticker' is everywhere to be seen. Every saloon keeper, dry goods man, grocer, and every club have their 'ticker.' To be without one would be to neglect the wants of their patrons. The men of the masses, who win their money by honest effort, stand about and listen to the 'tick!' 'tick!' 'tick!' from 'early morn to dewy eve.'" Although hyperbolic, Hatch explained that it is "the men of the masses" whom the ticker addressed, not only the financial baron and wealthy capitalist. David Hochfelder supports such a conclusion: "Within a few years of its introduction in 1867, the ticker could be found almost anywhere: in broker's offices, banks, hotels, restaurants, and even saloons and cigar stores" (2006: 340). He reports that in 1886 the Gold and Stock Telegraph Company operated "2,200 tickers scattered around the country" (2006: 430)—all disseminating quotations from the NYSE.[1] By the early 20th century, one finds comments indicating that the machines were available "wherever a subscriber care[d] to have them" (Duguid, 1926 [1904]: 94), and more dramatically to the point, "The whole American public has been educated to the 'ticker'" (Hirst, 1948: 94).[2]

A large part of this "public education" stemmed from the "bucket shops"—gambling establishments where customers with fewer means placed bets on the fluctuations of stock and commodity prices. The fluctuation of prices was "the lifeblood of the bucket shops" (Fabian, 1999: 191), and thus "the current of stock quotations brought in by the ticker" was their life-line (Anonymous, 1914). In the bucket shops customers avoided the high commissions of valid exchange brokers and could purchase less volume of stock. Because of this, these shops spread the activity of financial speculation to a larger public "playing the market" on a smaller scale. Importantly, transactions in these shops were never legitimate: "bucket shop transactions had no real effect on the market prices of stocks and commodities" (Hochfelder, 2006: 345). Orders made in these establishments never influenced the price of stocks and never appeared on the ticker. The bucket shops, in a move that would have made Jean Baudrillard grin, "outfitted their offices with the same tickers, blackboards, telephones, and reading matter found in brokers' offices" (Hochfelder, 2006: 343). Bucket shops simulated the apparatus of legitimate exchanges, and according to Ann Fabian, "in some cases the wire [of the ticker] ran only as far as the end of the rug" (1999: 192). Following the logic of the simulacrum, these bucket shops sparked decades of strife as "real" brokers attempted to differentiate themselves from the false.[3] Moreover, the bucket shops reveal an early example of network hacking, where some attempted to acquire official stock prices "by tapping into the wires of the stock exchange" (Knorr Cetina and Preda, 2007: 121).

Regardless of these contaminations, the bucket shops provided an outlet for the masses to be "educated to the ticker", to be exposed to the speculative infection which one commentator called "speculitis" (Ora, 1908: 159) and which was inevitably related to what another commentator, in 1926, called "tickeritis" (Harper, 1926: 10). Hochfelder summarizes:

> By broadcasting stock and commodity quotations to thousands of bucket shops, the ticker made speculation a popular activity. Whereas speculation had typically been the providence of the wealthy or well-connected, by 1880 ordinary men (and sometimes women) could step into a bucket shop and speculate in stocks or grain.
> (2006: 337)

In 1905, the Supreme Court decided that stock prices and their variations were the "products" of the speculators and brokers; thus, ticker tape machines could be further regulated and controlled. The "life-line" to the bucket shops was slowly cut off. Yet, while bucket shops withered during the first two decades of the 20th century, the legitimate stock exchange showed a marked increase in participants—participants already "educated to the ticker" and the general strategies of market speculation (Hochfelder, 2006: 357). Although one cannot claim saturation of the technology throughout society, many in the early 20th century were enticed by the "lure of the ticker" (Hochfelder, 2006: 339).

This lure turned luminous in 1923 when the Trans-Lux company installed a rear-projection apparatus on the floor of the NYSE. This machine printed quotations directly onto translucent film stock and projected enlarged quotations so that the crowds of brokers could observe price quotes from the floor (Preda, 2006: 767). The need for individuals to swarm the stock ticker—a common image in many photographs from the period—was disrupted by the appropriately named Movie Ticker. According to a company publication, *The Story of Trans-Lux* (1929), almost 500 such tickers were installed in "41 centers" in 1926, and over 1,200 installations "in 211 centers" by 1929. The projected tickers were installed on the floor of the NYSE, in brokerage boardrooms, and in executive offices. These tickers—some displaying quotations, others financial news—allowed information to "be easily read from any point in the room" (Burton, 1929: 10). The company publication notes that these electronic tickers "[freed] the members of the confusion, the delays, the physical and nervous strain that the ordinary ticker tape would impose" (Burton, 1929: 12). The "lure of the ticker" became more transparent as the "nervous strain" of the traditional ticker—supposedly caused by its materiality, its clattering, its messy and incessantly unfurling tape—was dispelled or at least politely removed. The ticker was thus perpetually present without its traditional distracting properties.

While a detailed recounting of the technical history of the ticker is beyond the scope of this essay, technological improvements appeared throughout the 19th and 20th centuries, often in terms of the speed of transmission. In the 1930s, the Teletype Corporation released a "black box" ticker system that could print 500 characters per minute (cpm), almost twice the speed achieved by previous tickers; these systems "became the standard for the next 34 years, until transmission speeds advanced up to 900 cpm" (Hale, 1995: 14). Electronic ticker devices began to appear in the 1960s, including the Quotron I, Quotron II, and Ultronic Systems Corporation's Stockmaster.[4] These electronic tickers stored information on magnetic tapes, allowing brokers to search stocks for price data.[5] As Kenneth Silber (2008) points out, in the 1960s "electronic boards and computer displays" were adopted because

> [t]he movement of print heads and tape reels [of traditional, mechanical tickers] could only be a bottleneck amid the growing digital networks. By the 1970s, mechanical stock tickers had mostly disappeared from financial circles, migrating in some cases to antique stores. The century-long run of ticker tape had come to an end.

Of course, these transformations occurred unevenly. The mechanical and electronic tickers that printed information to paper were not immediately discarded for screen-based displays: "Into the 1960s and early 1970s," writes Hale, "it was common to see traders and analysts scanning the paper ribbon" (1995: 14). In the 1970s and 1980s computerized display systems—"scopic systems", as Knorr Cetina and Preda call them—replaced the traditional delivery system of the ticker, but before returning to an analysis of the influence of the analog ticker on computerized systems, we must explore the ticker's importance during its historical peak.

Modernity and the Ticker

In discussing Thomas Edison—one of the key innovators in stock ticker technology—Jonathan Crary writes, "For Edison, cinema . . . had no significance in itself—it was simply one of a potentially endless stream of ways in which a space of consumption and circulation could be dynamized, activated" (1999: 31). The pre-cinematic ticker technology was another method of activating the flow of exchange; in fact, it visualized the flow and fluctuations of capital itself, creating a modern-conduit for observing the circulation of abstract data. This marked the emergence of the representation of capitalism in "real-time". The stock ticker was financial television where "one may get an almost complete picture of the market" (Atwood, 1917: 122).

If, as JoAnne Yates argues, the telegraph fueled the growth of markets by "reducing communication time and costs" and spurred the "vertical integration of firms" by extending their reach to distant markets, then the ticker spread the visual (and auditory) representation of this market (1986: 149–150). It called attention to the market, materializing it. Preda writes that the ticker "made the market in its turn visible as an abstract, faceless, yet very lively whole" (2006: 765). If the market was truly a miasma of traders clamoring in the stock pits, then the ticker was the "sensitive nerve" (as one commentator put it) that received and transmitted the data of these transactions (Conant, 1901: 701). With the emergence of the Trans-Lux Movie Ticker, the market and its continual flux were further projected for all to see, endlessly: the Movie Ticker "puts, and keeps, the Stock Exchange continually before their eyes" claims *The Story of Trans-Lux* (Burton, 1929).

In *Techniques of the Observer: On Vision and Modernity in the Nineteenth Century* (1990), Crary jettisons the use of the term "spectator" in favor of "observer" in order to bypass suggestions of "a passive onlooker at a spectacle".[6] The word "observer" invokes onlookers and also subjects active in their own obedience, in their own observance of societal rules and discursive practices. Instead of spectators as passive subjects, observers actively "[see] within a prescribed set of possibilities" (1990: 6). Historical and technological parameters act as conditions of possibility for the emergence of an active, and directed, vision. The observer implicitly remains as the mode of subjectivity operating in Crary's later work, *Suspensions of Perception: Attention, Spectacle and Modern Culture* (1999), which was an attempt to "examine some of the *consequences*" of his earlier mapping of modernity (1999: 4).[7] In his later work, attention became a key concept to understand the subjectification of the observer: "At the moment when the dynamic logic of capital began to dramatically undermine any stable or enduring structure of perception, this logic simultaneously attempted to impose a disciplinary regime of attentiveness" (1999: 13). Indeed, Crary mentions the stock ticker as a technology of "attentive consumption":

> Edison's first technological product, a hybrid telegraphy-stock ticker in the early 1870s, is paradigmatic for what it foreshadows in subsequent technological arrangements, including those of the late twentieth century: the indistinction between information and visual images, and the making of [a] quantifiable and abstract flow into the object of attentive consumption.
>
> (1999: 33)

The ticker operated as a modern technology of disciplinary attentiveness, calling one to pay attention to it, and through it, the market itself.[8] Yet, for Crary, both the concepts of attention and *suspension* are guiding threads in his analysis of modernity: attention is associated with waiting and contemplation while suspension is associated with interruption and distraction. Importantly, "Attention and distraction were not two essentially different states but existed on a single continuum" (1999: 47). Thus, attention could quickly turn into distraction and vice versa: "the two ceaselessly flow into one another" (1999: 51).

The technology of the stock ticker became a key site of this oscillating flow. The epigraph above provides an example where the ticker draws the attention of the writer's interlocutor, but this solicitation is simultaneously a distraction from the flow of conversation (Hatch, 1882). Even the sound of the ticker could both attract and distract. On the one hand, Albert Atwood wrote, "The louder the sound of ticking the more notice the machine attracts because loud ticking is a sign that the market is active" (1917: 122). On the other hand, the Trans-Lux company advertised their quote projections as important for "modern offices" where businessmen "dictate their letters, answer their telephone calls, hold their conferences without disturbing interruptions or the annoying clatter of a noisy ticker" (Burton, 1929: 20).[9]

Reading accounts of the ticker from the early 20th century, one encounters discourses of both attention and distraction, sometimes within the same work. For example, In *Truth of the Stock Tape* (1923), William Gann wrote:

> Tape reading requires *patience*, and the essence and value of it is *concentration*. There is no such thing as a man being born with a mind that can concentrate on 10 things at one time, much less 700. Then *success* depends upon *selecting* a *few stocks* and *concentrating* on them.
>
> (1923: 4)[10]

Yet, he also highlights the many disturbances that distract the tape reader in a crowded boardroom—conversations full of rumors and opinions, the distraction of the news ticker, etc. Thus Gann wrote, "With all these disturbances, there is not one man in a million that can concentrate enough to tell anything about what the stocks are going to do" (1923: 5). For those who cannot afford to have their own personal ticker, away from the distraction of others, Gann instructs, "Then it is *necessary* to *know how* to *read* the *tape without seeing it*, or without watching it all the time" (1923: 6). Perhaps what Gann envisaged here is a compromise, the observer concentrating over the tape one moment, and at another, avoiding the distractions emerging from the abundance of information on the tape itself and also the site of the ticker's reception.

For Crary, "the border that separated a socially useful attentiveness and a dangerously absorbed or diverted attention was profoundly nebulous" (1999: 47). Thus, attention was constantly troubled by either a fall into distraction (drift, daydream, reverie) or an overflowing excess (trance, the inactive stare, hypnosis), both categorized as absentmindedness by the norm of the "socially useful". Being put in a trance by the ticker tape is the direct result of the fluctuating phantasmagoria of capitalism. In his book *The Psychology of Speculation* (1926), Henry Harper wrote:

> The gyroscopic action of the prices recorded on the ticker-tape produces a sort of mental intoxication, which foreshortens the vision by involuntary submissiveness to momentary influences. It also produces on some minds an effect somewhat similar to that which one feels after standing for a considerable time intently watching the water as it flows over Niagara Falls. Dozens of people, without any suicidal intentions, have been drawn into this current and dashed on the rocks below.
>
> (1926: 11–12)[11]

This is an account of what Harper calls "tickeritis", and thus he recommends keeping one's distance from the ticker to reduce contagion. Nevertheless, the ticker operated as a disciplinary device that tuned one's attention to the continual presence of an abstract, financial market. The oscillation between concentration and distraction reveals the ticker as a key technology that calibrated the observer to the fluctuations of capital.

The Speculative Subject

The "mental intoxication" induced by the ticker tape is similar to the intoxication of the gambler, and indeed, gambling and stock speculation are endlessly collapsed and differentiated within discourses on speculation.[12] Applying a comment by Jean Baudrillard concerning "vertigo" to the experience of the gambler, Gerda Reith writes:

> Time freezes, and gamblers become absorbed in a total orientation to the immediate Here and Now. In this state they become creatures of sensation; seeing, but not really being aware of their surroundings; perceiving, but not truly cognisant of what is going on.
>
> (2006: 261)

The experience of the gambler tumbles into a perpetual "now", "a *constant repetition of a fleeting present*. In games of chance, the present is all-important" (2006: 269). The ticker tape provides a material instantiation of gambling with time. We have already seen how the ticker was the "life-line" of the Bucket Shops—the gambling dens that simulated real exchanges.

In discourses concerning the stock market, speculation is often placed on the side of reason and foresight while gambling is positioned as blindness. Albert Atwood wrote in 1917, "speculation presupposes intellectual effort; gambling, only blind chance" (1917: 292). Likewise, Thomas Gibson wrote in 1906, "gambling . . . is founded on blind chance, the equal possibility of certain events occurring or not occurring", while the speculator operates only "after careful investigation and study" (1906: 6–7). Atwood considers the success of the speculator as stemming from "reflection, foresight, forethought and knowledge of facts" (1917: 287).[13] Indeed, in the quotation above from Harper, he argues that the ticker tape "foreshortens the vision", it distracts the attention from "intellectual effort" which would extend analysis of the market by way of a longer-range thought. In an interesting image, Harper expands on this foreshortening:

> As a camera fails to record a true picture if placed in too close juxtaposition to the object, so in studying the ticker-tape one is restricted to a close-up view of conditions, resulting in a distorted gauge of values, for the figures recorded often mislead and confuse the attentive observer.
>
> (Harper, 1926: 12)

The ticker tape is *too close* to the market: its continuous unfurling blurs focus and restricts the view of the observer. In one sense, the ticker tape (considered as a whole) is a fluctuating, close-up image that is in danger of never being returned to a wider image where the quotations—and their meaning—can be reestablished in the scene of the market. In another sense, the isolated quotations that flow from the ticker tape are themselves like a series of close-ups that restrict outside perspective, similar, perhaps, to Carl Dreyer's film *The Passion of Joan of Arc* (1928). The threat when watching the ticker tape is that short-range thought will obscure long-range perspective:

> The business man in speculation will find it expedient to divorce himself from the alluring attractions of the ticker itself. Many traders whose long range views of values and approaching conditions are good, get their noses so close to the ticker as to shut out the true perspective.
>
> (Gibson, 1906: 47)

On the one hand, the examination of the ticker tape threatens blindness, a blindness that stems from closeness and a reliance on the short time span of the present, a blindness that is linked to pure chance and the realm of the gambler. The speculator, on the other hand, becomes a figure escaping chance through the examination of causes and determination.

The first definition of "speculate" in the *OED* Online is "To observe or view mentally; to consider, examine, or reflect upon with close attention; to contemplate." Speculation is a kind of immaterial seeing. More importantly, it marks a vision directed toward the future. Atwood provides the following description:

> The word speculation is derived from the Latin "speculare" meaning to spy out, look out, to observe. This Latin derivation affords the best idea concerning the meaning and real significance of speculation. It involves nothing more or less than the desire of the individual to overcome the domination of chance and to penetrate into the future with a view of profiting thereby.
>
> (1917: 286)

This mode of observing is directed outward and toward the future; indeed, the Latin word *specula* denoted a watch-tower or a look-out. Speculation is the desire for attaining an outside position from which to contemplate and analyze an object. One might add to Crary's category of the modern "observer" a mode of subjectivity aligned toward the concept of speculation. For Crary, attention was "a space in which new conditions of subjectivity were articulated, and thus a space in which effects of power operated and were circulated" (1999: 24). For the most part Crary limits his analysis of attention to the present moment; he does not have much to say about the *future* even though training the attentive subject also depends on eliciting concentration, extending perception from one instant to the next and into the future. Crary is more concerned with moments when attention turns to distraction or vice versa. Moreover, his examples of resistive modes of activity—reverie, daydream, trance, fixation, etc.—revolve on a present that escapes from the rationalized concatenation of events. Even trance, excessive concentration, devolves into a kind of timelessness where the future and past are ejected. As Atwood indicated, "to observe" is one of the definitions of "speculate", thus adding another temporal nuance to Crary's notion of the observer: one who not only attentively observes present norms and discursive practices but actively extends (or imagines) them into the future—colonizing and safeguarding the future.

For some financial analysts, the speculator was seen as socially useful in a world driven by capitalist markets, supply and demand, and the uncertain circulation of goods and capital:

> [The service of speculation] consists . . . in preparing for coming events. This preparation . . . forewarns every producer and holder of wheat that the price is going upward or downward in the future, according to the judgment of men who make it their business to study the conditions of demand and supply.
>
> (Conant, 1901: 702)

Furthermore, speculation "consists in the husbanding of the world's product against future needs" (Conant, 1901: 702). According to Conant, speculation and the prediction of future value help producers decide to raise or lower production, or even to switch to other "more profitable" forms of production. For Atwood, the future vision of the speculator "softens the intensity with which great economic forces work by lengthening the time over which their influences extend", which leads him to suggest that "[The speculator] is really a producer of 'time utility'" (1917: 297–298). Theoretically, the speculator helps to distribute goods, smooth circulation, equalize supply and demand, and even discover new markets. Time is usefully "smoothed" by linking the present in a conjectured concatenation with the future.

While part of the importance of the ticker concerned the progress toward "real-time" distribution of information, it was also a memory and storage device. Recall that Hatch explained that the ticker "machine is of little consequence, but the tally it keeps is a wonderful one". This tally provided a daily dump of information that could then be further compiled and analyzed. Preda details how the ticker led to the rise of "financial chartism", explaining that "the stock ticker enabled price data to be recorded in near real time. This minute recording led to a considerable increase in the available volume of data" (2007: 49). The massive accumulation of stored data allowed for the emergence of "financial chartists" who collected and analyzed past price fluctuations. "The aim of technical analysis", which Preda argues solidified in the 1930s, "is to forecast price movements on the grounds of detailed price charts" (2007: 58). Thus, the near real-time recording of data from the ticker was stored, charted, and processed for the assistance of future speculation. The stock ticker was one location where such a production of the speculative subject emerged. In discourses concerned with speculation, one discovers many recommendations to move away or beyond the constricting view of the tape. For example, Charles Dice wrote, "The successful sooner or later give up the close, confining work of concentration of attention on the tape for other types of trading quite as profitable, if not more so" (Dice: 1926: 288). It is not that attentive concentration disappears in these cases, but it is channeled in a different direction. Vision is turned away from the attractive watching of the tape, toward "financial chartism", and then toward the look-out position of the speculator. As Preda remarks, "technical analysis . . . provided market actors with an account of 'the market' as an orderly, totalizing phenomenon" (2007: 61). If the ticker drew attention to and materialized an abstract market, then chartism and speculation materialized an "orderly" market, rationalizing and extending it into the past and the future.[14]

The Ticker Tape Parade and the Materiality of Information

If modernity marks the process whereby "all that is solid melts into air", as Marx and Engels famously wrote, then the ticker was a modern technology of creating and circulating the wind. The ticker transformed material commodities into immaterial price data *and* materialized the immaterial by printing the price data on physical tape. First, the stream of ticker data becomes a commodity when it is rented to various subscribers. Fabian describes the long struggle over the ownership of prices between the farmers and the brokers, explaining, "The speculators argued that their products were price quotations and the markets themselves. They appropriated the language of production which had traditionally been the moral stance of the commodity producers" (1999: 196). According to the farmers, the speculators traded in "wind"—"wind wheat"—since the volumes of their traded speculations outstripped the actual volume of tangible product; moreover, these speculators were "parasites upon the social organism" since they produced nothing (unlike the farmers) and parasitically consumed the value of commodities for their own interests (Conant, 1901: 698). The speculators responded by slowly "appropriating the rhetoric of production and applying it, with all its historic moral weight, to the price quotations—the products—they said they made" (Fabian, 1999: 174). The struggle ended with the Supreme Court decision (mentioned above) that determined that price quotations were the products of the work of the traders and brokers

(Fabian, 1999: 199).[15] Second, ticker data were seen as a barometer of financial pressures and trade fluctuations: indeed, "These fluctuations simply reflect the ripples of opinion passing through the minds of the best judges regarding the value of such properties" (Conant, 1901: 703). If speculation concerned a "mental viewing" then the ticker tape materialized the mental views and immaterial thoughts of the speculators. The ticker tape substantiated the ideas driving the direction of the market, and if the prices occasionally fluctuated widely it was because current political, economic, or social events "flashed like a panorama across the minds of the brokers" (Conant, 1901: 704). The raw data of fluctuation that ticked across the tape became the physical material that the tape reader sculpted. Thus, the ticker can also be understood as a key harbinger of informational society: dematerializing reality through quantification and abstraction, producing immaterial information as a commodity, and giving weight and impact to informational forms.

It was within the air that the ticker tape found its most culturally celebrated role: the ticker-tape parade. The first parade where ticker tape appeared—as fact or legend would have it—was at the dedication of the Statue of Liberty in 1886, after which a few hundred such parades have occurred in New York City (Anonymous, 1886). Margaret Doty wrote of Admiral Byrd's return from Antarctica in 1930: "About fifty thousand rolls of ticker-tape were unwound and came twisting down on the explorers' heads, along with shredded telephone books and the contents of uncounted waste baskets" (1931: 150). In these parades, the growing, material accumulation of the burgeoning "information society" was ejected in the catharsis of a sky filled with debris. The farmers and workers who saw the stock brokers as trading in "wind wheat" might have seen their conclusions confirmed as the stalks of ticker tape blew like an urban, artificial dust bowl. The ticker tape parade commemorated the "progress" of modernity—the arrival of esteemed foreign dignitaries, successful exploration, victories in sports or in war—while releasing the future from the past in order to renew itself. As Walter Benjamin once wrote, information "does not survive the moment in which it was new", and thus these parades jettisoned the obsolete and became spectacular rituals allegorizing the dynamics of capitalism (1968: 90). They were just another day in the circulation of capital, a day to tidy up the office, to take out the trash, to clean house and make room for more modernity, more of the same, what Crary called, "a ceaseless and self-perpetuating creation of new needs, new consumption, and new production" (1990, 10)—and, certainly, new information.

It is not surprising that the "extinction" of the ticker tape parade was heralded in an issue of the newsweekly *Computerworld* from 1978, where, after witnessing a recent parade, author Tim Scannell conjectured that "traditional forms of the celebration are doomed to follow the same route as gas street lamps, coalpowered bed warmers and the Model T—thanks to computer systems" (1978: 5). At issue was the scant 135 tons of paper used in the parade, which Scannell compared to the 3,474 tons dropped during astronaut John Glenn's 1962 celebration. "City officials blamed the absence of ticker tape" and "the advent of computers and their 'paperless' processing", wrote Scannell, who then reported that spectators dumped "computer punch cards, rolls of bathroom tissue and several shoes" (1978: 5). By the late 1970s, the historical disappearance of the mechanical ticker, and its displacement by computerized quotation systems, was visually and materially evident even in the skies and on the streets of Manhattan.

The Tapering of the Ticker Tape

As the streamers of ticker tape disappeared from the skies, the dematerialized "form" of the ticker lived on, that is, as a stream of price data and trade volume. The 20th century witnessed the systematic expansion of the ticker-form, which migrated to the Trans-Lux projection systems, to electronic signs wrapping around city buildings, to television news programs where its flow served as an indicator of televisual "liveness", and then to the Internet news and financial tickers that appear on various websites. Users can even download a variety of stock ticker applications to run on their computer desktops or mobile devices. The modern technology of the ticker has replicated itself endlessly—at least in terms of its "paperless" form.

The mechanical ticker was eventually replaced by a variety of electronic and computerized systems. The Quotron I machine was produced by Scantlin Electronics Inc. (SEI) and appeared in 1960. It stored quotations on magnetic tape and allowed users to search for particular stock symbols and then print the latest quotations. Ultronics Systems Corporation released a more advanced service that used AT&T's Dataphone system; the Ultronics system stored stock information on a magnetic drum, and could broadcast the "previous day's close, open, high, low, net change, and volume" to a client machine (Phister, 1989: 111). In response, SEI introduced the Quotron II in late 1962 (which also used AT&T's Dataphone system), though SEI's service stored information "directly in computer memory" and thus increased retrieval speed (Freedman, 2006: 221). Contemporaneous with these early shifts to computerized technologies, the NYSE introduced a mechanical, high-speed ticker in 1964 dubbed the "900" in the popular press because it could print up to 900 cpm (Vartan, 1964b: 69). This machine was developed by Teletype Corporation, which also produced the "black box" ticker from the 1930s that the "900" replaced. To my knowledge, this was the last mechanical ticker that was introduced before the shift to systems that utilized computer displays in the late 1960s and 1970s (Vartan, 1964a: 43, 1964b: 71).[16] It is important to note that the shift from mechanical to computerized systems was not only a discontinuous rupture but also a technical evolution driven by the need to increase transmission speed in order to compensate for the increased volume of trades, thus decreasing temporal lags in reporting the real-time state of the market. The speedier computer systems that arrived in the 1970s (reaching speeds of 36,000 cpm) intensified processes inaugurated by historical development of tickers that eventually "[reached] the limit of electro-mechanical processing ability" (Teweles and Bradley, 1998: 149).

Knorr Cetina and Preda (2007) call these screen-based financial services (such as the culturally recognizable "Bloomberg Terminal") that eventually replaced the ticker, "scoping systems", which aggregate, sequence, and even stream an assortment of financial information and data on a single computer display.[17] While a detailed history of the various digital technologies that replaced the ticker is beyond the scope of this essay, more pertinent is Knorr Cetina and Preda's argument that the ticker "pre-cast" these scoping systems with the latter operating as an intensification of the form that the ticker presaged. While the mechanical tickers "sequentialized the market, initiating a price-and-volume flow that became at the same time an information flow" (Knorr Cetina and Preda, 2007: 123), the computerized, scoping systems added further dynamic threads of information to this flow including news streams, financial data, "insider observations", conversations between traders that include both "personal talk" and "conversational dealings" aimed at particular trades, and so on (2007: 133). As the authors explain, "now the tickers' bare price-volume record of past transactions has evolved into a Bergsonian multiplicity of 'everything' relevant to the market" (2007: 131).[18] The tapering of the ticker tape did not result in its disappearance, but the information contained on the ticker gradually narrowed into one thread within a wider financial and social tapestry, one flow of information surrounded by others upon the digitized screen.

Knorr Cetina and Preda explain how these scoping systems create what they call an "intensity" of observation and "connectedness" to the market as traders sit before the screen, "close enough to feel every 'tick' of [the market's] movement, and to tremble and shake whenever it trembles and shakes" (2007: 132). This is an augmentation of a similar affect that historical discourses surrounding the mechanical ticker recorded. The authors explain that this intense, mediated connection to the market catalyzes what they call "*preparedness*—the readiness to respond reflex-like to trading challenges that appear onscreen", which is a form of gut-response where "the trader can spring into action fast and 'unthinkingly' when prompted" (2007: 132).[19] This observation also dovetails with historical accounts of tape reading. For example, a tape watcher reminisced in 1923:

> The reason for what a certain stock does to-day may not be known for two or three days, or weeks, or months. But what the dickens does it matter? Your business with the tape is now—not tomorrow. The reason can wait. But you must act instantly or be left.
>
> (Lefevre, 1923: 11)

Gann uses similar rhetoric, claiming that the tape reader employs "intuition" as "instantaneous reason" in order to act (1923: 2).[20] By 1964, with the introduction of the "900" ticker, this instantaneous reason and preparedness was increasingly necessitated by technology: as one tape reader remarked at the time, "That faster tape doesn't give me much time to reflect" (Vartan, 1964b: 71). One could read the appearance of scoping systems and financial displays as an intensification of the processes inaugurated by the mechanical ticker which, as Crary remarked, made "[a] quantifiable and abstract flow into the object of attentive consumption" (1999, 93). Indeed, Knorr Cetina and Preda's description of traders ensconced before their screens—"a full-time task"—could be understood as an augmentation of this attentive consumption, where these scopic tapestries incorporate the distractions (e.g., conversations around the ticker, streams of news, etc.) that beleaguered mechanical tape readers in the past, recasting them as new attractive streams within an assemblage instead of distracting disturbances (2007: 132).[21]

This intense focus on presence, the business of "the now", the traders' lightning, unthinking reactions, allows us to speculate on what might be lost as the mechanical ticker tapers. The idea of "tapering" functions as a metaphor for the historical decline of the mechanical ticker, channeling connotations of diminishing, dwindling, and thinning as one approaches the end: the traditional price and volume data recorded by the ticker gradually become a thin thread within a thick array of financial data on a Bloomberg Terminal. Yet, it is also time itself that tapers off, the thick duration of the Now, the present, reduced into evermore fine, discrete "microticks". Today, high-frequency trading (HFT) is, if not a household word, at least widely understood. "Stock exchanges can now execute trades in less than a half a millionth of a second", writes Nick Baumann,

> more than a million times faster than the human mind can make a decision. Financial firms deploy sophisticated algorithms to battle for fractions of a cent. [. . .] these programs exploit minute movements and long-term patterns in the markets, buying a stock at $1.00 and selling it at $1.0001, for example. Do this 10,000 times a second and the proceeds add up.
>
> (Baumann, 2013)

These algorithmic and automatic high-frequency trades now account for a substantial portion of market activity. As "event fields far below the threshold of human perception and responsiveness" these microtemporal, automatic trades starkly contrast with the unfurling tape and the "tick", "tick", "tick" of the mechanical ticker, which—even if never truly "real-time" and lagless—evoked a tangible present at the scale of human action (Toscano, 2013). Indeed, when the "900" mechanical ticker went on-line in 1964 one tape reader complained of "[missing] every third sale", a statement that marks the gaps that emerged when technical speeds outran human perceptual processes. The mechanical ticker, at its peak and on the precipice of decline, silently announced the tapering of temporalities accessible to the human sensorium. Even the "ticking" sound entered a stage of diminuendo with the "900", where "the tapping sound is softer and switches alternately from a dissonant hum at lower speeds to a blended clacking sound at top speed" (Vartan, 1964b: 71).

This must not be read as nostalgia for a lost analog world, for a simpler, more tangible market displaced by an unknowable and immaterial financial world. As Alberto Toscano (2013) reminds in his ruminations on HFT, "attending to its operations dispels the equation of financialisation with immateriality and frictionlessness".[22] There is always the "revenge of matter and space", as Toscano quips, even when faced with what seems their erasure. Recall that even the mechanical ticker tape augured both the dematerialization of material commodities (into information) and the materialization of information (into commodities) that could be read through the tape, charted, and grasped. In doing so it instructs us that HFT has its own material scaffolding and impacts to which one must now attend. While this chapter has tugged at only a few threads that run through the historical brocade that was the stock ticker, clearly it is not "a machine of little consequence" as Rufus Hatch held in 1882. Rather, it harbors intricacies that will further enrich our understanding of both the analog and the digital, from real-time to lag and latency, from storage and memory to the emergence of information and the network society.

Notes

1 The precise number of tickers operating in New York City and around the country in the late 19th century is difficult to know. For another, more detailed list of stock ticker dissemination numbers, see Preda, 2006: 764.

2 A few humorous depictions from the early 20th century demonstrate the anxieties surrounding the ticker's dissemination of price data to the larger public. A cartoon from 1914 associates the ticker with the larger public, showing a "museum attendant" speaking to three clownish caricatures from "society" as they stare in astonishment at a stock ticker device. The caption reads, "These instruments, known as stock-tickers, *were* in use in Wall Street up to the year 1914. They were abandoned when the public got out of the market, and they are now very rare." Another humorous column tells of a "bond-salesman for one of the big-houses" traveling to Kansas in order to meet "prosperous farmers." The bond-salesman, thinking he can pull-one-over on the rube from the country, agrees to buy a stock from a farmer because he saw its price rising on a ticker before he drove out to the farm. In the end, the farmer himself has a stock ticker in his home office and pulls-one-over on the city-slicker: the farmer knew the stock price had dropped before the bond-salesman arrived (Anonymous, 1912).

3 See Brace, 1913: 235–241.

4 See, for example, Wagner (2014) and Bloom (2017).

5 Knorr Cetina and Preda discuss "how the ticker became replaced in stock exchanges" by these electronic systems, focusing on the Stockmaster and a collaboration between Reuters and Ultronic in 1964 to produce an international quotation system: "Ultronic 'absorbed' ticker-tape signals from all major stock exchanges and other markets by feeding them into master computers that processed the material and relayed it to subsidiary computers and finally to the offices of brokers" (2007: 128).

6 Preda explicitly links the ticker to Crary's notion of the "observer": "The observer of the market was the observer of the tape" (2006: 767).

7 For Crary, the onset of modernity—which he analyzes as beginning within the 1820s and 1830s—creates a new set of conditions from which a modern configuration of the observer emerges. During the 19th century the observing subject is shaped by a historical process whereby "capitalism uproots and makes mobile that which is grounded, clears away or obliterates that which impedes circulation, and makes exchangeable what is singular" (Crary, 1990: 10). Throughout this process, vision is unanchored from its stabilizing referent and perception is cast into a vortex of distraction.

8 Following Crary, Preda writes that "The ticker firmly bound investors and brokers to its ticks. Constant presence, attention and observation were explicitly required by manuals of the time" (2006: 786).

9 Such a projection, integrated into the workspace itself, is an early form of what Crary calls a "technological arrangement" that subjects one to "a permanent low-level attentiveness that is maintained to varying degrees throughout large expanses of waking life" (1999: 77). The Trans-Lux tickers removed the distracting seams of the work day, to make time more fluid: "This device puts the Stock Exchange right before [the businessmen's] eyes and keeps it there. They follow the course of trading without ever leaving their desks, without stopping, from time to time, to consult the ticker tape or to phone their broker, without a second's intrusion upon their pressing business routine" (Burton, 1929: 20).

10 Another example: "The tape reader spends every day at the ticker watching the tape with most careful attention. [. . .] The tape reader, to be successful, should devote his whole time to studying the tape" (Dice, 1926: 280).

11 See Preda for another example of "ticker trance" which aligns nicely with the "excesses" of attention: "I might be talking at the moment his eye began to pick up the tape again, but until he finished he was a person in a trance. If, reading the tape, he observed something that stimulated his mental machinery, I might go on talking indefinitely; he wouldn't get a word of it" (quoted in Preda, 2006: 768).

12 In his essay "Socialism and the Intellectuals" (1900), Paul Lafargue wrote, "The big capitalists interest themselves only in the operations of the stock exchange, which afford the delights of gambling; they dignify these by the pompous name of 'speculations,'—a word formerly reserved for the highest processes of philosophical or mathematical thought."

13 Interestingly, Gabrielle Brenner and Reuven Brenner call speculation "betting on an idea" (1990: 91).

14 In 1908, Reyam Ora wrote an article entitled "Speculitis", which, although never defined, really meant betting on the market "without results", gambling without knowledge of the system. Interestingly, the cure for "speculitis" is posited as knowledge and success: "*Absolute knowledge of when and how to play is the remedy; success at speculation, the cure.* To get that knowledge; to be a successful speculator is a matter of labor and research" (1908: 161). Speculitis, improper speculation as gambling, is cured by proper speculation—labor, research, and an imagined "absolute knowledge".

15 See also Hochfelder, 2006.

16 One can also see the tension between the mechanical and the electronic in the competition between the Trans-Lux Corporation and Ultronics Systems Corporation to market large, display boards for the "900" ticker. Trans-Lux continued with the analog approach that it had developed over the years to "record the transactions on a transparent tape for use in a projector", while Ultrasonic used an electronic system, the Lectrascan, which "[displayed] fixed characters that change at set intervals" on an electronic display (Smith, 1964).

17 For Knorr Cetina and Preda, a scoping system "assembles on one surface dispersed and diverse activities, interpretations and representations which in turn orient and constrain the response of an audience" (2007: 126).

18 Knorr Cetina and Preda leverage a kaleidoscopic metaphor of a ceaselessly unrolling carpet to describe these scoping systems where "threads (the lines of text appearing on screen) are woven into the carpet only as we step on it and unravel again behind our back" (2007: 130).

19 Caitlin Zaloom uncovered a similar experience of "being in the zone" on the floor of the Chicago Exchange: "Traders most value a sense of total absorption in the market. In the 'zone' conscious thought disappears and an ultimate sense of presence takes over. They are able to act without explicit thought. Their senses are highlighted to the rhythms and sounds of the market and the flow of trades. Achieving oneness with the market can wipe away thoughts beyond the moment" (2006: 136).

20 Gann (1923: 2) wrote:

> What is *intuition*? [. . .] The best definition I can give of intuition is that it is *instantaneous reasoning*. It is something which tells us when we are right or wrong before we have the

time to reason it out. The way to benefit through intuition is to act immediately, and not stop to reason or ask why. That is what a good tape reader does.

Yet, later in his book Gann would explain that "The best way to read the tape correctly is to stay away from it" (p. 76), and indeed, his embrace of "instantaneous reason" was grounded on the incorporation of other information into the decision-making process, including the keeping of charts that stored the past price changes of a certain stock (p. 82).

21 Knorr Cetina and Preda's description of scoping systems as "centering" and as "an array of crystals acting as lenses that collect light, focusing it on one point" parallels this sense of merging multiple information flows into a single stream of attentive consumption (2007: 126).

22 David Golumbia also discusses and laments the opacity of high-frequency trading, calling for a renewed, democratic transparency that could counter the "concentration of power" wielded by financial actors in this area (2013: 295–296).

References

Anonymous. "The Sights and Sightseers." *The New York Times.* October 29, 1886.

Anonymous. "How Wall Street Gets its News." *The New York Times.* January 19, 1902.

Anonymous. "How Stock Exchange Transactions Are Reported." *The Ticker.* Vol. 3, No. 1, November 1908: 13–16.

Anonymous. "Puck in Wall Street." *Puck.* Vol. 71, No. 1843, June 26, 1912: 6.

Anonymous. "The Remembered Lure." *Puck.* Vol. 76, No. 1969, November 18, 1914: 6.

Atwood, Albert W. *The Exchanges and Speculation.* New York: Alexander Hamilton Institute, 1917.

Baumann, Nick. "Too Fast to Fail: How High-Speed Trading Fuels Wall Street Disasters." *Mother Jones.* Jan/Feb, 2013. <www.motherjones.com/politics/2013/02/high-frequency-trading-danger-risk-wall-street> Accessed February 21, 2017.

Benjamin, Walter. *Illuminations.* Trans. Harry Zohn. New York: Shocken Books, 1968.

Bloom, Zach. "The History of Stock Quotes." Eager.io, <https://eager.io/blog/stock-quotes-through-history/> Accessed January 15, 2017.

Brace, Harrison H. *The Value of Organized Speculation.* Boston: Houghton Mifflin Company, 1913.

Brenner, Gabrielle; and Reuven Brenner. *Gambling and Speculation: A Theory, a History, and a Future of Some Human Decisions.* Cambridge: Cambridge University Press, 1990.

Burton, Jack. *The Story of Trans-Lux.* New York: Trans-Lux Daylight Picture Screen Corporation, 1929.

Conant, Charles A. "The Uses of Speculation." *Forum.* Vol. 31, No. 6, August 1901: 698–712.

Crary, Jonathan. *Techniques of the Observer: On Vision and Modernity in the Nineteenth Century.* Cambridge: MIT Press, 1990.

Crary, Jonathan. *Suspensions of Perception: Attention, Spectacle, and Modern Culture.* Cambridge: The MIT Press, 1999.

Dice, Charles. *The Stock Market.* New York: McGraw-Hill Book Company, Inc., 1926.

Doty, Margaret. "Cordiality by the Carload." *Forum and Century.* Vol. 85, No. 3, March 1931: 148–153.

Duguid, Charles. *The Stock Exchange.* Ed. E. D. Kissan. London: Methuen & Co. Ltd., 1926 [1904].

Fabian, Ann. *Card Sharps and Bucket Shops: Gambling in Nineteenth-Century America.* New York: Routledge, 1999.

Freedman, Roy. *Introduction to Financial Technology.* Burlington, MA: Elsevier Inc., 2006.

Gann, William D. *Truth of the Stock Tape.* New York: Financial Guardian Publishing Co., 1923.

Gibson, Thomas. *The Pitfalls of Speculation: The Human Element in Stock Market Transactions.* New York: The Moody Corporation, 1906.

Golumbia, David. "High-frequency Trading: Networks of Wealth and the Concentration of Power." *Social Semiotics.* Vol. 23, No. 2, 2013: 278–299.

Hale, Sam H. "The Ticker Tape: Yesterday, Today and Tomorrow." *MTA Journal.* Fall–Winter 1995: 10–22.

Harper, Henry Howard. *The Psychology of Speculation*. Boston: Privately Printed, 1926.

Hatch, Rufus. "Other People's Money." *The New York Times*. November 13, 1882: 5.

Hirst, Francis W. *The Stock Exchange: A Short Study of Investment and Speculation*. London: Oxford University Press, 1948.

Hochfelder, David. "'Where the Common People Could Speculate': The Ticker, Bucket Shops, and the Origins of Popular Participation in Financial Markets, 1880–1920." *The Journal of American History*. Vol. 93, No. 2, September 2006: 335–358.

Knorr Cetina, Karin; and Alex Preda. "The Temporalization of Financial Markets: From Network to Flow." *Theory, Culture, & Society*. Vol. 24, Nos. 7–8, 2007: 116–138.

Lafargue, Paul. "Socialism and the Intellectuals." 1900. <www.marxists.org/archive/lafargue/1900/03/socint.htm> Accessed January 15, 2017.

Lefevre, Edwin. *Reminiscences of a Stock Operator*. Larchmont: American Research Council, 1923.

Ora, Reyam. "Speculitis." *The Ticker*. Vol. 2, July 1908: 159–161.

Phister, Montgomery Jr. "Quotron II: An Early Multiprogrammed Multiprocessor for the Communication of Stock Market Data." *Annals of the History of Computing*. Vol. 1, No. 2, 1989: 109–126.

Preda, Alex. "Socio-Technical Agency in Financial Markets: The Case of the Stock Ticker." *Social Studies of Science*. Vol. 36, No. 5, 2006: 753–782.

Preda, Alex. "Where Do Analysts Come From? The Case of Financial Chartism." *Sociological Review*. Vol. 55, No. S2, October 2007: 40–64.

Reith, Gerda. "The Pursuit of Chance." In *The Sociology of Risk and Gambling Reader*. Ed. James F. Cosgrave. New York: Routledge, 2006.

Scannell, Tim. "Ticker-Tape Parades Quickly Becoming Extinct." *Computerworld*. November 6, 1978: 5.

Silber, Kenneth. "The Ticker's Rise and Fall." *Research Magazine*. February, 2008. <www.thinkadvisor.com/2008/02/01/the-tickers-rise-and-fall> Accessed January 15, 2017.

Smith, William. "Ticker Displays Spark a Battle." *The New York Times*. June 28, 1964.

Teweles, Richard; and Edward Bradley. *The Stock Market: Seventh Edition*. New York, NY: John Wiley and Sons, Inc., 1998.

Toscano, Alberto. "Gaming the Plumbing: High-Frequency Trading and the Spaces of Capital." *Mute*. Vol. 3, No. 4, January 16, 2013. <www.metamute.org/editorial/articles/gaming-plumbing-high-frequency-trading-and-spaces-capital#sdfootnote1anc> Accessed February 21, 2017.

Vartan, Vartanig. "High Speed '900' Ticker Is Installed Here." *The New York Times*. June 23, 1964a.

Vartan, Vartanig. "Quotations on High-Speed Ticker Fly Past the Eyes of Tape Readers." *The New York Times*. December 2, 1964b.

Wagner, Ralph. "The Rise and Fall of the Ticker-Tape Empire." *CFA Institute Magazine*, September/October, 2014: 47.

Yates, JoAnne. "The Telegraph's Effect on Nineteenth-Century Markets and Firms." *Business and Economic History*. Vol. 15, No. 2, 1986: 149–163.

Zaloom, Caitlin. *Out of the Pits: Traders and Technology from Chicago to London*. Chicago: The University of Chicago Press, 2006.

7

THE OVERHEAD PROJECTOR

Visuality and Materiality

Josh Zimmerman, Ken S. McAllister, and Judd Ethan Ruggill

Do you remember this? The teacher flicks a switch and the room darkens. Another switch clicks and the classroom's overhead projector hums to life (its whining cooling fan clearly in need of a new bearing) and casts a 5-foot slightly-keystoned square of light onto the wall. The retractable projection screen hanging from the ceiling near the front of the classroom—just above the chalkboard and behind the sanitized roll-up anatomy chart—has been broken for months, so the wall is its imperfect replacement. After struggling to separate a stack of cellulose acetate sheets (a.k.a. transparencies) and keep them from slipping onto the floor, the teacher places the first one on the projector's glass surface, adjusts the ostrich-like focal lens, and suddenly the crudely drawn image of a cell's membrane and cytoplasm covers the wall, shaded here and there by stray chalk marks, eraser dust, and pieces of cellophane tape that are brown at the edges. For the next 30 minutes, the teacher uses a felt-tipped pen to draw various additional cell structures on the projected diagram, explaining the development and function of each in turn. When the bell rings at the end of class, the students disgorge into the hallway while the teacher turns the lights back on and switches off the projector, stowing its cord. Between classes, the instructor uses an old wet rag to wipe the marker off the acetate sheets, preparing them for the next session. This was standard operating procedure in high school Biology classes until well into the new millennium.

Given the many recent improvements in computing and projection technology, today's students are more likely to view visual aids such as charts and diagrams on laptops and tablets, or see them displayed in large format through such devices connected to a ceiling-mounted digital projector. Sadly, the screen at the front of the room is likely to still be broken. The overhead projector is, as are all the technologies detailed in this book, an apparatus that has largely receded from view. Its most useful attributes—relative ease of use and transportation, the capacity to layer together and alter images, and the ability to reuse the recording medium (in this case, cellulose acetate sheets)—have all been improved on by newer projection and presentation technologies. Yet, the overhead projector's path toward obsolescence—an arguable assessment, as we will show—has been a circuitous one, emerging from two separate but related trends. Additionally, the overhead projector does something newer technologies cannot: manifest the materiality of projected objects. This unique feature is one of the primary factors keeping overhead projectors from complete extinction.

In this essay, we explore the development and obsolescence of the overhead projector. We begin with an overview of the projector's major components and functions, and describe the projector's use in classrooms, in business, and for military training applications. We then discuss its fall from ubiquity when faced with competing digital projection systems (e.g., the document camera), including the functionally similar "opaque projector". It was this latter technology in particular that made possible the two pathways to obsolescence that the overhead projector ultimately took. Finally, we conclude with a discussion of what makes the overhead projector useful even today, despite its generally agreed-upon obsolescence.

Mechanism

The overhead projector is a relatively simple device. The body of the projector is a metal or plastic box, inside of which sit various electrical components that power a bright lamp, a cooling fan, and a mirror that bounces the light of the projection lamp upward. The top of the box is a clear plate of tempered glass that contains or sits atop a Fresnel lens.[1] The plate—sometimes called the "platen" or "stage"—is where transparencies and other objects are placed for projection.

Affixed to the side or top of the projector body is an arm, usually made of metal or plastic, that extends upward. At the top of the arm is another small metal or plastic housing, which contains a smaller Fresnel lens on its bottom and through which the main unit's projected light is condensed again. This light—now an image containing whatever is on the stage—then bounces off an angled mirror and toward the wall or screen. The effect of these multiple refractions is that the projected image is correctly oriented on the screen (that is, right-side-up and unreversed). Finally, the small box at the top of the arm usually has a lever or dial that can be used to move it nearer to, or further from, the stage on which the projected item sits, thus changing the focus and magnification of the item.

The relative simplicity of the overhead projector belies some rather impressive engineering. The device is the result of literally thousands of years of technological development. As Paul Gardener noted in his 1996 address to the Jerusalem International Science and Technology Education conference, the Fresnel lenses that magnify the image were originally designed for use in lighthouses. The reflective mirror that orients the projected image has its roots in ancient Greece and Japan. Even the copper in the machine's power cord can be traced back to crafting practices in 18th-century Germany (Gardener, 1996: 6). The overhead projector is thus not a technology per se but a "technological system," with each of its component parts having its own multifaceted evolution (Gardener, 1996: 5). Understanding the overhead projector as a system is key to understanding its effervescence and evanescence. Its history is one in which technological, scientific, and social forces converge in such a way that cheap, easy, transportable projection technology becomes common and valuable. Before recounting this history, though, it is essential to outline its context.

Why Projection?

One of the ways to map the life of a technological system is to identify the challenges that system was meant to address. Assessing the socio-cultural importance of such challenges and systems often requires an understanding of the era's needs when the technology was developed. For instance, devising a system to automatically open the doors at a supermarket would be considered by many as solving a practical challenge: making it easier to carry one's groceries to the car. Conversely, Hero of Alexandria's design for a system to automatically open temple doors when a steam valve was activated is generally considered more of a novelty.

And yet, the designers of these two systems were addressing the same challenge: how to open and close a door automatically.

Among the fundamental challenges of the overhead projector are projection (of course) and enlargement. For the latter, the image needs to be expanded with no resulting distortion; when this happens, the image can have a significant cognitive impact, not only because details become easier to see, but also because the image itself is commanding. Consider how the combination of a movie theater projector and screen works to give a sense of depth and connection to a space battle or close-up of an antique photo in a documentary. The bright, large images of an overhead projector had a similar (though admittedly less dramatic) effect in their day.

Overhead projectors also made it possible to view an image of whatever was placed on the stage from a significant distance (15–20 feet). Such spatial flexibility was valuable because it permitted more comfortable viewing of charts, diagrams, and blocks of text, especially by groups of people. A piece of sheet music in a music history course, for example, becomes much easier for the class to see and discuss if it can be enlarged and projected than if everyone is crowding around a table looking at it. In this way, the overhead projector contributed greatly to the accessibility of briefing, meeting, and class rooms. Concomitantly, the overhead projector reduced the cost of information distribution. Because the device used low-cost transparencies with images or text printed directly on them, only one copy of the material needed to be created for an entire group. And because transparencies could be made permanent (that is, the images could be fixed to the sheets and not easily wiped away), they could be used on multiple occasions, allowing substantial cost savings over similar distribution systems such as commercial printing, mimeography, and photocopying.

The overhead projector also met another challenge, namely, how to facilitate collaboration around a single image or document. With cloud-based computing, such activity is commonplace today, but before such technologies existed, the overhead projector was the go-to option for such work. Picture a business meeting in which each attendee has a photocopy of a chart with the previous year's sales numbers. The regional manager reviews the chart aloud and asks attendees to brainstorm on how to improve the numbers. If the goal of such an exercise is to have attendees work alone, then individual copies of the sales chart makes sense: each person can look at her copy, jot down notes in the margins, and try to work out some specific solutions. If the manager wants to have the group work collaboratively, however—perhaps with the idea that the company will fare better if its middle managers all pull together in the same ways, at the same time, toward the same objectives—then distributing individual copies of the chart could actually undermine that goal, dispersing rather than concentrating the brainstorming efforts. Projecting the chart and making it the single focus of everyone in the room, by contrast, might well unify the group's energy and ideas. And because each person cannot write on the overhead transparency at the same time—a feature that some of today's cloud-based applications frustratingly do not include—attendees need to communicate their ideas aloud, creating an environment where everyone is able (personalities and politics aside) to offer, support, question, and critique everyone else's contributions. The overhead projector, in other words, offered a new way to organize the workplace, making it possible for it to be a place where the social and technical barriers to collaboration and shared governance could be lowered with relative ease.

Finally, it is worth noting that early overhead projection not only enabled interaction among viewers, but also with the projected object itself. As familiarity with the overhead projector developed in the classroom, for example, it became clear that there was power in turning over the machine to students themselves. Allowing students to write directly on the transparency sheets—e.g., labeling the cell structures for their peer audience—offered the possibility for engaging students more deeply than if they simply worked alone at their desks. This interactive potential is still valued today. At a 2016 Tucson Museum of Art event, for instance, an overhead projector was featured as a tool for spontaneous artistic collaboration. Having found the projector in one of the Museum's education rooms, several adults and children took turns placing objects from around the room on the glass platen to see what they would look like when projected. The objects—most of which were not transparent and thus cast only tinted silhouettes—were arranged, shifted, flipped, and removed, a creative process that local artist and teacher Lisa O'Neill described as "gleeful . . . . Suddenly all the adults were kids again. [There was] genuine curiosity, excitement, in collaborating and seeing the creation change" (O'Neill). The resulting scenes were a mishmash of colors and shapes that could not be easily reproduced by another technology. The overhead projector provided the perfect ratio of real-time object and collaborator interaction. There were no software menus to navigate to add new shapes, no hotkeys to learn in order to change colors. Rather, the chosen objects were gently staged and the mirrors, lenses, and light did the rest.

That such wonder could be coaxed from so banal a machine might seem surprising, especially to readers who grew up watching their homework or business ideas being dissected publicly on one. Yet the ancestry of the overhead projector is steeped in magic and stagecraft, in trickery, terror, and titillation. It is to these origins that we now turn, not only to historicize the overhead projector, but also to recall its congenital charms.

Early Days

Most of the technologies that comprise the overhead projector are quite literally ancient. Generally speaking, projection technology is thousands of years old. The camera obscura was used in Ancient Rome, while in China and Japan magic mirror projection systems were in use as early as the 5th century BCE (The Magic Lantern Society). Such systems gave rise to three technologies in particular that were key to the development of the overhead projector: the magic lantern, the opaque projector, and the viewgraph.

Magic Lantern

One of the earliest mentions of a projection apparatus—that is, a mechanism that depends on a light source to transmit the image of an object to another surface—is from the early 1400s. In his *Liber Instrumentorum*, Giovani de Fontana includes a description and drawing of a handheld projection lantern casting the image of a devil on a wall, which de Fontana described as "*Apparentia nocturna ad terrorem videntium*" ("a nocturnal appearance for terrifying viewers") (The Magic Lantern Society).[2] Much like an overhead projector, de Fontana's lantern enlarged an image—often a painting on a glass sheet called a "slide"—and displayed it on a wall, curtain, window, or other surface. Unlike an overhead projector, however, de Fontana's magic lantern functioned primarily as a novelty for entertainment rather than as an educational apparatus, and the lantern required the labor-intensive, custom creation of carved screens or painted slides to produce an image of relatively low fidelity.

Opaque Projector

As its name implies, the opaque projector allowed for the projection of opaque materials (e.g., coins, mineral samples) onto a projection surface. Developed in the late 19th century, its key limitation was that the object to be projected needed to fit inside the apparatus (either through a slot or door in the machine) or on a glass projection surface (much like that of an overhead projector) that was then covered. Opaque projectors worked by shining a bright light on an object, usually from behind or above. This light then struck one or more mirrors before being focused through a lens that correctly oriented the final image vertically and horizontally. If an object could not readily fit inside the projector, it simply could not be properly lit, reflected, and projected.

The development of the opaque projector marked a milestone in the history of the overhead projector because it made the direct projection of an object's likeness possible. Educators, artists, and others no longer needed to create a custom slide or screen depicting the object they wished to project, as the magic lantern had required. Instead, the opaque projector allowed objects to be correctly and accurately projected and enlarged with minimal effort. Unfortunately, because objects needed to remain within the projector itself (or under a cover connected to it) in order to be clearly seen, they could not be moved or manipulated without compromising the image quality. Furthermore, the lights used to produce the image tended to be extraordinarily hot, posing a fire hazard and discouraging anyone from trying to reach in and adjust the object being projected.

Viewgraph

Configured similarly to the modern overhead projector, the viewgraph was invented in the 1870s and is generally credited to optician Jules Duboscq (The National Museum of American History). Duboscq was a prolific inventor and manufacturer of optical devices, including the phenakistoscope—an early version of the motion picture projector—and the colorimeter, a device that measures the intensity of color (Brenni, 1996: 12). Duboscq is also known for the improvements he made to David Brewster's stereoscope ("Remembering inventor and pioneering photographer Jules Duboscq"), as well as for his work interconnecting various well-known optical technologies in order to create new or much improved machines. Duboscq's inventions—and those of other optical engineers of his day—highlight how multiple routes were taken to solve similar sets of projection-related challenges. Not surprisingly, as projection technologies evolved, the development paths guiding them began to diverge into specializations: entertainment, science, education, and so on.

Duboscq's path was the most prominent, especially among scientists, and his viewgraph was quickly adopted by others. Duboscq was considered "an indispensable companion of the scientists who were presenting scientific public lecture-demonstrations," largely because of the impact that his projection system lent to the performance of science (Brenni, 1996: 12). A no less interesting path (though more limited due to expense and scale) was taken by Johann Meopmuk Czermak. Czermak's *spectatorium* was a purpose-built medical lecture and demonstration hall in which two limelight projectors could be used to project "photographs of embryos, polished sections of bones, a colored photograph of a dog's knee, a cross section of human skin tissue, as well as [a] still-contracting frog heart" (Schmidgen, 2011: 44). And Czermak was not alone in thinking on this scale. As Schmidgen notes, "[t]hrough movable screens and powerful projectors and by the use of flags, rolling tables, as well as entire series of specifically designed demonstration devices . . . physiologists immersed their audiences in a cinema of life" (2011: 46). Despite the differences in movability and cost, these early iterations of the overhead projector shared the same basic function of projecting an object for a group of viewers with the understanding that visual perception could help to create knowledge about a subject.

Science educators such as Carl Jacobj were similarly keen to create projection devices and spaces in an effort to teach a "growing number of students . . . the rapidly growing body of knowledge produced in and by the experimental life sciences" (Schmidgen, 2011: 49). However, Jacobj's "presentation of movement images . . . [was] not an end in itself. Rather, it served a specifically modern economy of time and attention" (ibid.). Unlike Czermak's lecture hall, where seeing the object was thought to produce knowledge in and of itself, Jacobj's projection classroom assumed that sense impressions from sight were a precursor to knowledge production, not knowledge production itself. For Jacobj and other educators who were faced with an expanding body of knowledge and a growing student body, the problem that needed to be solved was not simply how to show an object and thus produce knowledge but how to produce knowledge efficiently for a large number of students without the distracting theatricality attendant to Czermak's *spectatorium* and other cinematic experiences. The *spectatorium*, though technically accomplished and visually impressive, was neither efficient nor free of distraction. Moreover, building and maintaining such a space was simply too expensive and inflexible to address changing needs. Instead, educators of all types turned to the much cheaper, more mobile, and more flexible overhead projectors of the type developed by Duboscq.

The Development Years

While the late 19th- and early 20th-century machines of Czermak's and Jacobj's cutting edge instructional theaters were rare, expensive, and powerful, the basic magic lantern technology that enabled the projection of an image from a glass slide onto a vertical surface had been fully commercialized at multiple price points. According to a story published on September 25, 1904 in *The Chicago Tribune*: "A good, gas burning stereopticon with a single projecting lens can be bought as low as $15, and from that the price lists range as high as $300 [approximately $7,740 in 2017[3]] for the double dissolving lanterns with all the latest attachments for calcium, electric, and other lights even more costly" ("How To Make . . . ," 1904: 3). The relative affordability of this projection technology resulted in the ancillary industry of glass slide production, much of which was located in Chicago. By 1904, more than 10 million magic lantern slides were made and sold in Chicago every year (rivalling the more well-established hobby of "the snapshot habit") and had long since become a standard educational technology in public schools (ibid.).[4]

By the mid-1940s, overhead projectors had been adopted by the U.S. military for training purposes, and military academies used them extensively in the years following World War II (The National Museum of American History). Other military forces did as well; writing in 1975 about the Canadian military's use of the overhead projector during war gaming exercises, W. P. Doyle (1975: 42) notes how transparencies depicting sensor and radar data were projected onto a vertically mounted map in order to alter the shape of the battlefield in real time. In these contexts, as with the *spectatorium*, the overhead projector was used as a tool for knowledge production. The overhead projector allowed military personnel to create a visual representation of an otherwise distant or abstract concept or scenario. And, particularly in the case of exercises of the type Doyle describes, the visualized scenario could be altered quickly, easily, and repeatedly. Much as classroom educators had found that seeing impacted their students' learning processes (recall Jacobj), military educators, too, realized that significant advantages accrued to soldiers who, by means of an overhead projector, could both see and manipulate objects, ideas, and data in real time.

The decades after World War II saw a number of improvements in overhead projector technology. One of the most important came from corporate scientist Roger Appeldorn in the early 1960s while he was working for 3M. Appledorn helped develop a microreplication technology that covered surfaces "with millions of tiny structures that enable[d] the surfaces to behave in specific ways" (Baskin, 1998: 134). This newly developed technology was then applied to the lenses of overhead projectors to improve the clarity of the projected images. That said, 3M's impetus to improve image quality might have had less to do with improving the overhead projector itself and more with creating demand for the company's newly developed transparent acetate sheets (The National Museum of American History). Notably, while the 1950s and 1960s saw improvements to the design of the overhead projector, early versions were still much larger than the machines that would later be wheeled around on carts between high school classrooms. For instance, instructors in the 1960s at the North Carolina College of Textiles were using overhead projectors the size of large coffee tables on wheels (NCSU Libraries' Digital Collections: Rare and Unique Materials). Nevertheless, even in this bulky form, overhead projectors of the 1960s were capable of a mobility far outstripping that of the *spectatorium* and other earlier projection devices.

Riding High

The period from the 1960s into the 1990s represents the apex of the overhead projector in classrooms, with teachers quickly finding uses for the technology across a range of disciplines. Henry Bissex, for example, wrote in 1967 that the overhead projector offers the writing composition instructor "a kind of controllable blackboard" which "give[s] control" to the teacher through "constant visual" cues to the student (1967: 10). Bissex also celebrated the overhead's ease of use, lauding the fact that it enabled one to "erase more easily and quickly," the ability to "prepare handwritten sheets of film in advance," and the ability to quickly reorder a lesson by rearranging the order of presentation sheets (1967: 6). Like Czermak, Bissex placed great weight on the act of seeing, noting that overhead projectors "make visible *what* [the teacher is] talking about" (1967: 10). More succinctly, Bissex observes that "[s]eeing makes a difference if we are to teach writing" (1967: 10). Interestingly, Bissex's approach to using the overhead projector reflects Jacobj's earlier desire to use the machine to maximize efficiency, elaborating extensively in one article, for example, on how he was able to copy seven different pieces of student writing onto a single transparency.

Bissex's 1967 discussion of the overhead projector clearly indicates that many of the same educational concerns of the late 1800s had carried through to the new century. A distinct difference between the two eras, however, was that while the basic exigencies that prompted the creation of overhead projectors remained, projection technologies themselves continued to shrink, becoming smaller, more mobile, and cheaper. And while the overhead projector of the 1960s lacked the grandeur of Czermak's *spectatorium*, it more than made up for its humbleness with mobility and affordability. Still, even the modern overhead projector had serious drawbacks. The visual fidelity of images transferred to a transparency were likely to be somewhat poor, for example, and the projector would not easily display moving images. Instead, classrooms and boardrooms relied on film projectors to show higher quality and moving images. This required instructors to have access to two different pieces of technology, each with its own maintenance, operating, and storage requirements—all problems seemingly mitigated by the development of the digital projector.

Two Pathways to Decline

There are two primary technologies that led to the overhead projector's obsolescence: the digital projector and the document camera. The term "digital projector" refers to any projector that can take digital input from a digital device (e.g., computer or smartphone) and project that input onto a surface. The methods by which the image is projected vary (e.g., CRT, Laser, LCD, and DLT), as do the various bells and whistles of projectors (e.g., image resolution, built-in speakers, and WiFi capability), but the basic parameters remain the same across the technology class. And, because the digital projector shows exactly what is on a computer screen, it enables presentation software such as *Microsoft PowerPoint* to exist.

The document camera, in contrast, is more akin to a traditional overhead projector, as it can be used to project images of physical objects. The document camera also shares important characteristics with the opaque projector in that it can be used to project more than just transparencies. Unlike the opaque projector, the document camera does not require objects to be placed in a special compartment, allowing instead for objects to simply be placed anywhere on the open stage below the camera. Moreover, most document cameras use LEDs as a light source, a feature that obviates the risk of fire. The document camera also has the advantage that it *looks* like a standard overhead projector, which means that the typical anxiety elicited when new technologies are introduced is lessened. The key physical difference is that the overhead projector's small, arm-mounted mirror box has been replaced with a small camera that can be zoomed in and out from the projector's console. Like the digital projector, though, the document camera requires a computer to control its functions and generate the images it projects.

Taken together, the digital projector and the document camera show how developments that seem only somewhat related (such as the rise of digital cameras) can eventually lead to the decline of even the most entrenched technology. To illustrate this phenomenon, we now examine how the digital projector and document camera first overtook and eventually supplanted the overhead projector in class-, board-, and briefing rooms.

Cause of Obsolescence #1: The Digital Projector

The seeds of the overhead projector's fall were largely sown in the late 1980s. Companies such as InFocus had digital projectors available for the public as early as 1987 ("InFocus Systems, Inc. History"). At the time, overhead projection was still "the most popular form of visual support used by presenters," despite the clear appetite for new solutions and products (Macnamara, 1996: 67). Even when these innovations were available, they did not immediately displace the old technology. In 1987, for instance, InFocus Systems released "a CGA-compatible monochrome display panel which, when placed on an overhead projector, displayed the output of a personal computer onto the screen" ("InFocus Systems, Inc. History"). In effect, the overhead projector and its eventual replacement were integrated, with the latter initially being an add-on meant to expand the utility of the conventional machine.

Such early digital projectors were suited only for small spaces such as conference rooms and classrooms. For bigger spaces—for example, demonstration theaters like the *spectatorium*—companies such as Digital Projection International began working on large-scale devices that could be used for a wide variety of applications, including movie theaters and educational auditoria. Digital Projection International's research, combined with an extremely accurate and sensitive reflecting mechanism developed by Texas Instruments, allowed for the creation of more powerful projector prototypes as early as 1992 (Digital Projection International). As with the systems of old, these powerful machines had steep prices. The Focus System TVT 6000 portable color LCD projector (a projector very similar to the ones found in classrooms and boardrooms today), for example, cost $7,995 in 1992 [approximately $14,058 in 2017] (Busse, 1991: 21). Top of the line models pushed the price point even higher, with

Proxima Corporation's Desktop Projector 9200 running at $11,999 in 1997 [approximately $18,314 in 2017] (Poor, "LCD Projectors Take the Next Step"). Over time, though, the price began to trend downward. Smaller, more powerful models were down to $799 by 2005 [approximately $1,017 in 2017] (Poor, "Dell Gets Business Projection Right"), and as of 2017 mini-projectors can be had for as little as $50.

The growing prominence of digital (or data) projectors would have been readily apparent to anyone reading magazine articles or books in the early 1990s about business presentations. These same texts, however, also reveal that the overhead projector was not quickly or easily sidelined. For instance, Manchester Open Learning's 1993 book *Making Effective Presentations* has several pages of tips for designing overhead projection transparencies, as well as a list of the benefits of using transparencies (1993: 61). Computer-based presentations are mentioned, but even here the assumption was that readers would simply be looking at a computer screen for notes while still presenting materials situated on an analog overhead projector (1993: 71). Jim R. Macnamara's *The Modern Presenter's Handbook* (1996) extends this practice, spending nearly 20 pages on tips for designing overhead transparencies. Macnamara, however, also mentions that data projection units are available, though he does not dedicate time to explaining how to design for this emerging digital medium. Also in 1996, Marya Holcombe and Judith Stein's *Presentations for Decision Makers* mentions "computer presentation system[s]" (1996: 82) as an option, and point out that the increasing capabilities of personal computers have produced higher expectations for the "quality and readability of visual support[s]" (ibid.), including overhead transparencies. As the turn of the millennium approached, handwritten or typed (that is, produced with a typewriter) transparencies lost favor among consumers, replaced by an appreciation for the wider range of aesthetic choices made possible by computer programs such as *Wordstar, Microsoft Word*, and *WordPerfect*. This led to one of the stranger liminal moments in technological history: computers and digital printers being used to produce overhead projector transparencies.

By 1997, the business world had actively begun to shift away from the analog/digital threshold and to embrace a fully computational solution for projecting documents, images, and objects of all kinds. Spring Asher and Wicke Chambers' *Wooing and Winning Business* (1997), for example, accepts that many businesses are still using overheads but are moving toward newer presentational forms that have their own presentational bugaboos. In a telling chart, Asher and Chambers detail the pros and cons of various presentation technologies. While the overhead projector is simple and can be nearly cost-free to use, it is not particularly easy to alter on-site and can only reach a limited audience size (1997: 63). The digital projector's imagery, on the other hand, can be viewed by a wide range of audience sizes depending on the available projector and screen, and the presentation itself can be easily altered *in situ*. The authors warn, however, that viewers "may be more interested in computer tricks than in your information" (ibid.).

Just one year later, the idea that computer tricks were something to avoid in presentations was being replaced by the opposite proposition, namely, that presenters should take advantage of the digital trickery that computers and digital projectors increasingly enabled. William J. Ringle's *Techedge: Using Computers to Present and Persuade* (1998), for example, presages the hard shift toward the digital projector that was soon to occur. Ringle acknowledges that "[o]verhead projectors are by far the most common way to project text and pictures on a screen to share with an audience" (p. 10) and that "[t]hey are also one of the easiest tools to learn how to use" (p. 11). But he also notes that considerable experience and skill are necessary to produce what he calls "progressive disclosure" on an overhead projector (p. 130), a technique that Ringle argues is key to making a truly compelling, memorable, and effective presentation. Progressive disclosure, Ringle says, shows viewers select information from a transparency or slide, revealing additional material only at the appropriate time—a simple chore for digital presentation software, but awkward and limited with transparencies and an overhead projector. Ringle also catalogs a host of practical problems with transparencies that digital slides do not suffer, from infuriating static cling to the headaches produced from simply transporting transparencies from one place to another: smudging, erasures, creases, flaking, and so on (p. 12). Yet, Ringle is aware that designing slides on a computer is not necessarily simple or easy, and he dedicates an entire chapter of his book to developing presentations using a computer.

Even into the early 2000s, the overhead projector hung on as a viable information-sharing technology. James Wagstaffe's 2002 book *Romancing the Room: How to Engage your Audience, Court Your Crowd, and Speak Successfully in Public* contains a list of "eye candy" that includes "short video clips, filmstrips, LaserDisc, images, films, slides, overhead transparencies, computer displays and animations" (p. 87). Despite such pansophic nods to the ways presenters could enamor an audience, however, the turn toward the digital projector—and away from the overhead—accelerated. The year 2008 saw the release of Nancy Duarte's *Slide:Ology: The Art and Science of Creating Great Presentations*, an in-depth description of how to produce high-quality professional visuals for many different situations. Duarte's book mentions neither the overhead projector nor similar technologies such as the slide projector; only the computer and digital projector are viable presentation technologies for her. And in 2011, Garr Reynolds's best-selling *The Naked Presenter: Delivering Powerful Presentations With or Without Slides* cemented the mass abandonment of the overhead projector, stating resolutely: "Build visuals in software" (p. 53). To Reynolds and his large mainstream audience—from mid-level executives and teachers to career scientists and artists—there was no longer an alternative to the digital projector. It had become the *tantum lux*, the only light.

Cause of Obsolescence #2: The Document Camera

Curiously, the document camera is not mentioned in any of the texts we reference above. This could be because some of the earliest models of the camera were actually bundled with a digital projector (Poor, "Lights, Camera, Project!"). And yet, while there is significant overlap in the projection challenges that the digital projector and document camera address, one challenge is uniquely suited to the document camera: projecting a non-digital object such as a magazine page or video game packaging. Using a document camera allows for the physical object to be seen in full color and for the object's lighting to be adjusted much more accurately than was possible with an overhead projector. In recent years, document cameras have been decoupled from digital projectors, and their prices have subsequently dropped dramatically. As with digital projectors, document camera prices vary widely based on the manufacturer, model, and features, but decent units are available in 2017 for as little as $45.00. Such (relatively) simple devices are little more than a small digital camera attached to a base by a flexible arm and connected to a computer with a USB cable. Other versions of the document camera bear a striking resemblance to the classic overhead projector, with a stage on which to place items and arm-mounted camera housing suspended above. These more elaborate models have a correspondingly large price tag, with the ubiquitous classroom ELMO model costing more than $2,660.

Regardless of price point, however, the document camera manages to solve a problem that the digital projector cannot: interaction. Much like an overhead projector, the document camera can project a text that can then be written on or altered, either digitally with corresponding software, or directly by writing on the page itself. This unprecedented level of flexibility for the presenter—allowing her to simultaneously interact with the audience and the projected object—means that the constraints of shape, color, sturdiness, duplicability, and so on are increasingly bygone problems. Just as the opaque projector and the overhead projector saw their rise and development overlap due to a set of shared needs by presenters of the 19th and 20th centuries, so too did their decline: as presenters wanted to do more, their technologies needed to fill the bill or disappear. In their heyday, the overhead and opaque projectors solved many of the same problems; it can be no surprise then that the technology that displaced one of them—the digital camera—ended up displacing them both.

Materiality

The obsolescence of the overhead projector does not mean that the needs it was designed to address have been fully and finally mitigated. Nearly every piece of information presented in a classroom or a conference room can be sent out over a shared website or e-mail, ready to be viewed on the multitude of individual screens available today. But the one-to-many sharing of information has never been the chief desideratum grounding overhead projection technology. A minor one perhaps, but not the main one. By sending slides out to everyone's cellphone, tablet, and smartwatch, presenters might well share their information but in so doing they lose an opportunity to unite people around a common project, to focus their colleagues' attention on a vision, challenge, or project in need of collective more than individual insight. The act of projection focuses viewers' attentions to a single frame, a bright, vital image at the front of the room. In focusing on this frame, the audience—it is hoped—finds it easier not only to engage with the presenter and her or his ideas, but also with each other. This is, and has always been, the singular though non-obvious strength of the overhead projector: intentionally or not, it virtually ensures that viewers think together. Through its brilliant insistence that viewers share a cognitive focus, the overhead projector constructs a shared space of imagination, one that can be at turns clarifying and intimidating.

Of course, the complete attention of everyone in the audience is never guaranteed, even with so transfixing a technology as the overhead projector. For reasons often unavoidable, distractions and boredom are ever-present. Yet even though the overhead projector has been supplanted by newer technologies, to those who remember its unique charms—the way it made looking at a projected image into a multisensory experience of whirring fan, burning dust, and blinding light—the overhead projector made learning embodied. By shining light through, around, and across whatever had been set upon its stage, the overhead projector reinforced and amplified that object's materiality in a way that presentation software or document cameras cannot. Perhaps this is because, as Giuliana Bruno observes, when we are "[e]nveloped in this surface of screening, we become aware that, just as in the film theater, the effect of projection is not simply visual but environmental" (2014: 67). The act of projection, in other words, does not just change a space—it *creates* one. From the moment the overhead projector cart was wheeled to the front of the room, the lights dimmed, and the screen extended, the conventional classroom or boardroom was temporarily transmuted into a projection hall, a place where shadow, light, and lenses effected a set of edifying intimacies that were greater than the sum of their technical parts. To be sure, the content of a lecture, choreographed with a transparency's projected text and illustrations, were the manifest agents of learning in a room outfitted with an overhead projector. But at a less conscious level, other factors enabled by the machine also compelled engagement in the learning process: the magnification of the speaker's hands—a fresh hangnail indicating nervousness, a dab of paint on the knuckle signaling a life beyond the classroom—or the way the projected light revealed an ear's blood vessels or a small scar invisible under normal light. Such humble but human observations, uniquely made possible by the overhead projector, often proved to be just the relational linkage needed to change apathy into intrigue.

One might reasonably wonder if digital projectors and document cameras cannot achieve the same magic. After all, both throw light and both have—in different ways—the capacity to magnify and project. The manner of projection, however, matters. Digital projectors and document cameras build an image from light: the light projected *is* the thing being seen. The object is carried in, as Roland Barthes writes, "visible and unperceived, that dancing cone which pierces the darkness like a laser beam" (1986: 421). This new process of projection thus creates a duplicate of the projected item, a sculpted doppelgänger of light. The overhead projector, in contrast, creates its image through interference. The projection is the result of light blocked, bent, and scattered, as much shadow as cone and plane of light. Even the most mundane text—a bit of doggerel or tangent table—activates this optical transformation when recorded on a transparent acetate sheet.

It is perhaps these more ephemeral qualities that have kept the overhead projector from absolute oblivion. Amateur and professional artists alike still use the overhead projector in their work, as do scholars and others who either have no better means of projection, or who simply appreciate the charms and conveniences this obsolete technology retains. Without question, the overhead projector has lost much ground, largely because digital projectors and document cameras do most of the same work more effectively and flexibly. But there are certain things that the overhead projector does that these other devices cannot, and for as long as that remains true, we suspect that this once ubiquitous machine will continue to lurk at the edges of presentation spaces, in storage closets, and in study rooms, waiting to be rolled front-and-center, powered on, and pressed into use once again.

Notes

1 A Fresnel lens is a flat lens made up of a series of concentric rings of glass or plastic that, due to its unique configuration, works to condense light. In the overhead projector, this has the effect of collecting and concentrating upward most of the oblique light given off by the projector's omnidirectional lamp.

2 De Fontana's text provides only a description of the projection apparatus; it offers neither proof nor claim that De Fontana actually created the device, leaving the possibility open that the genesis of the magic lantern could be considerably older.

3 All adjusted figures are taken from the Bureau of Labor Statistics' (BLS) CPI Inflation Calculator, https://data.bls.gov/cgi-bin/cpicalc.pl. As the BLS did not collect CPI information until 1913, prices prior to then cannot be reliably adjusted to 2017 dollars and so have been left as is.

4 According to *The New York Times Magazine*, by the end of World War I, the Chicago Public Schools had more than 8,000 magic lantern slides available for their teachers. Considering that a set of 30 high-quality lantern slides could cost upwards of $30 [roughly $520 in 2017], such emerging projection technologies were clearly seen by many educational institutions as a worthwhile investment (Wilson, Orellana, and Meek, 2010).

References

Asher, Spring and Wicke Chambers. *Wooing and Winning Business: The Foolproof Formula for Making Persuasive Business Presentations*. John Wiley & Sons, Inc., 1997.

Barthes, Roland. "Leaving the Movie Theater." *The Rustle of Language*. Translated by Richard Howard. University of California Press, 1986, pp. 345–349.

Baskin, Ken. *Corporate DNA: Learning from Life*. Butterworth-Heinemann, 1998.

Bissex, Henry. "The Use of the Overhead Projector in Teaching Composition." U.S. Department of Health, Education & Welfare—Office of Education's Commission on English: College Entrance Examination Board, 1967. https://files.eric.ed.gov/fulltext/ED036531.pdf. Accessed 25 August 2018.

Brenni, Paolo. "XIII: Soleil, Duboscq, and Their Successors." *Bulletin of the Scientific Instrument Society*, 51, 1996, pp. 7–16.

Bruno, Giuliana. *Surface: Matters of Aesthetics, Materiality, and Media*. University of Chicago Press, 2014.

Busse, Torsten. "Focus releases top-notch LCD video projector." *Info World*. 25 November 1991, p. 21.

Digital Projection International. "More About Digital Projection." www.digitalprojection.com/emea/more-about-digital-projection/. Accessed 26 September 2016.

Doyle, W.P. "Some Aspects of War Gaming: The Night Battle." *Military Strategy and Tactics: Computer Modeling of Land War Problems*. Ed. Reiner K. Huber, Lynn F. Jones, and Egil Reine. Plenum Press, 1975, pp. 37–43.

Duarte, Nancy. *Slide:Ology: The Art and Science of Creating Great Presentations*. 1st ed., O'Reilly Media, 2008.

Gardener, Paul. "Viewing the Roots of Technology and Science: A Philosophical and Historical View." Jerusalem International Science and Technology Education Conference. 11 January 1996, Jerusalem. Keynote Address.

Hero of Alexandria. "Temple Doors Opened by Fire on an Altar," *Pneumatics of Hero of Alexandria*. Translated by Bennet Woodcroft, Taylor Walton and Maberly, 1971.

Holcombe, Marya W. and Judith K. Stein. *Presentations for Decision Makers*. Van Nostrand Reinhold, 1996.

"How to Make Magic Lantern Slides for Your Own Entertainment." *The Chicago Tribune*. 25 September 1904. p. 3. http://archives.chicagotribune.com/1904/09/25/page/52/article/garment-designing-for-art-students. Accessed 25 January 2017.

"In Focus Systems, Inc. History." *Funding Universe*. www.fundinguniverse.com/company-histories/in-focus-systems-inc-history/. Accessed 13 October 2016.

Jacobj, Carl. "Visual Instruction and the Projection Method." *Methods and Problems of Medical Education*, 6, 1927, 257–264.

Macnamara, Jim R. *The Modern Presenter's Handbook*. Prentice Hall, 1996.

Magic Lantern Society, The. "An Introduction to Lantern History." www.magiclantern.org.uk/history/. Accessed 28 September 2016.

Manchester Open Learning. *Making Effective Presentations*. Kogan Page Ltd, 1993.

National Museum of American History. "Overhead Projectors." *Mobilizing Minds: Teaching Math and Science in the Age of Sputnik*. http://americanhistory.si.edu/mobilizing-minds/overhead-projectors. Accessed 22 September 2016.

NCSU Libraries' Digital Collections: Rare and Unique Materials. "Instructor Using an Overhead Projector to Teach a Lesson." https://d.lib.ncsu.edu/collections/catalog/0020075#?c=0&m=0&s=0&cv=0&z=0%2C-664.0714%2C6298%2C5848.1429. Accessed 26 September 2016.

O'Neill, Lisa. Personal correspondence. 22 October 2016.

Poor, Alfred. "Dell Gets Business Projection right." *PC Magazine*. July 2005.

Poor, Alfred. "LCD Projectors Take the Next Step." *PC Magazine*. 18 November 1997.

Poor, Alfred. "Lights, Camera, Project!" *PC Magazine*. 7 April 1998.

"Remembering inventor and pioneering photographer Jules Dubosq." *Photography-news.com*. 5 March 2016. www.photography-news.com/2011/03/remembering-inventor-and-pioneering.html. Accessed 25 September 2016.

Reynolds, Garr. *The Naked Presenter: Delivering Powerful Presentations With or Without Slides*. New Riders, 2011.

Ringle, William J. *Techedge: Using Computer to Present and Persuade*. Allyn and Bacon, 1998.

Schmidgen, Henning. "1900 – The Spectatorium: On Biology's Audiovisual Archive." *Grey Room*, 2011, pp. 42–65.

Wagstaffe, James. *Romancing the Room: How to Engage your Audience, Court Your Crowd, and Speak Successfully in Public*. Three Rivers Press, 2002.

Wilson, Charles; Marvin Orellana; and Miki Meek. "The Learning Machines: 1870, Magic Lantern." *The New York Times*, 15 September 2010. N.p.

8

FLAMMABLE WORKHORSE

A History of Nitrate Film from the Screen to the Vault

Amanda McQueen

"This article has been written to remind all concerned that *old nitrate motion picture films are dangerous!*"[1] So declared John M. Calhoun, spokesperson for Eastman Kodak, in 1962, over a decade after cellulose nitrate film had been discontinued. True, nitrate film is highly flammable, and as it burns, it generates oxygen that makes the fire difficult to extinguish and gases that are hazardous to human health. Furthermore, nitrate film's chemical instability causes it to decompose over time, destroying both the celluloid base and the image-carrying emulsion. Yet for sixty years, nitrate film was the "workhorse" of the professional film industry, circulating the globe in enormous quantities.[2] One of the first semi-synthetic plastics, it offered the transparency, flexibility, and durability the burgeoning cinematic medium required to a degree unmatched by any other contemporary material.

From its origins as a widespread technology to its current status as an archival rarity, nitrate film has been simultaneously characterized by its inherent instability and by its almost uncanny ability to endure. The history of nitrate motion picture film I present here thus pays particular attention to how the material's unique physical and chemical properties have colored its reputation and shaped day-to-day working methods. Beginning in the mid-1800s with the invention of cellulose nitrate, I discuss the development of celluloid-based film stock; the corresponding implementation of fire safety regulations; the attempts to introduce a less hazardous alternative; and the changing status of nitrate film within archival collections. Dangerous but dependable, unstable but irreplaceable, nitrate film's contradictory qualities make it not only a fascinating part of film history, but also a key consideration for those seeking to preserve that history for the future.

The Celluloid Boom

A simplified account of the pre-history of nitrate motion picture film begins in 1846 when German-Swiss chemist Christian Friedrich Schönbein discovered cellulose nitrate—alternatively known as nitrocellulose or gun cotton—when he combined cotton cellulose fibers with nitric and sulfuric acids. This initiated a chemical reaction process called nitration, in which the hydroxyl groups (OH) in the cellulose polymer chain are replaced by nitrate groups (NO_2), with sulfuric acid acting as the catalyst.[3] This produced an entirely new material that burned rapidly and without smoke, and Schönbein immediately recognized its potential as a military explosive. However, cellulose nitrate was also strong, waterproof, and moldable, making it equally suitable for a variety of other purposes, including paint removers, lacquers and enamels, imitation leather, artificial silk and other filaments, microscopy slides, and surgical bandages.[4]

These non-military applications required a more stable material, and in the late-1840s, a refinement of the nitration process yielded pyroxylin, a form of cellulose nitrate with a lower nitrogen content that was non-explosive, though still highly flammable.[5] It was further work with pyroxylin that proved instrumental to the development of motion picture film. First, around 1850, Louis-Nicolas Ménard found that dissolving pyroxylin in ethyl alcohol produced a colorless, viscous liquid called collodion (discussed more below).[6] This was followed by the discovery of pyroxylin plastic. In the 1850s, English inventor Alexander Parkes found that by adding solvents or plasticizers, particularly camphor, pyroxylin was converted into a moldable mass. This material could be formed into strong, transparent objects using heat and pressure, and colored with dyes, fillers, or paints.[7] It was not until the 1870s, however, that pyroxylin plastic was successfully commercialized. John Wesley Hyatt and Isaiah Smith Hyatt, working across the Atlantic in Albany, New York, patented a refined plastic solution in which the pyroxylin and the camphor were more uniformly mixed, and with key advances in machinery, they were able to achieve high-quality, large-scale production. The Hyatt brothers called their plastic "Celluloid", and in 1872, founded the Celluloid Manufacturing Company in Newark, New Jersey. By the following decade, they had licensed over twenty firms in both America and France.[8]

Significantly, in 1887, a lengthy patent lawsuit involving the Hyatts, Parkes, and manufacturer Daniel Spill determined that the basic process of plasticizing pyroxylin with camphor could not be restricted.[9] This meant that despite "certain variations in the method of manufacture . . . all pyroxylin plastics are essentially the same as far as chemical composition is concerned."[10] "Celluloid" was thus one of many trade names under which the same basic material was sold; it was also called Parkesine, Xylonite, coraline, viscoloid, and fiberloid.[11] However, likely because of the Hyatts' early and continued dominance of the market—thanks to their superior, patented equipment and refined methods of manufacturing and waste disposal—"celluloid" soon became the generic term for pyroxylin plastic "in universal popular use".[12] It is also worth noting that celluloid production was controlled by a small number of companies. Given the highly flammable nature of the chemicals involved, it was a dangerous business, and factory fires shut down a number of burgeoning companies and dissuaded many from entering the field altogether.[13]

As Edward Chauncey Worden exhaustively documents in his two-volume *Nitrocellulose Industry* (1911), celluloid proved an incredibly versatile material, and by the late-1920s, annual global production had reached 40,000 tons.[14] Chief among celluloid's benefits was that it could be made to resemble ivory, tortoiseshell, coral, amber, and other natural materials that were becoming increasingly rare and expensive. Celluloid manufacturing was thus driven by the demand for "forgeries" of those "beautiful and scarce" materials that were traditionally used to make the "necessities and luxuries of civilized life", such as cutlery handles, spectacle frames, trinket boxes, hand mirrors, billiard balls, piano keys, and corset stays. Celluloid was also used to manufacture dolls and other children's toys, phonograph cylinders, printing blocks, syringes, surgical instruments, waterproofed textiles, adhesives, and dental plates.[15] The mainstays of the celluloid industry, though, were decorative combs and celluloid-coated shirt collars and cuffs.[16]

Celluloid's specific qualities—its transparency, strength, water resistance, and moldability—facilitated its widespread application despite its obvious hazards. By the 1910s, multiple government-sponsored studies had established that celluloid ignited at comparatively low temperatures; that it generated hazardous and explosive gases as it burned; and that once ignited, it was difficult to extinguish.[17] Fires in celluloid factories and workshops threatened significant loss of property, not to mention injury and death, and accidents also occurred with some frequency in private residences when celluloid objects—most commonly clothing buttons, kitchen knife handles, and hair combs—were placed too close to a heat source.[18]

Still, even though celluloid was recognized as "a very dangerous article of common use", its utility outweighed its flammability.[19] Celluloid was often "the most suitable material for the purpose", and its manufacturing would continue until competitive alternatives were commercially available.[20] By the early-1900s, in fact, work was already under way on non-flammable plastics to replace celluloid, including casein, cellulose acetate, and phenolic resins—the latter popularized in the 1920s by the plastic Bakelite.[21] In the meantime, common sense safety measures were implemented to mitigate celluloid's dangers, without unduly burdening the businesses that depended on it.[22] There was one celluloid-based field, however, that was consistently thought to require "separate consideration" and more stringent regulations, especially as it came to vie with combs and collars for the largest share of the celluloid market.[23] This field was motion pictures.

Developing Celluloid as a Photographic Base

Almost immediately after collodion's invention, photographers adopted it to adhere photosensitive emulsion to glass plates. This wet-plate photographic process offered improved exposure times, but the weight and fragility of the glass could be a hindrance.[24] With the invention of celluloid, an alternative base support material became available; it was transparent like glass, which was necessary for the development process, but it was also light, flexible, and unbreakable. If cast into thin sheets and coated with emulsion, it could thus replace glass altogether. By the late-1880s, celluloid photographic plates were commercially available in both America and Europe, though their adoption was rather slow, as many professional photographers preferred the quality of glass plate negatives.[25]

There was another avenue of photography, however, where celluloid had an undeniable advantage. Plate photography required returning to the darkroom after each shot, a practice that was increasingly inconvenient as photographers moved outdoors to shoot in natural environments. In the 1850s, experiments began with roll film—emulsion-coated flexible strips, held on spool-like holders, that allowed photographers to make multiple exposures in succession. In 1888, George Eastman released the first affordable, commercial roll film system. Like other contemporary roll film systems, the Kodak No. 1 camera used a paper-backed "stripping film". After exposure, the backing was chemically removed and the emulsion was secured to glass for development; this process was both labor- and time-intensive, and the film's quality was inconsistent.[26] Transparent celluloid, however, was durable and did not need to be stripped from the emulsion prior to development. With a slight alteration to the plastic's chemical composition, it proved ideal for roll film. Using amyl acetate—a solvent patented by John H. Stevens, chief chemist at the Celluloid Manufacturing Company—Henry Reichenbach, chemist at the Eastman Dry Plate and Film Company, in Rochester, New York, created a more dilute celluloid mixture. This made the celluloid more uniform, and flexible enough to wind tightly around the spools inside the camera.[27] The Eastman Company began manufacturing celluloid roll film in late-1889, using a process that flowed liquid celluloid across long glass tables, and then coated it with emulsion and cut it into strips after it dried.[28]

Eastman's only American competitor, the Boston-based Blair Camera Company, released its own camera and celluloid roll film in 1890. However, save for a brief period from 1892 to

1894, when personnel difficulties forced the Eastman Company to suspend production, Blair captured only a small share of the roll film market, despite a superior manufacturing process.[29] Blair purchased its celluloid base from the Celluloid Manufacturing Company, which created long strips by flowing liquid celluloid over a slowly rotating heated drum. This patented continuous casting process produced smooth strips of roll film up to 300 feet long; the joints in Eastman's glass tables, on the other hand, sometimes created flaws in the stock. Emulsion was then applied at the Blair factories using a similar heated drum apparatus.[30]

Photographic celluloid posed unique manufacturing challenges. It needed to be entirely free of impurities to ensure the clarity of the stock, and just the right consistency—viscous enough to flow evenly but dilute enough not to blister, crack, or trap bubbles. Furthermore, the emulsion had to adhere firmly to the celluloid base, but any change to the chemical make-up of one component, such as boosting the emulsion's light sensitivity, impacted its relationship with the others. Early celluloid film thus often suffered from streaking, uneven thickness, brittleness, spotting, halo effects, and, in Eastman's case, a "mysterious vine or tree-shaped flaw" caused by static electricity.[31]

Yet despite the highly variable quality of the product, the demand for roll film grew quickly, particularly among amateur photographers. In 1891, Eastman opened three new factories outside Rochester, and by 1896, the re-named Eastman Kodak Company was employing 700 workers domestically and had established an overseas branch in England. In 1893, Blair, too, expanded to England, founding the European Blair Camera Company, which manufactured roll film using the same methods and equipment as its American counterpart.[32] As the commercial availability of celluloid roll film coincided with the first experiments with motion pictures, moreover, both of these companies would soon find a new market for their popular photographic material.

Celluloid Goes to the Movies

In the late-1880s, Thomas Alva Edison commissioned W. K. L. Dickson to begin work on what would become the Kinetograph motion picture camera and the Kinetoscope peep-show viewing machine. In 1889, the Eastman Company began supplying Dickson with special orders of celluloid roll film tailored to the requirements of these apparatuses. To produce images of the right size and resolution, the film needed to be narrower than that for still cameras. It also needed to be thicker—and therefore stronger—so that it could withstand both the strain of the Kinetograph's intermittent movements, and repeated viewings in the Kinetoscope. After some trial and error, Dickson and Eastman landed on the 35mm film stock that would become the international standard gauge for professional filmmaking. In 1892, when Eastman Kodak was temporarily out of the roll film business, Dickson ordered celluloid stock with these same specifications from the Blair Camera Company.[33]

The first Kinetoscope parlor opened in New York City in April 1894, showing films shot at Edison's Black Maria studio in Orange, New Jersey, and printed on Blair stock.[34] The ensuing demand for moving images was soon beyond what the Edison Manufacturing Company could supply, spurring rival companies, both in America and abroad, to produce their own cameras, films, and peep-show devices. Film projectors like Auguste and Louis Lumière's Cinématographe and Edison's Vitascope (invented by C. Francis Jenkins and Thomas Armat) were the logical next step.[35] Before long, the enormous popularity of cinema around the world had created a corresponding need for large quantities of celluloid film.

Until 1896, the majority of motion picture film in the United States came from the Blair Camera Company, which still sourced its celluloid from the Celluloid Company in Newark. The situation was similar abroad, with European Blair supplying a number of early filmmakers, such as Robert W. Paul and Birt Acres in Britain, Oskar Messier in Germany, and the Lumière brothers in France. However, European Blair stock had consistent problems with emulsion adhesion, and although the company—which gradually separated from its American branch—continued producing motion picture film until 1903, its share of the market declined as filmmakers turned to other suppliers or became manufacturers themselves.[36] In 1896, for instance, the Lumières and sheet film manufacturer Victor Planchon joined to found the Société des Celluloses Planchon in Lyon to produce motion picture film for their patented equipment.[37]

George Eastman had little interest in motion picture film at first, preferring to concentrate on the robust still photography market, which his company again dominated after resuming roll film production in 1894. In the summer of 1896, however, at the urging of George Dickman, London office manager, and Henry Strong, company president, Eastman agreed to the commercial production of motion picture film. In August, the Edison Company switched its substantial business from Blair to Kodak, and by November, motion picture film was a regular part of the company's output. Then, in 1899, after several years of negotiations and legal disputes, Eastman Kodak purchased the Blair Camera Company, gaining access to its continuous casting method. This made Kodak the largest film producer in the United States, and effectively forced the Celluloid Manufacturing Company out of this lucrative field.[38]

As with celluloid more generally, motion picture film was challenging and dangerous to manufacture, and global output was controlled by only a few companies.[39] Eastman Kodak remained one of the largest and, over the years, developed a refined and systematic manufacturing process, carried out entirely at Kodak Park in Rochester. First, specially treated cotton linters were converted into cellulose nitrate in the Nitrating Plant. The fibers were torn apart, weighed, and stirred in a controlled mixture of nitric acid, sulfuric acid, and water. Successive hot water and butyl alcohol washes removed the acids and excess water, respectively, and after centrifugal drying, the dehydrated cellulose nitrate was taken to the Dope Department. There it was mixed with a solvent consisting of methyl alcohol, ethyl alcohol, and acetone, and then camphor was added to create a clear, uniform solution. In the Roll Coating Department, this dope was funneled through a narrow slit onto a large, rotating wheel to form strips. The solvents were evaporated with a stream of hot air, and the dried celluloid was then coated with gelatin, which helped the emulsion adhere to the base. In the Emulsion Coating Department, a partly immersed roller passed the base through a pan of melted emulsion. Chilling set the emulsion, and the coated film was then dried under careful temperature, humidity, and air velocity conditions. Finally, in the Finished Film Department, the film was slit by revolving knives, and perforated by a series of carefully calibrated dies and punches.[40]

Certain steps in this process—chiefly the removal of excess nitric and sulfuric acids—increased the stability of the celluloid base, but they did not reduce its flammability.[41] Yet for those employed in the motion picture industry, nitrate film was simply part of their daily lives, just like oil lamps, gas cookers, and gasoline-fueled automobiles, and they were not unduly afraid of it. Government authorities and regulatory agencies, on the other hand, whether because of "ignorance, or artistic, social, or cultural prejudice", often ascribed an acute danger to motion picture film, and its use was carefully regulated from the medium's very beginning.[42]

Regulating the Dangerous Cinematograph

On May 4, 1897, a fire broke out during a cinematograph screening at the Bazar de la Charité Fair in Paris. While refilling the projector with ether, one of the operators lit a match, igniting the illuminant. Fueled by the flammable film, the decorative drapes, and the tar-coated building, the fire spread rapidly, destroying the entire structure and killing over 100 people within minutes.[43] Although not the first cinema fire, the Charity Bazar tragedy firmly established nitrate film's dangerous reputation, and directly led to laws controlling its use, including the 1897 regulations governing public entertainments in the province of Zaragoza, Spain, and the London County Council Act of 1898.[44] Other national, state, and municipal regulations continued to be implemented, leading one manual for projectionists and theater managers to complain that "No business . . . is subject to so many rules, regulations, conditions and restrictions as the kinema [*sic*] industry".[45]

Those in charge of public safety claimed that motion picture film required heightened oversight for two reasons. First, in film's early years, the celluloid used for the base had a higher nitrogen content than that used in other celluloid goods, making it more flammable.[46] Second, reels of film—standardized by the early-1900s at 1,000 feet—presented "a much larger surface than any other celluloid article", while the "exigencies of the trade" required "the concentration of great quantities of film" at theaters and especially at film exchanges.[47] It was thus on these two sectors of the motion picture industry—exhibition and distribution—that regulatory agencies concentrated their attention.

The top priority was placing any machine running flammable film in a fireproof enclosure. Initially, these were wooden booths lined with tin, sheet iron, or asbestos.[48] By the 1910s, however, as purpose-built cinemas grew in number, more substantial fireproofing was required. Projection rooms had to be constructed of hollow tile, cement, or brick, with fireproof doors that closed automatically and adequate ventilation. The number and size of lens and observation ports were restricted, and all such openings had to be equipped with automatic fire shutters.[49] The impetus behind this construction was to wholly contain a fire within the projection room, thereby minimizing property damage and preventing audience panic, which often proved more dangerous than the fire itself.

Some laws also placed stipulations on the quantity of exposed film, audience capacity, type and number of fire extinguishers, and equipment permitted in the booth.[50] Almost all projection-related equipment had to be fireproofed, including rewinders, which could ignite films by friction, and storage cabinets, which needed to both contain a blaze and limit the amount of film destroyed should one reel catch fire.[51] Receiving particular attention was the projector, where fires were most likely to start. Laws like Britain's Cinematograph Act of 1909 and regulatory agencies such as America's National Board of Fire Underwriters (NFBU) required certain safety features on all professional projectors. These included the fire shutter, a small metal barrier that dropped between the film and the hot projector lamp to prevent the film from igniting while it was stationary in the gate; and fireproof magazines, iron boxes holding the feed and take-up reels that were equipped with fire rollers to smother burning film, and that would also contain a fire should it spread.[52] Facilities screening flammable film generally needed a license from a government authority, and by the 1910s, so did the men (and occasionally women) who projected nitrate film for public exhibition.[53] Depending on the city, a projectionist license could be contingent on age, citizenship, or completion of an examination or specialized training, and some places required two operators in the booth to ensure safety.[54]

Regulations governing film distribution focused on transportation and exchanges. Most countries prohibited nitrate film on public transport, and to ship by truck, rail, or boat, distributors often had to comply with very particular requirements.[55] Railway shipping containers in the United Kingdom, for instance, had to be constructed of wood-lined galvanized iron with folding seams and hinged, locking lids, on which "Kinema [*sic*] Films" and "Keep in a cool place" were painted in black letters of a specified size.[56] In America, specialized private truck delivery services handled most domestic film shipments, which had to be insured and packed and labeled in metal containers from the Interstate Commerce Commission.[57]

Of more concern, however, were film exchanges, the regional distribution hubs that trafficked films to and from theaters. Exchanges stored large quantities of film at any given time; in 1936, it was estimated that American exchanges handled 30,000 miles of film daily.[58] As exchanges were also responsible for inspecting and repairing prints in between play dates, they were full of damaged films and celluloid scraps that needed to be discarded. Groups like the National Fire Protection Association (NFPA) thus classified exchanges "as among the most dangerous of occupancies, from the fire hazard viewpoint", and in the 1910s and 1920s, they started to be targeted with some vigor by governments and regulatory agencies.[59]

Like projection booths, exchanges had to be constructed of fireproof material, and ideally would not adjoin any residential premises. When not in use, each reel was to be enclosed in a separate metal box, and films had to be kept in fireproof store rooms or vaults, often of a specified capacity. The NFBU recommended in 1910, for instance, that vaults be a maximum of 750 cubic feet, while the British Celluloid and Kinematography Film Act of 1922 allowed no more than 560 reels or 1 ton of film per store room.[60] It was strongly recommended that these vaults were vented to prevent the explosion of built-up gases, and equipped with automatic sprinkler systems and vertical rack partitions, "permitting the free passage of water between the [film] containers, thus checking the spread of fire".[61] Other furniture had to be minimal and fireproofed; film waste disposal cans had to be lidded and labeled; adequate fire extinguishing equipment was, of course, required; and smoking was, of course, prohibited.[62]

These regulations likely did help prevent film fires, or at least minimize the damage they caused. In January 1918, for instance, *Exhibitors Herald* touted the "value of all-steel fireproof booths" following a fire at a theater in Elwood, Indiana: "Instantly the whole interior of the booth . . . was a mass of flames", but "the operator escaped uninjured", the "auditorium was not damaged to any extent, and, although persons in the audience were somewhat frightened, none was injured and there was no panic."[63] The film industry also recognized that cooperating with regulatory agencies to reduce the threat of film fires could boost cinema's cultural reputation, proving to those moral watchdogs who denigrated the movies as a disreputable pastime that theirs was a professional trade.[64] To this end, Hollywood's trade organizations developed committees to oversee construction and work practices in film exchanges; trade magazines touted the benefits of fireproof projection booths; and projectionist handbooks insisted that the "well-trained operator" need not worry about film fires "if he has conscientiously done his duty and cheerfully complied with regulations both written and unwritten".[65] In short, fire safety in theaters and exchanges became just as much about professional standing as it was forced compliance with the law.

Still, there were complaints that some of these regulations evidenced little understanding of how the film industry operated. In the fourth edition of his *Handbook of Projection* (1923), for instance, F. H. Richardson insisted that imposing limitations on the size of projection room ports served "absolutely no good purpose, either in safety or anything else", as smaller ports did little to impede the passage of smoke and only made it "impossible for the projectionist to secure a really good view of his screen" and thus do his job.[66] Moreover, even in the earliest days of cinema—when exhibition took place at fairgrounds, town halls, or other make-shift venues; when projection technology was comparatively crude; and when regulation was minimal—film fires were not as common as might be assumed. In many cases, furthermore, such as that at the Charity Bazar, the fire was caused not by the film itself, but by other flammable substances or human carelessness.[67] Nevertheless, the theoretical danger of nitrate film meant that almost as soon as it was adopted by the motion picture industry, efforts were being made to replace it.

Non-Flam Film

As early as 1897, researchers and chemists were trying to reduce the risk of film fires by dealing with the flammable stock itself. Most early attempts either to chemically treat the celluloid to reduce its flammability or to replace the base with another material seem to have amounted to little, but by the 1910s, cellulose acetate had emerged as a promising alternative.[68] In a process similar to that of nitrate film, cellulose acetate film stock was manufactured by plasticizing a cellulosic material—in this case, cellulose fibers treated with acetyl acid. The difficulty was achieving the right degree of acetylation—that is, in attaching the right number of acetyl groups to the cellulose polymer backbone. Fully acetylated cellulose, or cellulose triacetate, was difficult to cast into film strips and the stock was troublesome to splice. Removing acetyl groups to create cellulose diacetate resulted in a material that was easier to manufacture and splice, but that was also weaker, less rigid, and less heat resistant.[69]

The first cellulose acetate stocks, which appeared on the market around 1909, were predominately cellulose diacetate, and were quickly rejected for their inferior quality.[70] Acetate film readily became brittle; it warped and buckled under the heat from the projector; and it was not durable enough to suit the industry's needs.[71] An average nitrate print could withstand an estimated 200 projections before needing to be replaced, while a diacetate print was worn out after only fifty; switching to acetate would thus require producing four times as many prints for distribution.[72] In 1937, Kodak released an improved acetate base that used propionic acid as a solvent.[73] Though acetate-propionate was stronger and less subject to brittleness than diacetate, it was still about half as durable as nitrate.[74] Until cellulose acetate's physical properties could compete with those of cellulose nitrate, lack of flammability was not reason enough for the professional film industry to adopt it.

Strengthening producers' and distributors' resistance to safety film was its cost; it was about 12.5% more expensive per foot than nitrate, no matter the currency.[75] In addition, because non-flammable film was unregulated, it could be screened in unlicensed venues without a licensed projectionist, giving many in the exhibition sector further cause to object to it.[76] For these reasons, repeated attempts—whether by governments, regulatory agencies, or film manufacturers—to force a switch to acetate consistently failed. When Eastman Kodak and the Motion Picture Patents Company tried to contractually implement non-flam film in 1909, producers demanded a return to nitrate within two years.[77] When Pathé, seeking to promote its new safety stock, succeeded in outlawing nitrate film in France in the 1920s, the industry's continuous requests to defer the law's enforcement resulted in the issue simply being dropped.[78] And when New York State proposed prohibiting nitrate film in churches and schools in 1923, protests by the local projectionist union defeated the bill.[79]

Though rejected by the professional film industry, non-flam film was suitable for the home and amateur markets. Home cameras and projectors were commercially available by 1898, allowing people to bring nitrate film right into their living rooms.[80] Up until the 1930s, furthermore, toy projectors and film frame viewers were popular divertissements for children, and frequently came supplied, respectively, with short loops of nitrate film and packets of individual nitrate film frames.[81] Many government regulations, however, such as the Cinematograph Act of 1909, did not apply to private residences, and while some restrictions were placed on the sale of nitrate film to anyone under the age of sixteen, it remained fairly easy for amateurs and minors to obtain flammable film.[82] Thus certain manufacturers, particularly Eastman Kodak and Pathé, made a concerted effort to keep nitrate out of non-professional hands. As George Eastman acknowledged in a letter to the Edison Company in 1912, "in our opinion, the furnishing of cellulose nitrate for [amateur use] would be wholly indefensible and reprehensible."[83] Kodak and Pathé's amateur camera and projector systems, including the Edison Company's Home Kinetoscope (1912), the Pathé-Kok (1912), the Pathé Baby (1922), the Cine-Kodak (1923), and the Kodascope (1923), exclusively used small gauge safety film, as did these companies' respective toy projectors, the Kodatoy and the Pathé Kid.[84]

Until the late-1940s, then, safety film was almost entirely restricted to amateur filmmaking and niche markets, such as educational and scientific films, though there was a sharp uptick in the professional and semi-professional use of 16mm safety film during World War II.[85] Then, in March 1947, Eastman Kodak announced the development of an acetate stock that could replace nitrate, and by the following year, the company was conducting the requisite tests.[86] This cellulose acetate had an intermediate acetyl content, which made it physically stronger than diacetate, but still eliminated the problems with manufacturing and splicing that had plagued triacetate. A report in the *Journal of the Society of Motion Picture Engineers* in 1948 offered detailed comparisons between the new "high-acetyl" film, nitrate, and acetate-propionate, and concluded that on every significant point, particularly those connected to durability, the new acetate performed either as well as or close enough to nitrate that it would be suitable for professional use.[87] The speed with which Kodak got this improved safety stock to market is somewhat remarkable—at least compared to the time it took to develop acetate-propionate—and there is some evidence that the company utilized research taken from the Nazis after the war. Still, high-acetyl film was adopted quickly and widely. Now that acetate was on par with nitrate, it was in the film industry's best interest to switch, from both an economic and public relations standpoint: safety film could reduce overhead costs and halt the continuing negative publicity about nitrate fires. In 1950, with 85% of American films being produced on acetate, Kodak discontinued nitrate stock altogether.[88]

Of course, nitrate film was not immediately eliminated from professional use. Not every company ceased production at the same time, and not every country phased out nitrate prints at the same rate, so that nitrate film lingered in countries such as Poland, Japan, and Norway into the 1960s.[89] Even in countries that halted nitrate film manufacturing fairly quickly, such as the United Kingdom, which stopped production in 1951, existing nitrate prints remained in circulation for at least a year.[90] Once the conversion process was under way, furthermore, raw nitrate stock could be purchased cheaply, encouraging countries such as Francoist Spain to buy large quantities at discounted prices.[91] Eventually, however, nitrate film became obsolete, a relic of an earlier age, and those responsible for protecting the public from its fire hazards were no longer producers, distributors, or exhibitors, but archivists.

Nitrate in the Archives

As archives became the primary repositories of large quantities of nitrate film, they also became the sites of the most severe film fires. Fires like those at the Cinemateca Brasileira in 1957, the Canadian National Film Board Archives in 1967, the National Archives in Suitland, Maryland, in 1978, Mexico's Cineteca Nacional in 1982, and at the New York Historical Society in 2003, destroyed thousands of feet of unique heritage film. Many archive fires also resulted in significant property damage, personal injury, and death.[92] Indeed, the "special risk" posed by storing motion picture films due to "the inflammable character of the material used" was partly responsible for delaying concerted film preservation efforts, as administrators were reluctant to house celluloid alongside other precious artifacts.[93]

By the 1930s, though, there was a wider acceptance of cinema's historical and cultural value, and dedicated film archives were starting to establish long-term storage methods.[94] As John G. Bradley, Chief of the Division of Motion Pictures and Sound Recordings at the National Archives in Washington, D.C. explained, "No longer will the cellar or the abandoned garage be good enough. . . . photographic film, particularly nitrocellulose film, demands basic planning."[95] Naturally, fireproofing was central to such planning, and archives—often with the assistance of organizations like the Society of Motion Picture Engineers or the Kodak Research Laboratories—conducted extensive research on storage cabinet and vault construction.[96] When selecting permanent storage sites, furthermore, archives had to consider whether the general public would be put at risk in the event of a fire, and more than one nitrate collection has been forced to move due to population growth or changing regulations.[97]

Nitrate's flammability was only half of the preservation equation, however. Archives also had to reckon with the material's intrinsic chemical instability. Cellulose nitrate readily decomposes by way of two primary mechanisms: hydrolysis and chain scission. In the former, water breaks the bonds attaching the nitrate groups to the cellulose backbone—a process called denitration—and in the latter, the cellulose polymer chain itself breaks apart into smaller units. Exacerbated by heat and moisture, this polymer degradation process is autocatalytic, generating byproducts, including nitric acid, that further catalyze the decomposition reactions.[98] As a result, there are marked changes to the film's physical properties, which archivists commonly describe using a five-stage model. First, the image fades, the celluloid yellows and brittles, and the film begins to emit a noxious odor. The emulsion then becomes sticky, and subsequently blisters and bubbles, as the odor becomes more pungent. The film then congeals into a solid mass, sometimes forming a "viscous froth", and finally, it disintegrates into a brown powder.[99] (Celluloid objects undergo a similar decomposition process.)[100]

Professional film industries were aware of, but not terribly concerned about nitrate's instability, as most films, especially during the silent period, were destroyed long before they began to decompose on their own. Films were of little value to studios once they reached the end of their commercial run, save for any money that could be gleaned from their constituent elements. The silver in film emulsion always yielded a few dollars, and the celluloid base could be sold for a variety of purposes: coating for patent leather shoes, film leader, and reportedly even sausage casings and chewing gum.[101] Old nitrate prints were also sometimes burned to create fire and smoke special effects in new film productions, as in *We Can't Have Everything* (DeMille, 1918), *Faust* (Murnau, 1926), and *The Squatter's Daughter* (Hall, 1933).[102]

Those seeking to preserve film heritage, however, were tasked with determining how best to prevent this irreversible decay. By the 1910s, cold storage was known to delay the onset of decomposition, and by the 1930s, best practice was vault storage at 40–45 degrees Fahrenheit, which was considered a "manageable" temperature, but still "sufficiently low to prevent appreciable decomposition of the film base".[103] Other recommendations included storing films at an intermediate relative humidity (RH) to prevent brittleness; thoroughly washing films of residual chemicals that might trigger decomposition; and using vented cans to allow corrosive gases to escape.[104] Stored films were also to be inspected at regular intervals, so that deteriorating films could be duplicated before too much damage had occurred.[105]

Still, archivists had no way of anticipating when a film might begin to decompose, nor could they ascertain how long it would take to disintegrate entirely. Various tests were thus developed to try to identify when a film was in the earliest stage of decomposition, namely the Punch Test and the Alizarin Red Test, both of which involved heating a small sample of film in a test tube with a chemically sensitive paper strip. The fact that these tests were destructive, however, made them controversial in certain archival circles.[106] The five-stage model, outlined by the *Journal of the Society of Motion Picture Engineers* in 1950 and widely publicized by Kodak, was of some help, but it was not tied to any concrete timeline. It did, however, emphasize that films in stages three, four, and five were more flammable and should be destroyed.[107]

By the 1970s, nitrate film was becoming a burden to many archives. It was difficult to store, it would inevitably decay, and it would only get more dangerous as it aged.[108] So when the International Federation of Film Archives (FIAF) stated that nitrate film would last on average fifty years, and that the vast majority would be gone by the year 2000, a number of archives initiated copy-and-destroy policies.[109] Declaring that "Nitrate Won't Wait", these preservation campaigns asserted that "unless rescued in time, all of the films on nitrate stock will certainly crumble to powder if they do not spontaneously ignite".[110] By emphasizing the risk nitrate posed to both public safety and film heritage, archives were able to solicit funding to transfer nitrate films onto acetate stock.[111] The nitrate original was then destroyed.

Unfortunately, many of these copies were made hastily and poorly, and it turned out that cellulose acetate was not as stable as archivists initially believed. Not only did it decompose, but it also tended to do so more quickly than nitrate.[112] Subsequent research by the Image Permanence Institute estimated that, if properly stored, a nitrate print could last hundreds of years, while an acetate print would only last fifty.[113] Belatedly realizing that nitrate *would* wait, many archives—though certainly not all—decided that preservation was preferable to destruction.[114] Recognizing the role played by both heat and moisture in the decomposition process, contemporary best practices recommend storing nitrate collections at 2 degrees Celsius and 20–30% RH; under these conditions, it is believed that decay can be halted almost entirely.[115]

This does not mean, however, that the archival storage of nitrate film has become easier, as nitrate film is still thoroughly regulated. The film must be stored in fireproof vaults meeting particular specifications. Packing for shipment must be done by trained personnel or outside contractors, and many companies refuse to transport it. Disposal must also be handled by specialized companies.[116] These regulations can be irksome for large institutions, but they are particularly challenging for small ones, many of which might not have the resources to comply.[117] A better understanding of the relationship between nitrate film's stages of decomposition and its combustibility might lead to revised and relaxed regulations, but that relationship is still unclear. Recent physiochemical analyses of heritage nitrate film by the Wisconsin Nitrate Film Project, an interdisciplinary grant project funded by the National Endowment for the Humanities, found that the five-stage model does not fully correlate to nitrate's physical, chemical, and flammability properties. This means that nitrate film's visible characteristics—yellowing, emulsion bubbling, etc.—do not reliably indicate its stability or danger.[118] As this sort of research requires destroying the film, the available samples for further studies are limited.

Conclusion

While many film historians and archivists still view nitrate motion picture film chiefly as "a difficult and expensive fire hazard", others praise it as a "technological success story", the very thing that "made cinema possible".[119] Those who have seen a nitrate print projected—which can be done in select, highly fireproofed booths—swear by its superior sharpness, texture, and tonal warmth, while some have romanticized nitrate film precisely because its very chemical nature makes it "fragile, dangerous, precious, threatened [. . . and] forbidden".[120] These somewhat contradictory opinions about nitrate film are nothing new; they have been present through every phase of the technology's history. The film industry clung to nitrate film tenaciously because of the incomparable strength of its celluloid base. Simultaneously, the material's extreme flammability spurred continuous attempts to regulate, outlaw, and replace it. And nitrate film's potential hazards, chemical instability, and surprising longevity have all influenced, and at times altered, archival practices.

Though the conversion to cellulose acetate might have been delayed in part by commercial concerns and battles over industrial control, it is also the case that cellulose nitrate film possessed physical properties uniquely suited to cinema's needs and unmatched by anything else. This made its tendency to burn an easily overlooked inconvenience. It is also likely that if celluloid had not been available, cinema's pioneers would have found other ways to advance the medium. Still, the history of cinema is inextricably tied to the history of this unstable, yet durable material, and if predictions about nitrate film's lifespan prove true, this flammable workhorse will be responsible for preserving film history for future generations.

Notes

1 J.M. Calhoun, "Old Nitrate Films Are Dangerous!" *International Projectionist* 37, no. 5 (May 1962): 8. Emphasis original.

2 Sam Kula, "Mea Culpa: How I Abused the Nitrate in My Life", in *This Film Is Dangerous: A Celebration of Nitrate Film*, ed. Roger Smither and Catherine A. Surowiec (Brussels: FIAF, 2002), 168.

3 Alfonso del Amo, "A Brief History of Plasticized Cellulose Nitrate or Celuloid [*sic*]", *Journal of Film Preservation* 60/61 (2000), 42–44.

4 Deac Rossell, "Exploding Teeth, Unbreakable Sheets, and Continuous Casting: Nitrocellulose, from Gun Cotton to Early Cinema", in *This Film Is Dangerous*, 37; Edward Chauncey Worden, *Nitrocellulose Industry: A Compendium of the History, Chemistry, Manufacture, Commercial Application and Analysis of Nitrates, Acetates, and Xanthates of Cellulose As Applied to the Peaceful Arts, with a Chapter on Gun Cotton, Smokeless Powder and Explosive Cellulose Nitrates*, two volumes (London: Constable and Company, Ltd., 1911), 260–567, 794–827.

5 Amo, "Brief History", 44.

6 Ibid.

7 M. Kaufman, *The First Century of Plastics: Celluloid and Its Sequel* (London: Plastics Institute, 1963), 21–27; S.T.I. Mossman, "Parkesine and Celluloid", in *The Development of Plastics*, ed. S.T.I. Mossman and P.J.T. Morris (London: Royal Society of Chemistry, 1994), 11–15.

8 Kaufman, *First Century*, 33–38, 41; Mossman, "Parkesine and Celluloid", 15–21; Rossell, "Exploding Teeth", 39.

9 Kaufman, *First Century*, 38–40.

10 H.N. Stokes and H.C.P. Weber, *Effects of Heat on Celluloid and Similar Materials*, Technological Papers of the Bureau of Standards No. 98, Washington, DC, 1917, 4.

11 Worden, *Nitrocellulose Industry*, 448–451.

12 Paul C. Spehr, "Unaltered to Date: Developing 35mm Film", in *Moving Images: From Edison to the Webcam*, ed. John Fullerton and Astrid Söderbergh Widding (Sydney: Aura/John Libbey & Co., 2000), 5; Stokes and Weber, *Effects of Heat*, 4.

13 Rossell, "Exploding Teeth", 39.

14 Kaufman, *First Century*, 45.

15 Worden, *Nitrocellulose Industry*, 568–793.

16 Kaufman, *First Century*, 41–42, 48–49; Rossell, "Exploding Teeth", 40.

17 Stokes and Weber, *Effects of Heat*, 5–39; Celluloid Committee, *Report of the Departmental Committee on Celluloid*, London, 1913, 15–29.

18 Robert Friedel, *Pioneer Plastic: The Making and Selling of Celluloid* (Madison: University of Wisconsin Press, 1983), 96–98; Clyde Jeavons, "Playing with Fire", in *This Film Is Dangerous*, 237; Celluloid Committee, *Report*, 5; Rossell, "Exploding Teeth", 40.

19 W.G. Kubler Ridley, *The Common Hazards of Fire Insurance* (London: Sir Isaac Pitman & Sons, Ltd., 1922), 50.
20 Celluloid Committee, *Report*, 4.
21 Friedel, *Pioneer Plastics*, 98–100; Kaufman, *First Century*, 55–69.
22 Friedel, *Pioneer Plastics*, 96–98.
23 Ridley, *Common Hazards*, 47; Friedel, *Pioneer* Plastic, 95.
24 Rossell, "Exploding Teeth", 38.
25 Friedel, *Pioneer Plastics*, 91; Spehr, "Unaltered to Date", 5; Rossell, "Exploding Teeth", 41.
26 Rossell, "Exploding Teeth", 42; Friedel, *Pioneer Plastics*, 91–93.
27 Rossell, "Exploding Teeth", 42–43; Friedel, *Pioneer Plastics*, 93–94.
28 Spehr, "Unaltered to Date", 6; Rossell, "Exploding Teeth", 42.
29 Rossell, "Exploding Teeth", 44; Spehr, "Unaltered to Date", 9, 13.
30 Rossell, "Exploding Teeth", 42; Spehr, "Unaltered to Date", 10, 24 (note 33).
31 Rossell, "Exploding Teeth", 42; Spehr, "Unaltered to Date", 6, 9.
32 Friedel, *Pioneer Plastic*, 93–94; Spehr, "Unaltered to Date", 9, 13, 18; Rossell, "Exploding Teeth", 45.
33 Spehr, "Unaltered to Date", 10–15.
34 Rossell, "Exploding Teeth", 44.
35 Spehr, "Unaltered to Date", 15–17.
36 Spehr, "Unaltered to Date", 13, 15–17; Rossell, "Exploding Teeth", 44–45.
37 Spehr, "Unaltered to Date", 17; Rossell, "Exploding Teeth", 45.
38 Rossell, "Exploding Teeth", 44; Spehr, "Unaltered to Date", 18–19, 24; Friedel, *Pioneer Plastic*, 95–96.
39 Rossell, "Exploding Teeth", 46.
40 E.K. Carver, "The Manufacture of Motion Picture Film", *Journal of the Society of Motion Picture Engineers* 28, No. 6 (June 1937), 596–603. (Hereafter *JSMPE*)
41 John Reed, "Nitrate? Bah! Humbug! A Personal View from an Archive Heretic", in *This Film Is Dangerous*, 220.
42 Roger Smither and Catherine A. Surowiec, "'Calamity Howlers': An Introduction", in *This Film Is Dangerous*, 423–426.
43 "Calendar of Film Fires", in Roger Smither and Catherine A. Surowiec, eds., *This Film Is Dangerous*, 431.
44 Kula, "Mea Culpa", 166; Jorge Martin Neira, "Fires in the Spanish Cinema", trans. Dwight Porter, in *This Film Is Dangerous*, 474; Vanessa Toulmin, "Phantom Fires: An Evaluation of the Evidence for Nitrate Fires in Fairground Cinematograph Shows", in *This Film Is Dangerous*, 109.
45 J.H. Hutchison, *The Complete Kinema Manager* (London: Kinematograph Publications, Ltd., 1937), 126.
46 Sylvia Katz, "The Degradation and Disappearance of Cellulose Nitrate Objects", in *This Film Is Dangerous*, 198.
47 Celluloid Committee, *Report*, 11–12.
48 F.H. Richardson, "The Projection Room and Its Requirements", *Transactions of the Society of Motion Picture Engineers*, November 1918, 29.
49 Richardson, "Projection Room", 31–35; A. Humphrey Williams and Alfred Harris, *The Cinematograph Act, 1909, with Notes and an Appendix of Statutes and Regulations*, London, 1913, 60–61.
50 Neira, "Fires in the Spanish Cinema", 474; Williams and Harris, *Cinematograph Act*, 60–64; John B. Rathbun, *Motion Picture Making and Exhibiting* (Chicago: Charles C. Thompson Company, 1914), 175.
51 Roger Smither, "Unseen Showmen and Unsung Heroes: Projectionists in the Nitrate Era in the United Kingdom", in *This Film Is Dangerous*, 269; Jean-Louis Bigourdan, "From the Nitrate Experience to New Film Preservation Strategies", in *This Film Is Dangerous*, 56.
52 Williams and Harris, *Cinematograph Act*, 62; James R. Cameron, *Motion Picture Projection,* 4th edition (Manhattan Beach, NY: Cameron Publishing Co., Inc., 1928), 154.
53 Williams and Harris, *Cinematograph Act*, 10–19; Eileen Bowser, *The Transformation of Cinema, 1907–1912*, History of the American Cinema, Vol. 2 (Berkeley: University of California Press, 1990), 13.
54 "London County Council Regulation, Section XIX", in Colin N. Bennett, *A Guide to Kinematography (Projection Section) For Managers, Manager Operators, and Operators of Cinema Theatres* (London: Sir Isaac Pitman & Sons, Ltd., 1923), 172; F.H. Richardson, "Operators' Column", *Moving*

Picture World, July 30, 1910, 249; Smither, "Unseen Showmen", 276–278; Ivo Bloom, "Dutch Flames and Flickers", in *This Film Is Dangerous*, 479.

55 "Throw Me Out", in Roger Smither and Catherine A. Surowiec, eds., *This Film Is Dangerous*, 311; Kula, "Mea Culpa", 169.

56 "Summary of Railway Regulations for Carrying Film Programmes", in Bennett, *A Guide to Kinematography*, 177.

57 Henry Anderson, "Fire Safety in the Motion Picture Industry", *Quarterly of the National Fire Protection Association* 30, No. 1 (July 1936), 26. (Hereafter *QNFPA*)

58 Henry Anderson, "Fire Prevention in the Motion Picture Industry", *JSMPE* 27 (December 1936), 665.

59 "Film Exchange Fire Prevention Results", *QNFPA* (January 1926), 224.

60 "Fire Underwriters Make Film Rules", *Nickelodeon*, September 15, 1910, 159; "Celluloid and Kinematograph Film Act, 1922 (Abridgment)", in Bennett, *Guide to Kinematography*, 173–176.

61 George A. Blair "Reducing the Fire Hazard in Film Exchanges", *Transactions of the Society of Motion Picture Engineers* 2, No. 11 (October 1920), 56.

62 "Celluloid and Kinematograph Film Act", 175; "Fire Underwriters", 59; "Fire Prevention in Film Exchanges", *Safety Engineering* (November 1925), 263–264.

63 "Fireproof Booth Prevents Big Blaze", *Exhibitors Herald*, January 19, 1918, 35.

64 Bowser, *Transformation of Cinema*, 75.

65 "Film Exchange Fire Prevention Results", 224–228; Anderson, "Fire Prevention", 665–670; "Mock Lays Out Pantages Projection", *Exhibitors Herald*, January 8, 1921, 582; Bernard E. Jones, ed. *Cinematograph Book: A Complete Practical Guide to the Taking and Projecting of Cinematograph Pictures*, Revised edition (London: Cassell and Company, Ltd., 1919), 171.

66 F.H. Richardson, *Richardson's Handbook of Projection*, 4th edition (New York: Chalmers Publishing Company, 1923), 312.

67 Toulmin, "Phantom Fires", 109–110; Kula, "Mea Culpa", 166; Smither and Surowiec, "'Calamity Howlers,'" 423–424.

68 "Safety Film: False Dawns", 319–320; Leo Enticknap, "The Film Industry's Conversion from Nitrate to Safety Film in the Late-1940s: A Discussion of the Reason and Consequences", in Roger Smither and Catherine A. Surowiec, eds., *This Film Is Dangerous*, 203.

69 Charles R. Fordyce, "Improved Safety Motion Picture Film Support", *JSMPE* 51 (October 1948), 332.

70 Bigourdan, "Nitrate Experience", 55.

71 "False Dawns", 321, 323.

72 Anderson, "Fire Safety", 26–27.

73 Enticknap, "Film Industry's Conversion", 204.

74 Fordyce, "Improved Safety", 331–332, 341.

75 "False Dawns", 321–323.

76 Enticknap, "Film Industry's Conversion", 203–204.

77 "False Dawns", 320–323.

78 "False Dawns", 324.

79 David Pierce, "Legion of the Condemned: Why American Silent Films Perished", *Film History* 9, No. 1 (1997), 12.

80 David Cleveland, "Don't Try This at Home: Some Thoughts on Nitrate Film, with Particular Reference to Home Movie Systems", in *This Film Is Dangerous*, 192.

81 Stephen Herbert, "Projecting Nitrate", in *This Film Is Dangerous*, 105.

82 Williams and Harris, *Cinematograph Act*, 46; Jeavons, "Playing with Fire", 239–241.

83 Qtd. in Enticknap, "Film Industry's Conversion", 204.

84 Cleveland, "Don't Try This at Home", 194–195; Herbert, "Projecting Nitrate", 105.

85 Cleveland, "Don't Try This at Home", 193; Enticknap, "Film Industry's Conversion", 204.

86 Enticknap, "Film Industry's Conversion", 208.

87 Fordyce, "Improved Safety", 332–349.

88 Enticknap, "Film Industry's Conversion", 206–209.

89 Jan Slodowski, "Poland's Worst Nitrate Film Disaster", in *This Film Is Dangerous*, 472; Hisashi Okajimi and Yoshiko Irie, "Kyoto Tales and Tokyo Stories: Incidents in Japanese Film History", trans. Awake Saito, in *This Film Is Dangerous*, 485; Vigdis Lian, "The Last Fire in a Commercial Cinema?" in *This Film Is Dangerous*, 489.

90 Cleveland, "Don't Try This at Home", 197; Smither, "Unseen Showmen", 287.

91 Enrique Blanco, interviewed by Rosa Cardona Arnau and Jennifer Gallego Christensen, January 2000, trans. Dwight Porter, in *This Film Is Dangerous*, 309.

92 "Calendar of Film Fires", 446–451; Leslie Augenbraun, *Film Photography (aka Nitrate) Project at the New York Historical Society Final Report* (New York: New York Historical Society, 2006).

93 "Historic Films: The Difficulty of Preservation", *London Times*, November 28, 1916; Stephen Bottomore, "'A Fallen Star': Problems and Practices in Early Film Preservation", in *This Film Is Dangerous*, 187.

94 Heather Heckman, "Burn After Viewing, or, Fire in the Vaults: Nitrate Decomposition and Combustibility", *The American Archivist* 73 (Fall/Winter 2010), 494.

95 John G. Bradley, "Changing Aspects of the Film Storage Problem", *JSMPE* 30 (March 1938), 304–305.

96 Bigourdan, "Nitrate Experience", 56; Bradley, "Changing Aspects", 305–312; J.I. Crabtree and C.E. Ives, "The Storage of Valuable Motion Picture Film", *JSMPE* 15 (September 1930), 289–303.

97 Clyde Jeavons, "24 Years to Safety", in *This Film Is Dangerous*, 393; Paul C. Spehr, "The Library of Congress and Its 'Nitrate Problem'; or It was Necessary to Destroy the Nitrate in Order to Preserve It", in *This Film Is Dangerous*, 230.

98 Vance Kepley, Mahesh Mahanthappa, Kathleen Mullen, Mary Huelsbeck, and Amanda McQueen, "Investigation of Cellulose Nitrate Motion Picture Film Chemical Decomposition and Associated Fire Risk" (White Paper, National Endowment for the Humanities, 2015), 19.

99 Heckman, "Burn After Viewing", 491.

100 Katz, "Celluloid Nitrate Objects", 198–199.

101 Pierce, "Legion of the Condemned", 6; Bottomore, "Fallen Star", 188; Blanco, Interview with Arnau and Christensen, 310; "A New Film Fancy", *New Orleans Times-Picayune*, reprinted in *QNFPA* (October 1925), 129.

102 Roger Smither, "Fiery Tails", in *This Film Is Dangerous*, 449–507.

103 Bottomore, "Fallen Star", 186; Crabtree and Ives, "Storage of Valuable Motion Picture Film", 1930, 290.

104 Committee on the Preservation of Film, "Report of the Committee on the Preservation of Film", *JSMPE* 20 (June 1933), 524; Crabtree and Ives, "Valuable Motion Picture Film", 299, 302–303.

105 Crabtree and Ives, "Valuable Motion Picture Film", 303.

106 Harold Brown, "Trying to Save Frames", in *This Film Is Dangerous*, 99–100; Roger Smither, "Henri Langlois and Nitrate, Before and After 1959", in *This Film Is Dangerous*, 248–249.

107 Heckman, "Burn After Viewing", 498–499; James W. Cummings, Alvin C. Hutton, and Howard Silfin, "Spontaneous Ignition of Cellulose Nitrate Films", *Journal of the Society of Motion Picture and Television Engineers* 54 (March 1950), 271–274.

108 Bigourdan, "Nitrate Experience", 57.

109 Jeavons, "24 Years to Safety", 394.

110 John Culhane, "Nitrate Won't Wait", *American Film* 2 (1977), 54–55.

111 Kula, "Mea Culpa", 168.

112 Bigourdan, "Nitrate Experience", 55–57.

113 P.Z. Adelstein, J.M. Reilly, and F.G. Emmings, "Stability of Photographic Film: Part VI—Long-term Aging Studies", *SMPTE Journal* 111 no. 4 (April 2002), 142–143; Adelstein et al., "Stability of Ester Base Photographic Film Part IV—Behavior of Nitrate Base Film", *SMPTE Journal* 104 (June 1995), 369.

114 Heckman, "Burn After Viewing", 489–490.

115 Bigourdan, "Nitrate Experience", 61.

116 See, for example, National Fire Protection Association, *NFPA 40: Standard for the Storage and Handling of Motion Picture Film* (Quincy, MA: NFPA, 2016); International Federation of Film Archives, *Handling, Storage, and Transport of Cellulose Nitrate Film* (Brussels, FIAF, 1991); International Organization for Standardization, *ISO 10356- Cinematography—Storage and Handling of Nitrate-Base Motion Picture Films* (Geneva: ISO, 1996).

117 Kepley et al., "Investigation of Cellulose Nitrate", 5–11.

118 Kepley et al., "Investigation of Cellulose Nitrate", 16–38.

119 Leo Enticknap, "Nitrate's Still Waiting", *The Velvet Light Trap* 64 (Fall 2009), 86–87.

120 Dominique Païni, "Reproduction . . . Disappearance", in *This Film Is Dangerous*, 173.

9

FAREWELL TO THE PHOSPHORESCENT GLOW

The Long Life of the Cathode-Ray Tube

Mark J. P. Wolf

How many evenings, in how many homes in America and across the world, have families sat in their family rooms, the lights down low or off, basking in the flickering, ever-changing glow from a cathode-ray tube, in the form of a television? Or the glow of a computer monitor, or a television used to play video games? It is difficult to imagine life in the latter half of the 20th century without the cathode-ray tube (CRT), especially because so much of the 20th century—its news, entertainment, sporting events, the moon landing—were brought to so many via the CRT. As the premiere imaging device of the 20th century, the CRT was used in both raster and vector monitors for radar, oscilloscopes, television, computer graphics displays, video games, home computers, surveillance monitors, medical imaging, and more. The technology involved in the CRT has come to define mediated imagery, and how we perceive it, becoming a part of the cultures it conveys. Today, as flat-screen technologies have so quickly begun to replace bulkier CRTs, the cathode-ray tube's long-held role as the dominant moving image display has come to an end.

In order to discuss the CRT, one might ask, what is a cathode ray? In 1869, German physicist Johann Wilhelm Hittorf was experimenting with electricity and vacuum tubes, and discovered energy rays extending between negative electrodes (referred to as "cathodes" by Michael Faraday) and positive electrodes (referred to as "anodes") inside evacuated glass tubes. Seven years later, another German physicist, Eugen Goldstein, called these rays "cathode rays", and extended the study of them; thus, the tubes used came to be known as "cathode-ray tubes". Goldstein and other experimenters, such as Heinrich Hertz, thought that these rays were a type of wave, like light, while other chemists and physicists, including William Crookes and J. J. Thomson, believed they were streams of particles. In 1895, chemist Jean Baptiste Perrin set out to determine whether cathode rays were negatively charged. While his initial experiments did not conclusively prove this, a modified experiment by Thomson did manage to show that the cathode rays and the negative electrification followed the same pathways, and that the rays were indeed negatively charged. Crookes had discovered that cathode rays could be deflected magnetically, and Thomson went on to show that they could be electrostatically deflected as well. The particle theory seemed to be winning.

If cathode rays were made of particles, what kind of particles were they? German physicist Emil Wiechert measured the mass-to-charge ratio of the particles, declaring that "we are not dealing with the atoms known from chemistry, because the mass of the moving particles turned out to be 2000 to 4000 times smaller than the one of hydrogen atoms, the lightest of the known chemical atoms",[1] and atoms were thought to be indivisible.

A direct determination of the velocity of the particles was needed to complete a convincing argument, and while Wiechert was finally able to measure it in 1898, another physicist, J. J. Thomson, had already done it a year earlier, in October of 1897, using the "Braun Tube", a type of CRT.

CRTs were used to study cathode rays, but Karl Ferdinand Braun, a German physicist working at the Physics Institute of the Strasbourg University found another use for them in 1897. The electron beam was typically focused through a hole in the anode, magnets could deflect the beam, and the far end of the tube was coated with phosphorescent materials that would glow briefly when hit by the electron beam, so that one could tell where the beam was aimed. Braun put these elements together to produce the first oscilloscope, which moved the electron beam up and down, allowing changes in alternating currents to be visualized. Interestingly, Braun never patented the CRT, as he wanted it to be available to other scientists for research.[2]

Sir Joseph John Thompson, then, is credited with the discovery of the charge-carrying particles that made up cathode rays, which came to be called *electrons*, the first subatomic particles to be discovered, which also opened up the entire realm of subatomic particle physics. So the CRT, used as a particle accelerator, was instrumental in the discovery of the electron.[3] Over time, beams of electrons would no longer be called cathode rays, but the name "cathode-ray tube" would continue to be used.

CRTs for the Lab and Home: Oscilloscopes and "Magic Eye Tubes"

Another German physicist, Jonathan Zenneck, added a horizontal beam deflector to Braun's tube in 1899, allowing the beam to be moved about vertically and horizontally, improving Braun's tube and replacing the small rotating mirror that Braun used to expand the beam's movements into a second dimension; instead of scanning a line, the tube could now produce two-dimensional figures.

The oscilloscope made it possible to study current and voltage waveforms, and their sums, and Zenneck was excited by the possibilities it held. "It was just what I had long wanted," he wrote, "an instrument with which one could *see* what was happening in electrical circuits. In the days that followed I made it a sort of game to find as many applications as I could."[4] The oscilloscope would become a staple in the laboratories of the electronics industries as its uses grew in number. John F. Rider's book *The Cathode-ray Tube at Work* (1935) describes numerous types of CRTs and their applications, including circuit testing, the alignment of tuned circuits, transmitter adjustments, frequency analysis, the checking of audio-frequency amplifiers, and many other practical applications.[5]

The oscilloscope was not the first instance of the projection of a live, moving image; camera obscuras from antiquity onward had done so for centuries. The oscilloscope's image, abstract though it was, was the first *interactive* moving image, which could be adjusted and changed as it was occurring. While oscilloscopes could be purchased by the public, the first CRTs in people's homes were also interactive ones, in the form of "cat's eye tubes" or "magic eye tubes", as they were called. Invented in 1932 by Allen B. Dumont and commercially released in 1935, it was a small tube found in radios which indicated signal strength. The tube was adjustable, with green light filling in more and more of the circular screen as the signal grew stronger, allowing users to have a visual, interactive image of how well they were bringing in a radio signal. Thus, these radio "magic eye tubes" were often the public's first encounter with an interactive screen, whose abstract imaging aided in the accurate tuning of radio signals. These CRTs would remain in use in radios into the 1950s, and were also used in some tape recorders to indicate the recording level. They were finally replaced by semiconductor circuitry and optoelectronic displays.

The idea of an interactive moving image would become the basis of computer graphics, video games, the Internet and World Wide Web, tablet computing, and all other devices with interactive imagery. The oscilloscope, then, changed the role of imagery and introduced the potential for interaction, and it was the CRT that made this possible.

More CRTs for the Home: Television

> "Well gentlemen, you have now invented the biggest time waster of all time. Use it well."
>
> —*Isaac Shoenberg, following the demonstration of electronic television in 1934*[6]

Before its interactivity could be fully explored, the CRT had to be improved as an imaging device. The fact that the electron beam could now be used to quickly scan a two-dimensional area made it of interest to inventors working with scanning technologies. The concept of scanning had been around since 1880, when English inventor Sheldon Bidwell had built a "scanning telephotograph" that could transduce an image into an electrical signal, send it over a telegraph, and have it printed back into an image at the other end. Other inventors made similar inventions, for example, Alexander Graham Bell's photophone, also of 1880, or Elisha Gray's telautograph of 1888, that could send a person's signature over the telegraph. These machines, however, could not send images fast enough to be of use to moving image inventors; most of them had their hopes on the spinning disk patented by Paul Nipkow in 1884. Though Nipkow himself never built a system using his disk, it seemed the best, and perhaps the only, way to quickly scan and convert images to signals, along with the use of selenium photocells for transduction. Inventors in several countries competed with each other to try to develop a system for the transmission of images, but solutions to the problems involved, which included transduction and the synchronization of the machines sending and receiving, kept eluding everyone.

Finally, in 1906, German scientists Max Dieckmann and Gustav Glage considered using a CRT for the transmission of line drawings and hand-written materials. They attached a pencil to sliders on the x-axis and y-axis of the tube, which in turn controlled the strength of the electromagnets deflecting the beam. This meant that the motion of the pencil would be duplicated by the motion of the beam, which could be used to draw on the tube's small screen. In 1907, Russian inventor Boris Rosing took the idea one step further, adding a camera and rotating mirrors that scanned an image onto a photocell, and another deflector that would vary the brightness of the beam, so that it would lighten or darken the phosphors on the screen accordingly, resulting in a small, blurry grayscale image.[7]

The problem of transducing and sending the images still remained, and although some inventors would develop electromechanical television systems using Nipkow disks, their limitations soon became apparent. It was in 1908 that a British electrical engineer, Alan Archibald Campbell Swinton, suggested using two CRTs, one for sending and one for receiving, and simply synchronizing the deflection circuits of both tubes. Thus, it was the transduction of the image into a signal, not its display, which posed the most difficult challenge. Although Swinton did not have the answer, inventors began looking for a way to develop a CRT camera.

Rosing perfected his display system in 1911, patented a CRT with magnetic deflection coils around it, and also achieved the first distant transmission of images, though it used a mechanical device as a transmitter. And mechanical television technology in general, the ideas of which had been around since the invention of the Nipkow disk, was making headway. In 1923, British inventor John Logie Baird achieved the first instantaneous moving television image; he gave a public demonstration of his mechanical television system at Selfridge's department store in London in 1925, demonstrated a color television system in 1927, and achieved the first transatlantic television transmission in 1928. From 1929 onward, the BBC did broadcasts using Baird's 30-line television image, and by 1936, they had raised the standard to 240-line transmissions. But Baird's system's use of an intermediate film process, requiring images to be shot on film, developed, and then scanned, and the immobility of its cameras, doomed the system when it was compared to the all-electronic television systems that were continuously improving, and in 1937 the BBC's mechanical television broadcasts ceased.

Electronic television research, using CRTs, continued into the 1920s, alongside work on electromechanical systems. Rosing's assistant, Vladimir Zworykin, worked in the Russian Signal Corps and for Russian Marconi, and then left Russia for the United States in 1918. He worked at Westinghouse on electronic television systems, and returned to the problem of using CRTs as both transmitters and receivers. One of the main obstacles was the transduction of light (from the image) into electrons (from which a signal could be made). With no means of storing the light or the charge from images, Zworykin worked at improving the output of photoelectric cells, which convert light to electricity.

The first CRTs to capture images were image dissectors, for which Philo Farnsworth in the United States and Max Dieckmann and his student Rudolf Hell in Germany applied for independent patents in 1927. Image dissectors focus an image onto a photosensitive plate, which emits negatively charged photoelectrons; the plate is then manipulated so as to scan it several times a second across a tiny aperture which receives the negatively charged particles in a stream, turning the image into a signal. But without a stored charge, use of the dissector was limited, since most of the charged particles were lost and only a tiny amount were captured, meaning that the dissector would only work with extremely strong illumination. Farnsworth attempted to solve the problem with an "electron multiplier" that amplified the signal, but it was not reliable enough to be used.

Along with the refinement of CRTs, stored-charge technology was developing, with patents from Hungarian inventor Kálmán Tihanyi, Americans Harold McCreary and Farnsworth, and Zworykin's team. Zworykin had demonstrated a working camera tube in the fall of 1925, but the small, fuzzy and distorted image failed to impress Westinghouse executives.[8] Farnsworth transmitted silhouettes and then images with his system in 1928, even concluding that 30 frames a second would work well with the American 60-Hz power standard, and that a 400-line image would be competitive with motion pictures. But the shortcomings of Farnsworth's system kept it from being able to outdo the experiments with mechanical television of the day; by 1929, RCA was doing limited television broadcasts with a 60-line, 20 frames-per-second system that reached around 200 hobbyists with receivers.[9]

By that time, Zworykin was leading the team working on electronic television at RCA, and through the use of a number of innovations and patents purchased by RCA from their inventors, his team produced an improved CRT receiver, which became known as the Kinescope. On July 15, 1930, RCA held a contest between its two television development groups, and Zworykin's Kinescope was able to demonstrate a larger, brighter, and higher-resolution picture than any Nipkow-disk-based system, signaling the end for mechanical television research and ending its funding.[10] Within a few months, all RCA television research was under Zworykin, and its budget was no longer split between the two approaches.

Zworykin continued work on a CRT-based camera tube, and in 1931 RCA filed its first patent for what came to be known as the iconoscope. They had finally come up with a design that used stored charges and in which the scanning beam and the light from the image were on the same side of the photosensitive surface. Two years of further refinements later, the iconoscope was presented to the public in June 1933, and the first fully all-electronic television system was complete. Television also introduced raster scanning, in which the entire screen is scanned for every image, as opposed to the imaging done by oscilloscopes and later by vector scanning, where most of the screen is not scanned, and only the lines appearing on the screen are drawn, one line at a time.

Over the next few years, further refinements to CRTs, used by both cameras and displays, would increase the quality of video transmissions. In 1933, RCA produced a 240-line image at 24 frames per second, in 1934 this increased to a 343-line interlaced image at 30 frames per second, and by 1937 their television was one of 441 lines and 30 frames per second. RCA sought approval from the Federal Communications Commission (FCC) for a standard. In 1936, the Radio Manufacturers Association (RMA) proposed a standard of 440 to 450 lines, with an image with the same aspect ratio as film, 4:3. At the end of 1937, the FCC accepted this as a provisional standard. It would not be until 1941 that a more permanent standard would be introduced, when the National Television Systems Committee (NTSC) would send the FCC its recommendation for the standard of 525 lines, 30 interlaced frames, and FM sound, over a 6 MHz bandwidth. Of the 525 lines, 483 would make up the visual image, with the remaining lines used for vertical synchronization and retrace. To check if sets had their imaging set correctly, test patterns were shown on-screen at the end of the broadcast day (see Figure 9.1).

The NTSC standard would remain in place into the 21st century, and after a brief disruption during World War II, the television industry grew rapidly during the late 1940s and early 1950s. CRT technology would continue to advance, as inventors struggled to solve the problems involved with color television. Baird had demonstrated another system in 1940, which used two electron guns and a screen with cyan and magenta phosphors side by side; each gun hit a different set of phosphors to produce a color image. Mechanical wheels made a brief reappearance with red, green, and blue colored gels that spun in front of the screen image timed with corresponding black and white images (similar to the thinking behind the three-strip Technicolor process, which passed a color image through red, green, and blue filters to three black and white filmstrips). The FCC held color television demonstrations in 1948 to compare systems, forming the Joint Technical Advisory Committee (JTAC) to study them. A color system would have to fit its transmission in the same 6 MHz bandwidth, and have to be compatible with existing monochrome televisions. Some considered methods involving combining images from three CRTs, but these were impractical. While the deliberations went on, CBS used a mechanical color system for some of its broadcasts from 1950 to 1953.

The system finally accepted as the solution involved Alfred Schroeder's CRT which had three electron guns placed close together at the far end of the tube, one for each of the red, green, and blue signals. All three beams were deflected together by a single magnetic deflection system, and would scan a shadow mask on the inside of the CRT's screen. The shadow mask, an idea by German inventor Werner Flechsig, only allowed the electron beams to pass through certain precisely-aligned holes, after which they would hit the phosphors on the inside of the CRT's screen. The phosphors were arranged in triads of red, green, and blue,

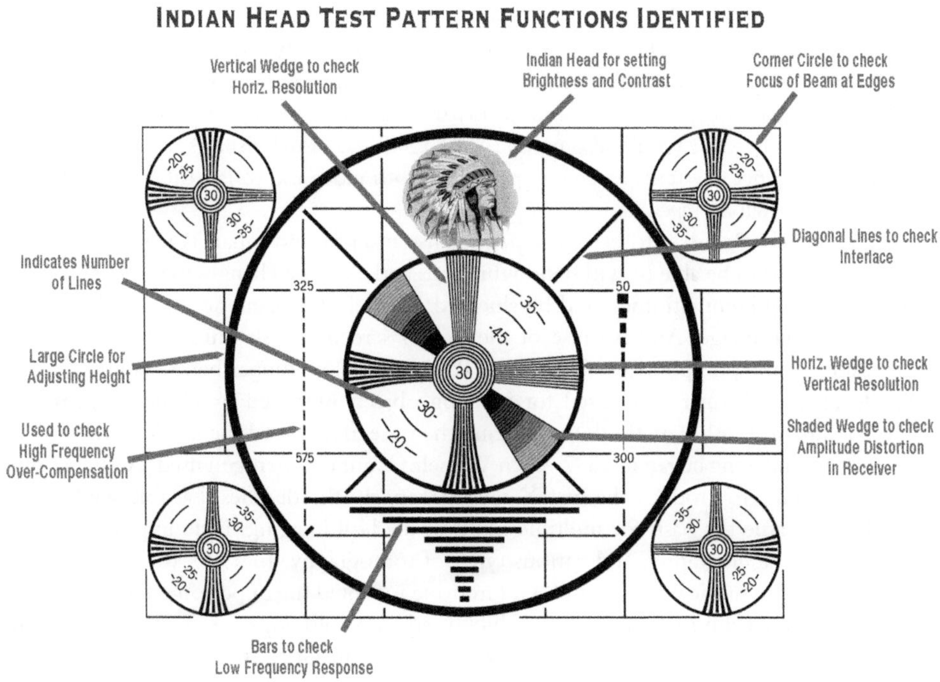

Figure 9.1 An explanation of the various features found on one of the Indian head test patterns used on early television. Public domain from Wikimedia Commons.

and because the three beams passed through the holes at slightly different angles, each hit only one color, and the varying intensities combined to produce a color image.

The CRT technology used in television sets would continue to improve, as would those used in camera tubes; for example, the image orthicon tubes developed in the 1940s that were far more light-sensitive and could even record by candlelight, or the vidicon tubes developed in the 1950s that NASA would use for remote sensing on unmanned space probes. Larger tubes would be made, as televisions offered larger images, as well as smaller tubes, for tiny televisions like the 1959 Philco Safari II-2010, a portable, battery-powered transistor TV set.

As television technology began to spread to the mainstream public after the establishment of the NTSC standard in 1941, the CRT was making its way into other arenas, including new uses in medicine and in the military and World War II.

CRTs for Information Display: Military and Medical Imaging

The United States military had been experimenting with pilotless aircraft after World War I, and began using radio-controlled planes during the 1930s. In 1941 they installed a CRT, an RCA camera tube, in a drone and a CRT receiver in the control aircraft from which the drone would be controlled. The camera in the drone would send an image to the receiver, allowing the drone to be steered and directed remotely. Eventually, the operator would be able to remain on the ground instead of in a plane following the drone, and in the early 1940s, the military in both America and Germany were experimenting with building cameras into guided missiles that sent visual feedback to operators on the ground.

Radar displays first used oscilloscopes, and the earliest displays from around 1940, known as A-scopes, only showed the range to targets instead of the direction. Later displays produced a two-dimensional image, including the plan position indicator (PPI) type of radar, in which the radar sweeps around continuously from a center point, refreshing the screen as it goes. Eventually sonar would also come to use CRTs as visual displays, in much the way that radar had. Since the term *oscilloscope* preceded both radar and sonar, their visual displays also became known as *scopes*.

New types of CRTs had to be developed for use in radar and sonar. Radar systems built into airplanes had to be able to withstand vibrations and extreme changes in temperature and air pressure at different altitudes, and be shielded from electromagnetic interference which could distort the image. And because of the slower screen refresh time in both radar and sonar scanning, the phosphors used inside a CRT had to glow longer before fading.

While radar and sonar were used for imaging objects obscured by clouds, fog, murky water, and distance, other methods for imaging the unseen involved the inside of the human body. Medical imaging began in 1897 when Wilhelm Conrad Roentgen used a CRT known as a Crookes tube to produce the first X-ray imagery. In the decades that followed, X-ray technology became the basis for multiple kinds of medical imaging, including fluoroscopy, mammography, angiography, and various types of tomography. In 1940, medical sonography began when acoustical physicist Floyd Firestone invented the Supersonic Reflectoscope, the first ultrasonic echo imaging device, which used an oscilloscope for its imaging; and the first use of ultrasound in a clinical setting was in 1956, in Glasgow, Scotland.[11] Later on, television technology replaced the oscilloscope, in magnetic resonance imaging, invented in 1971 by Paul C. Lauterbur, as well as in positron emission tomography, which appeared in the 1990s. CRTs have also been adapted to show feeds from microscopes (such as in the scanning electron microscope), telescopes, spectroscopes, laparoscopes, tachistoscopes, and other forms of optical-based imaging and computer imaging, including televised surgery, which began in 1947. Not only did the CRT make such imaging techniques possible, but it played a role in the development of the computer itself, and was even used as one of the earliest forms of electronic computer memory.

CRTs as Computer Memory: The First Random-Access Digital Storage Device

Electron beams hit the phosphors on the inside of the CRT's screen, causing them to glow, but if their energy level is higher, electrons are released out of the phosphor, a phenomenon known as *secondary emission*. The charged spots that appeared on the screen can then be detected by a nearby metal plate. Spots, and their absences, were read as zeroes and ones, producing a form of binary storage which could also be read by the electron beam. In 1946 and 1947, Professor Freddie C. Williams and his student Tom Kilburn applied for British patents for the Williams tube (sometimes also called the Williams-Kilburn tube), a type of CRT used for binary data storage, which was the first random-access digital storage device (see Figure 9.2). The number of spots that could be displayed on the screen depended on the size of the spots and the screen itself. In November 1946, Williams was able to store a single bit, and with the help of Tom Kilburn, the tube's capacity was increased so that by November of 1947 a single Williams tube could hold as much as 2048 bits of information (that is, 256 bytes), arranged in a 64-by-32 bit array.

The Williams tube was used in the Manchester Small-Scale Experimental Machine (SSEM), which ran its first program on June 21, 1948, and became the first stored-program computer.

Figure 9.2 A Williams tube, with a detection plate that closed over its screen (left), and the on-screen memory pattern of dots that represented bits (right). Images from the University of Manchester.

Once it was successful, it was developed into the Manchester Mark I, which became operational in April 1949, and which became the prototype for the Ferranti Mark I, the first commercially available general-purpose electronic computer, released in February of 1951. Williams tubes were also used in the IAS machine of 1951; the MANIAC (Mathematical Analyzer, Numerical Integrator, and Computer; or Mathematical Analyzer, Numerator, Integrator, and Computer) computer of 1952; and in 1953, the UNIVAC 1103 and the IBM 701, the company's first commercial scientific computer, which used 72 Williams tubes.

Williams tubes were relatively inexpensive, small, and faster than mercury delay line memory (which was also not random access), but they needed continual refreshing to keep their data, and they tended to be temperamental and not always reliable. For the Whirlwind I computer, which became operational in 1951, the engineers considered Williams tubes but decided to go with another form of CRT-based memory developed by the MIT Radiation Laboratory, a dual-gun tube that did not require refresh cycles. When the U.S. Navy and the Massachusetts Institute of Technology wrote their 1952 report on the Whirlwind I computer, they identified the tube memory as "the most important factor affecting reliability of the Whirlwind I system".[12] This referred, however, to the limitations of the system, and the Whirlwind had to employ a system that checked for hardware problems before they affected accuracy. Research on magnetic-core memory began in the 1940s, and by summer of 1953, the Whirlwind I was the first computer to have it installed, replacing its tube memory. Not only was magnetic-core memory more reliable, it was also faster: according to a Project Summary Report of 1953, core access time was only nine microseconds, compared to 25 microseconds for tube access.[13] From that time onward, Williams tubes and other forms of tube memory became obsolete as magnetic-core memory grew to become the main type of computer memory used, until it was replaced by semiconductor memory and integrated circuits around 1974.

Although the Whirlwind I had its tube memory replaced, there was another CRT that remained in use by the computer, and in a much more prominent place. It was the first computer to use a CRT as a data display, and the first appearance of computer graphics.

CRTs for Interaction: Computers, Art, Video Games, and Beyond

To the general public, the CRT was a technology that was used passively as the receiver of television broadcasts; its interactive uses, present since the first oscilloscopes, took longer to gain public notice. Computer graphics would bring new attention to the CRT as well as to the computer itself. The Whirlwind I used a large oscilloscope for its output, on which it could display text and show the trajectories of rockets and bouncing balls. It could also pick up planes by radar and display their positions on-screen. The Whirlwind I became the prototype of the SAGE (Semi-Automatic Ground Environment) computers used in a nation-wide ground-based air defense system, each of which used dozens of round, 48-inch vector-based CRTs to display information. A single SAGE computer could run 50 monitors and track 400 airplanes.[14] The CRTs used by SAGE computers included "Typotron" tubes capable of displaying 25,000 text characters per second, and they were also interactive; when an operator aimed a light gun at a spot on the screen, the computer would perform a particular operation.[15] For example, the light gun could be fired at a target on the screen, telling the computer to track the target.

The first light pen was developed for the Whirlwind I in 1955, and allowed a user to interact directly on the CRT's screen. The light pen senses the changing brightness in the pixels of a CRT's screen, and the exact timing of the changes is compared by the computer to the CRT scanning, which can match the times and determine where the light pen is pointing on the screen. Thus, a light pen can be used to draw vector imagery directly on the screen, and any CRT can be used since it is the computer that detects the input and matches the timing. A light pen was also used in Ivan Sutherland's program *Sketchpad* of 1963, which allowed the user to create and manipulate three-dimensional graphics on-screen. The program is credited for advancing not only human–computer interaction, but also mathematical visualization and three-dimensional computer graphics, which began to be explored in the 1960s.

The interactive possibilities for creating and manipulating computer graphics were not only explored by science, industry, and the military, but also by artists. In 1947, *Popular Science Monthly* featured a one-page article, "Even Necktie Designers Can Use Electrons" which read:

> Fabric designers in search of inspiration may well turn to a television receiver, General Electric engineers suggest.
>
> When they test cathode-ray tubes for modern sets by applying varying voltages, the engineers have found that the flying electron beam often sketches traceries of striking beauty upon the luminescent viewing screen. Ranging from random squiggles to intricate geometrical lacework, the designs offer fascinating possibilities for dresses, scarves, ties, and draperies.
>
> Small, portable test sets, currently available, are suited to apply this new scientific aid to art. Simply by twirling a dial, a seeker after new patterns can conjure up an infinite variety. Through the magic of electronics, the marvelous symmetry of motions resulting from the complex interplay of natural forces becomes visible to human eyes. Captured by photography, . . . the fleeting images may be preserved for subsequent transfer to fabric.[16]

Intrigued by the article, mathematician Ben F. Laposky took on the challenge, and in 1950 began making a series of photographs of oscilloscope patterns, which he called "Oscillons". His imagery was reproduced in over 250 books and magazines, and in a one-man 50-photograph exhibition, "Electronic Abstractions" at the Sanford Museum in Cherokee, Iowa, in 1952. The exhibition travelled the United States and France, introducing the world to a new kind of abstract art.

Soon after Laposky's work appeared, animators began using oscilloscopes to create abstract animation. Oscilloscope-based imagery was used by Norman McLaren in *Around and Around* (1950); by Hy Hirsch in *Divertissement Rococo* (1951), *Eneri* (1953), and *Come Closer* (1953); and by Mary Ellen Bute in *Abstronic* (1954) and *Mood Contrasts* (1956). Oscilloscope art combining sound and image continues to be explored today by artists including Jerobeam Fenderson, Vincent Adoxo, and Steve Bliss.

CRTs, in the form of television sets, also began appearing in artwork in the late 1950s. In 1958, Wolf Vostell's work *Transmigration* combined a slashed canvas with a flickering television screen. In 1963, Nam Jun Paik's first solo show, *Exposition of Electronic Music – Electronic Television*, at Galerie Parnass in Wuppertal, West Germany, featured 12 television sets whose images were distorted by magnets. Paik also was among the first users of the Sony Portapak, the first portable video recording equipment available to consumers which appeared in autumn of 1965. The Portapak allowed images to be created on CRTs, and was the beginning of video art. Video art includes videos made by artists, experimental video practices in which video imagery is manipulated (as with the Paik-Abe Video Synthesizer and others), and video art installations. For the latter, video's real-time recording and playback capabilities, as closed-circuit television, allows the audience and the CRT to appear on-screen live recursively, sometimes with startling effects, as in Bruce Naumann's *Video Corridor* (1968–1970).

While the CRTs found in interactive video art were limited mainly to the museum, another form of interactive art on CRTs was developing during the 1960s and became even more mainstream. The first patent for an electronic interactive game was United States Patent #2,455,992, "Cathode-Ray Tube Amusement Device", which was filed on January 25, 1947, and issued on December 14, 1948, to Thomas T. Goldsmith, Jr. and Estle Ray Mann. It was for an analog device that controlled the electronic beam of a CRT in an interactive fashion (it did not involve a video signal, however, and what it describes could therefore arguably be denied the status of a "video" game). As the patent describes it:

> In carrying out the invention a cathode-ray tube is used upon the face of which the trace of the ray or electron beam can be seen. One or more targets, such as pictures of airplanes, for example, are placed upon the face of the tube and controls are available to the player so that he can manipulate the trace or position of the beam which is automatically caused to move across the face of the tube. This movement of the beam may be periodic and its repetition rate may be varied. Its path is preferably caused to depart from a straight line so as to require an increased amount of skill and care for success in playing the game.[17]

The patent goes on to describe the circuitry in detail, referencing diagrams included in the patent. Apparently, the patent's authors did not build or exploit their invention, nor did it appear to be an obstacle to later video game developments. In 1951, Ralph H. Baer independently had the idea of playing games on a television set, and pursued the idea, producing a series of prototypes in the late 1960s. Games were also being programmed in university computer labs, such as William Higinbotham's *Tennis for Two* (1958) which used an oscilloscope for its screen, and *Spacewar!* (1962) written by Steve Russell, J. Martin Graetz, and Wayne Wiitanen for Digital Equipment Corporation's PDP-1 computer and its CRT display. *Spacewar!* inspired inventors to come up with a game that could be offered to the public. In September of 1971, Bill Pitts and Hugh Tuck used a PDP-11 computer set up in Stanford University's Tresidder Student Union to debut *Galaxy Game*, the first coin-operated video game. Due to the technology involved, the game was not commercially feasible, despite its popularity. Another *Spacewar!*-inspired game appeared in arcades two months later, *Computer Space*, created by Nolan Bushnell and Ted Dabney. *Computer Space* was thought to be too complicated to be operated by bar patrons, and also failed to find commercial success. In 1972, Baer went on to create the Magnavox Odyssey, the first home console video game system, and the same year, Bushnell and Dabney's arcade video game *PONG* (itself inspired by the Magnavox Odyssey) was a hit, and both the arcade video game and home video game industries were off to an explosive start.

Early arcade games used monochrome raster CRTs, although some vector-based games were produced from 1977 to 1985, and after 1980 virtually all games used color monitors. Many arcade games used the same kind of CRTs used for televisions, with a 4:3 aspect ratio, but positioned their screens in game cabinets with a "portrait" orientation, in which the screen is taller than it is wide, rather than the "landscape" orientation used by broadcast television. The presence of CRTs in people's homes, as television sets, made it less expensive for video games to enter the home, since no new display device was needed; in this sense, the CRT's ubiquity helped home video games to spread quickly, and their availability also helped arcade games to be produced in greater numbers.

Another use for CRTs that arose in the home during the 1970s was that of the home computer monitor. Like home video game consoles, many home computers used CRTs as their displays, and most even doubled as game-playing machines, since games were one of the reasons to buy a home computer. In this way, the CRT was crucial to the introduction of computer technology to the public, both in the arcade and in the home, and its entrenched presence made this entry seem natural, as consoles and computers offered new ways to use television sets (along with other new devices, such as video cassette recorders (VCRs) and later, laserdisc players).

Outside the home, CRTs continued their expanding role in art, science, industry, and government, and found new uses during the 1970s and 1980s, in closed-circuit television (CCTV) and surveillance video systems (which also became available for the home), kiosks in shopping malls, and automated teller machines (ATMs) used by banks. While closed-circuit surveillance video was used in Germany as early as 1942, and 1949 in the United States, home security systems were patented in 1969, and by the 1970s, video cameras had become smaller and inexpensive enough for their use in surveillance to become more widespread, especially with the appearance of the VCR in 1976, which provided a way to record video feeds. CRT touchscreen technology developed in the 1960s and 1970s, and in 1983 Hewlett-Packard released the HP-150, a home computer with a touchscreen CRT monitor. During the 1980s, touchscreen CRTs were used in ATMs, and even in cars; General Motors' Buick Riviera (1985–1989) and Buick Reatta (1988–1989) came with an "Electronic Control Center" (ECC) which featured a small touchscreen CRT that could be used to control the car's features; but it was not popular with consumers.

With all of the CRT's many uses, the CRT industry was booming in the 1980s, and in 1988 Sony produced the world's largest color CRT, with a 45-inch diagonal span; Mitsubishi experimented with 61-inch CRTs, but they were not commercially viable.[18] Even larger images could be had, with projection CRTs, which used an intense beam to produce an extremely bright image on a small, concave CRT screen housed in a chamber behind a focusing lens which projected the image from the CRT. With new technologies such as the Kodak Ektalite screen, these CRTs could project images as large as 72 inches.

By the early 1990s, there were seven producers of color television CRTs in the United States, and 30 producers of other kinds of CRTs.[19] No monochrome screens were being produced in the U.S., nor were any color screens smaller than 19 inches (measured diagonally) produced.[20] According to a 1995 Industry and Trade Summary, the period of 1989 to 1993 saw U.S. consumption of CRTs increase by 17% (from $2.1 billion to $2.4 billion); U.S. production of CRTs increased by 33% (from $1.8 billion to $2.4 billion); U.S. imports of CRTs increased by 24% (from $664 million to $822 million); and U.S. exports increased by an annual rate of almost 22% (from $352 million to $769 million). By 1994, CRT manufacturers were working at capacity for large-screen production, limited only by the glass supply.[21]

The same report reveals that in 1992, Japan was the world's greatest producer of CRTs, producing roughly 53% of world output, and the second largest was South Korea at 20%, while the U.S. produced 17%, Germany 8%, and Italy 6%.[22] The demand for CRTs in developing countries was also on the rise, as television, home computers, video games, and other technologies spread around the world during the 1980s and 1990s. The worldwide market kept growing throughout the 1990s; Fuji Chimera Research reported the CRT monitor market to be 57.8 million units in 1995, 67.1 million in 1996, and Stanford Resources found this increased to 84.2 million units in 1997.[23] In 2000, worldwide sales of CRTs rose to 90 million units, and in 2005 peaked at around 130 million units; but sales fell back to about 90 million the following year.[24] The year 2005 would remain the pinnacle of CRT production, for another technology was growing and about to successfully challenge the CRT and end its decades as the world's dominant screen technology.

LCD Screen Technology Eclipses the CRT

Although flat-screen technology would eventually replace the CRT, it also traces its origins to the CRT. The idea behind flat-screen technology had been around since General Electric had proposed such technology for its radar monitors, but the company did not pursue the idea. And when the first flat-screen television finally did go into production, appearing in 1958, it was a new type of CRT, also called the Aiken tube. During the mid-1950s, electrical engineer William Ross Aiken had invented a CRT with a flatter, more compact shape, working out the problems involved, such as the deflecting and focusing of the beam. Aiken built a prototype and took it to his former employer, Kaiser Aircraft and Electronics, who planned to use it in military aircraft cockpits. As an article in the January 1958 issue of *Popular Mechanics* described it,

> For the TV viewer at home, the new picture tube may result in new designs for sets, with screens mounted in any wall or hung like picture frames. The picture tube, only 2⅝ inches thick, is made of two rectangular pieces of plate glass with about an inch of space between them. The edges are sealed with powdered-glass solder to hold the vacuum. The surface of the thin tube is the equivalent of a 21-inch conventional screen. In the thin tube, the electron beam is injected at the bottom of one side. Deflection plates along the bottom edge bend the beam upward between the front and back glass walls. The inside of the front wall is coated with a new transparent phosphor which is said to improve the contrast. The thin TV tube is also reported to have sharper focusing properties. A new method of printing electrode elements on the inside surfaces of the glass eliminates the need for assembled metal parts. Printed circuits are used in the tube controls. The thin tube will replace many of the instruments needed for the blind flying of an airplane and can be operated by a small electronic computer.[25]

While some tubes were produced for military use, troubles with production, financing, and patent protection kept the Aiken tube from successfully being used in commercial televisions.[26] That same year, 1958, the Philco Predicta television also appeared, which was attached to a pole extending from its cabinet, allowing it to swivel. The Predicta's CRT was slimmer than other models of its day, but not nearly as thin as the Aiken tube. Despite its aesthetically daring design, the Predicta was a commercial failure and even drove Philco into bankruptcy.

The plasma display was the next attempt at a flat-screen technology, appearing in 1964. These displays, however, were much more expensive than CRTs, and monochrome, displaying pixels of a single color on a black background. Some were used as computer monitors, or as wall-mounted displays as in stock exchanges. Finally in 1992, Fujitsu released a 21-inch full-color plasma display for consumers, followed by a 42-inch display in 1995. In 1997, Philips also had a 42-inch full-color plasma display on market, which sold for $14,999, which included in-home installation.[27]

During the 1980s, liquid-crystal display technology was developing, with the first commercial LCD-based television being offered by Sharp Corporation in 1988. Still, LCD technology lagged behind plasma technology, and continued to do so, into the 1990s. LCD displays had a greater range of applications and sizes, which gave them an advantage over plasma displays when it came to economies of scale. The slow development of high-definition television standards also prolonged the time the two technologies had to compete.

After 2000, when HDTV was finally becoming a reality, large-screen sales picked up and it soon became apparent that the changeover from NTSC television to HDTV would occur in the near future. By 2006, large-screen television prices had been dropping as much almost 25% every year for several years, and plasma screens had come down considerably (for example, Panasonic's sold for $3,500 and Dell's for $2,600). And prices were still falling. According to a *New York Times* article of the day,

> All this spells good news for anyone thinking about upgrading from the old cathode-ray TV to screens that are 40 inches or larger in the three most popular formats. Full-featured plasma TV's [*sic*] with 50-inch screens that sold for $20,000 five years ago could edge close to $4,000 this season. A liquid-crystal-display version and a rear-projection TV with a digital light-processing chip will be considerably less, closing in on $1,800.[28]

In late 2006, the prices of LCD screens became competitive with those of plasma screens.[29] And in the following year, sales of LCD screens would overtake those of plasma screens.

According to DisplaySearch (owned by financial services company IHS Markit), worldwide shipments of LCD-based TVs finally overtook those of CRT-based TVs at the end of 2007, a move that surprised many. In the first quarter of 2006, CRTs had almost an 80% share of the market, while LCDs had less than 20%. By the first quarter of 2007, CRTs had only around 60% share, while LCDs had climbed to just over 30%. And by the end of the fourth quarter of 2007, LCDs surpassed CRTs, with LCDs reaching 47% market share while CRTs dropped to 46% share.[30]

LCD flat-screen technology had a number of advantages over CRTs: LCD screens are truly flat and have sharper images than CRTs; LCD screens can be much larger in size and have no glass and thus no glare; LCD screens take up less space (they are thin enough that they can even be hung on a wall like a painting); they weigh less than CRTs; and LCD screens are more energy efficient. Yet, LCD screens have some disadvantages: they sometimes have "dead pixels", pixels within the image that no longer function (remaining dark and unlit) whereas CRTs do not; the viewing angle might be more limited on an LCD screen; CRTs can produce true black, whereas LCDs cannot; CRTs can be used at any resolution up to the native one supported, but LCDs lose picture quality at resolutions other than their native resolution; and LCD screens are currently limited to 8-bit color whereas CRTs can reproduce colors equivalent to 32-bit color.[31] Despite these problems (which might be reduced or even eliminated as LCD technology advances), the LCD screen's overall quality, along with all of its other advantages, allowed it to surpass the CRT, once LCD screens were mass-produced enough for their prices to be affordable and competitive with CRTs.

Although they were no longer the dominant imaging technology, CRTs were still well-entrenched as late as 2008, and their reliability and long lives might have suggested a gradual decline and slow fade-out; but 2009 saw a big change in the world of television that sped up their demise and increased their obsolescence.

The Decline (and Death?) of the CRT

On June 12, 2009, high-power, over-the-air analog television broadcasting in the United States came to an end. The changeover date had originally been February 17, 2009, but when it became apparent that millions of people were not ready, Congress passed the DTV Delay Act, which changed the date to June. Other countries had already changed over to digital-only television broadcasting, beginning with Luxembourg in 2006, and even in the U.S., digital signals had been in use since July 23, 1996, when WRAL-TV in Raleigh, North Carolina, made the country's first digital television broadcast.[32] Throughout the 2000s and 2010s, most of the other countries of the world would also undergo the changeover from analog to digital television, though some have not planned to do so until into the early 2020s (for example, Brazil, which has scheduled the changeover for 2023).

The change to digital broadcasting did not force the immediate scrapping of all older televisions, but NTSC CRTs would now need a digital-to-analog converter box in order to receive broadcasting. The National Telecommunications and Information Administration (NTIA) in the U.S. Department of Commerce gave out coupons good for up to two $40 converter boxes per household, to ease the transition and allow families to continue using their analog television sets. Both the original changeover date and its delay were the subject of controversy, and some broadcasters began to broadcast simultaneous digital and analog signals, while others refused to pay the extra cost of simulcast and shut down their analog broadcasting on the original end date.

The makers of CRTs also tried to change with the times, with high-definition CRTs capable of receiving broadcast HDTV signals; but some sets weighed over 200 pounds, and were still not as flat as LCD screens, despite deflecting the electron beam by as much as 125 degrees to make the cabinet flatter. In the end, flat-panel display technology developed beyond what CRTs were capable of doing, and with advantages that outweighed whatever objections consumers might have had.

Although some companies tried making high-definition CRTs, most would soon cease producing CRTs altogether. In 2001, LG Philips Displays Holding ceased production of CRTs at their plant in Dapon, Taiwan, and on May 23, 2003, the company announced that it was closing two of its plants where CRTs were manufactured, in Newport, Wales, and Southport, England.[33] Mitsubishi also closed a CRT plant in 2003, in Mexicali, Mexico, and the following year, Victor Company of Japan closed its CRT plant in Oyama, Tochigi Prefecture.[34] In 2006, Thomson Consumer Electronics closed its Picture Tube Development

Group in Lancaster, Pennsylvania, and that same year another CRT plant, owned by Matsushita and Toshiba, closed in Troy, Ohio. Sony had ended CRT production in Japan in 2004 but continued producing them in other facilities in Singapore and Malaysia; then, on March 3, 2008, Sony announced that they would cease all production of CRTs.[35] Sony's plant in Westmoreland, Pennsylvania, was one of those that were closed, and was the last one in the United States, which ended production as of February, 2009.[36] And on September 28, 2012, Cathode Ray Technology in the Netherlands, the last CRT factory in Europe, finally closed.

As consumers switched over from CRTs to LCD and OLED screens, the disposal of old CRTs became a problem, due to the lead and other toxic metals present in the glass of the CRT. The answer seems to be recycling programs that will try to recover and use materials from old CRTs. In England, the Waste Electrical and Electronic Equipment (WEEE) Directive was begun in 2007, which sends old CRTs for reuse or recycling. The old sets are dismantled by hand, and cut in half due to the lead on the back side and phosphors on the screen, which are processed in different ways.[37] In 2012, the recycling company ECS Refining in Santa Clara, California developed a $2 million program for sorting glass from CRTs, in an operation in Stockton, California.[38] Originally, the glass from old CRTs was recycled into materials for new CRTs, but once CRT production ceased, some recyclers started storing the material rather than recycling it, leading to a crisis. According to the 2012 report, "U.S. CRT Glass Management: A Bellwether for Sustainability of Electronics Recycling in the United States", around 660 million pounds of material had already accumulated in U.S. landfills, and the same report estimated that the transport and recycling of these stockpiles would cost between $85.8 million and $349.8 million.[39]

Although in decline, CRTs are still common and numerous, and it is still too soon to say whether the nostalgia for them will see the CRT make a comeback as a luxury item, the way vinyl records have seen renewed interest among collectors. Might those who never experienced the phosphorescent glow of the gently curving glass screen be intrigued enough to revive the technology? How many of today's CRTs will be preserved for future generations? Will media connoisseurs of the future desire to watch early television programs and play early video games the way they were originally seen and played? Already there are video game players who prefer the CRT's quicker response time over other imaging technologies. A user going by the name ShortFuse posted the following on "The Lag Thread" at Smashboards.com:

> With Melee and Brawl, we already have a standard, and that standard is sub-1ms lag from CRTs. Smash4 will be a bit of a mess unless we standardize exactly what latency we're going to play with. Hopefully, OLED displays will be cheaper and more available by the time Smash4 comes out since they will have sub-1ms lag. If not, then, we might just have to play at 480i/480p CRTs or all agree to play on a certain monitor.[40]

Almost a year later on the same thread, another user, Kadano, verified the differences in lag time with precise data:

> It's not a myth. Prad.de measured CRT vs. LCD display latency with a $50k oscilloscope and these are the results:
>
> CRT: 670 ns = 0.67 μs = 0.00067 ms for a generic higher-end PC monitor
>
> LCD: 3000000-5000000 ns = 3000-5000 μs = 3-5 ms for the fastest ones.[41]

As with so many other technologies, the specific capabilities of the CRT are becoming better appreciated just as the technology has begun to disappear. Though it is difficult to imagine CRTs becoming as rare as mechanical televisions are today, the speed at which the changeover and abandonment of the CRT has taken place might suggest otherwise. But even if the CRT ends up as little more than an anachronistic museum piece, a curiosity from a long-gone era, it will still represent the way moving images were most commonly displayed during the latter half of the 20th century, as well as the start of so many imaging technologies and media, and the beginning of humanity's intimate involvement with the moving image in the home, where its luminescent glow held so many in its sway and enchantment.

Notes

I would like to thank Matt Knutson for pointing out the quotes by Shortfuse and Kadano regarding lag time and the CRT (see endnotes 40 and 41). All web pages were accessible on August 7, 2018.

1 Emil Wiechert (1897), "Ueber das Wesen der Elektricität", Schriften der physikalisch-ökonomischen Gesellschaft zu Köningsberg, Sitzungsberichte 38 3, quoted in Per F. Dahl, *Flash of the Cathode Rays: A History of J. J. Thomson's Electron*, London, England: Institute of Physics Publishing, 1997, page 154.

2 Cutler J. Cleveland and Christopher G. Morris, *Handbook of Energy: Chronologies, Top Ten Lists, and Word Clouds*, Waltham, Massachusetts: Elsevier, 2014, page 689.

3 The anode, to which the electrons jumped from the cathode and which attracted them, would often have a hole in the center so that electrons would pass through, creating a focused beam. Because the anode sped up the electrons on their way from the cathode, the CRT is considered a type of particle accelerator.

4 Jonathan Zenneck, "Zum 50jährrigen Jubiläum der Braunschen Röhre", Berlin, Physikalische Gesellschaft, 1947, as quoted in Friedrich Kurylo and Charles Susskind, *Ferdinand Braun: A Life of the Nobel Prizewinner and Inventor of the Cathode-Ray Oscilloscope*, Cambridge, Massachusetts: The MIT Press, 1981, page 92.

5 John F. Rider, *The Cathode-ray Tube at Work*, New York: John F. Rider, Publisher, 1935.

6 According to www.terramedia.co.uk/quotations/Quotes_S.htm.

7 See Alexander B. Magoun, *Television: The Life Story of a Technology*, Westport, Connecticut: Greenwood Press, 2007, pages 10–11.

8 Ibid., page 22.

9 Ibid., pages 43–44.

10 Ibid., pages 57–58.

11 Tanya Lewis, "5 Fascinating Facts About Fetal Ultrasounds", *Live Science*, May 16, 2013, available at www.livescience.com/32071-history-of-fetal-ultrasound.html.

12 M.I.T. Project Whirlwind, Summary Report #31, 1952, page 6. Institute, Archives and Special Collections, M.I.T. Libraries, Cambridge, MA. Quoted in *The Computer Museum Report*, Winter 1983, page 13, available online at http://gordonbell.azurewebsites.net/tcmwebpage/exhibits/whirlwindwinter1983.pdf

13 M.I.T. Project Whirlwind, Summary Report #35, 1953, page 33. Institute, Archives and Special Collections, M.I.T. Libraries, Cambridge, MA. Quoted in *The Computer Museum Report*, Winter 1983, page 13, available online at http://gordonbell.azurewebsites.net/tcmwebpage/exhibits/whirlwindwinter1983.pdf.

14 Stan Augarten, *Bit by Bit: An Illustrated History of Computers*, New York: Ticknor & Fields, 1984, page 206.

15 See "SAGE Computer System (developed in the 1950s, operational by 1963)", available at https://redlegagenda.com/2016/06/12/sage-computer-system-developed-in-the-1950s-operational-by-1963/.

16 "Even Necktie Designers Can Use Electrons", *Popular Science Monthly*, December 1947, page 115.

17 United States Patent 2,455,992, "Cathode-Ray Tube Amusement Device", issued to Thomas T. Goldsmith, Jr. and Estle Ray Mann, Application January 25, 1947. Available online at www.google.

com/patents?id=n-NZAAAAEBAJ&printsec=abstract&zoom=4&source=gbs_overview_r&cad=0#v=onepage&q&f=false.

18 About Sony's 45-inch monitor, see Alexander B. Magoun, *Television: The Life Story of a Technology*, Westport, Connecticut: Greenwood Press, 2007, page 173. As for Mitsubshi's 61-inch monitor, according to *Techwalla*,

> Mitsubishi was experimenting with larger CRT models and produced only a few 61-inch CRT TVs. The technology proved to be too unstable and the picture tubes on these TVs did not last long. As a result, they were discontinued quickly. The florescent contents of the CRT picture tube are highly dangerous and manufacturers were not able to make larger models that were safe.

See *Techwalla*, available at www.techwalla.com/articles/the-biggest-crt-televisions.

19 According to John W. Kitzmiller, "Industry and Trade Summary Television Picture Tubes and Other Cathode-Ray Tubes", Office of Industries, U.S. International Trade Commission, 1995, page 1. Available at www.usitc.gov/publications/332/pub2877_0.pdf.

20 Ibid., pages 3–4.

21 Ibid. Consumption, production, and import figures are from page 11, while export figures are from page 12. The reference to 1994 was from page 11.

22 Ibid., pages 6–7.

23 Environmental Protection Agency, *Computer Display Industry and Technology Profile*, December 1998, EPA 744-R-98-005, page 3. Available at www.epa.gov/sites/production/files/2014-01/documents/computer_display_profile.pdf.

24 According to David L. Chandler, "CRTs Going Down the Tubes? Hardly: Surprisingly, Old-style Television Sets and Computer Screens Are Still in Demand—To Make New TV sets", MIT News Office, February 2, 2010, available at http://news.mit.edu/2010/crt-recycle.

25 "Thin Tube Foretells Wall TV and Sky View for Air Pilots", *Popular Mechanics*, Volume 109, Number 1, January 1958, page 104.

26 William Ross Aiken, "History of the Kaiser-Aiken, Thin Cathode-ray Tube", *IEEE Transactions on Electron Devices*, Volume 31, Issue 11 (November 1984), pages 1605–1608.

27 According to "Plasma Display", *Wikipedia*, available at https://en.wikipedia.org/wiki/Plasma_display.

28 See Damon Darlin, "Falling Costs of Big-Screen TV's to Keep Falling", *The New York Times*, August 20, 2006, available at www.nytimes.com/2005/08/20/technology/falling-costs-of-bigscreen-tvs-to-keep-falling.html.

29 Ibid.

30 Darren Murph, "Worldwide LCD TV Shipments Surpass CRTs For First Time Ever", *engadget.com*, February 19, 2008, available at www.engadget.com/2008/02/19/worldwide-lcd-tv-shipments-surpass-crts-for-first-time-ever/; and "DisplaySearch Reports Q4'07 Worldwide LCD TV Shipments Surpass CRTs for First Time; TV Revenues Reach a Record High, Up 10% to $33B", *Businesswire.com*, February 19, 2008, available at www.businesswire.com/news/home/20080219005566/en/DisplaySearch-Reports-Q407-Worldwide-LCD-TV-Shipments.

31 See "CRT vs. LCD", Computer Hope, available at www.computerhope.com/issues/ch001394.htm; and "LCD vs CRT Pros and Cons (A Quick Overview) Version 0.534", *Bootstrike.com*, February 1, 2008, available at http://bootstrike.com/Articles/LCDvsCRT/. Both web pages accessed January 26, 2017.

32 "History of WRAL Digital", WRAL.com, May 9, 1996, updated July 14, 2014, available at www.wral.com/history-of-wral-digital/1069461/.

33 Amy Bennett, "LG Philips to Shut Down CRT Plant in Taiwan", *IT World*, July 5, 2001, available at www.itworld.com/article/2796061/hardware/lg-philips-to-shut-down-crt-plant-in-taiwan.html; and "LG.Philips Displays to idle over 900 As It Closes Two CRT Plants", *EE Times*, May 23, 2003, available at www.eetimes.com/document.asp?doc_id=1136842.

34 "Mitsubishi to Close CRT Plant", *Telecompaper*, June 25, 2003, available at www.telecompaper.com/news/mitsubishi-to-close-crt-plant-391794; and "Victor to Close Tochigi CRT Plant", *The Japan News*, July 4, 2004, available at www.japantimes.co.jp/news/2004/07/04/national/victor-to-close-tochigi-crt-plant/#.WKyZKk0UX1I.

35 "Sony to Stop Making Old-style Cathode Ray Tube TVs", *Marketwatch.com*, March 3, 2008, available at www.marketwatch.com/story/sony-to-stop-making-old-style-cathode-ray-tube-tvs.

36 Christopher MacManus, "Sony Closes Last TV Manufacturing Facility in USA", *Sony Insider*, December 14, 2008, available at www.crtsite.com/page3-2.html.

37 Dean Evans, "What's Happening to All the CRT TVs?", *Techradar.com*, February 6, 2009, available at www.techradar.com/news/television/what-s-happening-to-all-the-crt-tvs-525649.

38 Reed Fujii, "State Proposal to Allow CRT Glass into Landfills Imperils $2M Stockton Plant", *Recordnet.com*, January 22, 2012, available at www.recordnet.com/article/20120122/A_BIZ/201220301.

39 "U.S. CRT Glass Management: A Bellwether for Sustainability of Electronics Recycling in the United States", *Transparent Planet*, pages 31–32, available at http://transparentplanetllc.com/wp-content/uploads/2013/11/CRT-Glass-Management-Report-2-5-12.pdf.

40 ShortFuse, posted on "The Lag Thread", October 18, 2013, available at https://smashboards.com/threads/the-lag-thread.341361/.

41 Kadano, posted on The Lag Thread, October 7, 2014, available at https://smashboards.com/threads/the-lag-thread.341361/. Kadano compares lag time for different CRTs at: https://smashboards.com/threads/work-in-progress-perfect-setups-tv-monitor-console-capture-device.355292/page-7#post-21307864.

10

THE MOVIOLA AND OTHER ANALOG FILM EDITING MACHINES

Lori Landay

It lingers in the corner of the big film classroom, its soothing retro green at the edge of my visual field as I teach. Over a decade ago, the Moviola was still at the front of the room, and I probably moved around it in a way I thought was dramatic, and although I hope not, might have even caressed it once or twice. I confess: I love the Moviola, its funky upright shape, the place of privilege for film reels, and the way it involved the whole body, from eyes, to hands, to feet, and pains in the neck. It reminds us of a time before digital production whooshed away physical objects and skills into intangible code. Today, my students use the Moviola as a prop in their final movie project. Last year they cast it as a dream machine, but it used to be a machine that editors used to make dreams.

Analog editing is in its sunset phase, with the majority of film editing now being done digitally. "Moviola" often refers to any kind of machine for viewing and editing films, but it is a patented name of a specific type of editing device, produced by the Moviola Manufacturing Company, established in 1919. Moviola (primarily upright, with a single screen) and Steenbeck or KEM (flatbed, with dual screens and two soundtracks) machines were used for editing celluloid film. The Moviola enabled editors to run the film, stop it on a frame, and mark where they wanted to cut with a grease pencil. Analog editing was linear, and the editor used foot pedals to run the film either forwards or backwards, like thread in a sewing machine. Mechanical editing machines were also important for sound editing. At the dawn of sound film in the late 1920s, the sound Moviola made it possible to edit sound and picture separately in parallel.

Analog film editing on the upright Moviola and flatbed Steenbeck machines dominated film postproduction until the shift to digital nonlinear editing in the mid-1990s. Innovations in software as well as hardware made the move to digital editing of feature films possible. It happened quickly, with only three feature films edited digitally in 1994, but hundreds edited digitally in 1995, with an equal number edited digitally and mechanically. Nonlinear digital editing programs such as Avid, Media 100, Premiere, and Final Cut Pro were based on the language and metaphors of the analog technology they intended to replace, like many other digital media software applications.

The benefits of nonlinear digital editing abound: nondestructive editing, being able to try out many different ideas, faster turnaround for optical effects and transitions, and the computer keeps track of the editor's decisions. Editors have to keep the big picture of the entire film in mind at the same time that they focus on the single frame. They need to remember shots and frames that are almost exactly the same yet have subtle differences, and have confidence that when they cut the film, they have chosen the right place.

Because undoing a physical splice means tearing them apart, making a new cut, and resplicing, sometimes with a hunt for some of the frames previously removed (the "trim"), every act of analog editing has higher stakes of time at the least (as opposed to hitting an undo command). Arranging and organizing strips of celluloid physically fostered the essential skills of editing, including visualizing the sequence, scene, act, and film in one's head. Knowing where a piece of film is in physical space, in which bin or on which film hook on the editing bench, or one of several around the editor's neck, is different than remembering a file name or recognizing a thumbnail in a digital editing program. Digital nonlinear editing relieves some of the pressure on the editor, but also might make it harder to conceive of the film as a whole. What is definitely lost are the physical, tactile experiences of handling and splicing celluloid, and seeing light through it, all of which give the editor a visceral experience. What might also be lost because of the transition to digital editing is collaboration, because traditional editing rooms had multiple editing machines for assistant stations, on which assistants could work on their own machines at the same time as the editors, as opposed to waiting for their turn on the sole digital editing machine. Average shot lengths have decreased in the 21st century; is this because digital editing makes it easier to cut more? And is editing itself now facing the same specter of obsolescence as 360-degree video and virtual reality threaten continuity editing with immersive storytelling in 3D environments?

What Is the Moviola? The Flatbed? What Is the Purpose?

The Moviola is a machine that was used for editing celluloid film, primarily from 1924 to 1970, with some editors using it for longer. "Moviola" is the trademark name of the upright model invented by Iwan Serrurier and made by the Moviola Company in America. "Moviola" is also a generic term used for editing machines, specifically upright machines used in analog methods of bench editing (in contrast to computer-based editing).

Moviolas run film with foot pedals and have a four-inch wide viewer that magnifies the film frame in it. There are two kinds of Moviolas: with and without arms that hold film reels. The "Cutter Moviola" does not have arms and is used to cut film strips that are assembled into sequences that are wound onto reels and run through the machine with arms. The sound Moviola became available in 1930. It had a sound head equipped with an optical track reader comparable to the one on a film projector, and it ran both optical soundtrack and picture track either independently or locked together. Sound mixing developed in the early 1930s as it became possible to synchronize multiple sound tracks. When magnetic sound was introduced in the early 1950s, a magnetic head replaced the optical head.

In the film postproduction pipeline, editors and directors screen the "rushes" or "dailies", the raw footage that has been developed from the previous day's shoot. Editing begins with a work print made from the original camera negative of the film shot during production, and magnetic film (known as "mag") onto which the quarter-inch or DAT audiotape of the production sound has been transferred. Moviolas and other editing machines have rollers through which the picture track and the mag track (sound) are threaded; when the machine is engaged (with a foot pedal on the Moviola), it runs the picture track through the picture head and the mag track through the sound head, keeping them in sync. The editor can see the picture in the viewer and hear the sound through the speaker. A counter indicates film length in feet and frames. An editor runs the tracks, pulls the handbrake at the right place, marks the film with a grease pencil, and removes the film from the Moviola to cut and splice it on the editing bench, and builds the film by joining small pieces into larger arrangements that become scenes, sequences, acts, and then the film.

A flatbed also has a roller/picture head/sound head system, but is like a Moviola which is unfolded, flattened, and expanded with more options and more features. There are three sound heads and one picture head, with larger viewers, more speeds, much less noise, and visible controls, and the splicer is incorporated into the flatbed. Flatbeds are known by their trade names; KEM, Steenbeck, and—after 1970 when the Moviola Company made one of their own—the Moviola flatbed.

The purpose of the Moviola is to function as a tool for editing film. Editor Edgar Burcksen explains, "What editors do is artfully take advantage of something that is already hard-wired in our brains, compressing the redundancy and seemingly uneventfulness of our daily lives, thoughts, fantasies and memories into digestible bits with a beginning and an end" (Burcksen, 2017: 15).

Editing is the element of filmmaking that is unique to movies; other aspects are also found in other cultural forms, such as narrative structure in novels, *mise-en-scène* in theater, composition in the frame in photography, and the combination of music, sound effects, and dialogue in radio. The dominant style of editing has been continuity editing, which makes cuts "invisible", so the viewer does not notice them, even though the film is comprised of fragments—different shots from dissimilar camera positions, likely filmed on different days, and certainly not in sequential order, the way a play is in live theater. The style of editing which has predominated since the beginning of the 20th century gives the spectator a smooth ride that appears to be seamless, as unbroken as how we experience the world around us through our eyes and ears from within our bodies, but with the compression of time and diversity of physical (and often subjective) point of view that cinema affords. Continuity is established and maintained through editing shots that have consistent lighting, blocking (placement and movement of people and objects in the shot), and sound design so that the filmgoer is always experiencing the film from a stable viewing and listening position. The style of editing is "invisible" because it hides the potentially disorienting consequences of cutting between different camera positions and shots of different subject matter.

Working on the Moviola Made Editing Physical and Social

Editing with the Moviola, and to a lesser extent the flatbed, is an intensely physical process. The editor works with the machine, the governor of its noise and moving parts. When editor Steve Cohen described the experience of working on the Moviola, he emphasized how embodied it was:

> It was an amazing machine and incredibly freeing creatively because it was so simple and so physical. You stopped the film with your body, what could be more intuitive than that? Your editing was all about your reflexes and your nervous system.
>
> (Buck, 2015: 23)

Other editors remember the visceral nature of working on the Moviola. Walter Murch recollects, "One of the things I always liked about the Moviola is that you stand up to work, holding the Moviola in a kind of embrace – dancing with it, in a way" (Murch, 2001: 74). Editor Mark Goldblatt uses a similar metaphor: "I'd throw a clip in the air, and my assistant would catch it. I'd ask her for this close-up or that one. The film was alive. I'd throw it around my neck, keep it in my mouth. It was kind of dance" (Kirsner, 2008: 78).

The great Murch, who edited sound and picture on *Apocalypse Now* (1979) (recall the opening scene, with the helicopter and ceiling fan sound and image editing) was an early adopter, advocate, and philosopher of digital editing. His ideas about editing, developed primarily in lectures and interviews, are essential for anyone interested in understanding or practicing editing. He elaborates:

> The Moviola system "emulsifies" the film into little bits (individual shots) and then the editor reassembles it out of those bits, like making something out of clay. You take a little bit of clay and you stick it here and you take another little bit of clay and you stick it there. At the beginning of the process there is nothing in front of you, then there is something in front of you, and then there is finally the finished thing all built up out of little clay bricks, little pellets of information.
>
> (Murch, 2001: 75)

Editing on the Moviola was not only physical and noisy, but the technique also involved more than one person. In the film editing room, the senior editor worked with assistants and apprentices, who learned how to edit and editing room behavior and culture by being in the same physical space as their mentor. Assistants looked over the editor's shoulder, and handed editors pieces of the "trim" that had been cut and set aside in making the first assembly cut. Editors asked assistants, or editors from the next editing room over, to take a look and give their opinion. Assistants were tasked with organizing and repairing film (some of which was inevitably damaged by going through the Moviola), joining sequences, finding and fetching pieces of film, and sometimes doing a rough assembly of a scene. Editors, assistants, apprentices, directors, and cinematographers all screened the edited footage together in screening rooms regularly, because it was the only way to see it together, fostering a sense of collegiality and community through shared experience and discussion. Assistants had to work at the same time as editors, unlike now when the computer in the editing suite is maximized by having assistants and editors take shifts on the machines, putting the people at each level in a more isolated situation. Although editing with a Moviola came down to the editor, the grease pencil, and the Moviola, and was solitary in that way, the experience was done in a context full of sound, movement, and other people.

Editors working with their subordinates in editing rooms next to other editing teams in the film studio were another cog in the economy of production of Classical Hollywood Cinema. During the 1930s and 1940s, the heyday of the studio production era, directors were not typically involved in postproduction, and began their next studio assignment directing another film while the editing was done by the editorial department, with creative choices overseen by producers and studio executives. As full-time salaried editor positions in the studios were replaced by freelance single-film contracts by the end of the 1960s, when most film studios had divested from in-house postproduction, editors gained more creative control, more responsibility for managing the staff and workflow that the studios had previously taken care of, and also entered into new interpersonal relationships with directors. For an editor, a good working relationship with a director can still mean years of steady work. As Benjamin Wright explains in his summary of editing and special effects during the "auteur era" of 1968–1980, when editing came to be seen as an artful, expressive practice instead of just a technical trade, the role of the editor also expanded so that "the editor's professional identity as a problem solver is tied to both artistic and social aspects of client management that ultimately contribute to career advancement and reputation within the industry" (Wright, 2016: 107). As Gabriella Oldham puts it,

> Hand in hand with the overwork, however, does come a persistent sense of isolation—not just isolation imposed by outsiders who do not understand what

> editors do, but isolation even within the editing room. The capacity of the software to complete an unprecedented range of tasks—such as designing special effects and opticals, laying in sound and music tracks, or correcting color—has given the editor greater autonomy than ever before. Where previously it was critical to hand off the film to specialists like sound and music editors and to laboratories, the so-called "picture" editor is now able—even expected—to carry out many of these tasks alone. There are fewer hands to help along the way, so the job occasionally feels lonelier.
>
> (Oldham, 2012: 22)

The Moviola, with its one small viewing screen visible to only one person standing over it, did not foster director–editor interaction as much as editor expertise. With the shift to the flatbed, there were increased opportunities for the director to be involved in the editing process.

Inventing the Moviola

The Moviola was a machine that transformed film editing. Before the Moviola, editors used scissors and glue. They held strips of celluloid film up to a light, used jeweler's loupes and other magnifying glasses, and some editors could move the film from one hand to the other quickly enough to be able to see "moving" pictures. Editors were known as "cutters" who, before the Moviola was readily available in 1924, controlled film reels with hand cranks to move the film through a box topped with frosted glass and illuminated from underneath by a light bulb. Margaret Booth, one of the most influential of the many women film editors who established careers in the silent era that continued into sound film, remembered, "Before we had Moviolas for editing, we would run the pieces of film through our fingers. I would count, as if I were counting music, to give the scene the right tempo" (Hatch, 2013).

Editors sometimes fashioned their own viewers by modifying the intermittent mechanism in old projectors. William Hornbeck, who edited film from 1917 into the 1960s, when he became vice-president in charge of editorial operations at Universal Studios, recalled, "We constructed a sort of Moviola in 1921, but it was very crude and terribly noisy" (Thompson and Bordwell, 1983: 40).

The Moviola was invented in 1917 by Iwan Serrurier as a home movie projector to be sold to the public. Serrurier, an immigrant from Holland, was an electrical and mechanical engineer, interested in electricity and the movies. He settled in Pasadena, California, where he worked for the railroad. His interest in film led him to an idea about a home projector, like a Victrola phonograph, but for motion pictures. In 1919, Serrurier registered "Moviola" as a trademark and was awarded a US patent for a "Picture Projecting Apparatus". It was too expensive for mass adoption when it was released in 1920 at the price of $600, the same as a new car. He only sold three machines.

When a film editor at Pickford Fairbanks Studios showed Serrurier how editing was done with a magnifying glass, scissors, and a light well, and then run in the production room, Serrurier realized he could modify his home projector into an editing machine. By the end of the next weekend, Serrurier had created a personal editing system where cutters (as editors were called) could watch film rushes at the speed she or he wished, stop it at precisely the desired frame, mark it, cut the film with scissors and pass off the strips to an assistant to cement (or later tape) them together on a splicer. And all powered by a sewing machine motor. He sold four machines in less than a month. In 1924, Serrurier received a patent for the "Film Editing Machine" and the Moviola Midget editing machine was released.

Soon, both major and minor studios ordered so many Moviolas that Serrurier contracted with the Mitchell Camera Company to meet demand.

The next iteration after the Midget was fixed to a table, had bins to hold film, and had a foot pedal to control the motor in addition to the hand crank. During this time, the silent film era, editing was picture only. The Moviola Company's ability to adapt for editing sound during the industry's transition away from silent film cemented the editing machine's place in the filmmaking process. Adding an additional head for sound meant that dialogue could be perfectly synced with picture.

Editing Machines, Rhythm, and Thought

In addition to the physical and social aspects of working on the Moviola, there were aesthetic and cognitive styles fostered by the practice of film editing on the Moviola. The comically gruesome scene in Ethan and Joel Coen's movie about the Hollywood studio era, *Hail, Caesar!* (2016) depicts the peculiar rhythm of the Moviola that almost strangles C. C. Calhoun, the chain-smoking editor played by Frances McDormand, with her own scarf. But the pace of the physicality of working on the Moviola also promoted what some editors experienced as a contemplative practice. Legendary Hollywood editor William Reynolds, who won Academy Awards for editing *The Sound of Music* (1966) and *The Sting* (1973) in a career that lasted over sixty years, preferred the Moviola because

> even though there is a lot of taking film in and out of the Moviola mechanically, you're thinking about what your next step is while you're doing the mechanical things. I just found that wasn't true on the KEM, and after I did a couple pictures that way I went back to the Moviola.
>
> (quoted in Keil and Whissel, 2016: 8)

Other editors, who started on flatbeds and transitioned to computers, found a similar dialectic between physical actions and thought. For example, prize-winning documentary editor and director Alan Berliner recalls,

> the simple act of rewinding, fast-forwarding, or searching for a shot on a roll of film. Back in the day, there was something special about the quality of concentration involved in gazing at a stream of imagery as it quickly zoomed by on the flatbed—just fast enough, but also just slow enough to clearly "register" each image. On the surface, the act of rewinding film would seem passive, almost meaningless—something that would not appear to have anything to do with the cerebral part of editing. But when editing went digital and images could be found instantly, in microseconds, we lost some of the possibility for serendipitously chancing upon and stumbling across shots we never imagined could or would work in our film.
>
> (quoted in Oldham, 2012: 256)

The Three-Headed Monster and Television Editing

The Moviola was also significant in television production. One of the television programs in which audiovisual conventions of television style, storytelling, and production were established was *I Love Lucy* (1951–1957), and the Moviola played a role in that, too. Mark Serrurier, Iwan's son who helmed the Moviola Company after he graduated from Caltech in

1954, designed the "Three-Headed Monster" that *I Love Lucy* editor Dann Cahn used to cut the footage from the series' convention-setting simultaneous three-camera shoots in front of a live studio audience. Although *I Love Lucy* was not the first television program to combine multiple cameras, 35mm film, and a live studio audience, the people they brought together, such as production manager Al Simon, who was one of the pioneers of multiple cameras in television, to film legend Karl Freund to figure out the lighting and cinematography, and editors Dann Cahn, A.C.E. and Bud Molin all contributed to the aesthetic and technical standards that still govern television programs. Although editor Cahn called the custom Moviola the "three-headed monster", it actually had four heads, three for picture and one for sound. Nicknames aside, Cahn developed a process for editing on the monster that cut long editing times for a rough cut for an episode down to about a day and set precedents for television workflow. *I Love Lucy*'s success enabled Desilu to purchase the RKO Studio and expand production of television series even more, cementing the place of the editing and other equipment, techniques, and resulting audiovisual codifications in television history.

Moviolas and Flatbeds

The fundamental techniques of analog film editing remained stable from the introduction of the Moviola in the 1920s through the adoption of flatbed machines in the late 1960s to the rise of digital editing in the 1980s. The introduction of sound to film naturally had significant implications for technical and aesthetic aspects of editing, but the ability to synchronize sound and picture made the Moviola and other editing machines more essential to working with celluloid.

Although some editors stuck with the Moviola until the end of their careers, many shifted to the flatbed. The Moviola process, like any workflow, facilitated certain stylistic attributes and made others more difficult. Editors had a single point of focus in the viewing window, and a degree of privacy and concentration while working that alternated with more public and social exhibition of cut sequences in screening rooms. More like a motorcycle, perhaps, than a car, the Moviola responded physically to the editor's reflexes when she or he stopped the film to mark it or cue up the right frame in a shot. That singular control and visceral relationship with the machine vanished with the shift to the Steenbeck and KEM tables, which had their positive aspects that did not necessarily extend into the nonlinear computer editing systems (NLE) that replaced them. The flatbed made it easier for more than one person to see the footage, encouraging editors and directors to work together to use editing as an expressive tool. Many editors preferred the bigger screen, bigger spools of film, and room for three or more film reels with their own picture heads with which editors could compare different takes simultaneously to facilitate shot decisions. Flatbed editing tables also enabled much faster speeds of advancing and rewinding the film, five times faster than on the Moviola, and with much less risk of ripping the film. Replacing adhesive glue with tape to physically join shots also refined the editing process and gave editors more freedom, because editors could redo a splice without losing any frames, a development that was widespread by 1950.

Nonlinear Editing Systems: Editing in the Digital Age

Nonlinear editing systems (NLEs), such as the Avid, Final Cut, or Adobe Premiere, also use the tools and terminology of analog editing, such as the razor icon to "cut" footage and the term splice in Adobe Premiere. Early versions of Apple's iMovie application relied on similar metaphors, but more recent iterations abandoned professional editing nomenclature inherited from analog and the first NLEs for more language like "trim video" that is more intuitive to non-professionals and people born after the dawn of the digital age. With this semantic shift, analog editing loses one of its last traces in popular culture. In professional discourse, however, much of the lexicon from analog editing has been preserved.

Nonlinear editing systems mimic one of the key features of the Moviola workflow. As Murch explains,

> Computerized digital editing and, strangely enough, good old-fashioned Moviola editing with an assistant, are both random-access, non-linear systems: You ask for something specific and that thing—that thing alone—is delivered to you as quickly as possible. You are only shown what you ask for. The Avid is faster at it than the Moviola, but the process is the same.
>
> (Murch, 2001: 76)

Overall, the speed, flexibility, and convenience of digital editing is welcome among editors. It is part of the democratization of media-making; if a person can edit their own or someone else's video with a mobile device they have access to whenever they want, that is a drastically different circumstance from the days when there was a rigidly observed hierarchy of senior editors to apprentices in the film studio, with tightly controlled access to Moviolas and screening rooms.

There are losses, however, as well, and they extend beyond nostalgia for a physical activity or beloved and sometimes adversarial machine. NLEs mystify the editing process again, not by physically obscuring the viewing of the film in the Moviola, but through technology that many directors did not have, certainly in the mid-1990s when digital editing gained widespread adoption. There is more speed with digital editing, but also less necessity to review a reel of film, and much more footage with multicam shoots and digital video. Being able to save every version during the editing process encourages experimentation, and new "shots" can be made in postproduction with reframing and digital image editing, but can slow down the editing process. The loss of the physical aspect of analog editing means a detachment from the embodied process of filmmaking in the celluloid era. More choices and easier workflow do not necessarily make for a better end result.

There are many reasons why the average shot length (ASL) has decreased over time, and NLEs are one of them. As Paul Schrader explains:

> Why are there more cuts? Because film styles change, because our ability to process images has accelerated, but also because we can. The technology permits it. Each advance in editing makes it possible to edit more rapidly and with greater complexity. Now we need all that footage, because we've retrained viewers to expect all those cuts and multiple points of view.
>
> (Schrader, 2014)

In the big picture of the filmmaking process, the change from analog to digital editing means a paradigm shift from an emphasis on the relationship between shots to transforming what is in the shot. Editing now encompasses visual effects (VFX), or rather, digital postproduction includes editing, VFX, the addition of computer-generated imagery (CGI), color correction, lighting, the compositing of greenscreen and other footage, and editing within the frame on a production with live-action footage shot with a camera. On a movie with heavy CGI, SFX can be involved in most if not all of the shots. For example, in *Star Wars:*

The Force Awakens (2015), around 2,100 of the film's 2,500 shots involved digital effects, according to the film's visual effects supervisor Roger Guyett (Dent, 2016). When 80% of a movie is comprised of digital effects, isn't it more accurate to describe it as an artifact made with virtual assets (settings, characters, objects), captured by the "virtual kino-eye" of synthetic cameras than the live-action filmmaking of the studio era? Or, as new media theorist Lev Manovich posits:

> Cinema traditionally involved arranging physical reality to be filmed through the use of sets, models, art direction, cinematography, and so forth. Occasional manipulation of recorded film (for instance, through optical printing) was negligible compared to the extensive manipulation of reality in front of the camera. In digital filmmaking, shot footage is no longer the final point, it is merely raw material to be manipulated on a computer, where the real construction of a scene will take place. In short, production becomes just the first stage of post-production.
>
> (Manovich, 2001: 303)

Whither Editing in VR, 360-degree Filmmaking, and Immersive Media?

Practitioners and purveyors of virtual reality (VR), 360-degree media, and other forms of immersive media mostly agree that the conventions of continuity editing we inherit from film, video, and television do not work in the new media. In these new forms, the audience's sensory input is completely immersed in the media, with a headset that blocks all other visual input, and headphones that provide sound. That means that the camera position in the experience is your head, and you are in the middle of whatever place and situation you have been dropped into, with spatial audio soldering your experience in a specific place in a soundscape. You can look up and down, all around, continuing to hear sound from the correct place even as you turn your head, and depending on what the creators have made, your sensory input could be incongruous with your body's proprioception and sensation of being still. If VR and immersive media (which includes 360-degree video, 180-degree video, ambisonic audio, immersive journalism, and will include some augmented reality (AR) and mixed reality (MR or XR)), are going to be used for narrative experiences beyond one location at one time, some kind of editing will have to take place, but the invisible style of continuity editing does not stay invisible, and can be jarring or even nauseating.

Jessica Brillhart, principal filmmaker for VR at Google from 2009 to 2017 and now independent filmmaker, builds her philosophy of "probabilistic experiential editing" on Walter Murch's insights about film editing, titling her posts "In the Blink of a Mind" as a reference to Murch's *In the Blink of an Eye*. Probabilistic experiential editing means guessing where the viewer/visitor's attention will be, and using it as the point of transition.

In cutting frame-to-frame in film, editing creates a path through the film for the viewer, but Brillhart maintains,

> in VR, the frame is where the visitor engages. There are worlds of frames that need to be considered. If we can make good predictions and craft with potential engagement in mind, then we can pull someone through a full experience in a way that isn't jarring or strange, but personally gratifying. We allow the visitor to move through time and space.
>
> (Brillhart, 2017)

Brillhart recounts a story about how her first attempt at editing a VR experience looked fine on the computer screen, but when she watched it in the headset, it was terrible. That occurrence prompted her to search for metaphors to clarify what editing could be in immersive media. Instead of linear frames in a row, like in a strip of film, or a series of frames in a storyboard, she realized that VR, which always has the person wearing the headset at the center, was more of a series of concentric circles:

> It [should be] like a ripple effect—like a drop in a bucket, and then a ring around that, and a ring around that. [The editing] was really about rotating those rings to corral people through the general idea of a story, or an experience. Then, I thought, if I know that [viewers] are going to be doing certain things, I can edit for that sort of experience.
>
> (Macaulay, 2015)

The rules of continuity editing, based on not crossing an imaginary 180-degree line drawn through the middle of the scene, already challenged in post-Hollywood cinema style, are indicators of viewers' habits, if not obsolete. However, the aspects of human vision, such as having a preferred viewing zone of about 30 degrees in which shapes are easily recognized without head or much eye movement, on which editing conventions like the 30-degree minimum difference of camera position between cuts are based, of course are still crucial for constructing new 360-degree experiences.

Because VR and immersive media are relatively new, editing might now be experienced as more jarring than it will be in the future when people are used to it. We may be in a parallel moment to another change in relative size of the screen and the viewer: the advent of widescreen formats like Cinemascope. When they first came out, directors and cinematographers opted for more long takes and fewer cuts because they believed viewers would find it too jolting to see frequent cuts in such a large image field. However, as we know, ASL did decrease, and the very quick pace of cutting that debuted on television in the MTV era in the 1980s transferred to the bigger cinema screen, similar to how using the synthetic camera for flyovers and other kinetic movement initially seen in video games is now a staple of digital filmmaking aesthetics.

As VR devices become better at showing camera movement without causing motion sickness, or viewer/participants become used to the sensation, or new ways of experiencing some VR in which embodied experience is connected to the virtual experience, the virtual kino-eye might be able take full advantage of all three of the XYZ axes, and there might be shorter ASLs and different editing rhythms. I can imagine a combination of the person experiencing the immersive media being able to take on different roles with different degrees of agency, so there might be long takes in scenes or environments that the visitor (to use Brillhart's term) explores, triggering encounters when she or he moves toward a character or object, and also parts that are more time-based, perhaps in a scene with ambisonic audio in which the participant is able to choose the points of focus, akin to a 360-degree version of Mike Figgis's experimental film *Timecode* (2000) on DVD, with the audio controlled with the DVD remote.

There is another way of dividing up the 360 degree space of VR: only use the half of it that is in front you. Google's VR180 is designed to make immersive content creation easier by removing some of the biggest obstacles to VR video: that there is nothing outside the shot, getting audience attention in the right place if they are turned away from it, and the stitching and camera tripod removal required in postproduction. VR180 sets out an accessible workflow for immersive video: point and shoot 180 degree 3D stereoscopic video cameras priced to entice consumers, video streaming from the YouTube site that can be seen as flat monoscopic video without a headset, and experienced immersively with Google Cardboard or Daydream headsets that use a mobile phone, PlayStation VR, and standalone VR headsets like the 3DOF (degrees of freedom) Oculus Go (2018). That the 180VR demo at the Google booth at SIGGRAPH 2017 featured a father and son team who have a popular YouTube series, "What's Inside?" in which they cut stuff in half, signals 180VR's emphasis on vloggers and other YouTube content that has a low barrier of entry.

Cameras that shoot 180 degrees of video might provide a hybrid of immersion and traditional cinematic spectatorship that can preserve the core techniques of editing as they have been established since the advent of the Moviola: continuity editing, eyeline match, form cuts, match on action, and perhaps experiment with new ways of enforcing and breaking the 180 degree rule. Although 180VR is being shaped as a content creators' form, it seems likely that full 360 degree immersive experiences, which can be experienced on the web with WebVR, will also be increasingly in reach of YouTubers and vloggers, and whatever platforms and forms come next. The standalone 6DOF Oculus Quest headset with touch controllers (Spring 2019) is also a mixed reality device that can incorporate the physical environment.

Just as the Moviola fostered a kind of editing with one set of stylistic, workflow, and aesthetic ramifications, the flatbed another, and digital NLEs yet another, editing immersive media will likely involve new practices, tools, styles, and conventions. There are already, in 2018, programs to use in VR such as Quill, Google Blocks, Tilt Brush, AnimVR, and Medium to draw, paint, sculpt, model, and animate, and build in other programs such as Unity and Maya, as well as programs such as Mindshow for making real-time animation machinima.

In the virtual world High Fidelity, people can create their own non-player characters (NPCs) by recording actions, behaviors, and dialogue that can be played back at a later time, and even interact with objects in the virtual space.

> While we once saved moments with paintings, photographs, and video, we've now moved into the age of live streaming and are continuing to drive innovations around storing and saving memories and actions. Virtual reality enables us to take that to another level, where we can create versions of ourselves that persist in a digital space .
>
> (High Fidelity blog, 2017)

The playback is not video, but data capture of the avatar's movements, responses, and speech, which could be incorporated with interactive AI. What kind of editing will these new performances capture and personality assets inaugurate? What will it be like to edit wearing a headset not much bigger than the Moviola screen, moving the whole body in physical space with virtual hands grabbing and moving clips and assets? Will the editor finally be inside the film experience, and not outside of it, completing an arc that started standing over the Moviola, sitting down in front of the screens and reels of the flatbed, then whisked digital clips to and fro with the press of keyboard shortcuts, ever increasing the editor's postproduction domain? What is an editor in immersive experience, other than a guide—or a director of attention?

References

Brillhart, J. (2017). *These Uncomfortably Exciting Times | Filmmaker Magazine*, available at http://filmmakermagazine.com/101220-these-uncomfortably-exciting-times/#.WaMazzOZPEY [Accessed 2 Aug. 2017].

Buck, J. (2015). *Timeline Analog 1: A History of Editing 1860–1971*. [ebook] Tablo Publishing. Available at: www.amazon.com/Timeline-Analog-1-1860-1971-ebook/dp/B07DCX347D/ref=sr_1_6?s=digital-text&ie=UTF8&qid=1536934408&sr=1-6 [Accessed 11 Mar. 2017].

Burcksen, E. (2017). "The Blink of an Eye." *CINEMAEDITOR Qtr 1, Vol.* 67, p. 15.

Dent, S. (2016). *See How VFX Transforms "Star Wars: The Force Awakens" (updated).* [online] Engadget. Available at: www.engadget.com/2016/01/14/force-awakens-vfx-reel/ [Accessed 15 Jan. 2017].

Hatch, K. (2013). "Cutting Women: Margaret Booth and Hollywood's Pioneering Female Film Editors." In Jane Gaines, Radha Vatsal, & Monica Dall'Asta eds., *Women Film Pioneers Project*. Center for Digital Research and Scholarship. New York, NY: Columbia University Libraries. Available at: https://wfpp.cdrs.columbia.edu/essay/cutting-women/ [Accessed 11 Mar. 2017].

High Fidelity Inc. (2017). "Recording Yourself as an NPC." Blog. Available at: https://blog.highfidelity.com/recording-yourself-as-an-npc-3b241d4d66e2 [Accessed 17 Sept. 2018].

Keil, C., & Whissel, K. (2016). *Editing and Special/Visual Effects*. New Brunswick, NJ: Rutgers University Press.

Kirsner, S. (2008). *Inventing the Movies: Hollywood's Epic Battle between Innovation and the Status Quo, from Thomas Edison to Steve Jobs*. CinemaTech Books.

Macaulay, S. (2015). *Look Into the Cut: Jessica Brillhart on Editing VR | Filmmaker Magazine*. [online] Filmmaker Magazine. Available at: http://filmmakermagazine.com/96090-look-into-the-cut/#.WaMm_zOZPEY [Accessed 21 Jun. 2017].

Manovich, L. (2001). *The Language of New Media*. Cambridge, MA: MIT Press.

Murch, W. (2001). *In the Blink of an Eye: A Perspective on Film Editing*. Los Angeles, CA: Silman-James Press.

Oldham, G. (2012). *First Cut 2: More Conversations with Film Editors*. Berkeley; Los Angeles; London: University of California Press.

Polan, D. (2016). "Postwar Hollywood, 1947–1967: Editing." In C. Keil & K. Whissel eds., *Editing and Special/Visual Effects* (pp. 78–90). New Brunswick, NJ: Rutgers University Press.

Schrader, P. (2014). "Game Changers: Editing." *Film Comment*. [online] Available at: www.filmcomment.com/article/game-changers-editing/ [Accessed 14 Jan. 2017].

Thompson, K., & Bordwell, D. (1983). "From Sennett to Stevens: An Interview with William Hornbeck." *The Velvet Light Trap*, 20: 34.

Wright, B. (2016). "The Auteur Renaissance, 1968–1980: Editing." In C. Keil & K. Whissel eds., *Editing and Special/Visual Effects* (pp. 103–115). New Brunswick, NJ: Rutgers University Press.

11
ANALOG AUDIO SYNTHESIS
Oscillations, Traces, and Trajectories

Peer D. Bode

There is a beginning, middle, now, and next; beginning, now, middle, and unexpected; now, middle, beginning, and ruptured. Possibly more surprising than that events, moments, and times are linear, having a limited timeframe in a sequential additive progression, is the situation of events occurring in unpredictable ways. An argument could be made that sound-making and music-making are human occurrences, rich time-based forms, infinitely varied vibrational pitch events that participate in the varied paths of our human history. The sound arts as well as the visual arts, the time-based arts, are only deepening in their variety and their accessibility. The means to create and deliver these forms are changing. How does a culture develop and organize its material progress and evolution? Technological development sees periods of invention and innovation intertwined with periods of refinement and diversification. Technologies now are commonly need-tested and market-tested to determine viability. The pressure of innovation, improvements, and the pressure that comes from needing some degree of change, as well as repetition and ritual, are steered by many forces, not by industry alone. Change is with us, we cannot escape it. Having said that, contemporary acts of communication and development have long histories, the patterns of which and the lessons of which, can only be seen over time, at a distance. To be critical and to better understand the notion of obsolescence, the obsolete, and to use it productively, we need to consider that it functions within time-based forms. To consider and focus on varied timeframes alters the notion of obsolescence, its object, its impact, its usefulness. Let us start by enlarging our time-frame. By considering electronic music technologies and specifically analog electronic music technologies in the context of various historical conceptual trajectories, we can observe many seemingly diverse events, thick with living puzzles and queries, rich with historical connections. What follows is a directed weaving of concepts and narratives.

In his book, *Deep Time of the Media: Toward an Archaeology of Hearing and Seeing by Technical Means* (2006), author Siegfried Zielinsky suggests a lineage of historical technological development that is nonlinear, an archeology of technological development with many lines and branches of development, many dead ends, and many lines waiting, ready to begin again. The historical developments across overlapping timeframes he identifies are breathtaking. The time spans he uncovers are beautiful to consider. The histories are long. This long historical development leads right up to current analog electronic music technologies. Electronic development actualized, picked up speed in the early and mid-20th century, a time also of rich cultural innovation and invention.

In the present, electronic music and analog electronic music are more actively practiced and performed than ever before. There are, alone, over 1,000 Eurorack-style analog electronic music modules available as of 2017. The twisting and turning historical path of analog electronic music ideas and foundations are more clearly understood and embraced now with the benefits of our new technologies and networking capabilities. The larger electronic arts, which include sound and moving image and smart interactive objects and materials, are now de facto global practices. To understand this in its richness, I would suggest we start by evoking and engaging a number of notions and concepts to help us productively navigate the strands of this development.

The musician, composer, and theorist Anthony Braxton has established some useful and generous notions (Lock, 1988: 162). The terms he uses are Restructuralism, Stylism, and Traditionalism. He uses these terms to describe modes of cultural making, including music and sound-making. I suggest we consider these ideas to better understand creative and expressive practices, intellectual practices, and emerging potentials including the technological and technical practices of electronic music, which include those of analog electronic music. The restructuralists, he tells us, are those who engage in an evolution of ideas, forms, and materials. They find the ways to make change meaningful and significant, to make change relevant. Well, once something has been set into motion by the restructuralists, people usually take that information and use it . . . those are the stylists. There are master stylists too, but the masters are the ones who did not simply take without giving, who did not just play Charlie Parker's language and do nothing to it . . . stylists make works for the larger public. They transfer ideas to the larger forums. The traditionalists, on the other hand, move things forward by clarifying fundamentals and identifying the paths a lineage has moved in. Together, these three orientations form a productive dynamic that considers historical forms, contemporary constructs, and emerging experimentations all necessary for ideas and experiences to be life-bearing, nurturing, and empowering. Electronic music, media arts, and the moving sound, image, and object continuum are the manifestations of art thinking, of the crossing of culture and materialism. At this time, this later part of the 21st century's second decade, we might look to see how this thinking is further charged and opened with emerging notions of the new materialisms.

Useful notions of time are evoked by Siegfried Zielinsky, they are the ancient Greek ideas of time, *Kronos, Kairos*, and *Aion*. *Kronos* refers to sequential time, narrative development, linear time, past, present, and future. *Kairos* can refer to a season or an indeterminate time during which a significant event takes place. It can mean the right or opportune moment. In modern Greek, *Kairos* means "weather". *Aion*, the third term, is a Hellenistic deity, associated with time; it is the unbounded, the eternal. These are three very different notions of time; three ancient ideas of time varying as linear time, opportune moment time, and eternal time.

A significant composer of electronic music, Karl Heinz Stockhausen articulated a useful notion he called the "Moment Form". The Moment Form could be thought of as a solution to the challenge and problem of bridging the potential of the score, including the graphic score and the realization of complex electronic sound events. The Moment Form is a way of giving attention to, and focusing on, complex sound forms created by analog electronic and acoustic instruments. How does one organize and keep track of complex sound events, particularly those that cannot be properly represented in traditional scoring systems and also might be variably controllable? One solution is the graphic score. Another solution is that of the act of listening; to notice when a set of interacting sounds creates a sense of connectedness or a combined sense for periods of time. Like a high wire act, this can go on for some time. Then, one suddenly can hear a moment when the sounds, the interaction, the special mix of sounds, abruptly changes enough that the Moment Form has ended and another has begun. Organizing and connecting such varying moments is one strategy for perceptually mapping and constructing large, multiple, complex sound events.

Influential composer John Cage, often associated with strategies of listening, of the aleatoric and also of silence, described what he considered three historically significant eras of musical thinking; that of the composition, that of the performance, and that of the listening. His particular interest was our present and future era, challenging us to develop our listening abilities and skills.

The composer and performer Pauline Oliveros took this notion of listening further with her performative form she named "Deep Listening". Deep Listening, for her, was, in addition to a way of life, a way of activating a complex rich experience, that of listening together with performing as a compositional method. Deep Listening crossed the long tradition of improvisation with phenomenological sound listening strategies as key to the active performative composing event. This was performance grounded in interactive sound listening. The technologies she used to make sound were analog and then digital electronics, what she called her real-time "Expanded Instrument", an accordion with pitch-shifter foot pedal and delay foot pedal. The performer, the human body, was positioned as a real-time, sound listening, processing and reacting system. Oliveros's live performances and her subsequent recordings captured remarkable live listening-and-response dynamics in play.

The sound and video artist Nam June Paik, who studied philosophy and contemporary classical music as well as communication theory and electronics, realized that experiments in communication technologies and programming were key to making culturally significant and personally relevant contemporary art works. He made what could be called an electronic shamanistic art using sound and video electronic technology. Paik critically described the art writing of his time as "the fetishism of the idea" (Paik, 1973). He went on to say, "Sometimes I need red apple. Sometimes I need red lips" (Paik, 1964). His frustration concerned the state of critical discourse, at the time of new contemporary electronic media thinking, events, and art. In an interview by electronic sound and video artists Woody Vasulka and Steina, Paik, with some humor, described the moon as the oldest TV and video synthesizers as being thousands of years old. Videos made with video synthesizers were made with the fingers, turning knobs, not with the mouth, as was the case with traditional television. Paik was laying out an imaginary archeological history of electronic music and television. His own writings, informed by philosophy, electronics, and Dada practice, were often expressed in short, playful Haiku-like statements. He was the same artist who, in 1972 with his ground breaking broadcast television experimental art piece, "Global Groove", and his writings about the "Electronic Superhighway", pre-imagined the World Wide Web by some 20 years.

The above notions, considered together in a simultaneous or collage-like manner, open possibilities of varied musical thinking, sound and image thinking, what one could call event thinking. Such event thinking nourishes our listening as well as our imaging and thinking capabilities. These are possibilities, rich and productive assemblages, the thickening of narratives, the creating of new narratives, the opening of emerging moving sound and image opportunities. These ideas are critically useful in understanding and participating in the new media art works of the last hundred years, as well as those that are performed and presented every day. As Paik might have suggested, imagine these ideas as possible methods of cure.

Including the insights of psychoanalysis and philosophy into this query, Jacques Lacan and Alain Badiou both help us re-imagine and reconsider our narratives, our notions of history. As Badiou says, Lacan lays out his notion of the cure, the aim of which is to reopen the capacities of the subject, where the subject can lift back up and live again (Badiou and Roudinesco, 2014: 16). Being bogged down in the imaginary, one can once again exercise the power of symbolization. This involves the liberation of possibilities, giving innovation a chance, bridging from the litigious mind to the inspired mind. Alain Badiou builds on these notions, encouraging a re-engagement with histories, taking on, jumping in on unfinished stories, unfinished struggles. Badiou also encourages us to focus down on present formulations, supporting small and local innovations, not knowing ahead of time what is important about such experiments (Badiou and Milner, 2014: 126). He encourages vitality, hypothesis and propositions, reconsidering history, rebirths of history. In Badiou's possibilities, there is the projecting of the unimaginable.

Medievalist Gerry O'Grady, founder and director of the media arts center, Media Study Buffalo, friend and colleague of media theory iconoclast Marshal McLuhan, developed and embraced a methodology of media literacy that embraced combining theory and practice, studying media and media-making together, "para-deigma", from the Greek, that which is shown alongside. As an educator, O'Grady has inspired several generations of media makers and theorists to make media as a part of citizenship as well as of artistic aspirations and practices. Over the years, as we know, media practices have greatly expanded with many new techniques and strategies. Unfortunately, that part of media literacy that is media analysis and thinking about its potentials has lagged behind or become separated from media practice. In fact, media-making and thinking, and a new media literacy, as we might imagine, are still in their infancy or possibly adolescence, spread out across many countries and languages. The New Media Digital Humanities, as they are called, will be key to bringing together media's many disparate manifestations, including bringing the global academic media research contributions to the needy spaces of active public discourses.

The Advent of Musical Events

We know that sound and music predate recording technologies. Their early origins live in the voice, in the performative, the pre-historical, in the imagined; the human voice, our body-mind instrument sounding at the very earliest moments of our human history. Later, outside of our bodies, varying-sized stones were pounded, and could make predictably pitched sounds; not so very portable, but nevertheless solid sound-bearing matter, stone technology, mechanical mallet stone technology. Now antiquated, obsolete possibly, but ready to burst back into action when needed.

Early people who used the Lascaux Caves in what is now France, to protect their remarkable, sacred drawings of bison and deer, also no doubt made sound and in fact most likely made profound sound and image events with drumming, vocalizing, and visual shadow play. This would have been an early form of trance multimedia expression, an early image and sound proto-cinema. One might even imagine the cave drummers, vocalizers, and shadow players as so gifted in their abilities so as to be renowned in their early communities.

One of the earliest recorded narratives of sound and musical ecstasy and power is that cited in the early Greek literature of *The Odyssey*, attributed to the writer Homer, of Odysseus's ten-year return voyage from the decade-long Trojan War. The returning war hero asks his fellow seafarers to lash him to the sail mast, with wax inserted into their ears to protect them from the voices of the famous Sirens, those bearers of overwhelming voices and desire, an overwhelming cocktail of sound and vision possibly. The German materialist media theorist, Friedrich Kittler, toward the end of his career, shifted his media research to the subject of Love, taking on the story of the powerful Sirens and Odysseus of the epic poem as subject for his research. He revisited the Greek islands of the Sirens, the Li Galli Islands, to test whether one could actually hear and understand voices originating from the beach in the water, where Odysseus's ship might have passed. With an empirical test, a test of acoustics, Kittler confirmed the limit of *the* historical story of sound, female voices, desire, and possible disaster. This he did with the historical epic story of discovery and return to home. His acoustic test showed that the Sirens' voices as audible voices could not have physically reached the ship from the island shore. The sounds of the Sirens possibly but not their audible voices could have been carried to the passing ship. Voices could be heard, could have been heard, but not understood. As described by physics, acoustic vowel sounds carry far distances while consonant sounds do not. Lower frequencies carry across distances that high frequencies do not. The Sirens' voices could have been heard as sound, but not understood as language. Though *The Odyssey*'s truth in this case was challenged, the story's many other truths, lay in waiting, waiting to become real. Here we have events in the imaginary, ready to go, ready to come to actual life, awaiting the physical means, the technology to deliver the fearful, seductive, ecstatic sounds anew. By the 20th century, the imaginary seductive Siren sounds, via acoustics, could be better understood; they could also be re-imagined, transformed, and electronically evoked with new depth and uncanny presence. They could be electronically recorded as well. A long trip indeed. Kittler concluded that the voices of the Sirens, a mainstay of Western narratives, had indeed to be a partial fiction, though no less a significant fiction. The point was to materially test a culturally significant story, to test it as one would a physical medium such as phonography, photography, cinema, television, or their now analog and digitally evolved versions. As media create cultures, testing media is one significant way to further consider cultures.

After the epic of *The Odyssey*, the Greek philosopher Pythagoras, in the late 6th century BC, set important cornerstones for the future development of analog electronic music, including the relation of mathematics and musical sounds. He proposed that numbers and ratios were basic to the understanding of sound and music, further connecting that information to the understanding of the relationships and movements of the planets and stars. Tony Conrad, in the late 1960s, with a degree in mathematics from Harvard and a composer, filmmaker, media maker, and one of the founders of Minimal Music, made a series of sound pieces based on mathematically selected frequency combinations. One of these significant early minimal sound works, performance, and recording, he named, "Slapping Pythagoras". The piece was a strategic intervention into articulating, entering, and freeing new sound spaces. In so doing, he was also critiquing the far-reaching ideologies of control Pythagoras established, including those of blending notions of mathematics and mysticism.

Lists and Maps

David Ernst in his book, *The Evolution of Electronic Music* (1977), looks at the people, events, and moments along the way that pre-visioned and pre-auditioned the future to come, setting the stage for the future history of electronic music. Filling in a narrative with the time of *Kronos* (that is, sequential time), past, present, and future, Ernst gives names, developments, and dates. Siegfried Zielinsky describes the same events as a meandering story of stops and starts, accidents and discoveries, laid out for us in his book, *Deep Time of the Media*. This he describes as an archeology of media, the name of which he would later change to Variantology. Imagine an expansive map, a media archeology approach to these histories.

The simpler, sequential *Kronos* form might look like the grouping strata below. Many significant names and dates are there. The paths not seen are rich and often complicated. What follows is a useful, if incomplete, list of significant historical sound, music, and electronic instrument contributions as David Ernst and others have mapped out:

End of the 8th century BC Homer: *The Odyssey*.

570–495 BC Pythagoras: Proportions, Monochord, Quarter notes.

4th century BC Aristoxenos: Equal Temperament.

3rd century BC Ktesibios: Hydraulis pipe organ.

100–170 AD Claudius Ptolemy: Just Intonation.

1558 Gioseffo Zarlino: 19 divisions per octave.

1577 Francisco de Salinas: 19 and 24 divisions per octave.

1618 René Descartes: Scientific method.

1627 Francis Bacon: *The New Atlantis*, "Sound Houses", where all sounds exist.

1710 Conrad Henfling: 50 divisions per octave.

1722 Jean-Philippe Rameau: System of harmony based on overtone series.

1759 Jean-Baptiste Delaborde: The Electric Harpsichord.

1761 Benjamin Franklin: The Glass Harmonica.

1850 Henry Ward Poole: The Enharmonic organ with 50 divisions per octave.

1855 Moritz W. Drobisch: 43 and 73 divisions per octave.

1863 Hermann von Helmholtz: The text, "Sensation of Tone, Complex Tones Produced by Summing Individual Sine Waves".

1876 Alexander Bell: The Telephone.

1877 Thomas Edison: The Phonograph.

1897 Thaddius Cahill: The Telharmonium.

1899 William Duddell: The Singing Arc.

1909 Filippo Tommaso Marinetti: The Foundation and Manifesto of Futurism.

1915 Lee De Forest: The L-C vacuum tube oscillator.

1919 Leon Theremin: The Theremin.

1921 Charles-Emile Hugoniot: The "Hugoniot Organ", a photocell instrument.

1921 Joerg Mager: The Electrophone.

1923 Hugo Gernsback: The Staccatone.

1926 Leon Theremin: The Electric Harmonium with 1,200 divisions per octave.

1926 Hugo Gernsback: The Pianorad.

1927 Pierre Toulin: The Cellulophone.

1928 Maurice Martenot: The Ondes Martenot.

1930 Friedrich Trautwein: The Trautonium.

1930 Edouard Coupleaux and Joseph Givelet: The Coupleaux-Givelet Organ.

1932 Benjamin F. Miesner: The Electric Piano.

1934 Ivan Eremeef: The Syntronic Organ.

1935 Laurens Hammond: The Hammond Organ.

1937 Harald Bode and Christian Warnke: The Warbo Formant Organ.

1937 Homer Dudley: The Vocoder.

1938 Georges Jenny: The Ondioline.

1938 Harald Bode: The Melodium.

1939 Hammond Organ Co.: The Novachord.

1944 Percy Grainger and Burnett Cross: The Grainger-Cross Free Music Machine.

1945 John Hanert: The Hanert Electrical Orchestra.

1945 Hugh Le Caine: The Electronic Sackbut.

1947 Constant Martin: The Clavioline.

1947 Harald Bode, The Melochord.

1951 Earle Kent: The Electronic Music Box.

1955 Harry Olsen and Herbert Belar: The RCA Synthesizer.

1956 Raymond Scott: The Clavivox.

1957 Max Mathews: The Music 1 Program.

1959 Harold Rhodes: The Piano Bass.

1960 Harald Bode: The Audio System Synthesizer.

1963 Paul Ketoff: The Synket.

1964 Robert Moog: The Moog Synthesizer.

1965 Donald Buchla: The Buchla Synthesizer.

1969 Raymond Scott: The Electronium.

1969 Alan R. Pearlman: ARP Instruments, Inc.

1972 Harald Bode: The Bode Frequency Shifter.

1974 Dave Smith: Sequential Circuits.

1976 Tom Oberheim: The Oberheim Modular.

1967 John Chowning: FM Synthesis.

1977 Sennheiser Electronic GmbH & Co. KG and Bode/Moog Vocoders.

1986 Laurie Spiegel: Music Mouse.

1987 Joel Chadabe: M Interactive Composition.

1988 Miller Puckette: Max.

1990 Groupe de Recherches Musicales de l'Institut National de l'Audiovisuel: GRM Tools.

1991 Tom Erbe: Soundhack.

1995 Dieter Doepfer: The Doepfer Modular Eurorack Audio Synthesizer.

1996 Eric Wenger: MetaSynth.

1996 Miller Puckette: Pure Data.

1996 James McCartney: SuperCollider.

1999 Dominic Mazzoni and Roger Dannenberg: Audacity.

2001 Bernd Roggendorf, Robert Henke, and Gerhard Behles: Ableton Live.

2002 Curtis Roads: SweepingQGranulator.

2005 Michael Klingbeil: SPEAR: Sinusoidal Partial Editing Analysis and Resynthesis.

2017 Over 1,000 Euro Rack analog modular synth modules (modulargrid.net).

Across the last 100 years, we have seen Theremins, keyboard instruments, mechanical tone generators, electronic tone generators, disk and tape recording and manipulation, sound processors, and modular systems. We have seen vacuum tube amplifiers, transistors, integrated circuit chips, digital systems, programmable hardware and software systems, virtual instruments and plug-ins, circuit bending, and so on.

The 1950s and 1960s saw the beginnings of the early electronic music studio centers: the North West Deutsche Rundfunk Electronic Music Studio (Germany), the Columbia Princeton Electronic Music Studio (USA), Groupe de Recherches Musicales (France), Studio di Fonologia Musicale di Radio Milano (Italy), Swedish National Centre for Electronic Music and Sound Art (Sweden), Institute of Sonology (Netherlands), and NHK in Tokyo (Japan). With tape recordings in the 1950s and 1960s, there were opportunities to direct close attention and focus on detailed electronic and acoustic sound qualities as well as the resultant growth of rich sound archives. The close attention to sounds, via recordings, set the stage for the development of real-time performance of electronic sounds and durations. This took place within sound and music and also across media; for example, media artists Woody and Steina Vasulka moved from audiotape-based work to videotape-based work and then real-time recordings and performance.

In the 1950s, Robert Moog made Theremins. Shortly after hearing Harald Bode's presentation of his new modular sound processing system at the 1960 Audio Engineering Society conference, Moog began building his modular, voltage-controlled Moog synthesizer. Moog, with his interest in voltage control, could further the American folklore of the one-man band. With voltage control, multiple sound parameters could be altered live simultaneously with changing time-based waveforms. An individual could unleash a complex set of multiple variables, once the scene of the team of instrument players. On the West Coast, non-keyboard approaches to controlling large modular systems, such as voltage control as well as continuous ribbon controllers were carried out by Don Buchla.

In the 1970s, my audio and video colleagues were reading Bernie Hutchins's plans for building electronic synthesis modules, "Electronotes", out of Ithaca, New York. At the same time, PAiA Electronics sold mail-order audio synthesizer kits at low cost. I spoke to John Simonton, founder of PAiA Electronics, in Oklahoma City, on the phone in the late

1980s, inquiring as to details of an audio MIDI unit he was advertising. When I told him my name was Bode, he told me that he entered the audio synthesis tool business because he was inspired by my father, Harald Bode. Harald had proved that as an individual, one could build and sell electronic synthesizer instruments; one need not be a large corporation. These were clues of things to come.

A Personal Digression

I admit to digressing, wanting to re-contextualize and physicalize this text writing and reading activity. I have pulled various books and publications on electronic music and new music and music culture off the shelves. It is as the French theorist Roland Barthes described his own occasional engagement with his books and library, pulling books from the shelves, selectively placing them on the carpet, laying physically down next to them, considering bodily rolling over them. He picked them up and opened them to random pages. He would begin to read, needing only a few sentences or a paragraph and immediately remembering the atmosphere and resonance of the words and the stories he had read before, in some cases many years before. Here we see Barthes as the critical reader, literally, as the operator of the text. The images he describes are seductive as his writings are. I am transported and pleased and surprised how the texts and the pictures and diagrams I see on the pages I open, together are thoroughly inspiring and rewarding, yet once again. Electronic music, new music, and music culture are revealed as worlds of sounds and sensations, logics and sensuality, data representations, plots and diagrams, material realities, physics as material facts, sonic events as actualities, as well as excited spaces and sensuous geographies.

Histories

It seems one cannot help being drawn to considering the first, fundamental electronic music moments beginning in deep time, in the world's earliest reflections on the nature of voice and sound, of the relationships between sound frequencies and mathematical proportion, conspicuously evidenced with varying stone sizes and later lengths of strings, pulled tight and plucked. The one-to-one relationship between string length and frequency, note, and pitch could not be any more obvious and compelling. With the discovery of the relationship between varying lengths of strings and also pipes and varying sound pitches, the groundwork opened to rational, sensual, as well as mystical understandings of sound and then music. Those with mathematical skills could see deep into the possibilities of numbers and their relationship to sound. Multiple notes with their multiple string and pipe lengths could be measured, set, and reproduced. Complex sound waves could be created by simultaneously activating strings and pipes of varying lengths. Architecture and space could also become a metaphor of numerical proportions, celebrating worldly and unworldly sound.

Ptolemy and Pythagoras wrote early texts reporting these realizations. A romance with sound and mathematics was well on its way. Early texts described relations between sound and proportions and ratios as well as numerous tuning systems. In 1863, Hermann von Helmholtz published the celebrated "On the Sensations of Tone as a Physiological Basis for the Theory of Music", laying out a scientific and philosophically based understanding of sound, its cyclical nature, sound waves, and wave forms in the physical sense that had surprising properties via their fundamental frequencies that included the newly discovered partials, the multiples of the fundamental frequency.

> Raise the dampers of a pianoforte so that all the strings can vibrate freely, then sing the vowel *a* as in *father, art*, loudly to any note on the piano, directing the voice to the sounding board; the sympathetic resonance of the strings distinctly re-echoes the same *a*. On singing *oe* in *toe*, the same *oe* is echoed. . . . The vowel character of the echo arises from the re-echoing of those upper partial tones which characterize the vowels. . . . The musical effect of the resonance is compounded of the tones of several strings, and several separate partial tones combine to produce a musical tone of a particular quality.
>
> (von Helmholtz, 1863: 61)

He continues, "The upper partial tones corresponding to the simple vibrations of a compound motion of the air, are perceived synthetically, even when they are not always perceived analytically. But they can be made objects of analytic perception without any other help than a proper direction of the attention" (p. 65). The specific combination of fundamental tones and their overtones could create a sound wave of any tone. A world of sounds could be created with mixes of basic sound elements. The understandings of the physics and acoustics were there. The missing pieces, missing for a relatively short time, only decades, were the capabilities of creating sounds by electronic means. By tinkering and experimenting, and after numerous failed electrical circuits, the basics of stable electric oscillators became known, the creation of wave shapes became possible, and the triggering and combining of wave forms became the basis for a new sound world with new, evolving possibilities. Engineers and experimenters with varying mechanical and electrical and electronic experiences across Europe and in the United States set out to follow their curiosity and imagination, to reconsider what an electronic musical instrument might be. Tapping into the waves of progress that were driving industries and militaries to develop electronics, these inventors, electronic music pioneers, were often connected to research universities and research groups. A few were independents, and more recently, independent researchers and do-it-yourself approaches have led the way to innovations. Overall, they significantly shifted the electronic means to serve worlds of music, culture, art, and philosophy. They did not invent the electronics, but they did embrace the quickly evolving electronic tool kit and used its elements in new configurations to make new sounds and new spaces. Instruments were conceived and built. Musicians and composers took on the challenge of performing and composing with, and for, these new means; call them instruments, tools, systems, etc. The early literature, and popular press itself, oscillated between enthusiastic reporting of new instruments and negative dismissals of noise tools. Meanwhile, for several decades, composers were imagining their cultural instrument set expanding, their tone color experiments expanding, their music and culture space via radio expanding. The historical bridge between Pythagoras and Helmholtz and Varese, Eimert, Stockhausen, Ussachevsky, Cage, Cardew, Rowe, Oliveros, Conrad, Ashley, Carlos, Kraftwerk, and Eno is stunning. They explored mathematics, formal operations, multiplicities of infinities, sound clusters, pitch being replaced by duration, silence as a kind of presence, sound/music/moving image/film intersections and disruptions, set theory, and open systems with interchangeable sets. These included concepts of modularity, patching, and real-time generation and control of sounds, additive and subtractive processing, tracking sounds, triggered events, the sound signal, the physics of sound, vibrations, electronic principles, modulation of dynamics in registers of high, medium and low, and other developments such as psychoacoustics. Sound becomes access to, and articulation of, the real; analog electronics as physics and meaning, material meaning. The transformation of real events into sound. The actual, the real, the signal, sound physics; it is happening. I am hearing it. Hearing a bit. Thinking a bit. Being a bit.

Countering this, is the early Christian Gnostic dream of the self getting rid of the decay and inertia of material reality and floating from one selected embodiment to another. This has a contemporary ring to it. The bureaucratic scientific-technological is lurking. The recent cyberspace ideology, the dream of being freed from the attachment of the natural body, has historical roots. Now a rupture is taking place, the intervening of a new emerging non-essentialist materialism. The material, including the electronic as material, is being re-considered as multiple performance and event. The potentials of language and psychic subjectivity once again manifest. New musical orders, the known and unknown unfold. The intelligible and the illegible are productive partners in this moving sound and picture art. Woody Vasulka and Steina, new media artists, use phrases such as "dialoguing with the systems" to describe their real-time technology art interactions (Hill, 1995: 10). Media artist Ralph Hocking refers to the "artist as re-inventor of tools" ("Dialog with Ralph Hocking . . .": 25). He also, in conversation with the author, described, "machines for thinking with one's fingers", as key to his sensuous explorations. Hocking's media arts center, E.T.C., is a project of education, art, and technology.

We are currently experiencing a shift in the sensible, shifts in what is considered an event, access to what was earlier hidden and obscure, and now presents itself as articulate, as time-based experience and art. The time of historical development and preparation was long. The period of 20th-century experimental electronic implementation took place as Western music was evolving in parallel and absorbing new ideas, including Futurism, Dada, 12-tone music strategies, the aleatory, process sound, game explorations, improvisation, jazz, pop, electronic, and minimal music. Now we see electronics fueling global techno and rap music as well as electronics expanding as network infrastructure capabilities, AM/FM/shortwave radio re-surfacing as streaming/distribution genres, listening opportunities. Recording media, such as rolls, disks, vinyl, cassette tapes, CDs, new streaming archives, have evolved music/sound/genres/forms. Contemporary sound and electronics are significant components of moving image production, fine art screenings and installations, and moving image centers, as well as art centers, galleries and museums, black box and white box events, home theaters, and house music.

Amid personal computers, tablets and cell phones, the ubiquity and commercialization of electronics, the electronic game industry, and capital electronic investment and expansion, in all this flurry of electronic materialization and speculation, the fact remains of the great pleasure and curiosity involved in sound and sound experiences. The genius of sounds and electronics manifests forms, material actualities providing adventurous emerging experimenters, artist makers with the means to make practical and functional music for our time, regardless, or in spite of, or in tandem with large-scale industrial and military interests. Peter Wollen in his influential article "The Two Avant-Gardes" (1972: 92), named the two production systems that live side-by-side and yet often do not coincide, the co-op avant-garde and the commercial avant-garde. His articulation broke through what had been a long-running cultural divide or cultural bias of marginalization and censorship.

Tom Rhea, electronic musical instrument historian, states in an essay in "The Art of Electronic Music, The Instruments, Designers, and Musicians Behind the Artistic and Popular Explosion of Electronic Music", that the acceptance of electronic instruments historically happens across several stages (1984: 40). There is the initial period when the public is skeptical and which is fueled by resistance from the musical community. Then there is the moment of intense interest after the instrument is used successfully or is discovered by the musical community. Then there is the popularization by those devoted to the instrument or those marketing the device. The success, or lack of it, depends on the public and cultural response.

We might now add the cultures and subcultures that embrace and use the plentiful and wide array of electronic instruments, in their original configurations as well as in hybrid combinations and emulations. In the world of electronic tools and instruments, notions of variability and difference are alive and well. Surprisingly, unlike movie theaters, the evolution of concert halls and auditoriums has seen a surprising resistance to electronics in their basic accommodations and design. Furthermore, over one hundred years of electronic musical instruments and electronic music culture have not seen, apart from a few exceptions, the new instruments and imagination of our time manifest in the instruments of the symphony orchestra. It is curious that the Western canon of classical music that protects and celebrates the history of the remarkable adventurous and inventive musical ideas of the past, stops short when it comes to include our more recent electronic music history. One can imagine, as classical musical culture negotiates futurity, that electronic instrument sounds will be part of the future orchestra. As celebrated new composers increasingly embrace the larger instrument set thinking, we will see what has been a long-standing barrier come down, no longer desirable, productive, or livable, and electronics will be a common practice for our classical and contemporary music tradition.

Electronic media arts, or new media arts as it is often called in the current vernacular, is a vibrant if under-reported global contemporary cultural form; call it art, sculpture, moving image and sound, installation, performance or Internet practice. Call it the new essay, data visualization, or to use media visionary Nam June Paik's terms, "electronic serendipity" or "engineering accidentally". As photography once struggled to gain worldwide recognition as an art form, now electronic media arts finds itself beginning to settle into its status of the contemporary multicultural form in networking and clouding.

Electronic media arts have intermittently been an art-world darling (daring), participating in the larger administrative organizational dances of journalistic vollies (follies), tradeshow hits, and mass media probings. They are Internet social media fodder, going viral, hysterical advertising, and market-gorging. The sightings are regular, even if the understandings or meanings, are still unassigned; floaters and stem cells waiting to be more sensitively culturally coded. Cinema and audio, as art forms, once yearned for legitimacy and are now leading cultural and critical forces. Writer, photographer, and filmmaker Hollis Frampton once jokingly and challengingly quipped that filmmaking and its concepts and discourses were 20 years ahead of the fine arts and its writings. I believe this to still be the case. Surprisingly, now, we have confirmed the legitimacy of the cinema and audio with the establishment of future-looking academic new media arts programs, digital design programs, video game programs, all paralleled by newly arrived academic media theorists and historians. My own sense is that the electronic arts, in its varied manifestations, is poised to be the significant contemporary Lingua Franca, the contemporary present form of global human expression. Media's ubiquitous presence does challenge. Anthony Braxton's notions of Restructuralism, Stylism, and Traditionalism help us sort out and focus some differences, as well as provide important backwater spaces, percolators of experimentation and invention, small and large. They also provide necessary outsider positions, for example, where Edin Velez in his video sound work "Dance of Darkness" (1989) describes the importance of outsider spaces where Japanese Butoh dance and sound can continue to live and flourish.

Surprising to some, there are now thousands of analog electronic music sound recordings in existence. As they become available, they are inspiring analog electronic sound moments and events, across contemporary genres and forms including young DIY stars' practices. The DIY practices are nudging some electronic toolmaking companies to look at a new group of instrument users, who choose instruments for their productive hacking and modifying potentials. Some of those instruments allow for potentially open systems with clearly labeled circuit board soldering points for possible future circuit tampering and play, a new kind of hardware modularity, with open spots for control and interaction expansion. As companies such as Korg and others discover the value of the transformation potentials of their own hardware, the more we will see DIY-friendly boxes. At one time, the endless frontier looked like the sole purview of software development and coding, a proper and esteemed writing practice. Coders, carrying their badges of power and status, like designers, claimed a special place in the labyrinth of builders. Now hardware development is significant again.

We are in an era described by the reconstructuralists and new philosophers as the new materialism, reality, and art. We are at a moment with changed nuances for the notion of the obsolete, a term you can see I am happy to problematize. The shift is taking place; the need for a reconsidering of material reality, material researchers, a reorientation regarding the physical body, and the shift is profound. The former linguistic shift has left us short. The material reality shift we are seeing includes the world of analog electronics. We know that sound waves are real material phenomena transduced to electrical form, spectrums, signals ready to be plastically transformed and altered. After a generation of remarkable digital software and sound development, we now again see various analog hardware control and signal processing systems newly designed and manufactured. In numerous configurations, we see new stand-alone analog electronic music modules, hybrid analog and digital systems and historical analog instruments digitally modeled and reproduced.

We are now at another historic, technological moment regarding electronic sound generation and control. Systems once considered obsolete are again in cultural play and production. In this case, analog electronic music tools and instruments are rediscovered and reclaimed as functional, aesthetic and cultural necessities. We see lines of older material actions, analog electronic instruments, material signal-based events, electronic primitives, sine waves, square waves, and sawtooth waves, once again active, interesting, and even necessary. Is it interesting? Yes. What does it do? It opens our minds and imaginations.

Not only are analog electronic music modules being emulated in digital hardware and software, including digital plug-ins, we now are seeing the surge of a new generation of analog electronic music/sound modules being designed and built. The Eurorack, designed in the USA, is the format used for a vast array of modules. One of the most well-known companies in the Eurorack analog electronic instrument field is Doepfer Musikelektronik, by Dieter Doepfer. The early developments were inspired by the German electro pop band, Kraftwerk. As of 2017, there are over 1,000 analog electronic music Eurorack modules available on the market (www.modulargrid.net), manufactured by a large number of companies. Also, modular analog video synthesizer modules, Eurorack-compatible, are available by designers David Jones, D. J. Design Co., and Lars Larsen, LZX Industries. The electronic instrument tool set continues to grow. Possibly, the future will show more people performing electronics than viewing and listening to recordings.

The Belgian-Australian composer, musician, and multiple Grammy Award winner, "Gotye", Wally DeBacker, is involved in a project in Melbourne, Australia, which provides subscription-based access to a range of functioning historic analog electronic music instruments. The project is designed to allow contemporary composers and musicians access to historical studio instruments and ideas. There is also a significant collection of functioning electronic music instruments at the National Music Center in Calgary, Canada. In Amsterdam,

Holland, STEIM (Studio for Electro Instrumental Music), which has existed since 1969, is a significant center for research and development of new musical instruments. The Experimental Television Center (ETC), an education, art, and technology project located in upstate New York, established in 1972, fosters the development of artist-designed analog and digital sound and video instruments. And the on-line journal e*Contact!*, edited by Jeff Chippewa, is a comprehensive web-based source of information regarding electronic music and instrument culture and history (http://econtact.ca/index.html).

CD distributors over the past years have made available significant electronic music/new music releases. These include Deep Listening, Lovely Music, Columbia Princeton series, and the Stockhausen Verlag, including an array of distributors of Krautrock music. Additional scenes and movements surface today: the cassette underground, laptop music, and noise. With the availability of digital high-definition music formats, such as FLAC, a free lossless audio codec, the streaming of the immense archive of analog and digital electronic music will allow us to catch up with the discoveries of the recent musical past. This will no doubt impact the making of new music as well as new listening styles and, ultimately, new practices and understanding of sound, music, and cultural history.

The Obsolete

The term "obsolete" is a word that is commonly used, often casually spoken, and lightly considered. In fact, the term, the command, the slogan, is by no means neutral. It is culturally loaded. Is it a term to turn off debate? A term with a hidden back story, a term with a second term as secret life? Is obsolete, absolute? Is it a second-nature term, as from Roland Barthes's book *Mythologies* (1972 [1957]), "as we all know"? He sees contemporary mythologies as artificial intelligence stories told and retold uncritically by cultures and communities. "I resented seeing Nature and History confused at every turn, and I wanted to track down, in the decorative display of what-goes-without-saying, the ideological abuse which, in my view, is hidden there" (Barthes, 1972 [1957]: 11). "Ancient or not, mythology can only have an historical foundation, for myth is a type of speech chosen by history: it cannot possibly evolve from the 'nature' of things" (ibid.: 110). The chapter titles in the book *Mythologies* include "The Brain of Einstein" and "The Face of Garbo"; Barthes examines reappearing narratives that appear so natural and self-evident but are, in fact, constructions and fabrications; seemingly natural, but not natural at all. They are constructions, evolved cultures. Is the obsolete one of our contemporary mythologies, one of our shared partial narratives, functioning side-by-side with Einstein's Brain and Garbo's Face? Is the speculative or required contemporary mythology, that of thinking "outside of the box", sending us on journeys of additive strategies or exclusionary strategies, combining Einstein's brain with Garbo's face, or selecting neither, selecting what they are not? Which mythology is obsolete?

The obsolete, what is it, this often terrible judgment, this serious command? What other words does the term "obsolete" live with? Obsolete as the concrete performative, the judge's command, unconsciously linked and shadowed by the term "absolute", a performative command. We know of innovation and progress, experimentation, tinkering, and speculation. And we also know of financial processes such as expiration dates, shelf lives, quarterly growth statements; bureaucratic necessities challenging the access life of products. The obsolete refers to a stage of aging past a state of functionality. There is the interruption of industrial supply lines due to unavailability of parts. There is also the strategy of forced and planned obsolescence. A more general and vaguer use of the term, "Obsolescence" is used in a casual, vernacular way to refer to items no longer new, no longer this year's upgraded model, deeming older models as possibly undesirable and thereby granted the status of "obsolete". Is this more a function of status labeling than of functionality? It is often used as a strong and calculated strategic term of dismissal. As a judge in performative form announces a life-changing verdict, obsolescence is a verdict put upon an object or process or event or spoken from a position of presumed cultural power and influence. Is it really obsolete? Is it still usable? Does it still work? Can you still get the parts? Is it broken without the option of repair? What about systems of planned and predatory obsolescence, when companies are purchased and their products are discontinued? Is there a shelf life, a measurement of time, a temporal register? Can uses be obsolete? Industrial production suggests a sequence of new, broken, fixable, and unfixable. Painting is dead; in fact, every few years, painting is challenged and announced to be dead. Is it the same with obsolescence? As legend has it, Mark Twain is said to have stated, regarding his obituary reported in an American newspaper, "Reports of my death are grossly exaggerated."

The term "obsolete", whether correct or useful, is a term designating a symbolic position, presenting a judgment for a determined future, in the game or out of the game, center stage or relegated to the side. Meanwhile, the very notion of obsolescence could be considered an exaggeration, a judgment game, itself a side show, a distraction from the more interesting and more complex, difficult, and more nuanced consideration of cultures and needs. The increasingly celebrated market metrics and market rubrics do in fact hide and censor other questions and considerations. For example, consider human phenomenologies, experiences, and textures expanded by evolving philosophies, manifest in images, objects, and sounds, stationary and moving. World history is overwhelmingly filled with individuals moving these concerns forward. The issue is not whether something is obsolete or not, but rather if it is relevant and interesting. The unknown is an adventure.

Looking at the expanded timeframe suggested here, we see the long trajectory of particular ideas and concerns. Obsolescence of varying sorts comes into play inevitably as material strategies and material means change and evolve. What changes when we consider specific objects as instruments as opposed to utility tools? The drivers, the ideas and their manifestations, continue to move forward as they are necessary and achievable, with ruptures and surprises. Analog electronic music is an idea, long imaginable and now manifest and variably mutable in new forms.

Sound recording technologies are historically victims to the notion of the obsolete. We could say they have a short shelf life. Recording technologies such as the lacquered tube, vinyl disc, wire, and metal particle tape recording technologies in the form of open reel-to-reel, varying cassette formats, sound stripes on celluloid film, and high-quality sound on S-VHS video tape, all are now historical legacy materials, now considered obsolete. This is related to the moving image, with its optical lens-based image stored onto film and then lens-based electronic signals stored onto magnetic tape, onto DVD disks, and now onto portable memory chips or in vast memory chip arrays, we imaginatively call clouds.

Microphones have had a good run. The real acoustic sound continues to be transformed into electrical signals, the electro-acoustic.

We might consider multiple formats metaphorically as multiple languages. The contemporary analog-to-digital and digital-to-analog box is a converting, translating system. The current box converting digital HDMI to analog composite video and audio is a bridge between generational forms.

Physical vinyl records are again stimulating peoples' imaginations. They are again manufactured and distributed. The command of "obsolete" for vinyl records has been recalled. New releases of vinyl music are now accompanied by high-resolution digital files, downloaded from the Internet. The large format of vinyl records has added to the interest in the vinyl format. A younger generation is interactively engaged in the roughly 12-inch square visual format with custom covers, inserts, and decals. Radio amplifier tubes, once rare, are now manufactured again. Tube sound amplifiers are sought after.

Culture influences what is obsolete or not obsolete. The obsolete is not absolute. It is tentative, poll-tested, opportunistic, negotiated, not reversible; then surprisingly reversible. The obsolete status, following advertising pronouncements and economic contracts, is not a social contract. The obsolete is more varied and unpredictable than the Darwinian, survival-of-the-fittest bargain. Technologies do, in fact, become less and more desirable over time. There is a beginning, middle, now, and next. There is a beginning, now, and unexpected. There is also a now, a middle, a beginning, and a surprise.

References

Badiou, Alain; and Élisabeth Roudinesco (2014) *Jacques Lacan, Past and Present, a Dialogue*, New York: Columbia University Press. First published in French as *Controverse*, Éditions du Seuil.

Badiou, Alain; and Jean-Claude Milner (2014) *Controversies*, Cambridge, England and Malden, MA: Polity Press. First published in French as *Controverse*, Editions du Seuil.

Barthes, Roland, *Mythologies*, New York: Hill and Wang, 1972. First published in French by Éditions du Seuil, Paris, 1957. Translated by Jonathan Cape Ltd.

"Dialog with Ralph Hocking, Sherry Miller Hocking, Carolyn Tennant, Peer Bode and Pamela Hawkins", <www.experimentaltvcenter.org/sites/default/files/history/pdf/Tennantralph_2729.pdf>, 2005.

Helmholtz, Herman, *On the Sensations of Tone as a Physiological Basis for the Theory of Music*. Translated, thoroughly Revised and Corrected, rendered conformal to the Fourth (and last) German Edition of 1877, Dover Publications, Inc., New York 1954.

Hill, Chris, "Interview with Woody Vasulka by Chris Hill", 1995, Hill/Vasulka, *The Squealer*, Squeaky Wheel, Buffalo, New York.

Lock, Graham (1988) *Forces in Motion, The Music and Thoughts of Anthony Braxton*, New York: De Capo Press; an unabridged republication of *Forces in Motion: Anthony Braxton and the Meta-reality of Creative Music*, London, England: Quartet Books Ltd, 1988.

Paik, Nam June (June 1964) "Afterlude to the Exposition of Experimental Television", *Fluxus cc Five Three.*

Paik, Nam June (1973) "Videa 'n' Videology 1959–1973", *Global Groove 2004*. Deutsche Guggenheim, Syracuse, NY: Everson Museum of Art.

Rhea, Tom (1984) "The Art of Electronic Music, the Instruments, Designers, and Musicians Behind the Artistic and Popular Explosion of Electronic Music". Collection of articles originally published in *Keyboard Magazine* (GPI Publications, edited by Tom Darter) from 1975 to 1983.

Wollen, Peter (1982) "The Two Avant-Gardes". In *Readings and Writings, Semiotic Counter-Strategies* (pp. 92–104). London: Verso Editions and NLB. "The Two Avant-Gardes" was first published in *Studio International*, December 1975.

Zielinsky, Siegfried (2006) *Deep Time of the Media: Toward an Archaeology of Hearing and Seeing by Technical Means*, Cambridge, MA: MIT Press.

12

ARMCHAIR HARMONICS

Radio Remote Controls and the Historical Persistence of Push-Buttons

Brent Strang

The remote control, as a device, might well be on the road to obsolescence. Users, increasingly, are controlling their screens with laptops, game controllers, smartphone apps, voice, and gesture control. Perhaps it is time to empty our junk drawers and say good riddance to those dusty, unused, button-cluttered hunks of plastic. The remote control has long been an object of frustration, a cantankerous and uncooperative creature that multiplies rapidly and refuses to get along with others of its kind. But even as users become accustomed to new devices and habits, there is one feature of the remote control device that remains stubbornly resistant to obsolescence—the hard push-button.

For the past two decades, the peripherals company Logitech has been at the forefront of research and development of remote control technology for the consumer market. When I asked Ian Crowe, Logitech's director of product management for Harmony universal remotes, about the general trend toward streamlined touchscreen interfaces, voice control, and the elimination of hard buttons, he said he is not convinced that these trends are replacing the demand for traditional controllers. Physical button-placed remotes might not be "a permanent situation," he said, "but for as long as the people who are buying electronics are people who grew up with a certain way of doing things, they're going to continue using that, [and continue with] that way of doing things."[1]

Crowe's statement evinces Logitech's unique insight of how to capitalize on the historical persistence of users' habits, and how to proceed with the delicate art of introducing new design changes incrementally. Crowe is also implying here that buying and using cycles are generational, and thus, any major changes in remote control design would be generational as well. While there have been marked shifts in the remote control device's technology and functionality, as well as aesthetic alterations, certain fundamentals have remained more or less the same. From the first radio remotes in 1929 to Logitech's latest Harmony model, the remote control's evolution has retained a fairly conservative surface appearance: dark-colored and smooth-contoured, with hard buttons and rectilinear casings.

There is much packed into Ian Crowe's phrase, "a certain way of doing things". Something about it suggests more than a mere preference for hard-button remotes over soft-button smartphone apps or touchscreens. Consumers have maintained an association between remotes, buttons, and home entertainment for the better part of a century. During this time, remote control has insinuated itself into our everyday surroundings, impressed itself into tactile aspects of human apperception, and entwined itself with cultural meanings of instant gratification and technological fetishism. The remote control has come to occupy an intimately familiar place in our hand and in our subconscious; it has a structuring role in what we do and what we imagine we will do when we are presently consuming media or deliberating the purchase of new devices. This chapter seeks to locate the genesis of Americans' cultural bond with push-button remote controls, through study of a media device that vanished from our cultural memory long ago—the radio remote control.

The scope of this study begins with the first push-buttons on electrical devices and ends with the introduction of wireless radio remote controls before the outbreak of World War II. The main focus is the United States in the 1930s, specifically the technical, discursive, and cultural vectors that intersected and mutually influenced each other as the scene of remote control stabilized over the course of the decade. By 1942, a wartime ordinance ceased production of radios. When radio production resumed after the war, so, too, did television production. The scene formerly established by radio remotes would set the template for televisual remote control, not only in the way the remote control mediated spatial relations in the living room, but also the technologies within the devices themselves.[2]

It is important to state that radio remote controls were not initially popular or as ubiquitous as they became in the latter part of the twentieth century. In his study of the technical evolution of radio remote controls, Patrick Parsons estimated that only 4% of radios sold during the 1930s had remote controls.[3] With almost 49 million sets sold in the US during the 1930s,[4] we can extrapolate around two million US homes used radio remote controls. Even so, a thorough study of the origin of habits and practices needs to explore beyond the period of mass diffusion and ubiquitous use. Imaginaries precede habit formation. The socio-technical emergence of a phenomenon involves more than widespread practice, and more than technological innovation, it involves most fundamentally the coalescence of discursive formations—in other words, how popular culture begins to make sense of technical innovation.

Despite the relative rarity of radio remote controls in US homes, Americans knew very clearly by the end of decade what they were, and what they signified. The remote control was figured as a beacon signaling the future style of media consumption. It worked in concert with contemporary objects and innovations, from push-buttons, to station selectors, to chairside radios, that were all oriented toward the same *scene* that would come to dominate the living room, dens, and "TV rooms" of the twentieth century.

The following example is a synecdoche of how imaginaries pave the way to habit formation. In early 1932 *Radio Craft* reprinted a technical article by Dr. James Robinson describing his latest invention, the stenode radiostat. The device was expected to vastly improve the selectivity of radio and (it was hoped) television receivers, which were still in their primitive phase. In the bottom corner of the page, an illustration showed a man in his living room watching television with remote control (Figure 12.1).[5] Electronic television would not be mass manufactured in America for another seven years; nor would remote controls be sold with televisions for another seventeen. While the drawing incorrectly predicted "the television receiver of the future" as a tripartite extension of the radio chassis with separate screen and speaker, it accurately predicted the scene that would take place in virtually every American living room for the latter part of the twentieth century. Though it would be several decades before this habit took hold, the lucidity of the scene envisioned by the journal's editors is striking. It distills the intentionality of the radio and television industry's multiple actors and their collective resolution to establish armchair media consumption as an everyday practice in domestic space.

Before proceeding with an historical overview, I offer a brief qualification on this chapter's method. On numerous occasions I cite an invention, technical development, or use of terminology as a "first". However, this should not convey the impression of a deterministic

Figure 12.1 A 1932 illustration predicts the habit of televisual remote control while remote controls are still nascent, and several years before television is mass marketed in the US.

and linear history, where "great men" inventors break new ground and social history reacts in its wake. In almost every occasion when I note a key development as a first, there exist prior instances of patents, products, and people that remain obscured from that trajectory, even though they have contributed in some form.[6] Neither do I mean to imply that these trajectories are purely technical, as though emerging in a socio-cultural vacuum. Rather, the propensity for specifying dates and noting firsts is to explain push-button remote control as an incremental, material-discursive emergence. Dates and technical developments are therefore noted and contextualized alongside discourse analysis of trade press articles and advertisements. I proceed from this dialectic understanding of the socio-technical interface in order, finally, to narrate an account of how remote control coalesced in the cultural imaginary and, slowly, became a habit of everyday life.

Trajectories of Remote Control, Buttons, and Mechanization

The origin of remote control is commonly traced to Nikola Tesla. In 1898 he filed a patent for a "Method of and Apparatus for Controlling Mechanism of Moving Vessels or Vehicles", which he demonstrated with a four-foot radio-controlled boat at Madison Square Garden's first Electrical Exhibition. There were other examples of wireless command at a distance that had actually preceded Tesla's, including Marconi's demonstration with wireless telegraphy in 1896,[7] but historian W. Bernard Carlson claims that Tesla's presentation nonetheless popularized the concept in culture and engineering practice.[8] Still others have challenged whether any of these early examples should really count as "remote control" the way we understand it today because their technologies used only simple on/off commands.[9] By contrast, Spanish engineer Leonardo Torres Quevedo invented a device in 1902 that controlled vessels by sending and receiving coded signals from a specially developed set of codewords, each with distinct functions. Quevedo's device was thus able to transmit up to nineteen different commands to steer an unmanned aerostatic balloon. Still, neither Tesla nor Quevedo termed their invention "remote control": Tesla referred to his invention as a "teleautomaton" and Quevedo dubbed his the "Telekino". At the time, command-at-a-distance technology was known in the engineering profession as "telemechanics".

The earliest instance of the phrase "remote control" located through a Google N-gram search appears in a 1901 volume of the journal *Electrical World and Engineer*. General Electric developed "remote control switches" for use in central booths and sub-stations outfitted with banks of circuits for control of power generator units, city arc lamps, railway car motors, etc.[10] Throughout the decade, the phrase gained currency mainly in industrial settings, where dangerous machinery and high-potential current increasingly necessitated switchboards to control operation from a safe remove. A 1909 volume of *Electrical Review and Western Electrician* features a printing press motor controller (again manufactured by G.E.) with "remote push-buttons" to enable precise stopping or "inching" of a job to completion.[11] In the same volume, an article titled "Wireless Telecontrol" detailed experiments in France, Sweden, and the United States that attracted interest from the military.[12] The author described the inventors steering dirigible torpedoes and balloons with a "remote control"—which is perhaps the first time the phrase was used the way we have come to commonly understand it—that is, as a noun referring to a transmitter for complex control of a machine from a distance.

Meanwhile, the prevalence of the push-button coincided with the widespread introduction of electrical devices invented between 1880 and 1915, which included lights, office desk bells and doorbells, annunciators, flashlights (called "pocket lights"), fire alarms, and elevators. In her cultural history of the button, Rachel Plotnick maintains that, in everyday life, "push buttons achieved the most notoriety as a symbol of effortless machine interactions", as distinguished from the mechanical levers, spring mechanisms, and electrical switches that had been more commonly used to operate industrial machinery.[13] Though different social actors invested the button with different meanings at different times, Plotnick's central conceit is that the simplicity of the interface—its binary on/off functionality and ordinary material design—were responsible for the ease with which culture black-boxed the complex process of electrical wiring and transmission beneath the surface. Carolyn Marvin cites an anecdote from an 1895 magazine, *Science Siftings*, of a gentleman who complains to a lady that he does not "'understand how the incandescent light, now so extensively used, is procured.' 'Oh, it is very simple,' said the lady, with the air of one who knows it all. 'You just turn a button over the lamp, and the lights appear at once'."[14]

Sales and marketing literature employed the rhetoric of effortlessness and instant gratification extensively during this period. The push-button availability of electricity was mystified with magical topoi, most notably the figure of the genie. To promote "America's Electrical Week" in December of 1916, the prize-winning poster depicted a "Modern Aladdin" pushing a button to summon an illuminated genie from the darkness. A description of the poster in *Electrical Review and Western Electrician* read: "Gone is the ancient lamp. Now it is the gentle touch of a button and forthwith comes the Genie, Electricity."[15] In the same trade journal, authors often dubbed electricity and the telephone as the "genie of the wire".[16] Through such metaphor and rhetoric, combined with the deceptive simplicity of its appearance and mechanism, the button emerged as a paradigmatic emblem of reification. It worked not only to conceal the complexity of the underlying circuitry, wiring, and scientific processes, but also the social relations between household consumers, utility companies, and their regulators and laborers.

By 1917 less than 25% of American homes were electrified. There was, however, a dramatic increase in electrification in the years following World War I, reaching almost 35% by

1920 and over 68% by 1930.[17] The interwar years comprised what Siegfried Giedion dubbed the period of "full mechanization" in American homes. The dream of the kitchen with electric range, broiler, kettle, and saucepan, first demonstrated at the 1893 Chicago's World Fair, was now within the grasp of most families. As the makers of appliances reoriented for their massive push into the domestic sphere,[18] they followed the decades-old strategy of targeting the household's women.

The influx of "labor-saving" electrical devices served as a timely aid to America's ongoing "servant problem" and the steady transition to the servantless household.[19] As Ruth Schwartz Cowan observed, after World War I advertising imagery for household appliances eschewed servants, replacing them with "housewives, neatly manicured and elegantly coiffured".[20] The representation of domestic household appliances throughout the 1920s has been well documented by Sherri Inness in her book, *Dinner Roles*. Articles and advertisements in women's magazines and cookbooks commonly referred to "electric servants" and "push-button maids" to trope on the transition to the servantless household. A 1924 article in *Ladies Home Journal* titled "Aladdin's Newest Magic Is the Touch-a-Button Kitchen", used yet another topos common in the day to reinforce a prevailing theme in popular culture. The button was (ostensibly) eliminating labor and drudgery from everyday life, starting in the place where it matters most—the home. It is in this combined context of the rapid diffusion of electrical appliances and the increasing hours of labor for the households' women, that American domestic life experienced another change in the popularization of radio.

The First Radio Remote Control

In 1922, the same year that the *New York Times* declared radio the "most popular amusement in America",[21] the US Department of Commerce began gathering census data on consumer radios. Between 1922 and 1930, the number of American households with radio sets rose dramatically from 60,000 to 12 million—amounting to over 40% of the nation's homes.[22] Prior to this period, radio was largely the purview of the hobbyist and experimenter, who operated small, often homemade, crystal sets with headphones for two-way communication over short distances.

In order to stabilize radio as a medium for mass entertainment, fit for every home, marketing rhetoric had to eschew the longstanding association with amateur radio, with its gendered connotations of "manly" hobbyists, and messy technical apparatuses. Not surprisingly, the radio industry followed the same strategy that had been successful for kitchen and household appliances. Several magazine and trade press articles sought the advice of home economist Christine Frederick. In a 1925 *Radio Industry* article titled "Ten Suggestions to Help Dealers Sell Radio to Women" Frederick encouraged marketers to "take a leaf out of the book of washing machine, gas range, vacuum cleaner and other makers of the well-appointed home."[23] In 1922 newly promoted executive vice president of RCA David Sarnoff predicted that radio would become an essential household appliance in American domestic life. Comparing its necessity to indoor plumbing, he opined, "One constitutes a cleanser for the body and the other for the mind."[24]

Advertising pictured idealized settings with the family seated together in their living spaces, as radios themselves were increasingly designed to be "tastefully unobtrusive" and blend in with domestic furniture. As radio historian Louis Carlat puts it, these ads implied that "[w]ith only a slight adjustment of traditional furnishings and family habits . . . the radio could be integrated into the background of middle-class living space and life."[25] Over the course of the decade, both marketing and design were increasingly focused on differentiating "radio" from its forbearer, the "wireless". No longer a homemade tinkering device for men and boys to explore far away stations, radio's simplified tuning and ornate free-standing cabinetry was accessible to the whole family.

Despite the high percentage of American homes with radios by 1930, an overwhelming majority of these owners were "middle- and upper-class . . . white Americans, particularly native-born living in large cities."[26,27] A small segment of the population actually owned more than one radio, and it was toward these more affluent consumers that advertisers aimed their copy. Radio magazine ads illustrated couples entertaining guests in tuxedos and evening gowns. It was precisely this context of "conspicuous consumption" that Caetlin Benson-Allott argues set the conditions for the first consumer remote controls.[28] The 1929 Radio World's Fair at Madison Square Garden featured four radio receivers with push-button station selection that could be operated by remote controls, manufactured by Zenith, Kolster, Carter Radio, and Sleeper Research.

Together these manufacturers signaled the future of broadcast media consumption. It would no longer be an involved process of tinkering and constant adjustment while stations bleed in and out and over one another. Instead, stations would be homed in instantly at the touch of a button, even while seated at a comfortable remove from the receiver. No longer in thrall to the machine's whims, consumers would now exercise command at a distance through the magic of electro-mechanical robots. A few months before the fair, A. J. Carter, President of Carter Radio, blurbed about the state of the radio industry in *Radio Magazine*:

> Radio has been simplified to the point of one-dial control. The next radical change will be electric automatic tuning and remote control so that the stations may be selected at the radio set or at a distance by merely pressing a button.[29]

Even though they were cost-prohibitive for the masses, these high-end entertainment units were well worth their companies' investment. They were first of all, sold to upper-income consumers, where the profit margin was greatest. The 1929 Zenith 55A, for instance, cost $700—the equivalent of over $10,000 in 2017. Such high-end sets also functioned to help distinguish a company from the over six hundred radio manufacturers currently in competition by 1927.[30] But more to the point, they defined the image to which the average consumer would aspire and be able to afford in the coming years.

Delineating the Remote Control Device

Edgar Felix, a contributing editor to *Radio Broadcast* magazine, provided insight into how consumers initially felt about push-button tuning. In January of 1930, he wrote:

> The public has not found turning a dial to a desired setting, the correctness of which is easily checked by ear, such a trying operation that a mechanism of adjustable buttons, locking clamps, and flashing lights is anything to excite its enthusiasm. The public has recognized so-called automatic tuning of 1929 as the invention of despairing sales managers.[31]

Remote control, too, failed to excite the public upon its initial release: "Dealers report that the public estimate of remote control is that it is a $300 device enabling particularly lazy persons to press buttons at the end of a six foot cord rather than to reach for the tuning knob."[32]

It is striking to see how these two concepts, which contemporary users have taken for granted for decades, failed to excite the public in the first years of their introduction. But it is only in hindsight that the idea seems obvious, even brilliant, after having been habituated to remote control and having experienced the pleasure of instant gratification. To the buyer in 1930 not predisposed to this habit, the remote control was likely viewed as nothing more than an expensive gadget: the devices prominently featured at the World's Fair were all cost-prohibitive (costing as much as $300 equates to $4,400 in 2017!).

However, cost does not entirely rule out remote control's saleability, because cheaper options were then available. In the same issue of *Radio Broadcast*, an article describes in detail the operation of the Kinematic remote control, a device they claimed would only add $10 or $15 to a radio's price tag. We may deduce then, if anyone glimpsed the "obvious brilliance" of these concepts in 1930, it was a relatively small segment of the population—namely the industry people and inventors whose imaginations and livelihoods thrived by advancing radio's technical frontier. The author of the aforementioned Kinematic article (who is also the company's CEO) declared:

> The biggest feature of the Kinematic is not *seen* as much as it is *experienced*. That is, the device is habit-forming. Just, as from habit, you open your front door at night, and reach for the switch to turn on the lights, so you reach, with little direct attention, for the tuning knobs of the Kinematic, changing volume, tuning up or down to a program that is in keeping with your mood.[33]

Evidently, trade press articles needed to do a fair bit of description to sell prospective consumers and retailers on the idea of remote control, not only by detailing the operation, but by illustrating the domestic scene where habit formation takes place.

A similar article two years later elaborates the different functions of Stromberg-Carlson's latest innovation and the most advanced remote control of its day, the "Telektor" (Figure 12.2). Dubbed "the radio robot" by the author, the Telektor's twenty push-buttons controlled relays and motors to select stations, tune manually, adjust volume, and shift records in the phonograph. The article takes pains in describing the difference between buttons that manually tune by controlling motors and "must be pushed and *held down*" until the desired channel is attained, and buttons for station selection, which close the circuit of a relay and need "only [be] pressed for an instant".[34] The author marvels at how the one set is used for "cruising or hunting for stations", while the other is for saving your favorite stations.[35]

The Telektor article thus articulates a key distinction, for probably the first time in print, between two modes of push-button selection of media content. In the coming decades, the repetitive button pressing of these two alternatives—searching content or seizing upon it immediately—would lay the groundwork of habit and expectation and contribute to an ontological duality in media reception. Over the course of the century, users would come to expect this dual functionality on virtually all their media interfaces, from radio, television, DVD, and stereo remotes to the soft buttons on their browsers and digital music and photo libraries.[36] Button combinations labeled with active verbs such as scan vs. seek, fast forward vs. skip, shuffle vs. select, scroll vs. hyperlink, search vs. bookmark, signify the same bifurcative mode of consumption, despite the specificity of each particular function. The two paths mark the difference between "curious" versus "decisive" states of attention, between the user roaming, with weak or strong intentionality, versus knowing precisely what they want.

Before this cultural habit took hold, those with a vested interest in promoting remote control had to contend with numerous other meanings and associations for the phrase

Figure 12.2 Stromberg-Carlson's Telektor. (Photo courtesy of radiomuseum.org.)

"remote control" then in circulation. The phrase was mainly applied to industrial and military technologies in the first decades of the twentieth century. Starting in the 1920s, several other meanings sprang up in commercial and domestic settings. Radio stations referred to "remote control pickup locations" for recording live segments within range of their central broadcast antennae. Retailers installed "remote control" intercom systems in hotels to spread announcements and radio and phonograph music through their halls and function rooms. A 1924 article even promoted a program by Kansas State Agricultural College to broadcast classes by "remote control" from neighboring stations to farmsteads.[37] In 1930 a film titled *Remote Control* capitalized on the phrase's popularity and the cultural anxiety of radio invading people's personal spaces and influencing their thoughts.[38] The convoluted plot featured a gang leader posing as a psychic and directing the criminal activities of his minions through coded messages in his radio program.

Google N-gram shows the phrase sharply peaking through the 1930s and 40s, then declining and leveling off until the 1990s, when it surged again to the present. The profusion of different uses no doubt accounts for the early spike, while the leveling off phase corresponds to when "remote control" stabilized with its commonly accepted meaning—describing both the action and the object whereby a transmitter commands some electronic device at a distance. No doubt the phrase's initial popularity captured something stirring in the cultural imaginary, something expressive of the aspirations, expectations, and anxieties issuing from the confluence of broadcast radio and an emergent consumer culture. As households were busy outfitting their newly mechanized homes, they were acclimatizing to the notion of electricity coursing through their walls and appliances while radio waves permeated their domiciles. At the same time, they were getting a feel for the new interfaces—the dials, switches, and buttons that summoned power from unseen processes, which remained obscure and black-boxed. Remote control was an apt and suggestive phrase at this juncture. It described the process of harnessing remote EM waves and putting them to use in domestic space. It also suggested the luxuries and practical conveniences that an increasing number of people were coming to afford and expect. It also carried darker connotations: dirigible torpedoes, guided bombs, and indoctrination.

Within this landscape of meanings and uses, the radio industry advanced its own agenda for remote control through advertising and technical articles. Marketing techniques borrowed the familiar topoi of magic and instant gratification used earlier to market electricity and push-button appliances. Capehart Corporation dubbed their newest remote the "Aladdin Control".[39] Scott Corporation advertised their remote control with the motto, "Touch a button . . . there's your station!"[40] Promoters faced an obstacle, however, in the stigma, noted earlier by Edgar Felix, that such devices were conceived for the "lazy". The aforementioned Telektor article began boldly: "By the very nature of his existence, man is lazy. From time immemorial he has continually striven to minimize his daily work."[41] Before proceeding to situate the remote control within a long line of "so-called progress", the opening paragraph ends with a curious equivocation: "Whether laziness is a desirable characteristic or not depends entirely on the individual." Other promoters tried to mitigate the stigma by using similar tongue-in-cheek rhetoric to embrace laziness as the consumer's privilege. One author welcomed remote control as providing "a lazy man's paradise".[42] Another aftermarket remote option sold in the *Citizens Radio Call Book* with the title, "Remote Volume Control Is Ideal for the Lazy":

> If you are too-o-o-o-o tired to get up from your chair and turn off your radio set when the cooking expert gets to describing a favorite recipe for dill pickles, you may eliminate the pickles without leaving your chair if you fix up a variable resistor across the antenna and ground and run the two wires over to your chair.[43]

Scott Co.'s advertising took a more pragmatic approach by illustrating their remote in different settings, including the living room, office, hotel lobby, and retail store.[44] Claiming their remote was also "ideal for shut-ins" and the "bed-ridden",[45] one illustration depicted an infirm lady lying in bed, reaching for the remote.

Awareness of radio remotes was also bolstered by another device that has long since vanished from cultural memory—the automobile remote control. Custom searches through the 1930s' trade press magazines in americanradiohistory.com turn up roughly as many mentions for remote controls for automobiles as they do for home radios. For roughly twenty dollars, one could purchase and self-install an apparatus that would connect to the steering column from which they would be able to adjust the volume and tuning on their car radios. They were widely advertised from 1930 until the end of the decade because of the obvious safety hazard drivers courted when attempting to adjust their newly installed radios while driving. Though it was more of a safety device than an item of luxury and convenience, the plurality of home and auto remotes advertisements and technical articles strengthened the association of remote control with media devices.

In the absence of sales data, researchers can only glean the public perception of remote controls by commentary in the trade press. Though Edgar Felix dismissed it in 1930 as an expensive gadget, he also conceded the remote control's potential, provided it could be "intelligently merchandized".[46] In the next year, a *Radio News* writer commented that remote control was "not generally applied", but would inevitably be popular, especially to control volume, which was "a nuisance when the character of the program changes during the evening."[47] Articles between 1932 and 1934 indicated remote control had "increased demand",[48] had "long since passed the luxury stage",[49] and had even become "popular".[50] In these first four years, first- and third-party remote controls were prominently featured in the trade press, after which they began to disappear, resuming again in 1937 (for reasons later explained). Even so, "poor-man" options continued to be advertised from year to year. These were cheap, do-it-yourself installs or single-button kill switches. For $1.50 one could purchase and have installed, the "Blah Blah Eliminator" which would silence the volume on annoying commercials.[51] Time switches that turned off lights and appliances were also marketed as low-cost substitutes for remote controls. It is impossible to know how many of these after-market options were purchased, but given their consistent availability throughout the decade, we can reasonably assume that there was *some* middle-class diffusion, however little, which might have pushed the number of remote controls in American households somewhat above Parson's estimated 4%.

Settling into the Living Room

From 1930 to 1933, families were exposed to the idea of push-button station selection and remote control devices for radio, even if these luxuries remained out of reach for most. Much more affordable were the so-called "midget" radios and compact portable receivers, introduced in 1930 and 1932, respectively, and they could be moved from room to room. Portable receivers such as the Kadette were extremely popular at first, but the vogue waned somewhat by 1934, according to one author, who observed their impact on the future of remote control:

> [Compact receivers] brought attention to the advantages of being able to place the radio set in any convenient position where it could be reached and tuned with the minimum of effort, instead of having to locate it in the position, however inaccessible it might be, where it fitted in best with the other furnishings of the living room. It therefore seems safe to prophesy that receivers of the remote control type will develop a high degree of popularity because [they] . . . can be moved about from place to place . . . and thus always be kept within arms [*sic*] reach.[52]

The subject of this author's article was, incidentally, the functionality of Motorola's new "Lazy Boy" model remote control (so named three years after Edward Knabusch and Edwin Shoemaker dubbed their patented recliner the "La-Z-boy"). Evidently, Americans were getting familiar with the "lazy" trope and its associations with remote control and radio, as well as relaxation and the living room setting.

At this early stage in the remote control's conception, there were also competing visions of how to install remote control in the home. The traditional implementation was the armchair remote—a wired control that stretched to the radio cabinet—conceived for the living room setting. A second implementation dispensed with the cabinet and stored the chassis out of sight in the cellar or attic, connecting the desired number of speakers and remote control boxes by wire in various rooms throughout the house. The latter was introduced in a 1930 *Radio News* article, which claimed that the average family's listening habits had changed from "the old days" when people listened for one to two hours a day. Now, as people averaged between five and ten, the author declared, "[I]t seems illogical to limit radio enjoyment to the living-room."[53] In the following years, the makers of the Kinematic and the Telektor remote control promoted the idea repeatedly in the trade press, while boosting the added opportunities for the servicemen, architects, and builders involved in installation.

Indeed, the Kinematic remote control was marketed as a cost-effective alternative because it saved money otherwise spent on a cabinet in order to cover the expense of extra speakers and control boxes. To a large degree, radio makers were more in the furniture business than the radio business, and the quality of the radio chassis was often sacrificed for better craftsmanship of the cabinet. However, the boosters of remotely controlled "built-in radio" never cited the cost of installation in their articles, and, as it turned out, they were more successful in profiting from a few very affluent consumers than they were in offering a cost-effective alternative for the average family.

Just the same, one last big push to popularize built-in radio was featured in a four-part series of articles published by *Radio-Craft* in 1939. The magazine's editors oversaw the construction of an "average-priced" New Jersey bungalow with a built-in radio system manufactured by RCA that was supposed to be made affordable for the average American. Diagrams of the floor-plan indicated speaker units for every room (including the bathroom), each with its own push-button volume and tuning controls. Armchair remote outlets were also appointed for the living room, dining room, and kitchen. The author of the series, N. H. Lessem, boldly declared that the plan's widescale adoption would both "revitalize the anemic radio industry" and incentivize the building industries.[54] However, a key part of the plan depended upon the Federal Housing Administration's approval to allow the costs to be amortized within the homeowner's long-term mortgage. Soon after its inauguration as part of Roosevelt's New Deal, the FHA approved financing arrangements for public address systems in apartments, hospitals, schools, and other commercial buildings. The FHA was considering making similar concessions for built-in radio, provided installation could qualify in some way as "an integral part of the house". Hence, Lessem's insistence that radio is "as much an integral part of the house as plumbing, heating, and lighting systems".[55] After six months and four ambitious articles, it remained unclear whether the house's construction was ever actually completed.[56] In all likelihood, the FHA declined financing approval, since mention of built-in radio quietly vanished from future issues and popular discussion generally.

Without further research, one can only speculate as to why the FHA declined to finance built-in radio. It would have stimulated multiple industry sectors, and would have also (in keeping with the spirit of the times) democratized public access to a household convenience hitherto restricted to the rich. Surely Franklin Roosevelt, too, would have seen radio as "integral to the home", inasmuch as his fireside chats were integral to inspiriting the American public. But therein might lie the answer. The fireside chat invoked precisely that image—a family gathered together in one place. Perhaps it had something to do with preserving this image and disciplining the performance of the nuclear family: everyone assembled around father, who commands the dial or remote, together yielding to the nation's head patriarch.

Or perhaps the plan failed because of a recurrent complaint expressed by the public to the editors, which Lessem noted in his concluding article. The public took issue with the fact that only one program could be received at a given time, "thereby reviving the old family feuds (concerning program selection)".[57] Lessem tried to preempt this argument by advocating the purchase of additional "midget" receivers. This way, housewives could listen to their programs all day, as a midget could be "recessed into one of the kitchen cabinets"; meanwhile boys who liked to "listen to football scores or the Uncle Don hour" could keep a midget in their own room.[58] Aside from the obvious issue of cost—the average 1930s family could rarely afford two radios, let alone more[59]—Lessem's argument belies the unquestioned patriarchal privilege of who gets to decide what program blares through the family speakers, and when. (In the grammar of Lessem's article "He" who decides what plays for the entire family remained curiously unmentioned, though it was tellingly assumed.)

Whether the dream of a radio in every room died at Roosevelt's behest, the public's quibbles, or some combination of reasons, the matter determined in large part the future of remote-controlled media consumption in forthcoming decades. Remote control would not be associated with the freedom and independence of family members to tune in to a central hub from their separate rooms. At least until families could afford multiple radios and televisions, remote control would be associated with a centralized and distinctly patriarchal scene. It conspired with the electronic hearth to tether everyone together in one room, submitting to one will, belonging to whoever wields the remote as symbolic phallus.

In recent decades, things have changed substantially. Households with wireless connectivity and multiple screens have seen a gradual dissolution of their "home entertainment zone" as collective activity in these central spaces has declined. There has also been a trend, growing since the nineties, for custom sound installation throughout the house, with in-wall speakers or wireless "smart home" speakers controlled by digital assistants. While these trends seem natural, given the proliferation of mobile media devices and diffusion of wireless connectivity, one wonders how habits of media consumption, the home entertainment zone, and the "family scene" in general, might have developed differently had built-in radio come to fruition way back in the 1930s.

Built-in radio continued to exist toward the end of the thirties and after—just not in the form previously illustrated. Instead, the term was used to describe the installation of receivers into a wall, cabinet, or piece of furniture *within one particular room*. Some refrigerators were actually made with built-in radio, but most often it was installed in the living room, either recessed into the wall or built right into the easy-chair. Poring over the trends in radio styles, interfaces, and accessories over the decade, a picture emerges of how middle-class life habituated to a certain way of using and arranging their media devices. Either with or without the assistance of remote control, the user grew closer to their radios, both in proximity and in terms of intimacy, as visual and tactile elements of the interface became more appealing.

Several developments between 1934 and 1936 furthered the trend toward reclined consumption of radio. As economic conditions improved, free-standing console radios became fashionable again.[60] "Airplane dials" began to adorn the front face of cabinets in 1934. The tuning knobs rotated the needle slowly on these large round dials (often illuminated) to facilitate sharp-tuning into weaker station signals. An additional visual tuning aid appeared in 1935 called the "tuning eye" or "magic eye". These were glowing green cathode ray tubes that had a "pupil" and a gap in their "iris", which narrowed as stations were tuned in correctly. The popularity of airplane dials and tuning eyes marked a point when users became more visually drawn into the interface's technical dimension. Given the mounting anticipation of television, still a few years away, such innovations could be seen as a design strategy to entrance living room listeners with a proto-televisual appeal. A prime example was manufacturer Montgomery Ward's line of "Movie Dials" consoles, which displayed stations on a circulating 35mm filmstrip that was backlit and projected through a lens for the listener's visual entertainment.

Through the mid-thirties, compact tabletop and midget receivers continued to garner 75% of market share.[61] If listeners were not perched on their sofas, rapt by their new consoles' dazzling displays, they were likely reaching out and fiddling more with their tabletop radios. The closeness of these radios, placed within reach of one's easy-chair, no doubt encouraged more channel-cruising and tinkering with volume and tone. Manufacturers now started designing armchair radio sets of various descriptions: portables that would clamp onto the chair's arm, receivers that were installed directly into the chair, and, within a few years, a succession of "chairside" and "armchair" models with elaborate furniture styling. Though remote control was not the dominant interface during these years, it played an instrumental role within an ensemble of innovations to close the space between the user and the interface, further wedding their sensations and desires to the habit of information and entertainment consumption.

In the last few years of the decade, users would develop an even greater tactile intimacy with the interface through a series of new devices and innovations. In 1936 Grunow released the teledial, a circular disc that enclosed the airplane dial. Operating like a telephone dial, the user put their finger in one of fifteen holes, rotated the disc, and released when the desired station was centered at the bottom. The interface offered users a sensation that felt both playful and precise. Implementing a new technology called Automatic Frequency Control (AFC), the teledial reintroduced automatic tuning to the radio's interface. Automatic tuning through push-button station selection was rather rare before this date, appearing mainly on Zenith's early cam and lever mechanisms, as well as push-button remote controls. But these tuning technologies were rather clunky and imprecise. A common problem with push-button station selection was the inability of the apparatus to stop the condensers at the precise given setting, which worsened with repeated wear and tear.[62] For this reason, several remote control models opted for dials instead of buttons, so the listener could better fine tune from their chair. But this changed in 1936–37: AFC enabled buttons to "capture" the station, even if the preset setting was a little off.

Several manufacturers implemented the teledial interface on their 1937 and 1938 models, enhancing and remediating the airplane dials that were still in vogue. But an even greater number of manufacturers opted to use AFC technology to bring back push-buttons. Indeed the trade press declared at this point that push-buttons were "definitely here to stay".[63] They appeared on everything from consoles to tabletops to portables, and not least, remote controls, whose popularity began to rise again, as AFC made them more reliable. Remotes had also become more affordable. By the end of the decade one could purchase a high-end wired or wireless remote in a small handsomely crafted wooden cabinet for \$15–\$20. Mind you that still amounts to \$275–\$350 in 2017 dollars, so those brands remained out of reach for most middle-class consumers.

The final innovation of this decade is wireless remote control. Remotes' wires posed obvious obstacles for users throughout the decade, even though trade press articles rarely, if ever, discussed them. Children would trip on them, pets would chew on them, and early models could catch the carpet on fire, if their long cords generated too much heat. For these reasons, owners were continually advised to run the cords under the carpet, and flat ribbon-cable was soon implemented. To eliminate these issues and provide greater portability, Philco introduced the first wireless remote in 1938 (Figure 12.3), followed shortly after by Kadette, and several other brands in the following year. Most of these wireless remotes worked by transmitting radio waves that would either pulse in short bursts or carry a continuous pulse-modulated signal to be received and decoded by the console.

Nearly all contemporary wired and wireless remote controls used push-buttons, except Philco's "Mystery Control", which employed the teledial interface.[64] Philco's remote was almost cube shaped, enclosed in an 8 × 7 × 5 inch mahogany cabinet, with the dial mounted on a brass plate. The teledial itself was made of an early form of tenite, a cellulosic thermoplastic that was quite new and exotic (and, as collectors have learned, rather unstable and prone to warping over time).

The dial was a creamy marble color, smooth and softer to the touch than the hard, brittle bakelite buttons and knobs more commonly used. As it rotated and released, it made a clicking sound, familiar from rotary phones. The pleasing acoustic feedback and sumptuous interface were clearly designed to tempt the user to continually caress and fiddle with it, as they might a favorite toy.

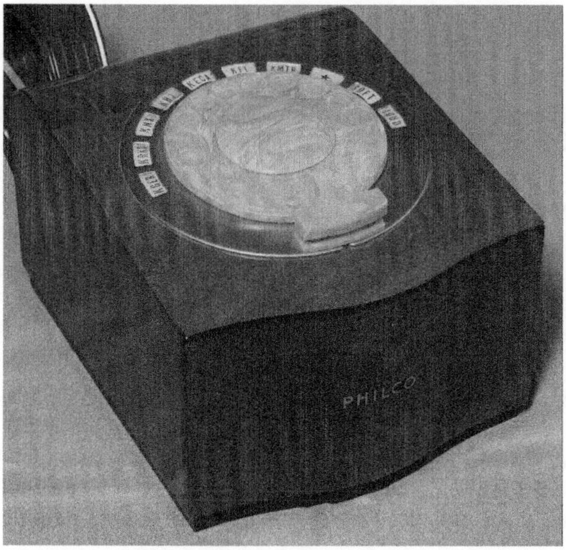

Figure 12.3 Philco Mystery Control's box-shape and teledial departed from contemporary designs such as Stromberg-Carlson's Telektor (Figure 12.2), which were conventionally rectilinear with push-buttons. (Photo courtesy of radiomuseum.org)

The Mystery Control was widely marketed as "the most thrilling invention since radio itself".[65] Whenever researchers refer to radio remote controls, the Mystery Control is often the only one mentioned or pictured. It is still coveted by collectors. It is a landmark device that heralded the future of portable remote control, and its design is unique from a present-day perspective. Perhaps this is why researchers and collectors rarely take notice of the era's numerous other radio remote controls, whose buttons and rectilinear designs resemble those to which we have become so accustomed. Yet both types of remote would have been equally novel to a user in the late 1930s discovering automatic tuning.

A Paradox in Obsolescence

The Mystery Control's fetishization thus demonstrates a paradox regarding vanishing media: while aspects of industrial design and technology may obsolesce, certain habits of operation such as push-button remote control remain deeply entrenched. The Philco Mystery Control belongs to a style and technology that has obsolesced many decades ago, and for these very reasons it attracts attention. On the other hand, button-laden remotes have long since become unremarkable, despite the many styles that have come and gone, from wired electro-mechanical to wireless RF remotes and ultrasonic clickers. Buttons have been increasingly culled from our consumer electronics in recent years, but there is still the sense that they are very much with us, even if interfaces have been reduced to a few all-purpose buttons, a single on-off or "home" button, or collapsed altogether into soft buttons on touchscreens. It is curious to observe how the rise of the popular "fidget cube" fad aligns with this trend in media device design: as smartphones replace our remote controls, and as smartphone manufacturers take away our devices' buttons, we seek out replacement buttons to stroke, caress, and fiddle with. The buttons on fidget cubes do not actually *do* anything when pressed, except satisfy our urge to press a button—and to release some unnamed anxiety by doing so.

We can locate the genesis of our ancestral bond to buttons in large part to radio remote controls. The remote habit itself did not become widespread until the 1980s and 90s, after the mass diffusion of infrared remotes trained the thumb's muscle memory for media. But a more fundamental conjuncture had already come together, settled its internal contradictions, and stabilized in the 1930s—this set the *scene* of push-button remote control. At the beginning of the decade the remote control device was merely a novel idea treading water in a sea of other uses and meanings. By the end of the decade remote control was clearly articulated to a certain embodiment: handheld with buttons, it triangulated the user, their furniture, and the media device within the centralized domestic space of the living room. Though guided along by the vision and determination of industry leaders, promoters, and inventors, it is as though the remote was preordained. As if some intractable force was always urging that users would get closer to their media, incrementally diminishing the gap between their desires and gratification through content delivery.

Notes

1 Strang, Interview with Ian Crowe, Director of Product Management for Harmony Remotes at Logitech.
2 The first television remotes followed a parallel trajectory to radio remotes with both expensive and after-market wired devices. The first wireless television remote, moreover, was directly inspired by earlier wireless radio remotes. Patrick Parsons discovered a 1942 RCA patent for a wireless, non-electric remote that struck vibrating reeds to send signals to a receiver. He draws a link between that technology and the first ultrasonic television remote invented by Robert Adler in 1956, whose patent cites the RCA reed remote patent. Parsons, "The 'Most Thrilling Invention Since the Radio Itself!'," pages 75–76.
3 Ibid., page 75.
4 Sterling and Haight, *The Mass Media*, page 360.
5 Robinson, "The Radio Craftsman's Page," page 469.
6 To give only two examples from this chapter, it is claimed that radio remote controls were first featured in the 1929 Radio World's Fair, but other, lesser known remote control devices had been on the market, such as the 1926 "Storad"—a battery eliminator and trickle charger with a single on/off kill switch that sold as "remote control"—and the 1927 "Thermiodyne"—advertised as "Like a Long Arm." it stretched a speedometer-like cable to the volume control (see "Marvelous New Storad," page 74 and "Like a Long Arm," page 329). Likewise, the 1938 Philco Mystery Control is claimed as the first "wireless" remote; however, a little-known company called "Remotrole" had invented the technology in 1927. See "Sets Tuned from a Distance by Latest Radio Invention" and "A Remote Control Without Wires."
7 In 1896 Marconi demonstrated the possibility of wireless telegraphy in a public demonstration where he "made a bell in a box ring by pushing a button in a different box, with no wires or cables in between." Yuste, "Early Developments of Wireless Remote Control: The Telekino of Torres-Quevedo," page 186.
8 Carlson, *Tesla*, page 229.
9 See for example Yuste and Palma, "The First Wireless Remote-Control: The Telekine of Torres Quevedo" and Yuste, "Early Developments of Wireless Remote Control: The Telekino of Torres-Quevedo."
10 "Remote Control Switches," page 38.
11 "A New Printing Press Motor Controller," pages 901–2.
12 "Wireless Telecontrol," page 210.
13 Plotnick, "Signal and Switch," page 31.
14 Marvin, *When Old Technologies Were New*, page 213.
15 "Prize-Winning Poster Selected for America's Electrical Week," page 321.

16 "Electrical devices and utensils will be omnipresent. Living room, kitchen, bed chamber, toilet, and den will be so equipped that the genie of the wire may ever be at the summons of its human master." Parker, "Increasing Current Consumption without Increasing 'Kicks'," page 99.
17 Wattenberg, *The Statistical History of the United States*, page 827.
18 Most every appliance had its origin in commercial use, and was redesigned for the household in later years. Giedion, *Mechanization Takes Command*, page 581.
19 According to Ruth Schwartz Cowan, "the proportion of servants to households in the nation dropped (1 servant to every 15 households in 1900; 1 to 42 in 1950)." Cowan, *More Work for Mother*, page 99.
20 Ibid., page 177.
21 Douglas, *Inventing American Broadcasting, 1899–1922*, page 303.
22 "Historical Statistics of the United States: Colonial Times to 1970, Part 2," page 796.
23 Boddy, *New Media and Popular Imagination*, page 37.
24 Horowitz, *His and Hers*, page 114.
25 Carlat, "A Cleanser for the Mind: Marketing Radio Receivers for the American Home, 1922–1932," page 129.
26 Ibid., page 120.
27 As reported in the 1930 US census, of the 2.9 million families of color only 13.4% of them owned radio sets (or, less than 2% of all US families who owned radios were non-white). By 1940 over 43% of families of color owned radio sets. "Census of the United States: 1930; Population Vol. 6: Families," page 33, and "Census of the United States: 1940; Housing Vol. 2," page 38.
28 For a concise history of the remote control that includes an analysis of gender and class discourse in radio remote control advertising, see Benson-Allott, *Remote Control*.
29 "As the Trade Thinketh . . .," page 31.
30 Douglas, *Radio Manufacturers of the 1920s*, Vol. 3, page xxiv.
31 Felix, "What Happened: Merchandising," page 130.
32 Ibid.
33 Sleeper, "A New Factor in Radio Construction: The Kinematic Remote Control," page 150.
34 "Telektor – The Radio Robot," page 689.
35 Ibid.
36 While adhering to this general scheme, obviously each of these interfaces has its unique trajectory, and the television remote is no exception. Through the 1950s and 60s, the channel up/down buttons were sufficient for the limited number of stations. As more channels became available through the 1970s, some infrared and wired CATV remotes began to put a button for each channel—effectively implementing the "favourite"—while still keeping the channel up/down buttons. With the mass diffusion of infrared remotes through the 1980s and 90s, the number pad was required for two- and three-digit direct entry. In recent years, due to the explosion of channels through cable and satellite, Logitech's Harmony models have done away with the number pad. Instead, their remote lets users choose their favorites by mapping an HBO or ESPN icon, for instance, directly on their touchscreen. Harmony's R&D has learned that while users still need the channel up/down to scan, they are overwhelmed trying to remember channel numbers and navigating the length of the onscreen menu guide.
37 Neely, "Every Kansas Farmstead Can Be a College Classroom," page 27.
38 Sedgwick, *Remote Control*.
39 Capehart Aladdin Control, Advertisement. February 1933, *Radio Retailing*, page 53.
40 "Scott Radio Control," page 546.
41 "Telektor – The Radio Robot," page 664.
42 "Take Your Choice of Remote Controls," page 625.
43 "Remote Volume Control Ideal for the Lazy," page 45.
44 "Scott Radio Control," page 546.
45 "Scott Designs Remote Control for His Screen Grid Super," page 57.
46 Felix, "What Happened: Merchandising," page 130.
47 Taylor, "Equipment Now and Then: Part Two," page 1074.
48 Geddes, "Trade Show to Reveal Improvements," page 991.
49 "A Remote Control Tuning Unit," page 333.
50 "New Methods Make Remote Control Popular," page 54.
51 "Be the Boss of Your Own Radio: Blah Blah Eliminator," page 638. In yet another demonstration of the parallel trajectories of radio and television remote control, the Blah Blah Eliminator was virtually

the same in kind and cost as the Blab-Off, a one-button remote to mute television that sold in 1953 for $2.98. Alpern, "Postwar American Television: Blab-Off."

52 "Remote Control (Motorola Model S-10)," page 116.

53 Taylor, "Practical Remote Control Systems for Servicemen and Experimenter," page 42.

54 Lessem, "This Home Wired for Radio: Part I," page 396.

55 Ibid., page 397.

56 In his post on the *Radio-Craft* articles, "When Radio in Every Room Was the Dream of the Future," *Paleofuture* editor Matt Novak was similarly convinced the home was never built, since there were no pictures showcasing the installation, and "Hugo Gernsback, sci-fi legend and publisher of *Radio-Craft*, was not known for doing anything understated in his magazines."

57 Lessem, "This Home Wired for Radio: Part IV," page 718.

58 Lessem, "This Home Wired for Radio: Part I," page 445.

59 Even though the cost of compact and midget radios had reduced to around $10 (equivalent to $150 in 2017), these were still the Depression years and such expenses were a luxury. In 1935 *Radio Engineering* magazine reported approximately 7% of American families with two or more radios. Since the upper-income bracket comprised about 12% of the population, two radios would have been out of reach for virtually all middle- and lower-income homes.

60 Sterling and O'Dell, *The Concise Encyclopedia of American Radio*, page 650.

61 Ibid., page 650.

62 A 1935 *Radio News* article glosses the remote control as a system with "complications," likely referring to these very issues, which could account for the downturn in the marketing of first- and third-party remote controls during the mid-30s. Borst, "A Home Built Receiver Designed for 'High-Fidelity' Reception," page 536.

63 "A Survey of the Latest Trends in Touch Tuning by Push-Button Control Systems," page 331.

64 Some recent publications have described Philco's remote incorrectly as requiring an AC power cord, thereby concluding it was not actually the first wireless remote. The remote was indeed battery-operated and wireless, as demonstrated in several YouTube videos, including SDSpike, *"Mystery Control" 1939 Philco Radio*.

65 "Philco Mystery Control . . . Most Thrilling Invention since Radio Itself," page 3.

References

"A New Printing Press Motor Controller." *Electrical Review and Western Electrician* 55, no. 19 (November 6, 1909): 901–2.

"A Remote Control Tuning Unit." *Radio-Craft*, December 1932.

"A Remote Control Without Wires." *Radio News*, February 1928.

Alpern, Laura. "Postwar American Television: Blab-Off." www.earlytelevision.org/blab_off.html. *Early Television Museum*, n.d. Accessed July 15, 2015.

"As the Trade Thinketh . . ." *Radio*, September 1929.

"A Survey of the Latest Trends in Touch Tuning by Push-Button Control Systems." *Radio News*, December 1937.

"Be the Boss of Your Own Radio: Blah Blah Eliminator." *Radio-Craft*, April 1932.

Benson-Allott, Caetlin. *Remote Control*. New York: Bloomsbury Academic, 2015.

Boddy, William. *New Media and Popular Imagination: Launching Radio, Television, and Digital Media in the United States*. 1st edition. Oxford; New York: Oxford University Press, 2004.

Borst, John M. "A Home Built Receiver Designed for 'High-Fidelity' Reception." *Radio News*, March 1935.

Carlat, Louis. "A Cleanser for the Mind: Marketing Radio Receivers for the American Home, 1922–1932." In *His and Hers: Gender, Consumption, and Technology*, edited by Roger Horowitz and Arwen Mohun, 115–38. Charlottesville: University of Virginia Press, 1998.

Carlson, W. Bernard. *Tesla: Inventor of the Electrical Age*. Reprint edition. Princeton, NJ: Princeton University Press, 2015.

"Census of the United States: 1930; Population Vol. 6: Families." Washington: Bureau of the Census, 1933.

"Census of the United States: 1940; Housing Vol. 2." Washington: Bureau of the Census, 1943.

Cowan, Ruth Schwartz. *More Work for Mother: The Ironies of Household Technology from the Open Hearth to the Microwave*. New York: Basic Books, 1983.

Douglas, Alan. *Radio Manufacturers of the 1920s*. Vol. 3. Sonoran Publishing, 1999.

Douglas, Susan J. *Inventing American Broadcasting, 1899–1922*. Johns Hopkins University Press, 1989.

Felix, Edgar H. "What Happened: Merchandising." *Radio Broadcast*, January 1930.

Geddes, Bond. "Trade Show to Reveal Improvements." *Radio News*, June 1932.

Giedion, Sigfried. *Mechanization Takes Command: A Contribution to Anonymous History*. Minneapolis: University of Minnesota Press, 2014.

"Historical Statistics of the United States: Colonial Times to 1970, Part 2." Washington: U.S. Department of Commerce, Bureau of the Census, 1975.

Horowitz, Roger. *His and Hers: Gender, Consumption, and Technology*. University of Virginia Press, 1998.

Lessem, N.H. "This Home Wired for Radio: Part I." *Radio-Craft*, January 1939.

Lessem, N.H. "This Home Wired for Radio: Part IV." *Radio-Craft*, June 1939.

"Like a Long Arm." *Radio Broadcast*, September 1927.

"Marvelous New Storad." *Radio Broadcast*, November 1926.

Marvin, Carolyn. *When Old Technologies Were New: Thinking About Electric Communication in the Late Nineteenth Century*. Reprint edition. Ill: Oxford University Press, 1990.

Neely, Henry M. "Every Kansas Farmstead Can Be a College Classroom." *Radio in the Home*, November 1924.

"New Methods Make Remote Control Popular." *Popular Science*, January 1934.

Parker, Charles A. "Increasing Current Consumption without Increasing 'Kicks.'" *Electrical Review and Western Electrician* 56, no. 2 (January 8, 1910): 97–99.

Parsons, Patrick. "The 'Most Thrilling Invention Since the Radio Itself!': The Evolution of the Radio Remote Control in the 1920s and 1930s." *Journal of Radio & Audio Media* 21, no. 1 (January 2, 2014): 66–79.

"Philco Mystery Control . . . Most Thrilling Invention since Radio Itself." *Saturday Evening Post*, October 1, 1938.

Plotnick, Rachel. "Signal and Switch: A Cultural History of the Push-Button Interface." Northwestern University, 2013.

"Prize-Winning Poster Selected for America's Electrical Week." *Electrical Review and Western Electrician* 69, no. 8 (August 19, 1916): 321–22.

"Remote Control (Motorola Model S-10)." *Radio News*, August 1934.

"Remote Control Switches." *Electrical World and Engineer* 38, no. 1 (July 6, 1901): 38.

"Remote Volume Control Ideal for the Lazy." *Citizens Radio Call Book Magazine and Technical Review*, March 1930.

Robinson, James. "The Radio Craftsman's Page." *Radio-Craft*, February 1932.

"Scott Designs Remote Control for His Screen Grid Super." *Citizens Radio Call Book Magazine and Technical Review*, March 1930.

"Scott Radio Control." *Radio-Craft*, April 1930.

SDSpike. *"Mystery Control" 1939 Philco Radio*. YouTube Video, 2011. www.youtube.com/watch?v=o608KIiBMqA&t=3s.

Sedgwick, Edward. *Remote Control*. MGM, 1930.

"Sets Tuned from a Distance by Latest Radio Invention." *New York Times*, September 25, 1927.

Sleeper, M.B. "A New Factor in Radio Construction: The Kinematic Remote Control." *Radio Broadcast*, January 1930.

Sterling, Christopher H., and Timothy R. Haight. *The Mass Media: Aspen Institute Guide to Communication Industry Trends*. Praeger, 1978.

Sterling, Christopher H., and Cary O'Dell. *The Concise Encyclopedia of American Radio*. Routledge, 2009.

Strang, Brent. Interview with Ian Crowe, Director of Product Management for Harmony Remotes at Logitech. Telephone, August 17, 2015.

"Take Your Choice of Remote Controls." *Radio News*, January 1930.

Taylor, S. Gordon. "Equipment Now and Then: Part Two." *Radio News*, June 1931.

Taylor, S. Gordon. "Practical Remote Control Systems for Servicemen and Experimenter." *Radio News*, July 1930.

"Telektor – The Radio Robot." *Radio-Craft*, May 1932.

Wattenberg, Ben J. *The Statistical History of the United States: From Colonial Times to the Present.* 1st edition. New York: Basic Books, 1976.

"Wireless Telecontrol." *Electrical Review and Western Electrician* 55, no. 5 (July 31, 1909): 210.

Yuste, Antonio Pérez. "Early Developments of Wireless Remote Control: The Telekino of Torres-Quevedo." *Proceedings of the IEEE* 96, no. 1 (January 2008): 186–90.

Yuste, Antonio Pérez, and Magdalena Salazar Palma. "The First Wireless Remote-Control: The Telekine of Torres Quevedo," 15. Bletchley Park, 2004.

13

STANDARDIZED FILM LEADERS

Matt Soar

A riddle: You will find me at the beginning, and often at the end, but never in the middle. My head is 20ft long, but my tail is shorter. I am occasionally blank, mostly silent, and very hard-working. I am meant to be partly read, and partly viewed, and am full of stories—but if you ever actually see the real me it is because someone screwed up. Even though I am designed to be invisible most people know me when they see me. What am I?

Why Film Leaders?

In 2011, I had the good fortune to be a participant in the Independent Imaging Retreat, a residential filmmaking workshop held annually in rural Ontario. The "Film Farm", as it is known colloquially, is hosted by the distinguished filmmaker and York University professor Phil Hoffman, who, along with a dedicated team of volunteers, gently encourages process and open-ended experimentation. Everyone at the Farm—from senior and mid-career artists to relative novices such as myself—shoots on 16mm film using wind-up Bolex cameras; processes by hand in the lower level of the barn; hangs the developed film to dry on an outdoor washing line; and edits, tints, and tones on the main level. Twice during the week, everyone gathers to screen their works-in-progress downstairs. Sitting together on lawn chairs, the flickering shadows of barn swallows occasionally pass across the makeshift screen.

One day I sidestepped most of these processes in favor of working with found footage: a physical, dusty bin, full of entangled, 16mm and 35mm film fragments. I decided I would juxtapose formally unrelated strips of film by having a go at handweaving them together. I cut along the horizontal framelines of one "scene", cut another scene vertically into long narrow strips, and then wove these threads through the cut framelines, and taped everything down. I remember being particularly drawn to a strip of film I found in the bin featuring large sequential numbers with an animated, rotating background, a curiosity I attribute partly to my earlier career in graphic design. This particular element offered an implied sense of urgency and authority, which I thought worked well with other fragments I wove into the center of the frame, such as sprocket holes and optical sound tracks. This particular fragment was of course a snippet of countdown leader, a piece of cinematic arcana that is surprisingly familiar to many people, given that the chief purpose of countdowns is to allow projectionists to cue consecutive film reels *without* drawing the audience's attention. As my opening riddle suggests, when they are used properly, they are invisible. Splicing several combined scenes together produced an exceedingly short non-narrative film which, with the later addition of a wonderful soundscape composed by audio artist Jackie Gallant, ultimately became *Lost Leaders #1* (2012), the first in an ongoing series of creative, archival, and scholarly explorations of film leaders as an historical, technical, and cultural phenomenon.

This was the beginning of my ongoing fascination with film leaders, the peculiar and particular artifacts I have come to understand as the "metadata" of the physical medium of film-as-film. My *Lost Leaders* creative experiments continue, and include: photomontage; stained glass; interactive narratives; cameraless animations; and microvideography. In a previous scholarly article (Soar, 2016) I presented an overview of four stages in the historical development of U.S. leader standards, covering the period from 1930 to the recent, rapid decline in the commercial development, printing, and projection of analog film. Unlike the earlier article, which prioritized the countdown section, in this essay I extend my research to explore some of the other common elements to be found on standardized leaders, and to consider the ways in which leaders have had an aesthetic impact on everything from experimental filmmaking to documentaries, movies, music videos, and book covers.

Until the recent, widespread adoption of digital projection technologies in the U.S. and Canada, movies were always physical things: narrow strips of nitrate, acetate, or polyester, sometimes miles long, cut into more manageable sections, wound onto reels, and distributed by studios to film exchanges and theaters inside protective canisters or cases. The recently arrived Digital Cinema Package (DCP) is a more controllable, durable technology for movie distribution and projection, offering (many would say) better picture quality, sound, and viewing. The DCP is a portable hard-drive which is shipped to the cinema in a protective case, contains the whole movie, and requires minimally skilled labor to load and play. Preparation time is negligible, and films need not be rewound. The Faustian deal struck with most theaters has been that the old analog film projectors had to be completely removed and even destroyed.

With physical film prints, "mutilation" (Soar, 2016) while in circulation was a very common problem, especially in the early days of popular cinema. Film exchanges and theater projectionists blamed each other for the most egregious damage, including excessive dirt, tears, scratches, damaged sprocket holes, missing scenes, added markings, punched holes, and even the use of pins or wire to fix breaks. Only as this emergent, booming industry began to mature, and the cost of such damage was fully realized, did new professional associations such as the Academy of Motion Picture Arts and Sciences (AMPAS) and the Society of Motion Picture and Television Engineers (SMPTE) develop and impose meaningful standards for the handling and care of films in general circulation.

The arrival of sound films presented an additional challenge: the synchronization of separate picture and audio sources. This was solved by the use of special timing marks on the area of the film reel immediately before the first frame of the movie proper, a section known since the early 1900s as the "leader". Leaders were already part of the lexicon of film printing, duplication, distribution, and projection but, as late as 1930, were still being officially defined as "blank". Archival evidence strongly suggests they were often anything but blank, even pre-standardization, since an area already attached to the filmed content provided an ideal space for identification, statements of ownership, and instruction.

The first U.S. leader standard, developed by an AMPAS technical committee and formally published in 1930, solved several problems at once. First, it specified an amount of leader that would act as protection for the entire film reel (8 feet, to be replaced if ever it got as short as 6 feet); second, it included an area devoted to identification, typically comprising one or more frames with the studio's logo, the movie title (usually handwritten), the reel number, and instructions for the projectionist such as sound level settings; and, finally, with the then-recent arrival of the "talkies", it provided a very particular countdown section to assist the projectionist in reliably cueing up and synchronizing the sound source with the picture. This might be a Vitaphone disc, or another, paired 35mm film reel containing the soundtrack ("track" for short). Although the idea of a countdown strikes contemporary ears as positively banal, it is worth taking a moment to explore some of its cultural and scientific roots.

Where Do Countdowns Come From?

In his *A Dictionary of the Space Age* (2009), Paul Dickson defines the "countdown" as "A step-by-step process that culminates in a climactic event, each step being performed in accordance with a schedule marked by a count in inverse numerical order . . . The act of counting inversely during this process" (2009: 20). He also credits Fritz Lang's highly influential, silent sci-fi melodrama *Frau Im Mond* (1929) as the place where the countdown "was first rendered graphically". Roberts (2015) goes further, claiming that "Lang is acknowledged to have invented the countdown sequence adopted by NASA when the U.S. space race began three decades later, by means of a montage of images and intertitles building to the moment of lift-off" (2015: 102).

The climactic launch scene in *Frau Im Mond* (1929) does indeed feature a dramatic graphical countdown from 6 to 1. The crew members lie nervously in their bunks as the commander shouts out his instructions, hand hovering over a large lever on the adjacent console. The title card reads "20 seconds to go – lie still – take a deep breath!"). Successive cards reveal larger and larger numbers: "Noch 6 Sekunden!" ("In 6 seconds!"), then 5, then 4, 3, 2, 1, and finally: "JETZT" ("NOW!").

It is certainly not the first ever countdown in fiction, however. Dickson mentions two stories written by George W. Griffith. In *The Crellin Comet* (1897), Griffith deploys a 10-to-1 countdown, and in *World Peril* (1910), he favors a 20-to-1 countdown. A simple N-Gram search of Google Books, which graphs the frequency of appearances of words and phrases in

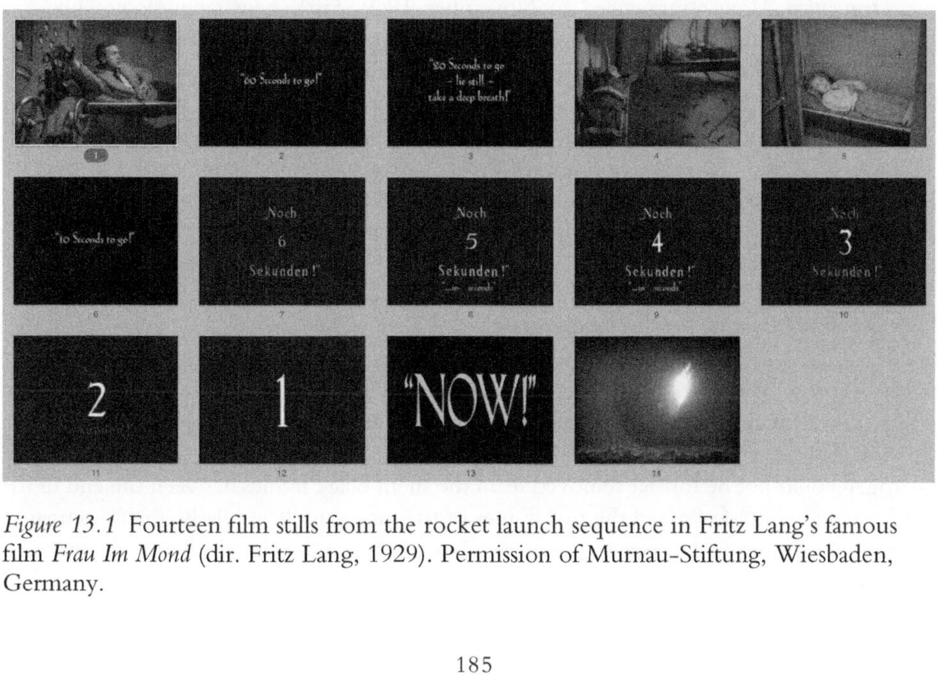

Figure 13.1 Fourteen film stills from the rocket launch sequence in Fritz Lang's famous film *Frau Im Mond* (dir. Fritz Lang, 1929). Permission of Murnau-Stiftung, Wiesbaden, Germany.

its vast collection (25 million and counting), shows that the word "countdown" is virtually unheard of before the mid-1940s, rises dramatically in the late 1960s and early 1970s, drops significantly from 1963 to 1974, and then begins a steady rise into the 2000s. The single most likely explanation for its sudden rise in popularity is NASA's Apollo program (1961–1972) and the associated "space race" with the Soviets, although there are, of course, plenty of confounding factors in the use of the word, such as in sporting events, space novels, and New Year's Eve celebrations, but also movies and TV series with "*Countdown*" as their title (at least 11 movies and nine TV shows, according to the Internet Movie Database).

Setting aside sporting events, Dickson (2009: 54) suggests that the modern application of the countdown began with "the first atomic bomb test (Trinity) in July 1945", while its initial use in relation to space travel was "the first American attempt to launch a Vanguard satellite in late October 1957". Roberts (2015: 105) and others have made direct connections between the research for Lang's film and the rise of the Third Reich: "in the years to follow, copies of the film were confiscated and withdrawn from distribution by the Gestapo, who—to the chagrin of Lang and other former VfR [Verein fur Raumschiffarht – the Society for Spaceship Travel] members—also seized the detailed five-foot spaceship model" (Miller, 1995: 95). Lang's chief technical adviser was Hermann Oberth, an engineer, member of the VfR, and author of an influential science textbook called *Die Rakete zu den Planetenraumen* [*The Rocket in Interplanetary Space*] (1923). Drawing on Geser (1996), Roberts also claims that "Oberth (along with a young scientist by the name of Wernher von Braun), took the plans for Lang's spaceship and went on to help the Nazis develop their own rocket programme during the years of the Third Reich" (2015: 105). Von Braun was a member of VfR, led the development of the V2 rocket for Hitler, and went on to direct the development of the Saturn V rocket for NASA, suggesting a direct "countdown" lineage from *Frau Im Mond* to the launch of the Apollo rockets.

What of the broader cultural context in the late 1940s and early 1950s? In his article on *Destination Moon* (1950), Schauer (2015) writes:

> Science fiction had burgeoning exploitation potential when production on *Destination Moon* commenced in November 1949. First, new technologies like rocketry and the atomic bomb were highly visible in American culture, and interest in popular science had reached new levels. In January 1949 *Life* magazine published a heavily illustrated article entitled "Rocket to the Moon" which argued that "engineers believe that a manned rocket . . . may get to the moon within the next 25 years." The article's illustrations seem to be a clear influence on the rocket design and lunar imagery of *Destination Moon*, as well as a competing film *Rocketship X-M* (1950).
>
> (Schauer, 2015: 7)

If rocket launches were a direct inspiration for the leader countdown, several important distinctions are worth pointing out: while a successful NASA launch countdown, for example, signals the imminent spectacle of sudden, rapid combustion and an awe-inspiring lift-off, a film leader countdown is by comparison profoundly anticlimactic, extinguishing itself visually after the count of 3 or 2. The vision of a scorching fireball of combusting liquid hydrogen could not be further removed from the silent black frames between the end of the visible movie countdown and the first frames of the movie itself; similarly, the accompanying roar of igniting fuel versus the discrete "2-pop", an audio "blip" often accompanying the visual count of 2.

The Historical Development of U.S. Leader Standards

The first U.S. standard, the Academy leader, was named after its creators, a technical committee of AMPAS. Although it was published in 1930 and revised multiple times, it did not actually become an official American Standard until 1947. In the intervening years, it became ubiquitous; so much so, that "Academy leader" has clearly become a generic term for any leader with a countdown. Unlike subsequent leader standards, the Academy leader was not drafted, printed, and distributed for use by studios and distributors. Rather, archival evidence strongly suggests that studios drew up their own versions based on the published guidelines, probably using their own art departments—the same folks who were likely responsible for props featuring type and lettering (e.g., newspapers, signs, etc.), intertitles (see Žilová, 2013), and publicity posters. Because the countdown figures were drawn and painted rather than typeset using a suitable display font, the tantalizing prospect emerges that we can potentially distinguish between instances, and the potential origins, of each studio art department's version of the Academy leader, based on the style of the drawn figures.

In what follows, some of the most distinctive sections that might typically be found on the "head" leader on the first reel of a 35mm movie are explored in turn, beginning with the head, and traveling along the leader until the beginning of the movie content; that is, the very first frame viewable to the theater audience. This is all assuming that the leader is intact, and the reel has been threaded and cued up by a competent projectionist who is alert and ready to open the projector's douser at the exact moment of the countdown's implied "zero".

Protective Bands

Film leaders since 1930 have had three declared functions: protection, identification, and synchronization. Much of the discussion leading up to the first de facto American Standard, the Academy leader (1930), focused on the widespread "mutilation" (Soar, 2016) of films in circulation. Blame was usually placed with projectionists and, less often, with film exchanges, which acted as intermediaries between studios and theaters. Aside from the cases and canisters in which film reels were transported, protection was achieved primarily through a stipulated minimum of 6 feet of blank film. Archival evidence suggests that efforts were sometimes made to add a protective wrapper of card or paper around each film reel. A short article in the May, 1947, issue of the *Journal of the Society of Motion Picture Engineers*, titled "A Proposed Film Lock and Identification Band" (Schwartz, 1947), proposes a narrowed section of the film leader,

> a tongue . . . with a groove or series of grooves formed in the body of the film adapted to receive the tongue, so that the outer strand of film may be locked on itself by inserting the tongue through one of the grooves [slots].
>
> (Schwartz, 1947: 473)

The tongue-and-groove arrangement stops the reel from unwinding, and suitable markings on this section provide for identification "so that there is no chance of misapplied data caused by separating a roll of film from its respective wrapper".

Identification

The "Academy Specifications for 35mm Motion Picture Release Prints" (Report of Standards and Nomenclature Committee, *JSMPE*, December 1930: 820) stipulate that the identification leader "shall contain not less than 32 frames in each of which is plainly printed in black letters on white background, type of print . . . part number . . . and picture title." (This 32-frame section is to be repeated at the end of the reel except the first part is "End of part" not "type of print".) Two feet into the synchronization section is "printed START (inverted) in black letters on white background ½ of frame height".

The architects of the "Society Leader", a.k.a. "All-Purpose Film Leader", took care to explain that, compared to the Academy leader (by this point an American Standard), "only additions have been made, and only such additions as cause no deletion of past features" (Townsend, 1951: 562). This is notwithstanding the fact that, overall, there is a great deal more graphical content than before, making the Society leader highly distinctive in appearance compared with its bland predecessor. While the identification section remains the same as before, the synchronization section begins with "PICTURE START", rather than "START", and is shown the right way up. Allowance is made for 35mm *and* 16mm film, as indicated by the audio cues.

The identification section for the Universal leader is by far the most comprehensive, compared to the two standards before it. The content begins with the phrase "SPLICE HERE" on the last frame of the protective section, "and an arrow pointing to the frameline between this frame and frame 1 of the identification section" ("American Standard Specifications for Leaders and Cue Marks for 35mm and 16mm Sound Motion-Picture Release Prints", *Journal of the SMPTE,* March 1966, page 222). Identification information is as follows: SUBJECT*, LENGTH*/ROLL*, REEL No.*/COLOR*/PICTURE, ASPECT RATIO*/TYPE OF SOUND*, HEAD, PICTURE, SMPTE UNIVERSAL LEADER, REEL No.*/PROD No.*/PLAY DATE*, and PICTURE*/COMPANY*/SERIES* (here, * indicates an elicitation of information with space for a response; forward slash indicates text appearing on the same frame). Except for HEAD and PICTURE (first mention) these are all specified as white letters on a black background (presumably so responses can be handwritten on the inter-negative, prior to final positive printing, or perhaps scratched onto the positive print). As with the Society leader, the synchronization section begins with PICTURE START. All the text appears to be hand lettered, rendered in a mix of plain sans serifs (some of it similar in appearance to Helvetica), with slab serifs (a.k.a. Egyptian) for PICTURE (first mention) and SMPTE UNIVERSAL LEADER, perhaps inspired by the typeface Rockwell. Allowance is made for 16, 35, and 70mm film, as indicated by the audio cues.

The Projection leader specifications move some information to the protective section: "logos, trademarks, part titles, or other extraneous materials, if absolutely necessary, should be placed in this section" (*SMPTE Standard for Motion-Picture Film – Theater Projection Leader, Trailer and Cue Marks* SMPTE 301–2005, page 2). The identification section includes much of the same information as the Universal leader, including the identification of the standard itself: SMPTE PROJECTION LEADER. Additions include: BEG at the very beginning; the language of the content; the lab name; and the frame rate. Allowance is made for 16, 35, and 70mm film, as indicated by the separate audio cues for each format. Almost everything appears to be typeset in Helvetica regular or Helvetica bold, with some of it optically distorted (stretched or compressed), a huge faux pas to most graphic designers, suggesting this leader was developed without the benefit of an art department.

Fader Setting Instructions

If the identification information on a typical "head" leader was of potential interest to everyone and anyone handling a given movie reel, the audio instructions were quite particular in terms of the intended audience. As an "SMPE Recommended Practice" published in the *JSMPE* in 1942 indicates, "The Fader Setting Instruction Leader shall consist of 15 frames located in the first 20 frames of the synchronizing leader . . . . The remaining frames may be used for whatever additional information *the studio may wish to transmit to the theater*" (*Recommended Practices*, May 1942, page 447, emphasis added). The accompanying illustration shows five different messages written out in 11 frames, with between one and three lines of text per frame. For example: REGULAR / SINGLE PUSH-PULL TRACK / PLAY / SOUND / 3 THREE / OR / 4 FOUR / db / ABOVE / *Insert Studio Name* / AVERAGE.

China Girls

In a recent article, film scholar Genevieve Yue writes:

> The China Girl, sometimes called the China Doll, China Lady, girl head, or any number of lab-specific nicknames like Ullie, Marcie, Shirley, and Lilly, was used from the late 1920s until the early '90s . . . . In Western nations, the China Girl is almost always female, young, conventionally attractive, and, despite the racial connotations of the name, white. In film laboratories, it is an essential part of quality-control processes, used to calibrate the desired exposure and color balance of film reels as well as the functionality of developing and printing machines. It has analogues in still photography and computer technologies as well.
>
> (Yue, 2015: 97)

Yue notes that the hundreds if not thousands of examples of anonymous China Girl models collectively stand as mute testimony to "the degree to which arbitrary preferences for *certain* [overwhelmingly white, occasionally light-skinned Asian] skin tones and a *particular* [*cis*-female] gender are made to conform to supposedly objective and neutral technological procedures" (Yue, 2015: 105). Interestingly, Yue lists a number of experimental films deploying China girls for critical and/or subversive ends, most prominently *Film in Which There Appear Edge Lettering, Sprocket Holes, Dirt Particles, Etc* (Owen Land, 1965), but also *Cosmic Ray* (Bruce Conner, 1961), *To the Happy Few* (Thomas Draschen and Stella Friedrichs, 2003), *China Girls* (Michelle Silva, 2006), and *Girls on Film* (Julie Buck and Karin Segal, 2005). See also Lorna Roth's essay (2009) on the cultural politics of "Shirley", an equally problematic analog of the China Girl, who is omnipresent in the medium of analog and digital still photography.

Countdowns

I have described the four de facto US leader standards in considerable detail elsewhere (Soar, 2016). These are: the Academy leader (1930/1945); the Society leader (1951); the Universal leader (1965/1966); and the Projection leader (1999). The last three were developed by the SMPTE, formerly the Society of Motion Picture Engineers (SMPE). Each version is distinctive graphically: over 80% of the countdown frames in the Academy leader are actually blank, while the numbers run from 11 to 3 inclusive, each one appearing upside down as the 16th frame (that is, 1 foot) after the previous one. The figures "9" and "6" have the words "NINE" and "SIX" directly underneath them to prevent confusion. Diamonds appear 20 frames before each number to provide visual cues for separate audio sources. Where the soundtrack was on an accompanying 35mm reel, the same leader would also be used.

The Society (a.k.a. "All-Purpose") leader was introduced to its intended audience in the journal of the SMPTE in 1951. Despite the evident optimism of its innovators, who were chiefly addressing the emergent needs of television broadcasters cueing up filmed source material, it was never taken up to any significant degree by the film industry, and was in any case rendered somewhat redundant with the introduction of video. As such, although it was slightly revised in 1953, it was still never ratified as an actual standard. The Society leader is perhaps the most technically ambitious and graphically sophisticated of the four "standards", and it is perhaps for this reason that it is over-represented in the derivative media discussed below, such as avant-garde films and music videos. (That, and its ready availability as 16mm found footage for a whole generation of experimental filmmakers, and as telecine source material for their students and protégés, who were coming of age as the music video industry began to take off.) Virtually every frame of the Society leader countdown carries a highly distinctive, repeated graphic motif, comprising two concentric circles, the inner one black, the outer one "white" (that is, clear). A double-headed black arrow points upwards and downwards to the edges of the inner circle, while another one points left and right to the edges of the outer circle. These are laid over a broader white/clear cross. The background of each frame is a mid-gray. The numbers are the right way up, and run from 11 to 3, every 16th frame. Each number is actually repeated or "echoed" twice: one frame before, and one frame after. The numbers 9 and 6 appear as text only. Dedicated frames mark the separate audio cue points for 16mm and 35mm film. Given its graphic complexity and the consequent difficulty of redrawing it accurately, the Society leader was produced, duplicated, and distributed by the SMPTE.

Given its extraordinary ubiquity in 16, 35, and 70mm film archives and in pop culture (see below), the SMPTE's Universal leader (1965/1966) countdown is fully deserving of its name. With numbers right side up, from 8 to 2, spaced every 24 frames (1 second), this leader is distinguished by its "clocksweep" (or as I prefer it, "radar" sweep) graphical animation. Two white concentric circles appear on a gray background with horizontal and vertical rules. Another black rule reaches out from the center to the edge of the frame and rotates through 360 degrees for every number in the countdown. Here, unlike the two previous designs, each number is visible for the duration of its given section, not just in one (Academy) or three (Society) frames. In the wake of the sweeping black rule is a darker tonal area, in the manner of a PPI (plan position indicator) radar screen. Since the number 9 has been eliminated, it is not necessary to prevent confusion between 6s and 9s. Even though the interval is now 24 frames instead of 16, because the 11, 10, *and* 9 have been dropped, this countdown is almost exactly the same length (hence duration) as the Academy and Society leaders. Again, the SMPTE produced the leader in bulk for distribution to studio labs.

The design of the SMPTE's Projection leader is largely a throwback to the 1930s: the gray background has gone, and the numbers run from 11 to 3, appearing every 16 frames, right side up. The word NINE appears to the top right of the figure 9; the word SIX appears to the bottom right of the figure 6. It appears to be too recent, and to my mind simply too bland, to be quoted in pop culture, despite its crisp design and liberal use of the much-lauded typeface Helvetica.

Turning convention on its head, I will now pass in silence over the actual movie content itself (title sequence, establishing shots, an unfolding narrative told through voiceover, action, dialog, etc.) to the end of the notional reel we have been traveling along.

Motor Cues, Changeover Cues

All four standards include a special mark added to the picture to signal the imminent end of the reel so that the projectionist can accurately cue the next one with minimal distraction for the audience (all assuming a two-projector set-up in the projection booth, rather than a single-projector system using a tower or platter, which involves temporarily splicing together all the movie's reels prior to projection). The special mark is a small circle either printed into, or punched through, the top right corners of four adjacent frames on the film strip. These are occasionally referred to in pop culture, if not professionally, as "cigarette burns" (see for example a brief scene in *Fight Club* (1999) in which one of the main characters, working as a projectionist, explains these cues, while also inserting single frames of pornography into family movies). The cue marks appear twice: the motor cue alerts the projectionist to start the motor on the second projector; the changeover cue signals to the projectionist to close the douser on the first projector and open the douser on the second projector. The time lag between each signal allows the motor on the second projector to achieve its steady working speed.

The Society leader largely carries over the Academy standard, while also adding a similar switching cue on the eighth frame before the picture (after the 3 of the countdown), for the benefit of television directors (Townsend, 1951: 564). Due to shifting conventions for measuring lengths of film in feet or seconds or frames, each of the other three standards defines the positions of the cues differently—even though film speed and frame size have remained consistent. The Academy leader specification stipulated that the motor cue and changeover cue be placed at 11 ft. and 1 ft. before the end of the picture, but by the time the Standard was established this had become 11 ft. and 22 frames; the Universal leader standard specifies 7 seconds and 1 second; the Projection leader standard specifies 172 frames and 18 frames. These are all roughly equal, since, for the historical time period collectively covered by the standards, there are 16 frames in 1 foot of 35mm film, and 24 frames per second. For each standard, then, the positions of the two cues are: 176 + 22 frames (Academy); 168 + 24 frames (Universal); and 172 + 18 frames (Projection).

Tail Leader, a.k.a. Trailer, a.k.a. Trailer Leader

In a short report titled "Five Recent American Standards on Motion Pictures" (*JSMPE*, March 1948) the technical committee was unanimous in endorsing the suggestion, offered by a representative from Technicolor, to use the term "tail leader" instead of "trailer": "in the experience of the American motion picture industry the word trailer had come to refer only to the preliminary advertising films which are exhibited in advance of the feature" (1948: 284). In contemporary parlance, too, the term "trailer" is used almost universally to describe short promos for upcoming feature films. Although the Academy leader specification of 1930 uses the term "trailer", the 1947 Standard does indeed prefer the phrase "tail leader". That said, the Universal standard reverts to "trailer" and the Projection standard uses the apparent oxymoron "trailer leader".

Immediately after the last frame of the picture is an opaque, blank section of film, stipulated as being 3 ft. in length (Academy), 87 frames (Universal), or 88 frames (Projection). This creates a visually neutral buffer between the end of the movie content and the identification section, giving the projectionist more time to close the douser. The Academy standard refers to this as the runout leader, the Universal and Projection as the runout section. In the Academy standard, this comes before the Identification Tail Leader and Protective Tail Leader; in the Universal standard, the runout section, identification section, and protective section are all part of the "trailer", and all part of the trailer leader in the Projection standard.

The Identification Tail Leader of the Academy standard is 24 frames in length, and specifies that 18 frames state "END OF REEL", the number of the reel, and the picture title, in black on white. This is repeated lengthwise on the remaining six frames, white on black. It is likely that the first section is intended to be viewed as projected, and the second section is to be read directly. For the Universal leader, there is a single frame reading FINISH between the runout and identification sections. The identification section is "3 ft. plus 5 frames" (53 frames) in length. It includes/elicits much the same information as the Academy leader, but is labeled SMPTE UNIVERSAL LEADER, and includes a single frame reading FOOT. The Projection leader specifies much the same information, but includes the word END in addition to FOOT.

The Tail End of the Leader Era

As physical film leaders drift toward analog obsolescence, becoming less and less a central part of filmmaking and projection, we would do well to note how they nevertheless linger in the shared imagination. In the remainder of this essay I consider two distinct registers in which leaders have persisted, and even flourished, beyond the rarefied, "hidden" realm of commercial and artisanal film production, distribution, and projection. Far from being entirely invisible to the public, as was always intended, the distinctive graphic language of leaders turns out to be pervasive in documentaries, movies, commercials and music videos, and on record covers, books, posters, and T-shirts. Below, I explore the ways film editors and designers for broadcast and print have sampled or quoted from the graphic language of countdown leaders, often with glaring inaccuracies. First, however, we must consider how leaders were initially drawn into the light by avant-garde/experimental filmmakers. Apart from being the first constituency to recognize their considerable creative *and* critical potential in an ongoing, reflexive critique of subjectivity, filmmaking, the movies, pop culture, and political life, it is vital to appreciate the fact that these filmmakers also understood the role leaders played as an integral part of the technical protocols (Gitelman, 2006) of cinema. After all, we cannot hope to effectively subvert a particular media form unless we fully understand its language.

Leader Countdowns in Avant-Garde Film Culture: Distancing, Irony, Dissent

In his discussion of the work of Canadian experimental filmmaker Arthur Lipsett, film scholar William C. Wees explains that, broadly speaking, "found footage films are composed of pre-existing footage, such as stock shots, archival materials, and extracts from previously released films." Likening the techniques of their makers to "the arbitrary relationships and dream-logic of Surrealism, the irony and iconoclasm of Dadaism, and the disjunctive conjunctions of collage and photomontage," Wees continues:

> As artist-archeologists of the film world, found-footage filmmakers sift through the accumulated audio-visual detritus of modern culture in search of artifacts that will reveal more about their origins and uses than their original makers consciously intended. Then they bring their findings together in image-sound relationships that offer both aesthetic pleasure and the opportunity to interpret and evaluate old material in new ways.
>
> (Wees, 2007: 3–4)

The synchronization sections of leaders (hereafter "countdowns") are present in abundance in the film work of many "artist-archeologists". Notable examples include: *A Movie* (Bruce

Conner, 1958); *Color Film* (Standish Lawder, 1972); *At The Academy* (Guy Sherwin, 1974); *Material Film* (Birgit and Wilhelm Hein, 1976); *Standard Gauge* (Morgan Fisher, 1984); *Academy Leader Variations* (David Ehrlich and collaborators, 1987); *To the Happy Few* (Thomas Draschan and Stella Friedrichs, 2003); *Girls on Film* (Karin Segal and Julie Buck, 2005); and *A Movie by Jen Proctor* (Jen Proctor, 2010). The deployment of this particular kind of recycled content is so plentiful that film scholar Federico Windhausen refers to these films collectively as "the film-leader-countdown genre" (Windhausen, 2008: 119).

Paul Arthur (2004: 75) writes:

> When Bruce Conner kicked off his 1958 collage classic *A Movie* with a burst of Academy leader . . . it was the opening gambit to one of the avant-garde's abiding obsessions: the retrieval and celebration of those aspects of cinema hidden from view, buried under the weight of commercial storytelling or simply marginalized as "unsightly" detritus.

Countdown leader is not only a key element of Windelhausen's "genre", then, but also emblematic of avant-garde filmmakers' "obsessions" more generally. "Hidden from view" is an especially important phrase, given the aforementioned distinction between the elements of film that are routinely familiar to theater audiences, and those aspects in plain view of anyone who has spent time sorting through garbage or film bins in search of re-usable footage, or loitering/working in labs and projection booths.

The use of found footage was often a necessity rather than a creative conceit. In Conner's case,

> his own poverty propelled him toward these materials, while the absence of a market made durability an afterthought. Economic necessity also dictated the form of his first film, *A Movie* (1958), which he created by selecting and splicing frames from condensed versions of feature films, documentaries, and newsreels that he found at a photo store. "I don't own anything except the splices," he said. "I put it together – and it was . . . a totally different thing."
>
> (Federman, 2016: 31)

Given the current context, two subsequent "classics", *At The Academy* (Sherwin, 1974) and *Academy Leader Variations* (Ehrlich *et al.*, 1987), are of particular interest, since the titles of both films refer to a single, specific source: the Academy leader, formally introduced in 1930, an American Standard in 1947. Sherwin writes, "the implication of the 'at' in At the Academy is of opposition, 'at' meaning against rather than just being there, i.e. against standard procedures in film, or the standardisation of anything (?)" (Sherwin, 2015). In his engaging account of the making of *Academy Leader Variations*, Ehrlich (2006) notes that "since 1978, each of my personal films had begun with a personal variation on the academy leader" (2006: 76); he continues, "the unifying element of [*Academy Leader Variations*] could be the structural theme of the academy leader" (2006: 76). The result is marvelously inventive.

Closer inspection reveals that Sherwin's film actually utilizes BBC leader (Sherwin, 2016), while Ehrlich's collaborative animation is clearly based on the Universal leader: although the signature "clocksweep" is not utilized in every section, every countdown begins at 8 and ends on 2, with an interval of roughly one second (24 frames) between each number. (The Academy, Society and Projection leaders all run from 11 to 3, with numbers placed at ⅔-sec intervals; that is, 16 frames or 1 ft.) The "burst of Academy leader" (Arthur, 2004: 75) at the

Figure 13.2 Two photos of the three-channel media installation *Three Screen Ray* (2006) by Bruce Conner. Courtesy of The Conner Family Trust and Kohn Gallery, Los Angeles.

beginning of Bruce Conner's *A Movie* (1958) is peculiar indeed, beginning on 9 (rather than 11) and ending on 1 (not 3). It is worth noting that Conner often used 16mm Society leader in his films and artworks, as in *Cosmic Ray* (1961), *Report* (1963–1967), a poster for the New York Film Festival (1965), and *Three Screen Ray* (2006) (see Figure 13.2).

While not detracting in any way from the impact and importance of these works, these errors do highlight the ways in which "Academy" leader is often used by makers, archivists, projectionists, and scholars as a generic term for any leader featuring a countdown. It also testifies to the lasting impact of the first leader standard in the U.S., which was the only one developed by AMPAS. The remaining three (Society, 1951; Universal, 1965; and Projection, 1999) were all designed and promoted by technical committees of SMPTE.

Leader Countdowns in Documentaries and Movies: Motivated and Misplaced

Documentaries

In Alex Gibney's scientology documentary *Going Clear* (2015), sections of Universal, Academy, *and* Society countdown leader make fleeting appearances. They are used to provide transitions either into archival footage from talking heads, or from archival photos to archival footage. In this context, the work being done is perhaps to flag graphically the switch to archival material for the sake of visual variety, as much as an aid to comprehension, but also to underscore a sense of its authenticity.

In *Lambert & Stamp* (2014), a documentary about the rock band *The Who*, BBC leader is used intermittently through motion graphics sequences (including the titles) as an atmospheric device. This aligns not only with the celebrated modish aesthetic of the band, but also with the underlying narrative of the film, in which the two eponymous filmmaker-impresarios would forever treat the band as film subjects at least as much as rock musicians. (In one memorable sequence, there's an argument over who deserved more credit for their rock-opera album *Tommy* (1969) and an associated movie script, with Stamp and guitarist Pete Townsend in flat disagreement.) The creative impulse to use leader in this way can be favorably compared to the tendency among the designers of movie titles sequences to use Courier—a typeface derived from typewriters—for films about writers or writing. Examples include *Adaptation* (2002) and *Capote* (2005).

Movies

The celebrated Oscar-winner *Cinema Paradiso* (1988) centers on a child growing up in a one-theater town in post-war Sicily, and his growing friendship with the Paradiso's old projectionist. Prior to each public screening, the town's priest previews each movie, forcing the projectionist to temporarily remove any ungodly content, which amounts to cuddles and kisses between leading characters. After a terrible nitrate fire, the child takes over as projectionist, but is ultimately encouraged by the old man to leave the town to find his fortune. Having become a famous film director, he returns to the small town for the old man's funeral, and receives a mysterious film canister from his widow. The final climactic scene is of the director, alone in a plush, private theater, viewing the film, which turns out to be a compilation of all the on-screen kisses the priest had censored. Appropriately enough, it begins with a countdown, the brief conversation between the director and the projectionist, its location in a *private* "movie industry" screening room, and the fact that the movie is in some senses *about* film projection, all providing a sufficient pretext for *Cinema Paradiso*'s audience to see the countdown too. It also provides a moment of narrative suspense: just what exactly are we counting down to? Finally, the countdown appears to be an Italian variant of the Academy leader, which would be entirely appropriate historically.

About half an hour into *The Artist* (2011), a delightful Hollywood movie about the arrival of sound films, the titular hero (Jean Dujardin) is collared on set by the studio head (John Goodman), and ushered into a private screening room. Cut to the flickering screen itself, where a rather jumbled countdown heralds a short sound test: a brief piece of film not intended for public consumption in which the hero's estranged co-star (Missi Pyle) can be seen and (by implication) heard, speaking and singing into a conspicuous microphone. (Dujardin's character reacts as if it is the silliest thing he has ever seen.) While the movie is in large part an immaculately observed period piece, the countdown we see here is excerpted from the Society leader, which was introduced 22 years after this scene supposedly takes place. (Thanks to Dick May for this insight.)

While the sound test film is full of portent regarding the momentous arrival of the talkies, the Society leader was formally proposed in 1951 to address an entirely distinct but no less momentous arrival: broadcast television. Other issues are apparent: a conspicuous "Picture Start" frame appears toward the *end* of the countdown, whereas it should actually be about 140 frames earlier; the sound test, beginning with a title card showing the purpose of the short film, the name of the actor, and the date, appears immediately after the 3, rather than at the implied zero. The countdown, therefore, is not actually performing its supposed job at all, which is to allow the projectionist to cue up a separate sound source and synchronize it with the picture, beginning at the implied zero frame. Although the countdown perhaps succeeds in signifying to both audiences (in the private screening room within the movie's narrative and, in theaters and living rooms since 2011) that "a film is beginning", the way it does so is almost entirely inaccurate. That aside, the other work being done here, as with *Cinema Paradiso*, is to signal that we are having a privileged viewing experience; that the curtain is being pulled back to reveal a further glimpse of the production process, a perspective that is consistent with the rest of the movie, where we see production meetings, conversations among extras, actors and directors, and successive takes of the same scene.

Leader Countdowns in Pop Culture: Derivative and Decorative

Music Videos

In a different register are the appearances of leaders in music videos. In the mid-to-late 1980s, Society leader was used at the beginning of music videos for REM (*Fall On Me*, 1986) and

The Tragically Hip (*Blow at High Dough*, 1989). Again, the leader is not doing any technical work at all, since it is integrated into the opening shots of a *video*, accompanied by additional text (REM) or greenscreen compositing (The Hip). While other examples surely exist, this creative approach takes us back to experimental filmmaker and artist Bruce Conner, who was given the dubious distinction of being the "father of MTV". The idea of an influential relationship between the avant-garde and pop culture is not a new one. As Dave Stewart, music producer and one half of Eurythmics, 1980s MTV darlings, has recently pointed out in relation to their hit *Sweet Dreams* (1983): "It was totally surreal: we were using [Luis] Buñuel, and Salvador Dali, and French filmmakers, Italian filmmakers, as references, it was like a sort of weird surreal vignette" (Episode 7 of *Soundbreaking: Stories from the Cutting Edge of Recorded Music*, US 2016).

TV Titles

NBC's *Dateline* (1992–present) often uses a modified Universal countdown with the familiar radar sweep, and descending numbers in the center of the frame (although these end on "1"). Given its routinely sensational journalistic approach, the countdown is chiefly a dramatic device, perhaps more akin to a timer on a bomb than a self-extinguishing technical protocol from the movies.

Commercials and Promos

A 2016 commercial for TGI Fridays, advertising their "endless apps" (that is, appetizers) weaves together typical food shots with a surprisingly adept riff on the Society and Universal leader designs. It begins with a Universal sweeping arm, a graphical and spoken countdown from 4 to 1 in NASA-style voiceover ("kitchens, we are a go"), Society crosshairs, a blur of one-letter-per-frame messaging, and fleeting graphics (ENDLESS APPS 2016 on its side, with two concentric circles and an arrow). All the graphics interspersed or overlaid on the kinetic food shots are in black and off-white, with slightly grainy edges reminiscent of film rather than video. Overall, it is a remarkably thoughtful pastiche, making the creative connection between film and rocket launch countdowns described above.

Air Canada's promo for its 2016 enRoute Film Festival ("Canada's Only Inflight Film Festival") is a much more diluted reference to the Universal leader. The countdown actually goes up, rapidly, from 1 to 10 (2016 marks its tenth anniversary), with a subtle, intermittent radar sweep in the background. It also includes a home movie projector in silhouette—the radar sweep becomes the projector beam—and a strip of film in red with white text and sprocket holes. The motion graphic clearly takes some of its visual cues from the language of film leaders, but they are probably lost on most viewers.

Graphic Design

The covers for LPs by the rock band Foreigner and Genesis keyboardist Tony Banks appear to show actual frames of Universal leader. Foreigner's album *Foreigner 4* (1981) shows a blowup of one frame of countdown which matches at least one version of the Universal leader *exactly*. Banks' album *Soundtracks* (1986) shows a series of thumbnails sampled from Universal leader with minor variations: there is evidence of 8 thru 3 (2 is missing); PICTURE START erroneously appears at the end rather than before the countdown; the 35mm SOUND START frame has been flipped and rotated and also appears *after* the countdown; and the graphical sequence begins with a flipped and rotated 16mm SOUND START graphic but with the text completely removed. While Foreigner's use of Universal countdown is entirely arbitrary, Banks' album features music from two movies, thereby providing an appropriate nod to work created behind the scenes.

When used as a graphical device for print jobs such as book covers, the designers responsible often take liberties in interpreting the leader standards. For the Filmspotting.com podcast logo, a countdown number on the Universal leader has been replaced by a stylized "F". The cover of Ronald Deutsch's film memoir *Inspirational Hollywood: Reflections on Life, Love and the Art of Filmmaking* (1997) wrongly shows adjacent Universal leader frames for 3, 4, 5, and 6 in a single strip; each number actually appears for 24 frames. Bruce Kawin's *Selected Film Essays and Interviews* shows a redrawn, stylized "9" without framelines, but with added vertical "scratches". The BBC Radio 5 podcast *The Film Programme* (1971–present) shows a strip of Universal leader in which an "8" and the two concentric circles have been redrawn, and the two fully visible frames are out of sequence. (Seen from the right side, the action goes from top to bottom, but in this case the "clocksweep" is in the wrong direction.) While graphic design often takes liberties with reference material—interpreting and adapting rather than quoting directly—we might reasonably question the lack of accuracy here, since all four examples are meant to signify the promise of uncommon insight into the world of film production.

Finally, a website called FilmLeader.co offers "apparel and design for film lovers". Among the offerings are button-down shirts, T-shirts, dresses, hoodies, and totes, all covered with Academy and Universal leader stripes. No background information is provided on these important historical artifacts; no acknowledgment is given to AMPAS or SMPTE.

Theorizing Leaders

In the introduction to her book *Always Already New: Media, History, and the Data of Culture* (Gitelman, 2006), titled "Media as Historical Subjects", Lisa Gitelman suggests that "it helps to locate media at the intersection of authority and amnesia" (2006: 6). Comparing the emergence of "scientific instruments" to that of "new" media, she writes:

> media become authoritative as the social processes of their definition and dissemination are separated out or forgotten, and as the social processes of protocol formation and acceptance get ignored. One might even say that a supporting protocol shared by both science and media is the eventual abnegation and invisibility of supporting protocols.
>
> (2006: 7)

I see this as referring not just to the relatively swift taken-for-grantedness of specific protocols and standards, and the cultural occlusion of their respective limitations and affordances, but also the ways in which digital protocols are often far less obvious and accessible than analog ones. While I often refer to film leaders as the "metadata" of film, it is far easier to locate, handle, and photograph 35mm leaders than, say, the file "header" information inside a DCP, or Digital Cinema Package, the portable hard drives that are the currently preferred way to distribute movies for screening at theaters and multiplexes. The case studies presented in *Always Already New* (Gitelman, 2006) strike me as a call to media historians to rediscover and interrogate these protocols.

One way to approach this task is to distinguish at the outset between "backstage" protocols that are intended to be hidden from—or at least *not learned by*—users or consumers (such as telephone switchboards or Internet Protocol addresses) and "frontstage" protocols—understanding area codes, dialing a number, or choosing a wifi network. This correlates with my proposed distinction between the *producerly* and *consumerly* paratext (Soar, 2016), the former being all the identification and synchronization information on the leader, and the latter being the opening titles and closing credits. The difference between the two is literally the width of a frameline, where the countdown ends (the final black frame implying "zero") and the first frame of the movie "proper"—which is also the very same point at which the projectionist opens the douser on the projector, allowing light (and hence moving image) onto the projection screen.

Jussi Parikka, in his exploration of the promise of media archaeology, notes the "centrality of the archive for media studies" (2012: 6). To marginalize (figuratively and literally) archival catalogs, film canisters, labels, inspection sheets, color timing cards, and film leaders as paratextual, is to do a disservice to the ongoing cultural work of engaging with "the movie". Further, it actively (if implicitly) devalues these vital cultural artifacts as trivial addenda. To take up Matthew Kirschenbaum's useful distinction, while "movie reels" have a *formal* materiality, all the other elements I would like us to consider add up to the movie's *forensic* materiality: the physical instance of *this* particular print sitting in my hands, with *these* scratches, in *this* canister, with *this* label and *this* leader. (One might extend Kirschenbaum's terms backwards: the movie-as-text could be considered an instance of formal *immateriality*.) What might it mean to refuse notions of supplementarity, paratextuality, marginality, if only briefly? Imagine, for example, a screening at which the audience is given a printed program featuring facsimiles of each of these elements, including a screengrab of the relevant catalog screen, and where the projectionist is asked to commit the heresy of revealing the leaders to the audience (if not at every reel changeover, at least at the beginning and the end)? Is this appealing, or asinine? And why, exactly?

References

Annual Report (1930). AMPAS.

Arthur, Paul (2004). (). *Film Comment 40*(1), Jan./Feb., p. 75. [Review of *()*, a short film by Morgan Fisher, US, 2003].

Dickson, Paul (2009). *A Dictionary of the Space Age*. Baltimore, MD: Johns Hopkins University Press.

Ehrlich, David (2006). International Collaboration Films. *Animation Journal 14*, pp. 76–78.

Federman, R. (2016). Bruce Conner: Fifty years in show business – a narrative chronology. In R. Frieling and G. Garrels (Eds.) *Bruce Conner: It's All True*. San Francisco, CA: SFMOMA/ University of California Press, pp. 16–269.

Five Recent American Standards on Motion Pictures. *JSMPE 50*, March 1948, pp. 282–289.

Geser, G. (1996). *Fritz Lang – Metropolis und Die Frau im Mond: Zukunftsfilm und Zukunftstechnik in der Stabilisierungszeit der Weimarer Republik*. Meitingen: Corian-Verlag H. Wimmerer.

Gitelman, L. (2006). *Always Already New: Media, history, and the data of culture*. Cambridge, MA: MIT Press.

Miller, Ron (1995). Silent Space: Early experiments in cinematic sci-fi. *Filmfax: The Magazine of Unusual Film & Television 49*, Mar./Apr., pp. 35–43, 94–95.

Parikka, J. (2012). *What is Media Archaeology?* Cambridge: Polity.

Roberts, Ian (2015). Primitive Miasmas and the Iconography of the Future in Fritz Lang's *Frau Im Mond* (1929). *Studies in European Cinema 12*(2), pp. 97–105.

Roth, Lorna (2009). Looking at Shirley, the Ultimate Norm: colour balance, image technologies, and cognitive equity. *Canadian Journal of Communication 34*(1), 111–136.

Schauer, Bradley (2015). 'The Greatest Exploitation Special Ever': *Destination Moon* and postwar independent distribution. *Film History 27*(1), pp. 1–28.

Schwartz, Gare (1947). A Proposed Film Lock and Identification Band. *JSMPE*, May, pp. 473–475.
Sherwin, Guy (May 5, 2015). E-mail correspondence with author.
Soar, M. (2016). The Beginnings and the Ends of Film: Leader standardization in the US and Canada (1930–1999). *The Moving Image 16*(2), Fall, pp. 21–44.
Specification for 35-Millimeter Sound Motion Picture Release Prints in Standard 2000-Foot Lengths I (1947). American Standard Z22.55-1947.
Standard Release Print Makeup and Practice (1930). AMPAS.
Report of Standards & Nomenclature Committee (1930). *JSMPE*, December.
Standards and Recommended Practices (1966). *Journal of the SMPTE*, March, pp. 222–228.
Standards and Recommended Practices (1970). *Journal of the SMPTE*, March.
Television Demonstration (1929). *Transactions of the SMPE.*
Thalberg, I. (1930). Technical Activities of the Academy of Motion Picture Arts and Sciences. *JSMPE* July.
Threading Leaders for TV Films (1956). *Journal of the SMPTE*, August.
Townsend, C. L. (1951). New All-Purpose Film Leader. *Journal of the SMPTE*, May.
Wees, W. C. (2007). From Compilation to Collage: The found-footage films of Arthur Lipsett/The Martin Walsh Memorial Lecture 2007. *Canadian Journal of Film Studies 16*(2) Fall, pp. 2–22.
Windhausen, Federico (2008). Historiographic Returns: Reviewing British avant-garde film of the 1970s. *Grey Room 30* (Winter), pp. 114–128.
Yue, G. (2015). The China Girl on the Margins of Film. *OCTOBER 153*, Summer, pp. 96–116.
Žilová, J. (2013). Intertitles: The visual force of writing. In J. Elkins and K. McGuire (Eds.) *Theorizing Visual Studies: Writing through the discipline.* New York: Routledge, pp. 143–145.

14

VINYL, VINYL EVERYWHERE

The Analog Record in the Digital World

Richard Osborne

Vinyl records differ from other media technologies in this book. Their obsolescence was planned and was perhaps expected but it has not happened. Quite the opposite, in fact: there have been a number of vinyl revivals, the strongest of which is currently taking place. In the U.S., sales of vinyl albums and EPs declined from their peak of 341 million units in 1978, to just 900,000 in 2006 (Hogan, 2014). The situation was similar in the U.K. 1978 was the peak year, with trade deliveries of 86 million albums; in 2007 only 205,000 vinyl albums were sold (Osborne, 2012: 1). The last decade has witnessed a transformation. In the U.S., sales of vinyl albums increased 15-fold between 2006 and 2017, when total sales reached 14.3 million units (Caulfield, 2018). In Britain, 4.1 million albums were sold in 2017, the highest figure for 21 years (BPI, 2018). This upward trend is expected to continue. Research and Markets have forecast an annual growth rate in worldwide vinyl sales of 55.15% between 2016 and 2020 (Houghton, 2016a).

This revival is not confined to a nostalgic, baby-boomer market. A U.S. survey suggested that while a quarter of vinyl listeners fall into the 55+ age bracket, one-third are aged 13–24 (Houghton, 2016b). Hipsters and alternative music fans are buying vinyl: the annual "Record Store Day", which celebrates vinyl and independent record shops, has done much to publicize the format's appeal. And yet the format is also mainstream. Britain's two leading supermarket chains, Tesco and Sainsbury's, have recently begun stocking vinyl albums. The biggest-selling vinyl album in 2015 was Adele's *25* (BPI, 2016). As this release indicates, new music is being manufactured on the format. New pressing plants are also opening up. In the U.S., Independent Record Pressing, Cascade Record Pressing, Third Man Records, and Disc Makers have each been established since 2014. Moreover, many of these vinyl records are actually being played. One factor that is often mentioned in reports on the revival is that some contemporary purchasers do not own record players; they instead buy "vinyl to own it as a piece of art" (Hall, 2006).[1] Nevertheless, while there is some evidence that this phenomenon is true, sales of turntables have increased considerably. For example, Amazon's top-selling audio product for Christmas 2015 was the Jensen JTS-230 (Owsinski, 2016).

The fall and revival of vinyl have much to do with digital technologies. The analog vinyl record was deliberately targeted by the digital compact disc (CD). Thomas Edison invented sound recording in 1877. Symbolically, the planned launch date of the CD was 1977, marking the end of a century of analog records (Osborne, 2012: 82). The CD was successful too, achieving world sales of 200 billion by 2007 ("Compact Disc Hits 25th Birthday", 2007). CD sales are in decline, however. They were first troubled by digital downloads;

now they are faced with digital streaming. In contrast, the vinyl record has been helped rather than hindered by these successor formats. Its analog qualities provide a complement and an alternative to intangible digital files.

And yet the affection for vinyl goes deeper. Vinyl is a multi-faceted product with diverse appeal. Many of its aspects were fetishized *prior* to the arrival of digital formats. In order to analyze these facets we need to explore the origins of the vinyl disc and the ways it has been perceived. We also need to go back beyond vinyl itself. Many important features of the analog record were in existence before vinyl was used for record manufacture. These include the groove, the disc shape, the label, the B-side, and the sleeve. As such, this essay will address analog discs more broadly, looking in turn at the way they have reproduced, shaped, described, and molded sound. It will also address differing types of vinyl records. Vinyl has sub-formats, most notably the 7-inch single, the 12-inch single, and the long-playing (LP) record. These sub-formats have differing qualities and attract differing adherents. They have each played a role in vinyl's growth and survival.

How to Reproduce Sound

The ability to preserve and reproduce sound was a long-held human desire. There were early attempts, such as the statue of Memon at Thebes, built approximately 1490 BC, or the talking head designed by the English monk Roger Bacon in the Middle Ages. In addition, there were fictitious reveries. Charles Grivel noted that "The wish for phonography more or less fills literature, from Homer to Rebelais to Porta to Becher, to Grunderl and Nadar, to Cros and Villiers" (1992: 39). However, it was not until the late 19th century, when Edison developed his phonograph, that this dream was successfully realized.

The technology was surprisingly simple. Sounds were made into a recording horn; these sounds would vibrate a diaphragm; the diaphragm would in turn vibrate a stylus; the stylus would etch the sounds into a rotated recording medium. To reproduce the sounds, this order was reversed. At the heart of this process lay the groove. This is where the sounds were captured and where they could be witnessed. The responses to this sound writing—the "phono graph"—have been varied. The groove provided an initial source of wonder; it is the reason why the record was considered to be outmoded; and it is one of the reasons why the record has survived. Key to its importance is the fact that it appeals to a number of senses: the groove can be seen, touched, and heard.

The legibility of the groove precedes its function as a sound carrier. The direct predecessor to Thomas Edison's phonograph—the phonautograph, invented by Edouard-Léon Scott de Martinville in 1856—was concerned only with creating a script of sound. It was designed, Scott claimed, to capture the "natural stenography" of noise (Sterne, 2003: 45). Edison's machine went further: it replayed the noises themselves. Nevertheless, an interest in the text remained. In December 1877, the phonograph was displayed to the staff of *Scientific American* (1845–present), the first people outside Edison's circle of employees to encounter the machine. They reported, "there is no doubt that by practice, and the aid of a magnifier, it would be possible to read phonetically Mr. Edison's records of dots and dashes" ("The Talking Phonograph", 1877). The groove fascinated because it was believed that its text could be decoded. Edison, for one, felt he could achieve this goal. He tested the quality of records by looking at them rather than hearing them (Israel, 1998: 436–437). These tracings inspired poets and artists. Writing in 1919, Rainer Maria Rilke recalled that "what impressed itself" on his memory following his first encounter with the phonograph "was not the sound from the funnel but the markings traced on the cylinder" (1954 [1919]: 51–52), while Làszlò Moholy-Nagy was so excited by this automatic writing that he wished to replicate it manually (1985 [1922]: 289).

Although a full comprehension of records' grooves has remained tantalizingly out of reach, their general patterns can be understood. They change with fluctuations in tempo and fluctuations in tone. This is more apparent with some sub-formats than others, however. The 78 rpm shellac disc was the dominant recording format from the second decade of the 20th century until the 1940s. It had large grooves, which enabled these changes to be easily noted and accessed. Consequently, when the LP was introduced mid-century, its narrow microgroove was bemoaned:

> It is going to be much harder now, if one wants for any reason to go back to a certain bar, to find the right spot and lower the needle without damaging the grooves . . . It will be impossible to mark up discs for demonstration, school or lecture use and certain broadcasting purposes. And L.P. records are a blow to people who, like myself, think they have at last mastered the art of reading music (the circular variety).
>
> (L.S., 1950: 4)

It would be another quarter of a century before the larger groove returned, via the 12-inch single, which was launched commercially in 1976. This sub-format had the same diameter as an LP but contained less music. Consequently, its grooves were more widely spaced. The introduction of the 12-inch single in the mid-1970s prompted a revolution in the art of DJing. The pioneering hip-hop DJ Grandmaster Flash re-introduced the idea of marking records. His "Clock Theory" involved drawing lines on a disc like the hands of a timepiece. He could then spin the records back to these segments with split-second timing. Flash's protégé, Grand Wizard Theodore, introduced further skills. He could accurately drop a record player's needle at the beginning of a percussive break:

> You watch the grooves, the thickest grooves are where the break part comes in. When the record rolls around at a 360-degree angle, you can pretty much see where it starts. You say, "Here it comes." I made sure that I picked up the needle at a certain point. I watch the record go round and round, then bam! It comes right in. I did it so many times that I came to do the needle drop. I developed a technique, and I didn't know what I was doing. I got this down to a science. I used to astonish myself.
>
> (Fricke & Ahearn, 2002: 62)

He also invented the most famous of all hip-hop's DJing innovations: scratching (manually moving a record's groove back and forth at speed to alter its pitch and distort its sound). It was the specific requirements of hip-hop and dance music DJs—their need to see and touch the grooves of a 12-inch single—that first ensured the continuation of vinyl in the digital world. As the sales of other sub-formats declined, sales of 12-inch singles remained healthy, thus keeping pressing plants in business (Davis, 1994: 18–19).

The digital campaign against analog discs was not centered on their ocular or tactile qualities. On the contrary, digital DJing equipment, such as Serato Scratch or Final Scratch, has sought to replicate the look and feel of vinyl records. Rather, it has been the audio quality of the groove that has been targeted. Shellac discs gained their best levels of reproduction when there was a close fit between needle and groove. Abrasives were added to the shellac mixture to help wear down the stylus, while the stylus itself would dig away at the groove. This battle resulted in close contact, but also fostered the distinctive "frying-bacon" sound of 78 rpm records. Shellac was eventually replaced by tougher and more durable vinyl records, but with these came the microgroove: styli now had to negotiate the most delicate of tracings. This new material was also static-electric prone and therefore attracted increased amounts of dust.

The CD's solution of contact-free reproduction had long been mooted. In 1917, Gresham L. George had written:

> That the superman of 2000 A.D. will be content to receive his aural titillation through the medium of a steel needle and shellac plate is, to me, inconceivable. Long before that, magnetised strands and selenium cells will have been exploited. By electro magnetion, or currents controlled by light – with in neither case direct contact between record and reproducer – will that fortunate descendant of the present scratch-enduring, squeak-abiding generation seek beatitude.
>
> (1917: 270)

The CD finally arrived in 1982. Philips and Sony, the co-developers of this format, focused their attacks on the fallibility of the vinyl groove. An early Sony advert pictured a vinyl record beside a CD and boasted, "no wow, no flutter, no wear, virtually immeasurable distortion, wide dynamic range and no surface noise" (Sony advertisement, 1982: 6–7). Philips's advertising claimed similarly: "No record or stylus wear. No dust, static or vibration problems. No surface noise" (Philips advertisement, 1983: 9).

These campaigns were effective. Ten years after its launch, the CD was the leading sales format in the U.K. (Osborne, 2012: 82). However, as vinyl records began to be superseded by this digital carrier, something curious happened: their patina began to be appreciated. It was worn as a badge of honor; held up as something that was missing from CDs. One boast was that the layer of detritus provided proof that the listener had been there:

> [I]n a world where it often seems important to assert that you were in on some things right at the start . . . those blemishes – scratches, bits of old candle, staunched beer spillages – are a crucial historical index. Records left their mark on you, and, in the spirit of co-operation, you left your mark on records. What hope for this two-way relationship if CDs have the gift of eternal youth?
>
> (Smith, 1995: 182)

Another claim was that background noise engendered a superior auditory ability: "We listened harder in those [vinyl] days. Music was made doubly precious by the thicket of noise from which it had to be plucked" (Eisenberg, 2005: 212). For others, the patina was part of the musical experience. The increased surface noise of the raised run-in groove, occasioned by its distorted shape, provided an anticipatory frame for the sound recording. This was followed by the pleasure of the steady crackle when the music was under way. The appreciation of background noise became obvious when artists began to sample worn vinyl records, a practice that rose in direct proportion to the dominance of the CD.

There was a sense that the true character of the analog format was to be found in its aging process. The artist Christian Marclay sought to highlight this phenomenon with his piece *Record without a Cover* (1985). This disc was issued sleeveless so that it could rapidly accumulate wear and tear. The recording itself consists of Marclay DJing with old and worn records, but also contains large expanses of "empty" grooves. As the record gets damaged it becomes difficult to tell which are its natural scratches and which come from other discs. For Marclay, the noise accumulated by records is as important as the noise incorporated in records:

> When a record skips or pops or we hear the surface noise, we try very hard to make an abstraction of it so it doesn't disrupt the musical flow. I try to make people aware of these imperfections, and accept them as music; the recording is a sort of illusion while the scratch on the record is more real.
>
> (Ferguson, 2003: 41)

The appeal of patina is wider than Marclay thinks, however. Many vinyl fans would echo Elvis Costello's sentiments in "45", his paean to the 7-inch single. The singer praises "Every scratch, every click, every heartbeat, every breath that I bless."[2]

How to Shape Sound

Once the groove was in existence it needed to be formatted. There were a number of possible solutions: to house the recording information on a continuous strip (as with a tape spool); to spiral it around a cylinder; or to spiral it on a disc. The pioneers of sound recording focused on the latter two options. Thomas Edison's early recordings were issued on cylinders. Emile Berliner, the inventor of the gramophone, chose to work with discs.

When it comes to sound quality, the cylinder is the superior choice. A standard rotation speed for a cylinder produces a standard surface speed for its groove. In contrast, when a disc turntable rotates at a regular speed, the analog groove has to cover a greater distance at the outside of the record than it does at the center. The faster the surface speed of the groove, the more space it has in which to hold recording information; the more space the recording information is given, the clearer its reproduction. Due to the lower surface speed at the center of a disc there is a decrease in quality as the record progresses.

The disc shape has advantages of its own. Being a flat object, it is more convenient to store. It can also be read and handled more easily, thus facilitating the sensory qualities outlined in the section above. In addition, it is two-sided. Doubling the surface area increases the amount of information that can be housed. It also enables the material on the two faces to be compared and contrasted. However, it was for reasons not inherent in the shapes of these formats that the disc eventually triumphed over the cylinder, becoming the leading format by the outbreak of World War I. Notably, the companies who manufactured discs hired celebrated artists, they had superior repertoire, and they marketed their recordings more effectively (Osborne, 2012: 35–43).

Because the analog disc had drawbacks, there were continual attempts to improve upon it. At first this remodeling took place *within* the parameters of the analog disc. There was a desire to find the optimum balance between sound quality, size, and duration. A record with a wider diameter would be able to contain more recording information. It would, however, be more cumbersome to store and would evidence a more pronounced decline in sound quality between its outer rim and its run-out groove. A faster revolution speed would alleviate some of the sound problems, but only at the expense of duration. Ultimately, the analog disc format set time limits for music. In the early years of sound recording, discs ran for no more than two minutes. By the middle of the 20th century, the majority of 10-inch, 78 rpm records were still restricted to a four-minute length. The 7-inch, 45 rpm discs introduced by RCA Victor in 1949 offered only a marginal increase upon this, running at a maximum duration of 5.3 minutes.

Attempts were made to create records of greater length, with the primary aim of housing longer pieces of classical repertoire. These items were generally ill-conceived, overly expensive, or inappropriately marketed, however, and it was not until the introduction of Columbia's 33 rpm, LP record in 1948, that some of these issues were successfully addressed. Each side of an LP could contain up to 25 minutes of music, while the narrow "microgroove" that was developed for this format offered sound reproduction of a high standard. This groove was nevertheless susceptible to dust and it required delicate handling.

Hence, format developments took place *beyond* the confines of the analog disc. The first alternative was tape spools. Reel-to-reel tape recording offered excellent sound quality, while the cassette tapes introduced by Philips in 1962 provided ease of handling. Both formats facilitated longer playing times than the analog disc. Moreover, such was the success of the cassette tape that it eventually superseded vinyl records, becoming the leading sales format in Britain by 1985. Cassettes were nevertheless considered as a complement to vinyl records; they were not intended to replace them.

The same is not true of the CD, which was marketed as superseding vinyl records in all respects. CDs would be of longer duration, but of a smaller, more convenient size. They would be of excellent, sustained sound quality and they would be less prone to damage and wear. As we have seen and will see, however, each of the supposed defects of vinyl records has also been viewed in a positive light, time constraints included. Although the parameters of vinyl formats are restrictive, they have become embedded in musical practice. The short burst of the 45 rpm disc has been idealized in the notion of the three-minute single. The 45–50 minute duration of the LP has also created a blueprint. Although the CD format increased the potential length of an album to 74 minutes, many artists continued to issue albums that lasted the timespan of an LP. These ideals provide one reason for the continued fetishization of vinyl records. If singles are still being made to a three-minute formula and albums are set at three-quarters of an hour, it makes sense to consume them via the formats that set these limits.

There are further aspects of musical repertoire that are most aptly consumed via vinyl formats, as it was these formats that helped to shape them in the first place. For example, the two-faced nature of vinyl singles helped foster the idea that the track on the B-side should be of a different but complementary nature to the one on the A-side (Osborne, 2012: 143–155). The B-side was sometimes of a different generic origin (as with Elvis Presley's Sun singles), it could have subterranean qualities (as with the wilder or more unguarded material that many artists have buried on their B-sides), or it could offer an inversion of the music on the other side (as with the dub sides of reggae singles or the remixes that evolved through dance music genres).

On a larger scale, artists have used the two sides of an LP record to create contrasts and continuities. Within rock music, Bob Dylan and the Beatles were pioneers of mirroring and inverting the two halves of an LP. Dylan's folk music apostasy is reversed on 1965's *Bringing It All Back Home* (which by a few months pre-dates his notorious Newport Festival appearance). On side one he goes electric; on side two there is just guitar and voice. Other rock artists, including David Bowie (*Low*, 1977), Joni Mitchell (*Court and Spark*, 1974) and the Beatles (*Abbey Road*, 1969), have produced albums with strongly contrasting sides, as have soul artists, such as the Isley Brothers, who commonly issued albums with up-tempo songs on one side and ballads on the other. The Beatles' *Sgt. Pepper's Lonely Hearts Club Band* (1967) is also dependent on the dual-sided nature of the LP. It is patterned like a theatrical production, beginning with an orchestra tuning up and concluding with an encore, and it is broken by an intermission between the two sides. However, it is the "Sgt. Pepper" reprise that makes the whole concept hang together: bridging and dividing the two halves, it summarizes the record and marks out the distance the listener has travelled. This use of the reprise has been much imitated. Neil Young used it in combination with an acoustic/electric split, book-ending at least three of his albums with different versions of the same song. Pink Floyd was also fond of this device: "Shine on You Crazy Diamond" opens and closes *Wish You Were Here* (1975); "Pigs on the Wing" performs the same task on *Animals* (1977).

These devices have been carried over to later formats: although one-sided, cassette singles and CD singles have both included B-side material. There was also, for a time, the idea of the download B-side (Osborne, 2012: 153). In addition, artists such as Julian Cope (*Citizen Cain'd*, 2005) have issued albums of contrasting material as double-CD sets, even though the music would fit onto a single CD. The idea is to replicate the two-sided nature of the LP. Other artists have sought to digitally replicate the intermission of an LP record. At the halfway point of Beck's *Odelay* (1996) CD there is the sound of a needle being picked up from a record, followed by the sound of the needle being put down again. Similarly, halfway through the CD of Robert Wyatt's *Cuckooland* (2003) there are 30 seconds of silence, put there "for those with tired ears to pause and resume listening later" (Wyatt, 2003).

Later formats also sought to replicate the varied appearances of vinyl sub-formats. Cassette singles and CD singles were both packaged differently from their album counterparts, and some of the earliest CD singles were smaller in diameter than CD albums. These differences were never as marked as those between the various vinyl sub-formats, however. Seven-inch singles, 12-inch singles, and LP records have different sizes, speeds, and sleeve design styles. Each of these formats has different sound qualities as well. Although these sub-formats have been utilized by all musical genres, they have also been articulated for particular musical causes. The LP format was developed for classical music and henceforth was suited to music that had mature and serious aspirations (successively, the earnest causes of jazz, folk, and progressive rock were each honed on the LP record) (Osborne, 2012: 94–115). The 7-inch single, meanwhile, has been variously associated with commercial music (because of its links with the charts and radio), with dance music (because of jukeboxes), and with DIY punk and indie music (because it is cheaper to self-produce a vinyl single than to record and manufacture an LP). Finally, the 12-inch single was developed by DJs and has remained the preserve of electronic dance music (it is louder, longer, and has better bass frequencies than the 7-inch single, and its wider grooves can be accessed and manipulated more easily).

The differences between sub-formats have proved crucial for vinyl's continued success. Vinyl is the only sound carrier with sub-formats that sound and perform differently from one another, and it is vinyl that provides the most effective means of putting musical preferences on display. Moreover, it is the taste preferences of different musical tribes that have kept vinyl in continuous production. It was the demand for 12-inch dance music singles that ensured vinyl's survival during the fallow period of the 1990s. The early 2000s saw a rise of garage rock bands, such as the White Stripes and the Strokes. Their espousal of the immediate and speedy thrills offered by the 7-inch single saw an upturn in sales of this format. The more recent vinyl revival has been based on LP sales, as a mature market has emerged for classic rock releases.

How to Describe Sound

An analog record has two main packaging features: the label and the sleeve. These devices have different emphases and provide one of the means by which sub-formats have been utilized to promote different types of music. They have also had an effect on the way that music is disseminated and perceived. They are required for a number of reasons. First, they make up for the fact that the groove cannot be read as text: the consumer requires further literature and imagery to describe the audio tracings. Second, they provide a means of marketing music. As they do so, new visual and tactile qualities are introduced to the consumption process. Third, these items make up for the design limitations of the disc. The label gains its position at the center of the analog record because there is a point at which the groove's spiral can no longer accurately reproduce sound. The record cover is required because of the delicacy of the groove.

Faced with the void at the center of his gramophone discs, Emile Berliner turned this flaw into an advantage. He used the space to inscribe the title of each recording as well as details of patents he had been awarded. Eldridge Johnson, the first head of Victor Records, went further. In April 1900, he wrote to his British colleague William Barry Owen, "Strange to say, one of our greatest difficulties has been the proper marking of these records. We never tried before to mark them properly, as if we were making them to sell" (Edge & Petts, 1987: 201). His solution was the paper record label, first witnessed on his "Improved Gram-O-Phone Records". The name of his record company was placed prominently on the top half of the label; the specifics of the disc (the name of the song, the artist, and other recording information) were outlined in smaller print and placed on the bottom half. This design standard has remained dominant throughout the history of analog disc manufacture.

The clearest outcome of labeling lies in the fact that the term "record label" became a synonym for "record company". Recordings have become strongly identified with particular manufacturers. Victor Records, for example, were particularly keen that their company name and masthead would be associated with the finest repertoire and with particular artistes. This marketing ideology was adopted by other major companies in the early 20th century but became difficult to sustain as their rosters grew in size and they sought to cover the market for all forms of music. As a result, the larger companies used labels to divide their music along generic lines, utilizing a range of label colors and sequences of catalog numbers to subdivide their repertoire. It was by these means that American record labels ghettoized "race" and "hillbilly" genres in the first half of the 20th century (Osborne, 2012: 50–61).

Labeling is also one of the more obvious means by which major companies can be distinguished from independent companies. This is beneficial for the latter, as some consumers identify positively with their business ethos. Independent companies can also find it easier to build a strong musical identity through labeling practices, retaining a close association with a particular type of music. Some label names, such as Motown or Blue Beat, have even become generic terms. In contrast, the music of major labels can be tainted by the presence of their company names, rendered "corporate" or "manufactured" because of their business ideologies. As a result, the majors have on occasion masked their involvement behind bogus independent label names (Osborne, 2012: 66).

These practices have been important because the label has been prominent. Shellac records were sold in brown paper bags with cut-away centers, which allowed the record label to show through. It was only in the late 1940s, roughly coinciding with the arrival of the LP, that records began to be housed in picture sleeves and the label was hidden from immediate view. In Britain, this luxury was not regularly afforded 45 rpm singles until the 1970s. Before then, vinyl singles took their design cues from shellac discs, being sold in paper bags with cut-away centers. The basic design for these bags was nearly always an extension of the record companies' logos. The dominance of the label name and iconography in this layout is one of the reasons why singles have been negatively considered more "commercial" than LP releases. Some practitioners nonetheless continue to see the virtues in this design. The cut-away layout can still be witnessed among 12-inch single releases, for example. Within dance music genres, the label name is often as important as that of the artist. Hence, this design scheme makes sense. This packaging method also highlights another icon of dance music scenes: the blank "white label" of promotional releases.

The record sleeve has the opposite virtues. It downplays the label name and instead puts artists at the forefront. Pictorial designs were first regularly used for album sets of shellac records, grouping together discs that would make up a classical composition or a theatrical show. These sets were also the first items to receive sleevenotes, which were introduced by Herbert C. Ridout of Columbia Records in 1925 (Osborne, 2012: 162). The first sleeve designer to achieve individual credit and recognition was Alex Steinweiss, art director for Columbia from 1939 to 1954. He persuaded company president Edward Wallerstein to allow him to illustrate the 1940 release *Smash Song Hits by Rodgers and Hart*. His bold graphics and illustrations not only revolutionized sleeve design, they also transformed record sales. Six months after designing the Rodgers and Hart sleeve, he repackaged a recording of Beethoven's *Symphony No. 9*, previously issued in plain grey cloth, and sales increased by 894 percent (McKnight-Trontz & Steinweiss, 2000: 31–32). Nevertheless, the biggest fillip for record sleeve design came with the introduction of the LP. This format offered a larger expanse for designers to explore. Additionally, because the music on this format was longer and more varied than that on shellac discs, it required cohesive illustration and elucidation.

There have been many trajectories within the history of sleeve design. One general factor, however, has been a movement away from scripture. Early LP sleeves were quite textural. The front covers would feature the names of the artists, the title of the album, the titles of some of the songs, the name of the record company, and possibly some boastful proclamations about the LP. The back covers would feature full track details, sleevenotes, and information about how to protect your records. In addition, they would sometimes advertise other products manufactured by the record company. However, when it was discovered that pictorial devices provided a better means of selling records, these texts began to diminish (Osborne, 2012: 165–166). Most sleeves came to be dominated by pictures of the artists or by abstract imagery.

The sleeve is one of the main reasons for the continued popularity of vinyl records. In part, this is because some of the covers have achieved classic status. This can be attributed to the skills of particular designers (ranging from Reid Miles for Blue Note to Vaughan Oliver for 4AD); it can be because a particularly striking image has been employed (the prism of Pink Floyd's *Dark Side of the Moon* (1973) or the radio pulses depicted for Joy Division's *Unknown Pleasures* (1979)); it can be because of the synchronicity between the music and the design (the psychedelic *Sgt. Pepper* sleeve; Jamie Reid's graphics for the Sex Pistols); it can even be because of a disjuncture between the cover and the text (the sleeve of Beach Boys' *Pet Sounds* (1966) looks increasingly dated while the music remains advanced).

What is also crucial is that vinyl sleeves have advantages over the packaging for other recording formats. In the first instance, they are bigger. The 12″ × 12″ frame has had sufficient impact to attract serious artists to its cause (Peter Blake, Richard Hamilton, and Andy Warhol all designed LP covers). Second, these sleeves can be made bigger still, folding out into gatefolds and posters if required. Third, different shapes can be utilized (although square is the most common, there have been circular and hexagonal sleeves). Fourth, these sleeves can accommodate other materials (inner sleeves, posters, singles, and various "free" gifts). Fifth, they can be made of a range of materials. By the early 1970s, album sleeves had incorporated "linen, sand grain, calf, imitation leather and embossing" (Britt, 1971: 25). Finally, whereas cassette and CD sleeves are enclosed in plastic cases, LP sleeves are open to the elements.

Although this renders them susceptible to damage, it also means that they encounter nostalgic aging processes. There is a pleasure in the patina of an LP sleeve, just as there is pleasure in the patina of the groove.

How to Mold Sound

One curious factor about the vinyl record is how late vinyl enters the frame. The groove, the disc shape, the label, and the sleeve were all in existence before polyvinyl chloride (PVC) was used for record manufacture. This material was introduced as a replacement for shellac, the substance that dominated record manufacture from the early 1900s to the middle of the 20th century. Shellac's downfall was not due to its sound reproduction. Despite its "frying-bacon" sizzle, shellac was capable of withstanding continued audio improvement: by 1945 Arthur Haddy of Decca Records had developed "full-frequency range recording", which had the ability to cover almost the entire range of frequencies heard by the human ear. Instead, its problem was geographical. Shellac comes from secretions made by the female "lac" beetle. Its main sources are the Malay Peninsula and French Indochina. The Japanese occupied both territories during World War II, consequently for the West, shellac was in short supply.

The Union Carbon and Carbide Company introduced PVC in the early 1930s. This substance was almost immediately employed in record production: the first vinyl discs were produced in this decade, housing transcription recordings of popular radio shows. The harder and finer material of vinyl allowed for closer groove spacing, thus providing the longer playing time that these discs required. Vinyl also produced less surface noise than shellac and gave a wider frequency response. At this time, however, it was more expensive to manufacture. Although the blockades of World War II helped to urge its commercial usage, it was only with the arrival of the LP record in 1948 that vinyl proved to be commercially viable. The LP's longer playing time justified its higher cost.

Nevertheless, this manufacturing material was not the most novel aspect of the LP record: by the time of the LP's introduction, consumers had become used to discs of different substances, including celluloid and laminated discs. More innovative was the microgroove system, which made increased playing time possible. Music journals and record industry literature originally distinguished LP and 45 rpm vinyl records from 78 rpm shellac discs by reference to their speeds, rather than by referring to the material from which they were made. Indeed, it was only following the introduction of these newer discs that shellac records began to be referred to as "78s". Correspondingly, it was only with the introduction of further formats—the tape cassette and the CD—that LPs and 45s were commonly referred to as "vinyl" albums and singles, instead of being distinguished by their disc speeds.

The point being stressed here is the lack of attention paid to the vinyl compound at the moment of its wider introduction. If anything, artists and record companies wanted the focus to be elsewhere. Concentration on the material aspects of the record could lead to accusations of plasticity (an attitude that can be witnessed in the Byrds' 1967 release "So You Want to Be a Rock 'n' Roll Star", which refers to the singles-making process as "sell[ing] your soul to the company / who are waiting there to sell plastic ware").[3] The history of vinyl production is curious and contradictory, however. By the 1970s, record companies were drawing attention to vinyl's qualities. By the 1980s, this plastic product was being described in organic terms. In fact, vinyl began to be thought of quite differently from how it was viewed when it was the leading music format.

Vinyl became ostentatious because this helped to sell records. There had been a long tradition of using the groove or the recording material to create quirky or distinctive discs.

As early as 1898, Emile Berliner was cutting records that featured two grooves on the same surface; the needle would then arbitrarily choose which one of two piano solos would be heard. The picture disc was also an early innovation, first appearing in the 1920s and used mainly for children's or advertisement records. In the 1930s the first square-shaped "discs" were manufactured. Moreover, neither shellac nor vinyl records are naturally black. As such, different colors have been used to demarcate prestige releases and different genres of music. RCA Victor's original 45 rpm releases, for example, were colored in this manner.

When these techniques were re-introduced in the 1970s, they were applied more individualistically and were used for the record companies' core musical products. The first vinyl picture disc was Curved Air's *Air Conditioning* (1970); color (or lack of) made a comeback with Faust's eponymous clear-vinyl LP (1972). Richard Myhill's "It Takes Two to Tango" (1978) claimed to be the world's first square-shaped single, while Alan Price's "Baby of Mine" (1979) was the first in the shape of a heart. During the punk and new wave era, a rash of colored vinyl records was released. The double groove also returned: in 1979 John Cooper Clarke issued "Splat" / "Twat", featuring parallel clean and rude versions of the same poem.

Vinyl also brought attention to itself because of its declining quality. For a period, it seemed as if the substance would suffer the same fate as shellac, its demise occasioned by its geographical source. PVC is derived from petroleum, for which the record industry has largely been dependent on the Middle East. The oil embargo instigated in 1973 by OPEC countries led to an increase in prices and a shortage of supplies. Record companies responded to this situation by reducing the amount of pure vinyl that records contained. In addition, records became thinner, with the weight of both singles and albums being reduced by about 8 percent. Despite consistent outcries from consumers and an increase in returned faulty goods, poor standards of vinyl production continued into the 1980s. It was against this background that the CD was developed and launched.

Vinyl enthusiasts responded to the introduction of the CD with an argument that has since become familiar. Despite the poor standards of vinyl then available, this substance was advocated as sounding "warm" and feeling "alive". In contrast, the CD was accused of having an "alien clarity" (Loder, 1991: 94). A year after the launch of the CD, it was reported that hi-fi enthusiasts found the new format lacking in "emotion" (Burbeck, 1984: 10). By 1987, the Smiths' lead singer Morrissey was claiming, "Vinyl, when rubbed vigorously against human skin, is passionately all-consuming"; in contrast he claimed that "Aromatically the Compact Disc has a vacuous 'Shake 'n' Vac' stench about it" (Martin, 1987: 21).

The irony here is that it was the arrival of the CD that enabled the analog disc to transform its image. It was only now that it could be elevated from assembly-line product toward something approaching an art object. Statistics helped. As the CD rose to dominance, it superseded vinyl as the music industry's mass-produced product. Vinyl, in contrast, came to look like a cottage-industry good. It was market forces that enabled artists such as Neil Young to advocate vinyl over the CD. Young waited until the early 1990s before launching his attack on digital reproduction. He then argued:

> With analog records the moment used to be captured – complete with the flaws, but you could hear it: all the *nuances*. With digital what happened was they removed everything that seemed like a flaw and all you have left is the *semblance* of sound.
>
> (Thompson, 2001: 205; emphasis in original)

The repositioning of vinyl would not have been possible were there not art as well as industry in a record's manufacture. A vinyl record *is* nuanced, its emphases dependent on the cutting engineer's skill. In contrast to a digital reproduction, an analog record usually operates at the limits of its reach: there is a desire to achieve maximum duration, volume, and depth of tone. Cutting engineers have to account for and compromise between each of these. Greater duration requires closer groove spacing, which results in lower volume. Ignoring this can result in over-cutting (this can contribute to the fuzzy "warmth" attributed to vinyl reproduction). Increased bass frequencies also require wider groove spacing; while excessively high frequencies can result in distortion and accelerated record wear. Engineers also have to accommodate for the decline in sound quality as the groove reaches the record's center.

As studio technology improved in the 1960s and 1970s, the dynamic range of tape master recordings increased. Vinyl did not keep pace with this change, however. Discrepancy between the two contributed to the increased number of faulty records in the early 1970s. Normally the dynamic range of the original studio recording has had to be compressed in order to achieve a successful transfer to vinyl. Again, this requires the skill of the recording engineer. Vinyl's warm analog glow is also the result of compression. What is fortuitous, according to remastering engineer Bill Inglot, is that vinyl "kind of mushes things just right" (Milano, 2003: 43). Another engineer, Martin Giles, believes that the aural limitations of vinyl have helped to ensure its survival. He argues:

> They talk about digital music sounding harsh or tinny or brittle in comparison and it is because you can't get away with that kind of top-end on vinyl, and you have to find other ways to cut it . . . In a way, vinyl won't let you get away with cutting unmusical stuff.
>
> (Woods, 2003: 15)

The CD had another effect on vinyl production: quality improved. In 1984, the Linn record label was set up with the idea of "making very high quality vinyl records to compete with CD" ("Linn Set . . .", 1984: 3). Thirteen years later, *Music Week* was reporting that "[t]he high quality associated with CD has inevitably meant that vinyl customers expect perfect replication"; the journal added, "Far from finding itself a casualty of the CD age, vinyl mastering has risen to the challenge" (Faux, 1997: 35). Record companies also began to use more vinyl. Toward the end of the 1990s, LP releases began to proudly advertise the fact that they were being issued on 180-gram vinyl as opposed to the 130-gram records that had previously become common. The manufacturers of these thicker discs reported a "surge in orders" (Tesco, 1999: 23). These records also boasted that they were "virgin vinyl pressings" and used "heavy quality" sleeves. Vinyl was no longer plasticware; it was sold by the gram as though it were fine food, and marketed as "luxury" issues in the manner of collectable art.

Digital downloading and digital streaming have brought another transformation. Increasingly, it has been vinyl, rather than the CD, that is held up as the physical product with which to counter the computer's free-flowing zeroes-and-ones. As a result, the artistry of music and the artistry of manufacture have become thoroughly confused. By 2005, Malcolm Swindell, music account manager at AGI Media, could claim without irony that "[t]he use of special packaging reminds people why they got into music in the first place" (Webb, 2005: 16). Similarly, Gordon Gibson of Action Records, speaking in 2007, could comment: "In every new batch of students there always seems to be a few that are actually into the music", pointing out that these are the people who are buying 7-inch singles: "They seem to be in it for the artwork – that's the main attraction nowadays" (Poole & Giacomantonio, 2007). Thus, we have reached the strange situation whereby "real" fans—those who are most passionate about music—are proving this point by investing in pressing and packaging. The 21st century has witnessed the advances of free music and the exorbitant vinyl release.

Conclusion

Looked at from one perspective, the vinyl revival could seem precarious. Vinyl is susceptible to changing tastes in music. Its current popularity is in part underpinned by the large market that exists for "classic" rock music. This market might not always exist and its consumers are certainly getting older. In addition, vinyl is a hipster's choice and is being bought in reaction to prevailing trends. It marks the purchaser as someone who is resisting digital dominance and wants a return to a slower, more tangible world. What happens if these trends change? And what happens if vinyl becomes so popular that it is no longer an alternative product? There are already signs of a backlash. As Record Store Day has become increasingly successful, long-term adherents have criticized its appeal to the "wanton masses" and its takeover by major labels (Hebblethwaite, 2014).

However, the vinyl record is multi-faceted and so is its revival. As I have outlined above, its sub-formats appeal to different musical scenes. Vinyl's regenerating ability is evidenced by the fact that its afterlife is now longer than its term as the leading sales format. In Britain, vinyl was at the summit for 27 years: it overtook shellac records in 1958 and was overtaken by the cassette in 1985. The year 2018 marks 33 years since the format was eclipsed. Within this period, there have been a number of different revivals—the 12-inch single in the late 1990s, the 7-inch single in the mid-noughties, the LP in the current decade—furthered by different musical trends.

It is also important to note that the appeal of vinyl does not reside solely in its analog obstinacy. It is true that some aspects of vinyl appreciation only came to the forefront after the introduction of digital formats. It was in response to the introduction of the CD, for example, that consumers began to talk with wonder about scratches and dust. It was this "alien" format that made vinyl seem warm. The advocacy of vinyl is not wholly reliant on digital dialectics, however. You can peel back its layers and witness a longer-lasting appeal. For example, here is Colin MacInnes, writing about purchasing records in the 1950s: "The disc shops with those lovely sleeves set in their windows, the most original thing to come out in our lifetime" (1980: 65). And here is Stuart Maconie, talking about his early encounter with records:

> The first objects that I can recall being imbued with anything approaching mystery and beauty, objects that I coveted with what clearly bordered on the fetishistic, were the vinyl records in the darkened, reverential space inside my nana's radiogram. . . . I remember almost every one; its look, its label design, even its smell.
>
> (2003: 13, 14)

Here is Twiggy, with a 1950s take on a supposedly modern phenomenon:

> My sister saved up all her pocket money and one day she came home and she had the record of "Diana" by Paul Anka. This is the first record that I'd actually ever seen in real life. It was very exciting. I remember sitting round our tea table and we all handed it round to each other and talked about it, because we actually didn't have a record player, so we couldn't play it. We could only touch it and look at it, but we all thought that that was fantastic. It didn't seem a weird thing to do.
>
> ("The People's History", 2016)

Finally, here is Rainer Maria Rilke, talking about the arrival of sound recording in the late 19th century: "What impressed itself on my memory most deeply was not the sound from the funnel but the markings traced on the cylinder; these made a most definite impression" (1954 [1919]: 51–52).

Vinyl has outlasted cassettes; its sales are going upwards while those of the CD are in decline; and it recently surpassed digital downloads in terms of sales income (Moore, 2016). At this rate, it looks set to outlast digital streams. When it comes to recording formats, the story of obsolescence lies elsewhere.

Notes

1 See also Green 2006; Bray 2008; Plummer 2009; Bignell 2011.
2 Elvis Costello, "45," written by MacManus. BMG Music, 2002.
3 The Byrds, "So You Want to be a Rock 'n' Roll Star," written by McGuinn & Hillman. Tro Essex Music, 1967.

References

Bignell, P. (2011). "Vinyl finds its groove with young music lovers," *Independent*. Retrieved September 18, 2018, from <www.independent.co.uk/arts-entertainment/music/news/vinyl-finds-its-groove-with-young-music-lovers-6261656.html>

BPI. (2018). "Rising UK consumption enjoys fastest growth this millennium." Retrieved September 18, 2018, from <www.bpi.co.uk/news-analysis/rising-uk-music-consumption-enjoys-fastest-growth-this-millennium/>

Britt, S. (1971, March 6). "What's new in sleeve manufacture," *Record Retailer*, 25.

Burbeck, R. (1984, March 31). "Give the CD a chance," *Music Week*, 10.

Caulfield, K. (2018). "U.S. vinyl album sales hit Nielsen music-era record high in 2017," *Billboard*. Retrieved September 18, 2018, from <www.billboard.com/articles/columns/chart-beat/8085951/us-vinyl-album-sales-nielsen-music-record-high-2017>

Compact disc hits 25th birthday. (2007). Retrieved October 28, 2016, from <http://news.bbc.co.uk/1/hi/technology/6950845.stm>

Davis, S. (1994, May 28). "Vinyl demands," *Music Week: Record Mirror*, 18–19.

Edge, R., & Petts, L. (1987, June). "The first ten-inch records, part 1," *Hillandale News*, 156, 201–205.

Eisenberg, E. (2005). *The recording angel*. 2nd edn. New Haven, CT; London: Yale University Press.

Faux, K. (1997, March 29). "Demand comes full circle," *Music Week*, 35.

Ferguson, R. (2003). The variety of din. In J. Huyn (Ed.), *Christian Marclay* (pp. 19–51). Los Angeles: UCLA.

Fricke, J., & Ahearn, C. (2002). *Yes yes y'all: The experience music project: Oral history of hip hop's first decade*. Cambridge: Da Capo.

George, G.L. (1917, November). "Stereophony," *Talking Machine News and Side Lines*, 15/74, 270.

Green, T. (2006). "Getting back into the groove," *Telegraph*. Retrieved September 18, 2018, from <www.telegraph.co.uk/culture/music/rockandjazzmusic/3655723/Getting-back-into-the-groove.html>

Grivel, C. (1992). The phonograph's horned mouth. In D. Kahn & G. Whitehead (Eds.), *Wireless imagination: Sound, radio, and the avant-garde* (pp. 31–61). Cambridge, MA; London, England: MIT Press.

Hall, J. (2006). *The old guard is getting streetwise*. Retrieved October 28, 2016, from <www.telegraph.co.uk/finance/migrationtemp/2950913/The-old-guard-is-getting-streetwise.html>

Hebblethwaite, P. (2014). *Is Record Store Day in crisis?* Retrieved December 9, 2016, from <http://thequietus.com/articles/15031-record-store-day-2014-problems-distribution-vinyl?utm_source=feedblitz&utm_medium=FeedBlitzEmail&utm_content=395530&utm_campaign=0>

Hogan, M. (2014). *Did vinyl really die in the '90s? Well, sort of* . . . Retrieved October 28, 2016, from <www.spin.com/2014/05/did-vinyl-really-die-in-the-90s-death-resurgence-sales/>

Houghton, B. (2016a). *Vinyl sales projected to grow exponentially through 2020.* Retrieved October 28, 2016, from <www.hypebot.com/hypebot/2016/09/vinyl-sales-projected-to-grow-expodentially-through-2020.html>

Houghton, B. (2016b). *Vinyl sales up another 17%, used sales surpass new.* Retrieved October 28, 2016, from <www.hypebot.com/hypebot/2016/07/vinyl-sales-up-another-17-used-sales-surpass-new-.html>

Israel, P. (1998). *Edison: A life of invention.* New York: John Wiley and Sons.

"Linn Set to Challenge CD Quality Claims with Vinyl Label." (1984, January 21). *Music Week, 3.*

Loder, K. (1991, November 28). "Songs lasting three minutes – and forever," *Rolling Stone*, 94.

L.S. (1950, June). "Decca issues L.P. records," *Gramophone*, 28, 4–5.

MacInnes, C. (1980 [1959]). *Absolute beginners.* London: Allison & Busby.

Maconie, S. (2003). *Cider with roadies: From school bus to tour bus without ever growing up.* London: Ebury Press.

Martin, G. (1987, February 14). "Out of the groove," *New Musical Express*, 20–21.

McKnight-Trontz, J., & Steinweiss, A. (2000). *For the record: The life and work of Alex Steinweiss.* New York: Princetown Architectural Press.

Milano, B. (2003). *Vinyl junkies: Adventures in record collecting.* New York: St Martin's Press.

Moholy-Nagy, L. (1985 [1922]). In K. Passuth (Ed.). *Moholy-Nagy.* New York: Thames and Hudson.

Moore, S. (2016). *Vinyl album sales outstrip digital downloads for the first time ever.* Retrieved December 9, 2016, from <www.nme.com/news/music/vinyl-album-sales-outstrip-digital-downloads-first-time-ever-1893095#KE1xmgPzBtHocz7b.99>

Osborne, R. (2012). *Vinyl: A history of the analog record.* Farnham: Ashgate.

Owsinski, B. (2016). *A turntable was Amazon's top selling holiday audio product.* Retrieved October 28, 2016, from <www.hypebot.com/hypebot/2016/01/the-turntable-amazons-top-holiday-audio-product.html>

"The People's History of Pop: 1. The Birth of the Fan." (2016). Dir. by James Giles. BBC.

Philips advertisement (1983, March 12). *New Musical Express*, 9.

Plummer, R. (2009). "Revived 45 heads for 60th birthday," *BBC News.* Retrieved September 18, 2018, from <http://news.bbc.co.uk/1/hi/business/7750581.stm>

Poole, D., & Giacomantonio, A. (2007). *Music: the 10 best independent UK outlets.* Retrieved April 24, 2008, from <www.independent.co.uk/student/student-life/music-film/music-the-10-best-independent-uk-music-outlets-5335136.html>

Rilke, R. M. (1954 [1919]). *Selected works: Volume I prose.* C. Houston (Trans.). London: Hogarth Press).

Smith, G. (1995). *Lost in music: A pop odyssey.* London: Picador.

Sony advertisement (1982, June 26). *Music Week*, 6–7.

Sterne, J. (2003). *The Audible past: Cultural origins of sound reproduction.* Durham, NC; London: Duke University Press.

Tesco, N. (1999, April 17). "Worth its weight in vinyl," *Music Week*, 23–24.

"The Talking Phonograph." (1877). *Scientific American*, 32/25, 384–385.

Thompson, B. (2001). *Ways of hearing.* London: Orion.

Webb, A. (2005, March 19). "Thinking outside the box," *Music Week*, 15–16.

Woods, A. (2003, July 19). "Never give up on a good thing," *Music Week*, 13–15.

Wyatt, R. (2003). *Cuckooland* (sleevenotes). Hannibal Records: HNCD1468.

15

DON'T TAKE MY KODACHROME AWAY

The Rise, Fall, and Return of Kodachrome Color Film

M. M. Chandler

Even if you never heard of a "Kodachrome" before, chances are you have seen one. From Gerald Sheedy's 1937 snapshots of the fiery Hindenburg explosion, to Abraham Zapruder's infamous 1963 home movie of President Kennedy's assassination, to Steve McCurry's haunting 1984 National Geographic photos of *The Afghan Girl*, Sharbat Gula, the tragedies and triumphs of modern history have been burned into our collective memory as color-saturated images captured on Kodachrome film.[1] Released by George Eastman's Kodak Company as the first color film stock in 1935, Kodachrome changed the nature and future of color photo-cinematic image-making. Throughout the 1930s–1980s, Kodachrome achieved widespread commercial success in both amateur and professional markets as a still photography and motion picture medium. Analog Kodachrome film began to fade into outmoded obsolescence when the market turned to inchoate digital imaging technologies beginning in the mid-1970s, which ultimately culminated in the discontinuation of its production and photo-lab processing by 2010. In many histories of technology, this would signal the end of the story, but this is not the case for Kodachrome.

The iconic images referenced above not only capture historic events, but also chart the life history and many of the major milestones for Kodachrome technology. In addition to chronicling the historic rise, fall, and return of Kodachrome, this essay will consider the lasting cultural work accomplished through this groundbreaking imaging technology. During the time of its popularization, Kodachrome's unique chemical processing and color-imaging properties intersected with a larger American cultural turn toward artificial materials, synthetic consumer goods, and escapist national rhetoric. The particular color palette offered through Kodachrome photography and motion pictures became intertwined with a forward-looking, futurist mindset that sought to escape the dark realities of the American Great Depression. Users developed a new way to look at the world through Kodachrome and, even more importantly, used its supernatural colors to fabricate a world view that saw things not as they were, but as they were desired and imagined to be. While Kodachrome no longer exists in its original form today, it has been revived as a nostalgic visual throw-back to an analog past and still functions as an escapist vehicle. However, instead of providing escapism through color into a brighter future, today's users summon Kodachrome's colors as a nostalgic pathway into the imagined rosiness of a past time. The year 2016 was, indeed, an important revival moment for Kodachrome and its iconic colors, as well as retro escapism: Kodachrome was revived as a digital photo after-effect in AlienSkin's latest editing platform, ExposureX2, and it became part of a larger nostalgia phenomenon spreading throughout American visual culture as most recently captured in Nadav Kander's Kodachrome-inspired portrait of President-elect Donald Trump for the cover of TIME magazine's "Person of the Year" issue.[2]

But how did an imaging technology once associated with progressive change and American dreams become the aesthetics of backward-looking memory? This essay seeks to answer this question by mapping and reframing our understanding of Kodachrome through an analysis of its technological history, color aesthetics, and its larger cultural work. Throughout its history, and especially today, Kodachrome offers more than just the means for making pretty pictures. Rather, it offers a critical lens through which we can re-examine darker aspects of American cultural history and discourse. Although technically obsolete, Kodachrome continues to haunt the present as a meaningful cultural force entangled with broader issues of artifice, escapism, nostalgia, and desires for mediated, rose-tinted memories that actively shape rather than simply reflect lived events. Kodachrome's past, present, and future are even richer than the photographs or motion pictures created with it, and this essay will illuminate the multifaceted, prismatic aspects of its colorful legacy.

Birth: From Black-and-White to Color, 1816–1935

Attempts to mechanically reproduce images in color date back to the invention of photography itself. One of the early innovators of photography, Joseph Nicéphore Niépce, wrote to his brother in 1816 that "I must succeed in fixing the colors; this is what occupies me at the moment, and it is the most difficult."[3] Hand painting, tinted lenses, organic dyes were all used to infuse color into black-and-white photographs, though with much effort and mixed results. The earliest commercially successful method for producing colored photographs was the Autochrome, introduced by the Lumière brothers in 1907. In this process, dyes made from potato starch grains were added to the developed print to give it color.[4]

Color motion pictures became a technical possibility with the introduction of Kodak's first motion picture color system, Kodacolor, in 1928. George Eastman unveiled the new Kodacolor system at a lavish garden party hosted at his Rochester estate, which was captured on Kodacolor film and remains in existence today as a 16mm home movie entitled *Kodacolor Party* (1928).[5] Initially marketed to amateur cinematographers, Kodacolor could produce the illusion of moving, color images by using an additive (also called lenticular) method. Essentially, Kodacolor's additive colorizing process was based on the Keller-Dorian method: a 1920s French innovation that used a complex array of lenses to separate red, green, and blue wavelengths before individually capturing them on black-and-white film. When these three color-coded strips were combined, or "added" back together, and played through a special projector outfitted with the same three color-splitting lenses, the final projected image would appear as a rudimentary color moving picture.[6] Though an important first step toward color imaging, Kodacolor was not ideal, especially not for everyday, amateur users: it was expensive, required additional lenses, and a special projector unit. Adding to these disadvantages was the fact that images could only be temporary projected and not permanently printed in color. As such, color motion picture prints could not be reproduced or kept in color, only visually projected and experienced that way in the moment.[7] Within professional filmmaking, similar color-splitting techniques were used to create colorful Hollywood films, as in the case of the "glorious" Technicolor technique perhaps best known for bringing the fantasy world of ruby red slippers, emerald cities, and yellow brick roads to life in MGM's 1939 musical spectacle, The Wizard of Oz.[8] Kodachrome, however, would emerge as a revolutionary new type of film stock—truly "a horse of a different color", in every way.

The Kodak Company hired two amateur photographers, Leopold Mannes and Leopold Godowsky Jr., to overcome Kodacolors pitfalls and develop a streamlined, more realistic color alternative.[9] Their work resulted in Kodachrome: the first color-capable film stock, launched in 1935, that did not require additive lenses or special projectors. At first, Kodachrome was only offered as 16mm motion picture film, but within a year Kodak ramped up production to include 35mm photographic film (costing $3.50 a roll, equivalent to approximately $54 US dollars today), and a smaller, cheaper 8mm motion picture film for home moviemakers. Unlike previous lens-based methods, Kodachrome was designed with three emulsion layers embedded within a single film strip. Each of these layers contained silver halide crystals sensitive to a different primary color wavelength.[10] Kodachrome was also different in that it did not contain dye couplers; rather, these were only added during the development process. This innovative change resulted in final images with increased image clarity, finer detail, and a high-contrast, broad-spectrum color palette that became Kodachrome's defining hallmark. Adding dye couplers during processing also made this step more complicated; numerous dye baths containing color chemicals and other fixatives unique to both the Kodak Company and to the Kodachrome product line had to be expertly orchestrated. Due to this complicated and proprietary process, Kodak required all users to mail their undeveloped rolls to an official Kodak processing laboratory. The cost of processing, along with a convenient envelope, was included in the price of the film. However, the 1954 court case United States v. Eastman Kodak Co., put an end to this monopoly, and independent facilities gained access to the means and materials to develop Kodachrome themselves. No other analog film stock, either before or after Kodachrome, would use the same dye set formulas which has contributed to the unique aura surrounding these truly one-of-a-kind images.

Consumers instantly heralded Kodachrome as the literal gold standard of color imaging, and marveled at its ability to not only produce an unmatched range of vibrant colors but to offer improved archival capabilities. Kodachrome products promised to retain the clarity and color of the final image for up to 100 years with proper storage. These archival abilities, combined with Kodachrome's novel aesthetics, captured the market and turned Kodachrome into an idealized color imaging and preservation material. Kodak's marketing for Kodachrome film repeatedly emphasized its ability to reproduce and permanently maintain color vitality.[11] Candy-colored photographs of ephemeral flower arrangements, impossibly verdant vegetation, and perfectly rosy-cheeked children proliferated the pages of equally color-obsessed magazines such as Popular Science and Life, all carrying the message that Kodachrome film could make everything seem "vividly alive" and even better than the real thing.[12] This promotional rhetoric was, of course, riddled with hyperbolic promises that no reader would have literally believed at the time. However, these early advertisement strategies do reveal palpable desires to overcome the limits of the natural world and reshape reality through new technologies. Humankind could, and would, venture to remake the world through artificial materials and colors that defied the possibilities and limits of nature, and Kodachrome would become part of this process.

Growth: From a Color Revolution to a War in Color, 1930s–1940s

Kodachrome's initial success as a color-imaging technology also intersected with popular interests in color aesthetics and newly coined notions of "The American Dream". James Truslow Adams gave birth to the concept of The American Dream shortly before the birth of Kodachrome. In *The Epic of America* (1931) he wrote:

> The American Dream is that dream of a land in which life should be better and richer and fuller for everyone, with opportunity for each according to ability or achievement. . . . It is not a dream of motor cars and high wages merely, but a dream of social order in which each man and each woman shall be able to attain to the fullest stature of which they are innately capable, and be recognized by others for what they are, regardless of the fortuitous circumstances of birth or position.[13]

In part, Adams' vision of The American Dream is grounded in the premise that a richer livelihood could be obtained through hard work and upward social mobility—accomplishments that could be actualized through the consumption of goods and other visual displays of status. While this does not account for all that Adams envisioned, these aspects would be pushed further to the forefront by surges of consumerism and desires for new, colorful consumer products through the 1920s and into the 1930s. Color aesthetics and consumer goods made of artificial materials fit into these growing trends and would come to symbolize the type of thriving American livelihood outlined in Adams' dream.

During the 1930s, color became an important and valued aesthetic within American culture, in large part because it was linked to progressive, modern living and served as an antidote to the dark realities of the American Great Depression and lead-up to World War II. With the exception of some photographers shooting in color, most government-funded Farm Security Administration photographers working during the Dust Bowl and Depression reproduced images of American life in the stoic gravitas of black-and-white. On the other side of the spectrum and as a bold alternative to this grim reality, a "color revolution" was steadily sweeping across America in the form of brightly colored consumer goods. As Regina Lee Blaszczyk notes in The Color Revolution (2012), through the 1920s and 1930s a host of products made of newly engineered, cheaper plastic materials offered both an economic and visual reprieve from austerity.[14] In the same year that Adams dreamed his optimistic dream for American social ascension, the Tennessee Eastman company began selling their first plastic-based, artificially dyed acetate yarns, textiles, and home decor products. Synthetic materials not only promised to capture and reproduce vibrant colors, but would do so in a way that lasted longer and resisted wear-and-tear, all while being more cost effective and economically accessible to less economically solvent consumers. Interestingly, part of the sales pitch for these new plastic products included claims that artificial materials were simply better than natural products; as the infamous 1935 DuPont slogan promised, plastic would yield "Better Things for Better Living . . . Through Chemistry".

In an article for *Popular Science*, proudly entitled "New Feats of Chemical Wizards Remake the World We Live In", Alden P. Armagnac detailed how modern chemists were re-fabricating and even replacing natural products with technologically engineered substitutes.[15] "[L]aboratory workers have gone the silkworm one better", Armagnac enthusiastically cheered in a section on the marvels of synthetic fibers, and the "lowly silkworm's" natural production abilities have now been outdone by chemists who could spin their own new, improved, and colorful versions of plastic silk.[16] Other engineered alternatives, many of which prominently featured color-rich plastics, invaded nearly every corner of consumer culture: "your home, your clothing, your car, and the whole world about you," Armagnac concluded, "are benefiting from the wizardry of [the chemists'] touch".[17] This Midas touch would turn the world not only into a better place, but also a literally brighter and more colorful one. Kodachrome film stock fit into this overarching narrative in both aesthetic and form: it was made of the same acetate plastic composition as early plastic silks as well as many other artificial consumer products, and its innovative colors were made possible by novel chemical baths and complex processing procedures.

One of the benefits offered by plastic products and chemical dyes was a new range of super-saturated and non-fading color surfaces. Rogers, Eastman, and several other companies launched widespread marketing campaigns leading up to World War II that fueled this emergent frenzy for color through promoting a host of color-rich products. By switching to synthetic threads, chemical dyes, and amassing personal archives of unfading color photographs, individuals could remake the world in their own preferred image, and it was one full of artificial colors. Photography historian Sally Stein has further noted that American consumer interests in color aesthetics developed into a "rhetoric of the colorful": a discursive trend that positioned color "as an all-purpose positive sign of progress".[18] Faded, dull colors were for a past time, the impoverished, and the dead whereas vibrant colors would become the hallmark of a reinvigorated era dreaming of a new, thriving livelihood. Kodachrome film stocks further contributed to ambivalent reproaches of nature and voracious desires for color by offering supernaturally vibrant images that, departing from Mannes and Godowsky Jr.'s original goal of representing colors more naturalistically, quickly veered toward a surreal color pallet that transcended the colors found in nature.

Color appearances, even if artificial in origin and unnatural in appearance, were taking hold within modern American tastes and Kodachrome became a new rhetorical tool used to support positive associations between color, vitality, hope, and moving toward a brighter future. As the Library of Congress describes in their exhibition "Bound for Glory: America in Color", this would be "the dawn of a new era—the Kodachrome era".[19] As with any act of rhetoric, however, Kodachrome products and advertisements also began to sell a deeper message: that artifice was preferable to reality, and nostalgic fantasies take precedence over truthful remembrance. These colorful images would couple with Depression-era escapism, wartime anxieties, and mid-century consumerism to mark a historic divide between "the monochrome world of the pre-modern age and the brilliant hues of the present".[20]

In contrast to the color craze happening within pre-World War II American consumer culture, the majority of images associated with World War II, both in the historical cannon as well as popular imagination, exist in the form of black-and-white photographs. However, beyond Joe Rosenthal's patriotic *Raising the Flag on Iwo Jima* (1945) or Toyo Miyatake's oppositional *Boys Behind Barbed Wire* (1942–1945), there exist a number of wartime images captured as uncannily rosy Kodachromes.[21] Between 1939 and 1944, U.S. government-employed photographers working with the Farm Security Administration (later renamed the Office of War Information) shot approximately 1,600 color photographs consisting mostly of 35mm Kodachrome slides or transparencies, plus a few 16mm motion pictures. While small in number compared to the other 171,000 black-and-white images produced by the Farm Security Administration/Office of War Information, the visual and rhetorical impact of these wartime Kodachromes are perhaps even more striking than their counterparts.[22]

The subject matter of these colored photographs varied as widely as their hues: some depicted rural life and farm labor throughout the United States, while others captured the industrial buildup of military equipment and other wartime mobilization efforts. This last subset of photographs benefited as much from Kodachrome's color-imaging abilities as Renaissance-era paintings did from new innovations in oil paint; only Kodachrome could have captured the metallic sheen of aircraft parts as they marched down the assembly line, or the menacing appearance of killer bee yellow fighter planes as they circled mid-air. Black-and-white photography could not have capture these material subtleties, plays of light, or psychological impressions, and only Kodachrome could render the range, clarity, and detail seen in these images.

The most striking images are not of the machines of war, though, but of the women behind their construction. Women take center stage in many of Alfred Palmer's FSA/OWI photographs of military manufacturing, and the women featured in his *Engine Inspector for North American Aviation at Long Beach, California* (1942) and *Operating a Hand Drill at Vultee-Nashville, Woman Is Working on a 'Vengeance' Dive Bomber, Tennessee* (1943) offer an especially meaningful look at Kodachrome's discursive power to alter meaning and shape perceptions through color.[23] Dressed in a short-sleeved red, black, and yellow plaid shirt with high-waisted light denim jeans, a Caucasian engine inspector from California leans into an open engine block to adjust one of its many chrome tubes. In the second photo, an African-American hand drill operator from Tennessee wears a vibrant, short-sleeved navy denim jumpsuit while she drills rivets into a slab of sheet metal. Even though her brown skin is mostly rendered in shadow compared to the white engine inspector, we can still make out her reflection in the bright, reflective metal surface. Both women have their dark hair neatly tucked into a deep crimson headscarf, knotted at the front. In their dress and dogged attitudes, the visual parallels to the iconic Rosie the Riveter are as clear as the colors are sharp. The garish blues, yellows, and reds that gleam from Palmer's photographs are especially similar to the popular propaganda poster of Rosie the Riveter crafted by J. Howard Miller in 1942. Responding to what appears to be a case of art/artifice becoming real life, a contemporary article in the Daily Mail proclaimed that these photographs show "a moment where Rosie the Riveter moves out of the propaganda poster and into [the] world".[24]

What these photographs truly accomplish, however, is the reverse: in an act of visual rhetoric facilitated by the color properties of Kodachrome, these images move the world into the realm of propaganda. In short, life became art, or rather propaganda, through these Kodachrome renderings. Though any visual mediations and acts of representation can undoubtedly function in these ways, Kodachrome enabled a unique visual sleight-of-hand through its bold and captivating colors. Blinded by color, viewers marvel at the surface and get lost in the spectacle of technology (fighter planes or Kodachrome, itself); they are displaced into a different time—past or future, never present—without having to take stock of the larger issues and murkier truths lurking beneath the image. By dressing-up these scenes of female and African-American labor in the glossy saturated hues of Kodachrome, Palmer blurred the harsh realities facing women at this time in American history when neither women in general nor African-Americans of either gender were afforded equal rights or unfettered opportunity. In her writings on gender, race, and labor in WWII-era American industry, feminist scholar Maureen Honey notes that while African-American women did accomplish some meaningful gains in the military labor force, these advances were sharply circumscribed and undercut by residing practices of legalized racial segregation and systemic discrimination.[25] Honey's research shows that African-American men and women only accounted for 6 percent of all employees in U.S. aircraft industries during WWII.[26] When workers of color were employed, they were often relegated to the most menial and physically dangerous positions. African-American women, in particular, found themselves last in line for safe and gainful employment compared to their white female counterparts. In fact, the same year Palmer photographed a woman of color operating a hand drill in Tennessee, a Western Electric plant in Baltimore built segregated toilet facilities after protests from white women workers—certainly a different scene and sentiment than Palmer creates in his vibrant and progressive-seeming photos. The realities of work and polarized experiences of white women and women of color in the military labor force are glaringly absent from the world of

Rosie the Riveter as well as these Kodachrome photographs. Palmer's photos may share the same cartoonish and up-key colors as Miller's Rosie poster, but her "We can do it!" slogan would only ring dimly true for many real-life Rosies.

Palmer gives us another color-rich view of a white woman worker in *Noontime rest for an Assembly Worker at the Long Beach, Calif, plant of Douglas Aircraft Company* (1942). Here, a chestnut-haired and rosy-cheeked young woman flashes a red-tinted smile as she sits awkwardly perched in front of a scratched metal nacelle. Her silky burnt-orange blouse matches the rough wooden beams she sits on, and her bright red socks provide a balancing counterpoint to her red lipstick. Completing her look and the parade of Kodachrome-rendered hues are her electric-blue trousers, greenish-brown loafers, and a shining green lunch box propped conspicuously in the foreground. Apart from the fuselage in the background, this photo seems only vaguely related to wartime manufacturing and labor. Rather, what seems to be on full, celebratory display is the range, depth, and quality of Kodachrome's color capabilities. Indeed, the final image seems more like an advertisement for Kodachrome than a picture of war. The featured "model" does not appear as a Rosie figure, but as a harbinger of another influential, imagined female icon: Shirley Page, a model who would appear in the Kodak Company's instructional photo printing cards and guidebooks for independent development facilities produced during the 1950s. "Shirley Cards", as these mass-produced photo cards came to be nicknamed, first appeared in the 1950s —a decade when Kodak sold almost all of the color film used in the U.S., but lost its exclusive rights to develop finished rolls in-house. These infamous cards depicting a young, white woman draped in vibrant clothing and surrounded by saturated consumer objects became more than just functional extensions of color photography processing: they ultimately reveal how industry guidelines designed around artificially engineered hues shaped what consumers' perceived to be "good" colors as well as "real" colors.[27]

Maturity: From Recolored Recollections to Chromatic Consumption, 1950s–1960s

When the U.S. federal government put an end to Kodak's monopoly on color film printing with the 1954 case United States v. Eastman Kodak Co., Kodak supplied independent processing labs with Shirley Cards in an effort to maintain color standards and uniformity across the skin tones, lighting, and the range of hues that would appear in the final print. In her article, "Looking at Shirley, the Ultimate Norm: Colour Balance, Image Technologies, and Cognitive Equity" (2009), media historian Lorna Roth chronicles how the entire color film industry calibrated their norms based on getting white skin to look "right".[28] Kodak's entire line of color film stocks were so skewed toward whiteness and away from browns or dark tones that filmmakers, wood furniture manufacturers, and candy companies all raised complaints.[29] French New Wave auteur filmmaker, Jean-Luc Godard, even refused to use Kodachrome when shooting in Mozambique on the grounds that the stock was inherently "racist".[30] Shirley Cards worked as an institutional mechanism to keep these color biases consistent and established aesthetic preferences for "good" colors, defined as bright white faces for people and bright rainbow tones for objects.

In addition to these racial implications importantly discussed by Roth, Shirley Cards calibrated and forcefully privileged color matrices that, when dutifully followed by processing labs all over the country and world, produced final Kodachrome images that shaped users' perceptions of the natural world and memory of once-lived events into artificially color-augmented images deemed "better" than reality. Bill Pyne, the general manager of Richard

Photo Lab in Hollywood, states that Shirley Cards and color correction processes in general are about establishing a sense that "'[t]his is the standard.' And truthfully, in the real world, there is no standard." Echoing Pyne, media historian Brian Winston writes that, "[c]olour photography is not bound to be 'faithful' to the natural world. Choices are made in the development and production of photographic materials."[31] Kodachrome's unique color properties and processes clouded viewers' abilities to distinguish between artifice and reality, while ultimately urging them to value the former over the latter. Kodachrome's slant toward hyper-realism and exaggerative coloration—especially its infamous bloody tomato reds, on full display in the socks worn in Palmer's *Noontime Rest for an Assembly Worker*—gave images an artificial appearance that surpassed anything in nature. Rather than being disparaged as tacky or dismissed for being unrealistic, though, the consuming public embraced, and commercial printers safeguarded, Kodachrome's artificial range of colors, ensuring its lasting and iconic appeal.

Kodak's advertising strategies for Kodachrome attempted to sell consumers on the premise that life and memories looked better in Kodachrome. Several ads throughout the 1950s claimed that "color snapshots tell the story best"; that Kodachrome is "the gift that keeps memories bright"; or that users could keep "memories fresh in snapshots".[32] Many domestic users even came to prefer these renderings to reality, saying they favored artificial Kodachrome appearances precisely because they improved upon real life and made events look happier.[33] "Kodachrome had a way of making shots look better than real life", claims amateur photographer, Charles Moore.[34] Tom Stone, a commercial photographer and camera store owner since the 1960s, further confirms that "people liked it because it made things look better than real".[35] Another amateur photography enthusiast, Mark Reed, reveals the success and impact of Kodachrome mediations upon one's recollection of lived events: "I remember the 4th of July, 1956 like it was yesterday", he recalls on his personal website, "[i]t was sixty-years ago, but I remember it in Kodachrome."[36] Reed's sentiments are especially curious in that they display the slippage between lived events and memory representations, reality and Kodachrome renderings. This slippage is brought into even sharper focus when considered alongside C. J. Bartleson's investigation of chromatic shifts within memory.

C. J. Bartleson was a scientist employed by the Kodak Company to research the relationship between perception, color, and memory. In his 1960 study, "Memory Colors of Familiar Objects", Bartleson showed that when asked to identify the color of a familiar object from memory, the vast majority of test subjects incorrectly gauged the color, with a penchant for aggrandizing saturation and amplifying brightness.[37] Grass was greener in memory than in reality, and red bricks were much more red in what Bartleson termed "memory colors". Across all participants, bright and snappy versions of familiar objects seemed the most "correct", even though these judgments were in fact incorrect and far exceeded the actual color-metric values for those objects. Bartleson's findings effectively dispelled the ubiquitous saying that memories fade with time; rather, it was clearly the opposite. This reversal also accounts for Kodachrome's wide popularity and success as a hyper-saturated imaging medium charged with the mission of keeping users' memories alive. Kodachrome's departure from naturalistic hues made it the ideal medium for not just capturing, but literally creating memories; its impossibly bright and saturated scenes did not capture reality, but did appease and seem right to how users saw things in memory, making it the ideal medium for memory construction.

Based on Bartleson's findings, Kodak doubled-down on their investments in saturated color schemes, forsaking the pursuit of realistic accuracy and instead pursuing distorted memory color ideals. Bartleson's findings established a crucial mandate within Kodak as well as the photo-cinematic imaging industry at large: that representations should carry the colors of memory and not reality in order to look "right" in users' eyes.[38] This version of "right", however, was not based in factual reality, but rather on how the mind's eye chose to recolor what it saw in hindsight, reshaping it into what the viewer would like to remember. Photography theorist and visual culture philosopher, Vilém Flusser, would later propose in Towards a Philosophy of Photography (1983) that the photochemical concept of "green" is not based on reality but some imagined image of "green" drawn from a fantastical world of gilded dreams and memories.[39] Foreshadowing Flusser's 1980s' theories, Kodachrome was already accomplishing this work in practice, leading enthusiasts such as Mark Reed to start using "Kodachrome" as a neologistic adjective to describe their memories.

The same color-distorting, memory-shifting work was also happening on a larger cultural level in post-WWII American visual culture. Photographers and the popular press utilized Kodachrome's revisionist memory colors to rescript how the American public would see and remember the atomic attacks on Nagasaki and Hiroshima. Wayne Miller's iconic shots of the icy blues and fiery reds of atomic explosions, for example, were printed as full color spreads in the May 3, 1954, edition of Life magazine, while also appearing as the final, jarring image in Edward Steichen's "The Family of Man" international photography exhibition held at the Museum of Modern Art in 1955.[40] These mediated, Kodachrome-produced views of nuclear war did not confront viewers with the harsh, gory aftermath of the bombs, but instead provided an aesthetic shelter from the tragic sights of scorched bodies in the displaced world of selective, recolored memories. Through the prism of Kodachrome, we see and remember the colors, not the carnage, and are allowed to visually revel in the "beauty" of the atom bomb divorced from its darker reality.

Commercial advertisers also stood to benefit from the color-shifting properties of Kodachrome. Similar to memory, advertising and marketing also perform acts of perception dilation, albeit with the added motivation of convincing consumers to buy their products. Throughout the 1950s and 1960s, Kodachrome photographs and color aesthetics proliferated in magazines and product spreads. Everything from fashion to food was given the Kodachrome treatment, even in ways that seemed to render food supernaturally inedible. A half-page spread for Prem premium canned ham, for example, features a zoomed-in Kodachrome photograph of a dinner plate generously portioned with golden potato chips, a fruit salad of glistening green grapes and red cherries, and thick slabs of pink modeled meat product bearing a striking similarity to the color and texture of chewed pencil erasers.[41] Even though the end results look more plasticine than epicurean, what advertisers were banking on and ultimately selling to viewers was the dream and Oz-like fantasy of a better life to be found, or rather bought, somewhere over the Kodachrome rainbow. Continuing the discourse begun in the 1920s and 1930s, such Kodachrome-rendered ads sold and became synonymous with "The American Dream", which was now being revived by post-war white middle-class suburbanites.

After World War II, American consumers were flooded with cheaply made commodities that traded in idealized images of "the American nuclear family"—defined as home owners, car drivers, Coca-Cola drinkers, and Kodachrome image-takers. Kodak's own marketing for Kodachrome products played into this narrative in two ways. First, it echoed marketing strategies for other consumer goods by promising that The American Dream could be obtained through domestic consumerism and the lived performance of "familial happiness" defined as children's birthday parties, drives to the countryside, marriages, graduations, and opening presents around the Christmas tree. Second, Kodak drenched their images in the glow of memory colors and visually offered up saccharine scenes as desirable, future memories that you could have too, if you shot them with Kodachrome film. Not only were advertisers selling American dreams through Kodachrome-colored magazine spreads, they were essentially selling The American Dream as Kodachrome, and Kodachrome as The American Dream. Kodak's marketing of Kodachrome, as well as the product itself, exploited the power of color to re-imagine a world that only exists in memory, and that was already a fabricated illusion disconnected from reality. Kodachrome sold consumers a glimpse of a happy future, but it was a future fashioned out of artificially constructed dreams and augmented memory colors.

Kodachrome's popularity peaked in the 1960s, thanks in large part to several technological updates that helped extend its market reach and armed users with even easier ways to catalog every highlight of their lives as still and moving color images. Kodachrome II came out in 1961, in a smaller, cheaper, and more user-friendly 8mm format. Kodachrome II also boasted a faster film speed, which made it more versatile for point-and-shoot photography and spontaneous home moviemaking. In 1963, Abraham Zapruder used 8mm Kodachrome II to shoot the most infamous home movie in recorded history. The "Zapruder film", as it came to be known, spontaneously captured the assassination of President John F. Kennedy on motion picture film and was subsequently pored over as visual evidence of the crime. However, the Zapruder film offered neither definitive truth nor an unambiguous view of who shot Kennedy (or how), and the film remains shrouded in clouds of doubt despite its warm, glowing colors. Beginning in 1965, a new silent motion picture format sold as "Super 8" repacked Kodachrome film into a plastic cartridge case, which users could simply slip into a hand-held camera.[42] Super 8's cartridge design proved to be a revolutionary update that made home moviemaking even easier for amateur and young users, who now did not have to handle or manually thread the film stock.

While the general public continued their fervent love affair with Kodachrome, a backlash emerged from professionals who railed against its vulgar color aesthetics. Indeed, Kodachrome was deemed "vulgar" by art photographers not just for its garish color schemes, but for its "low-brow" and "common" mass appeal. As a technology initially intended for home users and famed for a color palette at home in commercial advertising, professionals largely rejected Kodachrome film products. Exasperating this divide even further was Hollywood's inevitable embrace of Kodachrome. As the quintessential American dream factory that straddles art, commerce, and appeals to popular taste, Hollywood studios jumped on the Kodachrome bandwagon and extensively used it in publicity shots and to amp-up the larger-than-life star appearances of celebrities.

In an attempt to woo resistant, high-profile American art photographers including Ansel Adams, Paul Strand, Charles Sheeler, and Edward Weston, Kodak offered commissioned projects throughout the 1960s. Echoing the staunch perspectives of many, Walker Evans remained resistant and even went so far as to say:

> Many photographers are apt to confuse color with noise, and to congratulate themselves when they have almost blown you down with screeching hues alone—a bebop of electric blues, furious reds, and poison greens. Color tends to corrupt photography and absolute color corrupts it absolutely. Consider the way color film usually renders blue sky, green foliage, lipstick red, and the kiddies' playsuit. These are four simple words which must be whispered: color photography is vulgar.[43]

From Evans' perspective, the colors of Kodachrome were loud and untamed (with derogatory, racist connections to the anti-classical style of 1940's African-American jazz music, known as Bebop), corrosive and even potentially deadly. Encapsulated in Evans' chromophobic comments is a racist, classist, and elitist distaste for Kodachrome undoubtedly connected to its prevalence in popular and consumer culture. In an ironic twist, Evans' verbalized disdain and insults match the same words used to praise Kodachrome II within the realm of commercial advertising. In one particular advert, printed as a full back-cover spread in the March 1962 issues of U.S. Camera (the leading magazine for both professional and amateur photo enthusiasts in America), boxed rolls of Kodachrome II lay intermixed with other consumer products—literally consumables in the form of fruit pieces—and are described as capturing all the once-elusive "blazing scarlets, lush greens, [and] tart yellows" of life, "translating your desires into new color", and "giv[ing] you color so good you can taste it".[44]

Art photographers might have maintained a certain disdain for Kodachrome, but avant-garde filmmakers took a divergent stance and enthusiastically embraced it within their practices throughout the late 1950s–1960s. Kodachrome especially factored into the work of mid-century American Structuralist/Materialist filmmakers—a cadre of artists who, as film scholars Peter Gidal and P. Adam Sitney have described, were interested in exploring and expanding the form, structure, materials, and audio/visual language of the cinematic medium.[45] American Structuralist/Materialist filmmakers, including Ken Jacobs and Stan Brakhage, found a particularly apt tool in 16mm and Super 8 Kodachrome film; its unique range of color, visual texture, and lack of audio gave these filmmakers a dynamic range in which to question, subvert, and re-imagine both the ontological nature and function of color images within motion pictures. Their goals might have been different, and their end films quite distinct from each other, but both amateur home moviemakers and avant-garde art filmmakers shared a love for Kodachrome and embraced its unique, populist features as a means to catalog their life pursuits and express themselves.

Still photography also benefited from amateur-orientated improvements in the 1960s, especially in terms of decreased cost for Kodachrome transparency slides. With the addition of a cheap projector unit, American households became exhibition spaces for showing off family vacations, milestone activities, and other wall-sized, colorful testaments of their successful pursuit of The American Dream. There was still one integral way, however, in which Kodachrome did not, and could not, keep up with consumer desires for fast, instant imaging production. As previously detailed, Kodachrome required a complex, time-consuming development process—one that would never be able to make the "one-hour photo" deadline or reach the instant imaging capabilities customers began to demand by the 1970s.[46] Public desires were quickly turning toward faster and instant imaging options offered by companies such as Fuji or Polaroid. Polaroid, in fact, introduced the first instant color film in 1963, followed by the first fully automatic, folding camera with instant color prints in 1972.[47] In light of these new consumer shifts and market competitions, Kodachrome began its steady slide into obsolescence—a death march whose pace would only be quickened by the emergence of digital imaging technologies.

Death: From Digital Eclipses to Industry Sunsets, 1970s–2000s

The first few years of the 1970s still looked bright for Kodak: the company was attempting to expand their Kodachrome product line while American folk-singer, Paul Simon, released the heartfelt ballad "Kodachrome" in 1973. In one particularly poignant passage, Simon encapsulated the essence of Kodachrome as well as its nostalgic, memory-recoloring work:

> They give us those nice bright colors
> They give us the greens of summers

Makes you think all the world's a sunny day.
[. . .]
I know they'd never match
My sweet imagination
Everything looks worse in black and white.

Within the next few years, though, several pivotal changes would supply the first nails to Kodachrome's coffin, as well as fundamentally change the nature of analog image-making.

In 1975, Kodak unveiled the world's first digital camera prototype. Weighing 8 pounds, sporting the dimensions of a toaster, and only having the ability to take black-and-white low resolution (.01 megapixel) images, its nascent digital image did not hold a candle to Kodachrome or other analog film images. However, these first digital photos did mark an important turning point away from analog, a move that would continue to make successive, innovative strides over the following decades.[48] Kodachrome was beginning to slide into the register of nostalgic past-tenseness. Rather than appearing as glossy advertising spreads illustrating the heights of American dreams, by the 1980s it was recast as the go-to visual marker for the "olden glory days", as seen in the Super 8 footage used in the opening credits for The Wonder Years (1988–1993), a television series dedicated to wistfully looking back at American life in the 1960s.[49]

Another critical blow came in the form of digital alternatives to traditional photo printing practices. In 1990, Kodak announced a new Photo CD system that rendered typical picture development old-fashioned and obsolete. Rather than settling for small-scaled individual color prints, users could now store several photos on a Photo CD and display their images on television screens in a digital update of the projected family slideshow. Numbered were the days of tangible, paper-based photos, home movie reels, and the analog slideshow. Pressure from other companies, namely Fujifilm, pushed Kodachrome further to the margins. Fuji's new Fujichrome Velvia stock swooped in to eclipse Kodachrome's position within the waning analog market by offering a more sensitive color stock and cheaper processing method. The new millennium also brought competition from other digital camera developers, such as Nikon and Canon, who helped to steer the course of the market and visual culture into the digital horizon. Twenty-one years after Kodak's first digital camera and only six years after Photo CD, the first of the Kodachrome formats was discontinued: Kodachrome 120 was taken away in 1996, with Super 8 following suit in 2005, and 16mm Kodachrome in 2006. The last professional grade 35mm format of Kodachrome ended in 2009. Just shy of the product line's 75th birthday, all Kodachrome color film stocks were decommissioned in a move that Kodak self-described as "break[ing] one of the largest remaining ties to the era of pre-digital photography".[50] The Kodak Company itself would also file for bankruptcy, just three years later in 2012.

To fill the void left by Kodachrome's demise, some users moved laterally over to Ektachrome, an alternative color stock first made available in 1946 as a way to supplement Kodak's line of color film and to compete against Fuji's release of its first color reversal stock. Photojournalist Steve McCurry, for example, returned to Afghanistan in 2002 to find and rephotograph Sharbat Gula ("The Afghan Girl") with Ektrachrome instead of his original choice of Kodachrome. Other photographers, however, remained faithful to Kodachrome to the end. Facing a new lymphoma diagnosis, photographer Jeff Jacobson purchased one last cache of Kodachrome and turned his final shots into a book grimly entitled The Last Roll (2013).[51] Switching back to Kodachrome in order to pay his final respects, Steve McCurry also embarked on a pilgrimage to shoot his last roll of film. McCurry's journey resulted in both a television featurette produced by National Geographic as well as an on-line exhibit on his personal website—an ironic sign of the digital times.[52] Others photographers clung to whatever remained of vanishing Kodachrome, and have maintained their devotion to the iconic stock by stockpiling rolls of unused film in freezers, turning to auction sites like eBay to get their hands on what was still in circulation, or scrambling to get their shot rolls developed before December 30, 2010 —the day the last standing Kodachrome lab, Dwayne's Photo in Parsons, Kansas, was slated to end their processing services.[53] In the fullest form of obsolescence, Kodachrome is no longer being made and whatever rolls do remain can never be developed, rendering their captured contents sealed in the filmstrip-cum-tomb, preserved yet inaccessible, and ultimately gone for good.

Liz Coffey, an amateur filmmaker and archivist with the Harvard Film Archive, provided a heartfelt visual sendoff to Kodachrome in her short silent film, Funeral for a Friend (2006).[54] Shot on a now-rare roll of Super 8 film in 2006, Coffey's eulogy begins as a faux home movie for the film in commemoration of its 5th birthday, complete with intercut found footage depicting a real child's birthday in 1940. We fast-forward next to 1975 and Kodachrome's 70th birthday, which is visualized as a fast-paced montage of more sober scenes showing the exterior of a Woodman's Market, an eerily empty school bus with blackout windows, a woman removing her vest, and a quick flash of some kind of legal document. All of these scenes bear the telltale signs of analog film—a slight graininess and several bright blue streaks run down the frame from improper processing—as well as metaphorically hint at consumerism, transition, and change.

The second half of Coffey's 2 minute and 30 second film focuses on the death and burial of a pristine, red and gold box of Super 8 Kodachrome 40. Utilizing rudimentary stop-motion animation techniques, Coffey flings open the lid of a miniature toy casket, offering an implicit invitation for Kodachrome to assume its place. A box of Kodachrome 40 suddenly appears inside and through a series of rapid jump cuts, an ominous skeleton image is intercut with the Kodachrome coffin. A burial sequence follows next, with a coffin-side memorial service attended by a semi-circle of vintage analog cameras gathered to pay their final respects. After the surreal memorial, the cameras become pallbearers and carry the coffin through a field of grass to a derelict cemetery. A new, handmade tombstone appears among the chipped and faded stone headstone. Made of paper and black marker, the headstone has a floating skull on top, a row of iconic analog film notches along each side, and an inscription that reads:

> Kodachrome
>
> 1935–2006
>
> Killed by "The Market"
>
> R.I.P.

To end, Coffey makes one final visual proclamation, one that echoes the last rallying call of Kodachrome's many enthusiasts: "Viva Kodachrome!", which is spelt out in white alphabet blocks on a patterned, hunter green carpet. Indeed, Coffey's sentiments echo many of the reactions to the end of Kodachrome by those who loved it. While perhaps a bit melodramatic, these mournful reactions, often phrased in terms of loss and death, reveal the profound investment and formative admiration many held, and continue to hold, for Kodachrome. Kodachrome is dead, long live Kodachrome.

Afterlife: From Contemporary Resurrections to Political Critique, 2010 and beyond

Kodachrome might have been rendered obsolete as an analog film stock by the early 21st century, but continued enthusiasm from fans and capitalistic hope from Kodak are giving the defunct format new prospects for a second coming. With the teasing headline, "A comeback for Kodachrome? Maybe, Kodak says", Sea Lahman chronicles Kodak's latest attempts to capitalize on recent nostalgic technology trends and reboot Super 8 cameras, Ektachrome, and perhaps even Kodachrome film.[55] Super 8 and Ektachrome will be easier to resurrect: Kodak's new Super 8 camera model will blend the best of original analog with the latest in digital camera technology, while Ektachrome processing was never as complex or individualized as Kodachrome's.[56] Kodachrome film, however, might never be able to make the jump back from nostalgic remembrance to commercial viability. What it has been able to do, though, is make the jump from analog film to digital filter.

Kodachrome has, in fact, already returned this past decade as a revived relic and fetishized aesthetic within digital visual culture. In 2010, the same year the last independent Kodachrome processing facility stopped offering its services, Instagram debuted as a digital photo-editing and sharing application primarily designed for mobile smart phones. Seven years after its debut, Instagram remains a widely popular application that enables users to disseminate photos, many of which they have digitally re-imagined through filters that recreate the unique visual aesthetics of antiquated media formats—from the vibrant vermilion hues of Kodachrome, to the austere metallic tones of early Daguerreotypes. In 2016, the software company, Alien Skin, launched a new edition of their digital editing plug-in and app, Exposure X2. Touted as one of the most advanced products on the market, Exposure X2 can realistically "emulate the most iconic analog films, ranging from vintage Daguerreotype to modern portrait films like Kodak Portra", with the goal of visually harkening back to the analog past.[57] Alien Skin claims to have perfected their mimicry of vintage film stocks through scrupulous scientific analysis of analog formats and research into the chemistry of processing techniques; the company even claims to have analyzed "film grain under a microscope to get the proper characteristic look".[58] Like the chemists in 1936 who strove to "[go] the silkworm one better" with their uncanny artificial silk fabrics, digital software engineers are now striving to digitally go the analog film strip one better.[59]

Even after Kodachrome and other analog imaging products have been phased out, digital applications revive and profit from their most distinguishing characteristics and sell them to users as a way to repackage scenes from contemporary life and send them into the realm of nostalgia. Users of Instagram, Exposure X2, or the similar apps on the market are empowered to make their current lives look "better" by shrouding them in digitally produced, analog aesthetics—essentially turning what just happened into a "vintage memory". In a twist on Bartleson's memory colors research, today's digital image manipulators are once again using color to play with and shape their recollections. The aesthetics associated with fetishized "past-tense-ness", analog "realness", or an "aura of periodization", as Arjun Appadurai has termed it, are used to make the present seem more "real" or "meaningful" by recreating it through the analog colors of the past.[60] While these practices might seem to offer the promise of reviving historic imaging media and bringing increased awareness to them, detaching the visual qualities from the original format and re-appropriating them within digital platforms gives way to a type of ahistorical aestheticism that even goes so far as to erase the actual names of the analog source materials.

In the same way that nostalgia is disarticulated from reality, so is this form of digital revival: while these apps and filters offer the appearance of reviving historic imaging media and bringing increased awareness to extinct forms like Kodachrome, they ultimately detach its visual qualities from the original format and its historical moment, and re-present them as an ahistorical, renamed aesthetic. A filter channeling the aesthetics and qualities of Kodachrome, for example, becomes an Instagram filter simply renamed "Lo-Fi". The omission of the Kodachrome monicker is likely a consequence of copyright issues and trademark licensing, but the end result remains: even if the color appearances of Kodachrome seem to have come back from the margins of forgotten obsolescence, the name and form itself have ultimately disappeared and are replaced by a misleading departure from both the history and historic use of Kodachrome, which was never "low-fi".[61]

This departure from reality also characterizes how contemporary users try to make sense of Kodachrome and history today. When attempting to describe the resurfaced World War II Kodachromes of Alfred Palmer, a Seattle newspaper saw it necessary to remind readers that "Nope, it's not Instagram, it's Kodachrome", further proving Instagram's cannibalization and erasure of analog history.[62] Looking back into Kodachrome's own history, however, we find a similar phenomenon where users replaced reality with Kodachrome. Kodachrome enthusiast Mark Reed, for example, described his memories of childhood as "Kodachromes". Evoking Jean Baudrillard's postmodern theories of simulation and simulacrum, reality is turned into Kodachrome, which is then turned into Instagram, and each is mistaken for the other.[63]

A 2015 music video starring Dr. Teeth and a gaggle of multicolored Jim Henson Muppets (known as The Electric Mayhem Band) provides another example of Kodachrome's complex, time-warping reincarnation within the digital present.[64] In this cover of Paul Simon's iconic "Kodachrome", a swirling rainbow of colors and a parade of Muppet characters pose for digital snapshots and take their own smartphone selfies. We see Gonzo zoom in and focus his smartphone's camera on a group of Muppet chickens; when he snaps the photo, the image rather magically transforms into brighter, higher-contrast Kodachrome colors as if passed through an Instagram filter. The onscreen inclusion of various hashtag markers—including a "#noteasy" placed on top of an almost "toxic" green shot of Kermit the Frog—places us clearly within the linguistic realm of Instagram. Even in this tribute song for Kodachrome, we are experiencing and even confusing Kodachrome for Instagram and vice versa.

In another act of digital revival, photographer Nadav Kander selected a distinctive color scheme reminiscent of analog Kodachrome to create a subtly subversive portrait of President-elect Donald Trump for the cover of TIME magazine's 2016 "Person of the Year" issue. Evoking all of its historic associations with The American Dream and visual connections to nostalgic, mid-century Americana, Kander uses Kodachrome as a meta-commentary and apt aesthetic choice to represent a political figure who has risen to power by evoking rose-tinted images of past American grandeur (and whose own personal coloration is, perhaps, best described as garish burnt orange). Similar to Palmer's use of Kodachrome to color public perceptions of World War II, Kander conjures its high-contrast, past-tense palette to visually contextualize Trump's nostalgia-laced rhetoric, best summed up by his slogan "Make America Great Again". This rhetoric essentially proposes to move America forward by nostalgically bringing back past imagined glories. Journalist Jake Romm further notes how Kander consciously plays upon Kodachrome's "distinctive look [which] defines our common visual concept of nostalgia. . . . By reproducing a Kodachrome color palette, the TIME cover makes us re-imagine the cover as if it were an image from the era of Kodachrome's mass popularity."[65] Whether this brings the viewer back into a sober connection to atomic war and deadly civic injustices, or into a color-shifted dream of American greatness is the most pressing and open-ended question posed by Kander's contemporary revival of Kodachrome, and is a lingering question hanging over the history of Kodachrome film itself.

From its beginning, Kodachrome has functioned as a type of visual transportation tool, taking viewers out of the present and either into a brighter vision of the future or into a rosier version of the past. In the 1930s, a "rhetoric of the colorful" emerged in full force to visually pull Americans out of the Great Depression and into an upwardly mobile pursuit of "The American Dream". In the 1940s, the horrors of World War II and inequalities faced on the home front by women and people of color were "color-washed" away in photographs that bore the same garish hues as those used in pro-American propaganda posters. The 1950s brought an outburst of color-saturated marketing and advertisement campaigns selling the idea that American domestic bliss was achievable through consumption and consummated through the accumulations of colorful consumer goods, including Kodachrome film products. Kodachrome reached maturity throughout the 1960s–1970s, when new research provided tangible proof for the color-shifting nature of memory and how aggrandized pictures in the mind's eye are readily mistaken for reality. Despite beginning to fade from mass popularity in the 1980s and going technically extinct in the wake of new digital imaging technologies in the 1990s and 2000s, Kodachrome is also being digitally resurrected today, or at least its aesthetics. Platforms such as Instagram and programs like Exposure X2 have lifted the stock's groundbreaking, trademark color properties from its obsolete analog base and repackaged them for a new generation of users to digitally transport themselves into a candy-colored, "better" past world. Indeed, the legacy of Kodachrome has proven to be as prismatic and enduring as its most iconic images, and its ghostly presence is still recoloring our perception of the past and present. Kodachrome might be dead, but long lives its continued influence. Viva Kodachrome!

Notes

1 In 1937, Gerald Sheedy, a staff photographer for *The Daily Mirror*, used a mini camera loaded with Kodachrome to capture the red flames and orange smoke that enveloped the Hindenburg. In 1963, a private citizen and home moviemaker named Abraham Zapruder incidentally captured the assassination of President John F. Kennedy on 8mm silent Kodachrome II acetate safety film and a high-end 414 PD Bell & Howell Zoomatic Director Series camera. In 1985, photojournalist Steve McCurry captured the visage of Sharbat Gula, known then only as "The Afghan Girl" and face for Afghan refugees, with Kodachrome 64 and a Nikon FM2 camera.

2 Nadav Kander for *TIME*, Donald Trump. November 28, 2016.

3 Quoted in Henry Wihelm, "A History of Permanence in Traditional and Digital Photography: The Role of Nash Editions", *Nash Editions: Photography and the Art of Digital Printing,* Garrett White, editor (Berkeley: Nash Editions, 2007): 101.

4 For an accessible description of the Autochrome process, see Robert E. Martin, "Secrets of New Color Movies", *Popular Science* (Oct. 1928): 17–18; 153.

5 Alan Kattelle, *Home Movies: A History of the American Industry*, 1897–1979 (Nashua: Transition Pub., 2000).

6 For more on Kodacolor and amateur color filmmaking, see Charles Tepperman, "Color Unlimited: Amateur Color Cinema in the 1930s", in *Color and the Moving Image: History, Theory, Aesthetics, Archive*, Brown, Stree, Watkins, editors (New York: Routledge, 2013): 138–149.

7 Stephen A. Booth, "Kodachrome at 50", *Popular Mechanics* 163.1 (Jan. 1986): 46; 50.

8 *The Wizard of Oz*. Directed by Victor Fleming, Mervyn LeRoy, George Cukor, King Vidor, and Norman Taurog, MGM, 1939. See Richard W. Haines, *Technicolor Movies: The History of Dye Transfer Printing* (North Carolina: McFarland & Company, Inc., 2003), and Fred E. Basten, *Glorious Technicolor: The Movies' Magic Rainbow*, Ninetieth Anniversary Edition (New York: Easton Studio Press, 2005).

9 Stephen A. Booth, "Kodachrome at 50", *Popular Mechanics* 163.1 (Jan 1986): 46; 50.

10 Brian Coe, *Colour Photography: The First Hundred Years 1840–1940* (London: Ash & Grant, 1978): 121.

11 Eastman Kodak, "All the Wonders of Awakening Life" advertisement (1938); "Seems As If She Could Walk Right out of the Picture" advertisement (1939); "She's So Real You Want to Pick Her Up and Hug Her" advertisement (1939); "Life is a Movie. Get it with a Movie Camera and You Have a Lasting Record" advertisement (1940); "All This Beauty on Your Home Screen . . ." advertisement (ca. 1940s).

12 Eastman Kodak, "Bring 'em Back Alive with this Movie Camera" advertisement. *Popular Science* 130.6 (June 1937): 100.

13 James Truslow Adams, *The Epic of America* (Westport: Greenwood Press, 1931): 214–215.

14 Regina Lee Blaszczyk, *The Color Revolution* (Cambridge: MIT Press, 2012).

15 Alden P. Armagnac, "New Feats of Chemical Wizards Remake the World We Live In", *Popular Science* 129.1 (July 1936): 9–11; 109.

16 Ibid.

17 Ibid.

18 Sally A. Stein, *The Rhetoric of the Colorful and the Colorless: American Photography and Material Culture between the Wars.* Dissertation (Yale University, 1991): 11.

19 Library of Congress, *Bound for Glory: America in Color.* Online photography exhibition. September 8–January 21, 2006. Accessed Jan. 10, 2017. www.loc.gov/exhibits/bound-for-glory/credit.html

20 Ibid.

21 Joe Rosenthal, *Raising the Flag on Iwo Jima.* February 23, 1945, The Associated Press; Toyo Miyatake, *Boys Behind Barbed Wire.* 1942–5, Toyo Miyatake Studio.

22 Library of Congress, "Farm Security Administration/Office of War Information Color Photographs". Accessed Jan. 10, 2017. www.loc.gov/collections/fsa-owi-color-photographs/about-this-collection/.

23 Alfred Palmer, *Engine Inspector for North American Aviation at Long Beach, California.* June 1942, 4″ × 5″ Kodachrome transparency, Library of Congress. Alfred Palmer, *Operating a Hand Drill at Vultee-Nashville, Woman is Working on a "Vengeance" Dive Bomber,* Tennessee. Feb 1943, 4″ × 5″ Kodachrome transparency, Library of Congress.

24 "World War II in Kodachrome: Vivid color photos paint a moving picture of the 1940s American war effort." *The Daily Mail.* July 1, 2013. Accessed Jan. 10, 2017. www.dailymail.co.uk/news/article-2352173/World-War-II-Kodachrome-Vivid-color-photos-paint-moving-picture-1940s-American-war-effort.html

25 Maureen Honey, *Creating Rosie the Riveter: Class, Gender, and Propaganda in World War II* (Amherst: University of Massachusetts Press, 1985).

26 Maureen Honey, "African American Women in World War II", *History Now: The Journal of the Gilder Lehrman Institute.* Accessed Jan. 14, 17. www.gilderlehrman.org/history-by-era/world-war-ii/essays/african-american-women-world-war-ii.

27 Alfred Palmer, *Noontime Rest for an Assembly Worker at the Long Beach, Calif., Plant of Douglas Aircraft Company.* Nacelle parts for a heavy bomber form the background. October 1942. 4″ × 5″ Kodachrome transparency. Library of Congress.

28 Lorna Roth, "Looking at Shirley, the Ultimate Norm: Colour Balance, Image Technologies, and Cognitive Equity", *Canadian Journal of Communication* 34.1 (April 2009): 111–136.

29 Rosie Cima, "How Photography Was Optimized for White Skin Color." *Priceonomics* (24 Apr 2015). Accessed Jan. 1, 2017. www.priceonomics.com/how-photography-was-optimized-for-white-skin/.

30 Sean O'Toole, "Making, Refusing, Remaking: Adam Broomberg and Oliver Chanarin's Recent Photography." *Safundi: The Journal of South African and American Studies* 15.2-3 ("South African Photography: A Special Double Issue"): 2014. Lorna Roth, "Looking at Shirley, the Ultimate Norm: Colour Balance, Image Technologies, and Cognitive Equity", Canadian Journal of Communication 34.1 (April 2009): 111–136.

31 Brian Winston, *Technologies of Seeing: Photography, Cinema and Television* (London: British Film Institute, 1996): 96.

32 Kodak advertisements, ca. 1949–1950s.

33 *Popular Photography* 16.1 (Jan 1945): 17.

34 Charles Moore, "HDR Darkroom Pro Brings Kodachrome Quality to Digital Photography", Oct. 10, 2011. Accessed Jan. 15, 2017. www.macprices.net/2011/10/10/hdr-darkroom-pro-brings-kodachrome-quality-to-digital-photography/.

35 Tom Stone, "My Review of Kodachrome." *Digital Photography Review* (25 June 2009). Accessed Jan. 15, 2017. www.dpreview.com/forums/thread/2611767
36 Mark Reed, "4th of July, 1956." *Missing the Mark*. Accessed Jan. 10, 2017. www.markreed2.wordpress.com/2016/07/03/4th-of-july-1956/comment-page-1/.
37 C. J. Bartleson, "Memory Colors of Familiar Objects." *Journal of the Optical Society of America* 50.1 (January 1960): 73–77.
38 R. W. G. Hunt, *The Reproduction of Colour*, 6th Edition (Hoboken: Wiley, 2006).
39 Vilém Flusser, *Towards a Philosophy of Photography*, Anthony Mathews, Trans. (London: Reaktion Books, 1983).
40 "New 'Ivy' Pictures Show Fire and Ice", *Life* 36.18 (3 May 1954): 54; Edward Steichen and Ezra Stoller, *The Family of Man: The Photographic Exhibition* held at the Museum of Modern Art in 1955. Catalog published for the Museum of Modern Art by Simon and Schuster (New York) in collaboration with the Maco Magazine Corp.: 1955.
41 Prem advertisement, printed in *LIFE* 13.1 (6 July 1942): 54.
42 The original Super 8 film release was a silent format, but in 1973 a sound-on-film version was released.
43 Quoted in David Batchelor, *Chromophobia* (London: Reaktion Books, 2000): 12.
44 Kodak, advertisement for Kodachrome II printed in *U.S. Camera* (March 1962): back cover.
45 See Peter Gidal, *Structural Film Anthology* (London: British Film Institute, 1976). Peter Gidal, *Materialist Film* (New York: Routledge, 1989). P. Adam Sitney, *Visionary Film: The American Avant-Garde, 1943–2000,* 3rd Edition (New York: Oxford University Press, 2002).
46 Clair Suddath, "Brief History of Kodachrome." *TIME* (June 23, 2009). Accessed Jan. 17, 2016. http://content.time.com/time/arts/article/0,8599,1906503,00.html.
47 Polaroid, "History." Accessed Jan. 17, 2017. www.polaroid.com/history.
48 The Kodak Company, "Milestones." Accessed Jan. 17, 2017. www.kodak.com/corp/aboutus/heritage/milestones/default.htm.
49 Neal Marlens and Carol Black, creators. *The Wonder Years*. The Black/Marlens Company. 1988–1993.
50 Ibid.
51 Jeff Jacobson, *The Last Roll* (New York: Daylight Books, 2013).
52 "The Final Exposure (Episode 202)." Nat Geo's Most Amazing Photos, produced by Yvonne Russo and Hans Weise, edited by Carol Slatkin, *National Geographic*, Sept. 17, 2010; Steve McCurry, "Last Roll of Kodachrome" (2012–16). Accessed Jan. 1, 2017. www.stevemccurry.com/galleries/last-roll-kodachrome.
53 John DeFeo, "Kodachrome: Funeral for a Friend (1935–2010)." *The Street* (Dec. 30, 2010). Accessed Dec. 15, 2016. www.thestreet.com/story/10957423/1/kodachrome-funeral-for-a-friend-1935-2010.html.
54 *Funeral for a Friend*. Directed, filmed, edited, and produced by Liz Coffey, 2006. Accessed Aug. 26, 2018. https://archive.org/details/FuneralForAFriend.
55 Sean Lahman, "A Comeback for Kodachrome? Maybe, Kodak Says." *Democrat & Chronicle*. Jan. 26, 2017. Accessed June 10, 2017. www.democratandchronicle.com/story/money/2017/01/26/kodak-ektachrome-super-8-kodachrome-film-possible-return/96532280/.
56 As Lahman notes, the new Super 8 camera will use film cartridges that, when purchased, include the cost of processing, digital scanning, and upload to The Cloud. Coming full circle, this returns users to the early days of Kodak's "You press the button, we do the rest" business model and slogan.
57 Alien Skin Software company website. "Exposure X2" product page. Accessed Jan. 22, 2017. www.alienskin.com/exposure/creative.
58 Ibid.
59 Alden P. Armagnac, "New Feats of Chemical Wizards Remake the World We Live in." *Popular Science* 129.1 (July 1936): 9–11; 109.
60 Arjun Appadurai, "Consumption, Duration and History." *Stanford Literary Review* 10 (1–2, Spring–Fall 1993): 11–23.
61 Interestingly, a reactionary turn in digital image sharing has been to adopt a "no filter" policy, literally self-proclaimed with the inclusion of "#nofilter" to images, to make clear they have not been manipulated with the addition of any after-effect filters or lenses to change the color appearances of the snapshot.

62 "Time & Life: Rare Color Photos from WWII", Seattle Pi. Accessed Jan. 5, 2017. www.seattlepi.com/news/slideshow/Time-Life-Rare-color-photos-from-WWII-73604.php.
63 Jean Baudrillard, "Simulacra and Simulations", *Literary Theory: An Anthology*, 2nd Edition, ed. Julie Rivkin and Michael Ryan (Maiden, MA and Oxford, England: Blackwell, 2004).
64 "Kodachrome" music video. "The Muppets at YouTube Space LA", directed by Kirk Thatcher, performance by Dr. Teeth and the Mayhem Band, Dec. 26, 2015.
65 Jake Romm, "Why Time's Trump Cover is a Subversive Work of Political Art." *Forward* (Dec. 8, 2016). Accessed Dec. 10, 2016. www.forward.com/culture/356537/why-times-trump-cover-is-a-subversive-work-of-political-art/.

16

SHAKE IT LIKE A POLAROID PICTURE

The Rise and Fall of an Analog Social Medium

Sheila C. Murphy

People of a certain age know the move—the flick of the wrist as one shakes a Polaroid print in an effort to hasten its resolution into a photographic image. This essay primarily focuses on the consumer phenomenon of the Polaroid instant camera. While other Polaroid technologies were used extensively in art photography, the company's instant camera was a hit with consumers. Therefore, "Shake It" looks to the temporal and phenomenological dimensions of taking, viewing, and sharing the analog visual experience afforded by Polaroid's "instant" camera.[1]

The history of Polaroid devices grounds this contemplation of the temporality and physicality of the Polaroid. The Polaroid camera—so successful that it became synonymous with "instant pictures"—has an industrial history marked by rapid growth and equally rapid decline. In less than a century, Polaroid innovatively introduced a new technology of representation and then both this technology and the company passed into relative obscurity. This essay asks what drove the consumer desire for "instantaneous" images by considering the social aspects of Polaroid prints—the iconic small squares of film with a rectangular border that could be posted, drawn on, shared, and thrown away. What was the social logic in which Polaroids circulated? Was the act of taking a Polaroid perhaps more significant than the resulting image? Grounded in media history and theory, this essay is an account of how this gesture is part of the history of photography, dating back to when Polaroid was a company, a brand, a camera, an image, and an activity typified by "shaking it". If you are looking for a definitive history of the Polaroid Corporation or its cameras, that exceeds the scope of this essay but can be easily found.[2] Indeed, Peter Buse's excellent book on Polaroid, *The Camera Does the Rest: How Polaroid Changed Photography* (2015), covers the company's history in great detail but it also focuses on the more ephemeral experience of Polaroid—its playfulness and how instant snapshots promoted a particular relationship to the photographic image that did not exist before Polaroid pictures did.[3]

At the height of Polaroid's success in the 1970s, Christopher Bonanos describes how photographers shot over one billion Polaroid photographs per year. Formed in 1937, Polaroid was founded by Edwin Land and George Wheelwright to market Land's Polaroid polymer, a plastic sheet that could act as a polarizer or filter. Today, LCDs, microscopes, and sunglasses continue to use such filters. However, the corporation is most well remembered for the consumer products it produced, namely instant cameras and film. But by 2001, the Polaroid

Corporation had gone bankrupt and its assets were sold off, only to reform as a "new" Polaroid Corporation that later went into bankruptcy in 2008, the same year the company stopped producing its instant film.[4]

Shake It Like a Polaroid Picture . . .

Every event, or my memory of every important milestone of my childhood, includes my memory of a certain ritual. Whether I was opening a present, attending a live event, or performing at a school function, my Aunt Patty would be front and center with her Polaroid camera to document what was happening. Patty, who had Down Syndrome, *loved* to take pictures of her nieces and nephews. She also loved to see the "instant" results of her efforts. Decades later, cleaning out my grandmother's house, we found boxes and boxes of her blurry, imperfect Polaroid pictures. There, in the boxes, were both our memories and the physical evidence of my aunt's photographic practice. But what mattered to us were not the images themselves but the memory of having those photographs taken, when we posed for our aunt's Polaroid pictures, back in the 1970s and 1980s. In a certain sense, my childhood itself was shaped by Polaroid, in the many moments we posed and waited for its "instant" images to appear on the film. While this essay is not about my familial use of photography, it is about the Polaroid habits and technological practices that have now faded into memory and history.

While Polaroid signifies a corporation, a patented method for processing film, a line of cameras using that patented process to make "Polaroid pictures", and more, perhaps the most significant cultural legacy of Polaroid is the way it changed the understanding of time in relation to images. For, more than anything else, Polaroid speaks to a sense of "instantness" in photography that was once singular and unique—it sold a lot of Polaroid brand cameras and film and inspired competitors such as Fuji to enter the new instant photography marketplace, but now the instantness of a Polaroid seems almost slow—a temporal lag in a world where the instant is now priority and privilege. Indeed, if we understand Polaroid's instant in the context of the postwar era's love affair with mass production and the new, industrial products, evident everywhere from the damning critique they receive in Horkheimer and Adorno's work to Lynn Spigel's historical analysis of the suburbs' contingent rise with television to the mass-made "ethnic" tchotkes in today's West Elm catalog, mass, industrial, instant goods were, and remain, in demand. It is quite clear, then, that by making instant image production and processing cheap and available to consumers, Polaroid made its mark on how we perceived *time*, how we measured it in lengths like the duration of shakes it took to make an image appear on the trademarked square-in-the-rectangle of a Polaroid instant photograph. Today, that perception of duration stretches out to a good beat, as the nostalgic hook in a song about past fun, which declares "Nothing Is Forever".

There are many potential inroads to understanding the cultural relevance of Polaroid. One could write about the notion of the snapshot, as I have done here, or look instead at how Polaroids have long functioned in fashion photography as part of the process of setting up lighting and shots, where they are a kind of work product before final images are produced. In art photography, Polaroid has become synonymous not with instantaneity so much as with large format Polaroid Land 20 × 24 cameras used by artists as varied as Chuck Close, Robert Rauschenberg, Mary Ellen Marks, and William Wegman.[5] Such work was essentially foreclosed upon when Polaroid stopped producing its instant film in 2008 as a result of its bankruptcy. The film by documentarian Errol Morris, *The B-side: Elsa Dorfman's Portrait Photography* (2016) is also a history of this aspect of Polaroid. During the same period,

Polaroid introduced a consumer series of cameras. Beginning with the 1972 Polaroid SX-70 Land camera, an automatic, motorized, folding, instant color camera, the company had a hit technology in the marketplace and the 1970s and 1980s saw a series of cameras launched, included the OneStep, Spectra, and Sun lines. In the 1990s, the company began mining its own design history, producing new camera models that recalled the design and style of earlier models from the height of the OneStep's popularity in the late 1970s. Seen as disposable, Polaroid snapshots are often touted as an organizational tool, as this image of shoe boxes with Polaroid photographs of their contents demonstrates (see Figure 16.1). And, of course, crossing between art, fashion, and popular culture, Richard Avedon used a Polaroid to photograph Andy Warhol, who was also rarely seen without his Polaroid camera. Like his ubiquitous tape recorder, which Warhol referred to as his "wife", the artist used Polaroids in his daily life and his art work, as the basis for the portraits he is so well known for and in producing his Pop Art.[6] Clearly Polaroid, whether understood as a corporation or a format of photography or a set of social practices around instant images, was a powerful part of 20th-century visual culture.

In a literal sense, Polaroid exists because of patents held for technological innovations and processes. Indeed, Polaroid is known in intellectual property circles for its patents and its deft defense of those patents over the years. Throughout the 1980s, Polaroid and Kodak held a lengthy battle in court over Kodak's infringement of Polaroid intellectual property dating back to 1976. The settlement that Kodak was eventually ordered to pay Polaroid was

Figure 16.1 Shoeboxes labeled with Polaroids depicting their contents. Image credit: Brit. co, Lisa Raphael, "Get Ahead on Spring Cleaning with SERIOUS Shoe Organization Inspo", available at www.brit.co/diy-shoe-organization/.

for $909 million dollars and held the record as the highest patent infringement award until 2015, when a case over the design of blood vessel grafts surpassed the Polaroid v. Kodak settlement.[7] This is not just arcane trivia over who sells more cameras or more film, both seemingly moot points in this moment of mobile phone photography and digital visualities.[8] Kodak's role in photographic and cinematic history depends upon its corporate approach toward patents that George Eastman first purchased from an inventor, David Houston. Houston's initial patents were transformed into Kodak's early patents for motion picture film, while Eastman also registered Kodak as a trademark in 1888. Eastman established his business upon building trademarks and patents, legitimizing his practices in a competitive marketplace. For corporate leaders in the early part of the 20th century, filing patents and trademarks was one way to ensure success and build upon technologies and processes a company developed. Indeed, attorney Ronald Fierstein has studied the discourse of patents at Polaroid and other major 20th-century corporate inventors, resulting in his book *A Triumph of Genius: Edwin Land, the Polaroid Corporation and the Kodak Patent War* (2015).[9] Patents, trademarks, and actual innovations made for a successful company, one that could expand exponentially, extending its brand into far-reaching areas.[10] Edwin Land's vision for Polaroid was unrestrained. The company made and sold sunglasses, consumer electronics (still cameras, motion picture cameras, instant cameras), military applications of polarized lenses, and film (black and white, color, and instant). While patents insured a place in the competitive marketplace, Polaroid chased success wherever it could be found. This is evident when one looks at how Polaroid's most well-known and key product, the Polaroid OneStep Land Camera, was only invented in 1977, forty years after Land changed his company's name from the Land Company to the Polaroid Corporation. Both Kodak and Polaroid bear out a kind of "long tail" lesson in this regard. Today, these companies, each once a model of American capitalist success, are more known and successful on the basis of their nostalgic brands rather than for their current contributions to the industries that have emerged out of photography and filmmaking. Today, Kodak and Polaroid signify high-quality 20th-century innovations in motion picture and consumer photography. Neither corporation could easily pay out a near-billion dollar patent settlement, but both corporations' business models were reliant upon patents during the era when patent-driven, Fordist innovation dominated the marketplace. Both Kodak and Polaroid fit into Vivian Sobchack's model of photographic and cinematic screen presence in her 1994 essay, "The Scene of the Screen", in which Sobchack maps visual presence onto Frederic Jameson's eras of capitalism in *Postmodernism, or, the Cultural Logic of Late Capitalism* (1991). For both Jameson and Sobchack, an era's dominant aesthetics are defined by its economic model, such that "late capitalism" is represented by a multinational capitalist economy and a postmodern aesthetic.[11] Sobchack's "The Scene of the Screen" was first published in 1994, just three years after Jameson's book, a work that had a paradigm-shifting impact upon the humanistic study of culture. Today, there are obvious critiques and updates that must be made in order to utilize either essay well, particularly when it comes to each author's periodization of the postmodern and definition of the aesthetic, but Sobchack's application of Jameson to cinema and photography still instructs us in valid ways: under what Jameson calls "late capitalism" and others term the "multinational" era when corporations such as Polaroid and Kodak thrived and grew into new global markets, their corporate brands became collapsed with their technologies. Eventually, as we see today in the various attempts to purchase and utilize the Polaroid brand for novelty and nostalgia purposes, the brands themselves become the product. It is no longer about Polaroid producing the most ubiquitous instant camera, it is about Polaroid being understood as the brand of instantness.

The Instant Performance

While I began with describing my own relationship to the consumer Polaroid snapshot, as the subject of many a Polaroid picture taken by my aunt, I find myself returning again and again to Andy Warhol, a champion of Polaroid technology in his own art, life, and commissioned portraiture.[12] For Warhol, the instantaneousness and the portability of the consumer Polaroid fit in well with his other disposable analog devices, like his "wife", the nickname he gave to his handheld tape recorder he used for making notes and tracking expenses.[13] While he used the Polaroid and another instant camera format, the photo booth, in his art and studio to make "serious" work, like his self-portraits in drag (see Figure 16.2) or as the basis for large-scale screen prints, Warhol made just as much use of it as a consumer. Warhol once famously spoke about his photographic practices in a way that echoes much of what I have written thus far about the Polaroid and its relationship to time: "A picture means I know where I was every minute. That's why I take pictures. It's a visual diary." Today we read this sentence and are reminded of Instagram, but it is likely Warhol is referring to another diary, his personal diary that was later published but began as a way of tracking tax expenses. Having made a large-scale public art installation that was critical of then-president Nixon, Andy Warhol was routinely audited by the IRS every year from 1972 until his death in 1987.[14] For an artist trained and working in images, a visual diary makes a lot of sense.[15] And Polaroid's instant film made it convenient, portable, and sensible for Warhol. Indeed, while I am arguing that Warhol's use of Polaroid embraced the technology's most salient features in its period, others have repeatedly made the comparison of Warhol's Polaroid portraiture with Instagram, declaring, as CBS News does, "Before there was Instagram, there was Warhol."[16] Or take *WIRED*'s headline from 2015: "Andy Warhol's Celebrity Polaroids Are Better Than Your Instagrams."[17] Warhol's Polaroids, like much of his other work, blur the distinctions between art and snapshot, as in Figure 16.2. Here Warhol is photographed

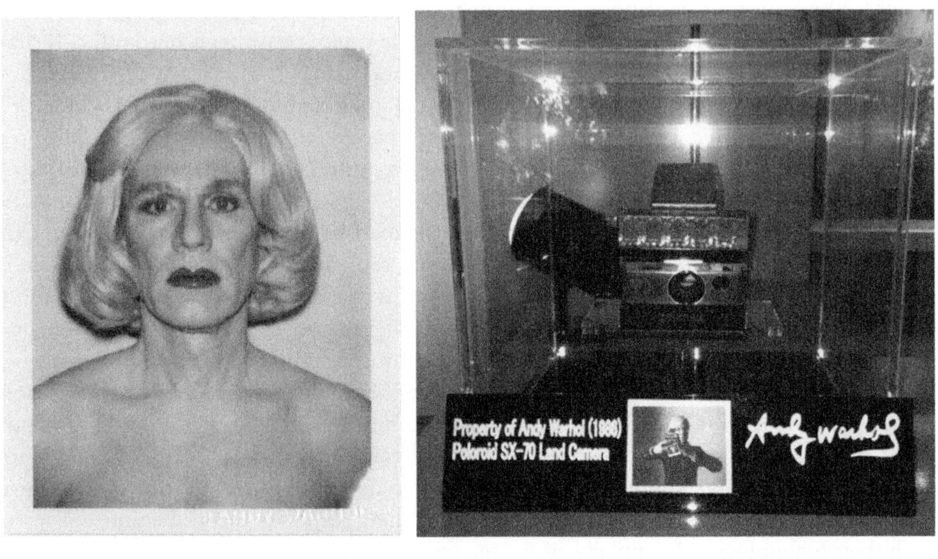

Figure 16.2 Left: Andy Warhol, *Self-Portrait in Drag* (1982). Right: The Polaroid SX-70 Land camera owned by Andy Warhol. Image credit: Ebay via PetaPixel, available at https://petapixel.com/2013/08/09/andy-warhols-1986-sx-70-polaroid-land-camera-selling-on-ebay-for-50000/.

in careful drag, with precise makeup applied and a neatly combed woman's wig. His naked shoulders and bare chest contrast with the garish make-believe of his frosty eye shadow and deep red lips. This is a typically opaque self-portrait of Warhol—his seemingly empty gaze tells us nothing. But the lower right corner *does* speak—in it one can see the flawed edge and chemical surface of a Polaroid print, now part of the art itself.

While Polaroid introduced its patented polarizing film process and several related products, including the line of Polaroid consumer cameras and instant film, the company had its shares of struggles and declared bankruptcy in 2001. The combination of several ill-timed product launches, including the home movie camera Polyvision, which was unable to compete with home video products that were on the rise in 1977 when Polyvision launched, weakened the company's standing in the market. But ultimately Polaroid's corporate history is best understood in the context of technological change and its effect upon a corporation's development. Throughout the 1980s and 1990s, there were significant changes in consumer photography—from how quickly film could be processed to the introduction of video and other new technologies—that Polaroid had to weather while still selling its product lines. By 2001, the company was bankrupt and its assets, including its brand name, were sold.[18] Eventually, after a series of litigations between interested parties, a "new Polaroid" corporation emerged in 2009. Now producing digital cameras with instant printing, Polaroid has returned as a company promoting the nostalgic, retro-futurism of its original commercial success producing instant snapshots.

It seems to make a certain kind of sense that in 2017, Polaroid has largely been reduced to a set of icons. As a consumer technology, Polaroid was a highly branded experience, requiring the purchase of a particular camera and type of film, both primarily produced by the company itself. Polaroid *was* instant photography, just as Kleenex was facial tissue or Xerox was copies. This new experience of "instant" photography was understood through a name with certain iconic associations with its camera technology and film. A Polaroid is a picture that one shakes in order to hasten the film's development; it is, oddly, an image linked to

Figure 16.3 Left: The Polaroid OneStep camera (image from Wikipedia). Right: The Polaroid OneStep camera used as a logo by Instagram (image from Instagram).

a gesture that developed around using the technology itself. A Polaroid is a square photographic image contained on a larger, slightly rectangular piece of polarized plastic paper film, with a decidedly larger white border on one side than on the others. Polaroid is a rainbow stripe on a camera. All of these icons now circulate more freely, mostly distinct from actual technologies, and signify Polaroid in the cultural imagination. While today's Polaroid still produces novelty nostalgic devices like the Polaroid Snap Camera and the Polaroid Snap Touch Camera, released in 2016 and 2017, respectively, the company's emphasis is upon exploiting its brand identity of instant photography as a novelty in the Instagram era. The Snap models are, like their analog predecessors, combination camera-printers, produced in cheery plastic colors. While the physical design resembles an on-line app logo, the camera is designed to produce nostalgia. One can shoot in black and white, color, or vintage sepia with a Snap or Snap Touch and the camera also allows users to print their images with the "Polaroid Classic Border Logo". So even though the Snap produces digital image prints, one can still invoke the idea and action of "shaking it like a Polaroid picture".[19]

It is hard to say at this moment if Polaroid has exceeded its original goal of producing polarized film products, since the market for those products has shifted so thoroughly with the introduction of digital technologies, or if Polaroid has succeeded far beyond its goals by becoming an icon of instantaneousness. In this way, Polaroid is not just an analog technology of a past aesthetic and economic moment, but it is also an image for going forward that frames one way social photography is imagined today.

Today, Polaroid (the corporation, the film, the cameras) is more memory (or capitalist-driven commodification of memory) than it is material, but its material effects on popular culture remain and even seem to grow.[20] The Polaroid Supercolor OneStep, a 1977 product produced in the Polaroid Land Camera series, was seemingly readymade for the "flashback Fridays" of Instagram, when the company's founder first debuted their original logo based upon the OneStep. Removing the slot on which the directly printed film slid out of the camera and replacing the OneStep camera name with Instagram, this logo directly connects the analog era of instant photography to the instant era of on-line shared visuality. Eventually Instagram stylized their logo even further, simplifying the visual design of the imagined technology (see Figure 16.4), but keeping the bright Polaroid rainbow and fixed lens, this new logo invokes seeing through its viewfinder and taking photographs through the camera lens. At the same time, the white plastic case of the OneStep is traded in for a two-tone tan device, the top of which has a leather-like texture—the flashback now goes further back into an era of cameras encased in leather goods. The image is simplified visually and it is more an evocation of photography than a depiction of an actual photographic technology. This Instagram has no way of actually producing photographs on paper like its Polaroid predecessors. Instead, it is all about making images, images which may circulate via the app and enter the social media sphere. "Chemical photography" is of the past, now only an icon representing an index that has been lost, that only exists in cultural metaphor as a memory or gesture. "Shake it like a Polaroid picture" becomes a quaint phrase, redirecting our attention to historical pastimes rather than to a present in which Polaroid becomes an unacknowledged icon of Instant-ness. Its specificity, as a medium, a technology, a brand, fades just like those chemical photographs of the 1970s and 1980s that yellow over time. But all is not lost; such gestures and approaches to technology are important. Shake the photo, jiggle the toner cartridge, blow the dust off the cartridge. These consumer behaviors are part of the history of technology too. Even in the "Insta" age that is all image, no print, we bring with us our own photographic habits. In 2017, Instagram has replaced its graphic once more and now the connections to photography, history and image-making are even less visually

Figure 16.4 Instagram's camera logo (left) and glyph icon (right). Image credits: Instagram.

delineated as the graphic has become a line-drawn, minimalist image that the company refers to as its "glyph icon" (see Figure 16.4).[21]

The photographic is no longer even in evidence, as the background of the "device" is now a gradient blur from yellow to purple color tones. The "rainbow" of photographic visuality has been reduced to the demonstration of a color spectrum. Indeed, in the glyph icon, the portmanteau of Instagram (instant camera + telegram) is a transformed image, now symbolic of a more nuanced, yet also opaque combination of technologies and communication modalities.[22] In this visualization of a corporation, a software application, and a social network there is no such thing as obsolescence, just layers of nostalgia, technological history, cultural gesture, and memory. With Polaroid, we see the media archaeological roots of visual, social media like Instagram in a technology that was simultaneously analog, "instantaneous", everyday, and visual.

Notes

1 Polaroid's name is drawn from its patented "polarizing polymer" process, which involves the incorporation of a polarizing plastic or polymer filter into film. This process is still used to produce liquid crystal displays, sunglasses, and other devices in addition to photography.

2 See analog enthusiast Florian Kaps's *Polaroid: The Magic Material* (London: Frances Lincoln, 2016) for a cultural history of Polaroid, as well as Christopher Bonanos's *Instant: The Story of Polaroid* (New York: Princeton Architectural Press, 2012).

3 See Buse, *The Camera Does the Rest: How Polaroid Changed Photography* (Chicago and London: University of Chicago Press, 2016) 4–24.

4 Rob Beschizza, "Gear: Polaroid Ends Instant Film Production," *WIRED*, 13 February 2008, available at www.wired.com/2008/02/polaroid-ends-i/ (accessed August 16, 2018).

5 Randy Kennedy, "Champions of a Monster Polaroid Yield to a Digital World," *The New York Times*, June 20, 2016, available at www.nytimes.com/2016/06/21/arts/design/champion-of-a-polaroid-behemoth-yields-to-the-digital-world.html (accessed August 16, 2018).

6 See Arthur Danto, *Andy Warhol* (New Haven: Yale University Press, 2009), pages 112–119 on the Mao portraits, among others. See also Richard Woodward, *Andy Warhol: Polaroids, 1958–1987*, edited by Reuel Golden (New York, NY: Taschen, 2015). This book has been marketed by Taschen with the tag line *Instant Andy: Before there was Instagram, there was Warhol.*

7 Ronald Fierstein, "Polaroid v. Kodak, Still the Champ," *IPWatchdog*, April 12, 2015, available at www.ipwatchdog.com/2015/04/12/polaroid-v-kodak-still-the-champ/id=56654/ (accessed August 16, 2018).

8 Brian X. Chen, "The Smartphone's Future, It's All About the Camera," *The New York Times*, August 30, 2017, available at www.nytimes.com/2017/08/30/technology/personaltech/future-smartphone-camera-augmented-reality.html (accessed August 16, 2018).

9 See Fierstein, *A Triumph of Genius: Edwin Land, the Polaroid Corporation and the Kodak Patent War* (New York: Ankerwycke Books, 2015). Ankerwycke is the publishing division of the American Bar Association.

10 Ronald Fierstein, "How the Inventor of Polaroid Championed the Patent," *The Atlantic*, February 19, 2015, available at www.theatlantic.com/technology/archive/2015/02/how-the-inventor-of-the-polaroid-saved-the-patent/385617/ (accessed August 16, 2018).

11 See Vivian Sobchack, "The Scene of the Screen: Envisioning Photographic, Cinematic, and Electronic 'Presence'," *Carnal Thoughts: Embodiment and Moving Image Culture* (Oakland, CA: University of California Press, 2004), pages 135–162; and Frederic Jameson, *Postmodernism, or the Industrial Logic of Late Capitalism* (Durham, NC: Duke University Press, 1991).

12 Warhol's iconic commissioned portraits were painted silkscreens of Polaroid snapshots. He approached these images as a business deal, charging a standard rate of $25,000 per portrait. In 2007, the Andy Warhol Museum launched its Andy Warhol Photographic Legacy project, distributing over 28,500 photographs to museological and educational institutions in the United States. Many of these images were Polaroid prints. See Tony Shafrizi, editor, *Andy Warhol Portraits* (London: Phaidon, 2007). http://warholfoundation.org/legacy/photographic.html (accessed August 16, 2018).

13 See Gustavus Stadler, "My Wife: The Tape Recorder and Warhol's Queer Ways of Listening," *Criticism* 56: 3 (425–456).

14 Victor Bockris, *Warhol: The Biography* (New York: Da Capo Press, 2009), page 358.

15 See Murphy, *Lurking and Looking: Media Technologies and Cultural Convergences of Spectatorship, Voyeurism, and Surveillance*. Doctoral dissertation, UMI Press, 2002.

16 www.cbsnews.com/pictures/instant-andy-before-there-was-instagram-there-was-warhol/ (accessed August 16, 2018).

17 Margaret Rhodes, "Andy Warhol's Celebrity Polaroids Are Better Than Your Instagrams," *WIRED* July 27, 2015. Both of these journalistic celebrations of Warhol and Polaroid were on the occasion of the publication of *Andy Warhol: Polaroids, 1958–1987*, eds. Richard Woodward and Rueul Golden (New York: Taschen, 2015).

18 See Peter Buse, *The Camera Does the Rest: How Polaroid Changed Photography* (Chicago and London: University of Chicago Press, 2016).

19 See Polaroid website: www.polaroid.com/products/polaroid-snap-touch-camera (accessed August 16, 2018).

20 While Polaroid exists today, it is a post-bankruptcy version of the company made up of assets acquired in a bankruptcy auction. The company was reformed by a small group of Dutch Polaroid enthusiasts known as the Impossible Company, which produced films made to be printed on Polaroid film before acquiring the actual Polaroid brand and assets at auction. In a way, Polaroid is now a fan-owned corporation.

21 See the Instagram corporate website: https://en.instagram-brand.com/assets/glyph-icon (accessed August 16, 2018). This site instructs web users on how to correctly utilize Instagram assets such as the glyph icon on their websites and applications.

22 See Adam Lang, "Why is it called Instagram?" *Rewind & Capture*, September 2, 2014, available at www.rewindandcapture.com/why-is-it-called-instagram/ (accessed August 16, 2018).

17

HOLLYWOOD IN A BOX

Time-shifting, Rental, and Videocassettes

Joshua Greenberg

Broader access to printed books quickly followed the invention of the printing press, and consumer sales of recorded music date back to the wax cylinders of the late 19th century, but for much of their existence the personal ownership of movies and television programs was restricted to a small group of industry insiders and enthusiasts. For those decades, most audiences' experience of moving images was passive, managed by theater projectionists and network broadcasters. It was not until the late 1970s that the introduction of the consumer videocassette gave average consumers the ability to actually hold a movie or TV show in their hands and the power to decide when, where, and at what pace to watch it.

For the last decades of the 20th century, the humble videocassette offered the ability to record broadcast or cable television, to capture family celebrations, and to bring mainstream Hollywood movies into the home, in all three cases stabilizing moving images for transportation across space and time and introducing a set of practices that changed audiences' relationship to those texts. Within a remarkably short time, it became commonplace to watch television programs on one's own schedule, even fast-forwarding through the commercials that had taken a captive audience for granted; meanwhile the rewind and pause buttons enabled home viewers to subject movies to a whole new level of critical scrutiny. The world of flexible, instantaneous, on-demand access promised by contemporary services such as Amazon, Netflix, Hulu, and iTunes has its roots in a black plastic rectangle with a "Be Kind, Rewind" sticker.

Videocassettes in Industry

Attempts to record and "time-shift" broadcast television signals go as far back as British television pioneer John Logie Baird, who recorded television signals as grooves on phonograph records in 1927. Not unlike his mechanical television, Baird's "Phonovision" never really took off, and for the first few decades of the television industry, the standard way to record a broadcast, either for archival purposes or to rebroadcast later, was to point a motion picture camera at a television receiver and capture the video image on film. Through the 1950s, the preferred system for doing so was the Kinescope, and the best "hot kines" systems could shoot a broadcast in New York and have the film developed and ready for rebroadcast on the West Coast within three hours.[1]

It did not take long, however, for magnetic tape recorders to approach high enough resolutions that they could replace Kinescopes and other film-based technologies. In the late 1950s, the Ampex reel-to-reel videotape recorder offered instantaneous recording and playback in vibrant color, and made its network debut by recording Nixon's 1959 Moscow debate with Khrushchev for later broadcast in the United States. The Ampex machines, along with subsequent recorders by RCA and Sony, raced through tape at speeds of up to 15 inches per second, requiring massive reels just to record an hour-long program, and the technical history of videotape in the 1960s essentially revolved around the question of how to slow down the recording speed (with the ultimate goal of shrinking the size of the tape reels).

As video quickly became a standard part of broadcast control rooms, some manufacturers began to position video for other applications. Thinking that a more-packaged technology might be easier for corporate training and other instructional contexts, Sony created a fully enclosed cassette that measured 6 by 10 by 3¾ inches (by 1971 they had revised their standard to ¾-inch-wide tape). The resulting videocassette recorder was the U-Matic, released in 1969 and able to play up to an hour of video with stereo audio. The U-Matic quickly became an industry standard, due in no small part to Sony's inter-manufacturer agreement which licensed the U-Matic technology to Matsushita and JVC (Victor Company of Japan).

Building a market for the new technology, however, was more difficult than establishing a technical standard. Sony's initial approach was a sort of traveling road show, a well-choreographed demonstration of the videocassette player to audiences around the country ranging from the Institute of Electrical and Electronics Engineers and Hollywood boardrooms to major politicians and the military. Though later models included recording capability, this initial Sony "U-Matic" itself was a playback-only machine, and while audiences tended to be very impressed with the new cassette technology, which seemed far more user-friendly than a hand-threaded projector or recorder, sales were lean. Facing failure, Peter Keane (the de facto "English language spokesman" for Sony, who had worked in the film industry since the early days of Technicolor and had been hired in large part because of his contacts in the film industry) developed his own pipeline of demonstration videos for the corporate and educational markets, mainly in order to show customers how simple it was to play back videotape compared with threading up a 16 mm film projector. Keane's hope was to develop "enough software to get the ball rolling", and by 1972 one educational film company had published a list of at least 50 or 60 programs available on ¾-inch tape.[2]

Interestingly, some of the earliest adopters of the U-Matic outside of the television industry were operators of adult movie arcades, who had traditionally relied on never-ending loops of film to give their customers their 25 cents' worth, but found the video recorder more economical. Though the conventional wisdom that pornography fueled the early video industry is most definitely true,[3] the earliest adopters were adult bookstores and arcades rather than home users.

Videocassettes in Homes

The first magnetic video recorder explicitly intended for home use appears to have been the Ampex Signature V, which measured over 9 feet long, included an Ampex reel-to-reel video recorder as well as a black-and white camera, television receiver, and home music center, and was featured in the 1963 Neiman-Marcus Christmas Catalog for $30,000.[4] Needless to say, the Signature V (whose purchase price included a visit by an Ampex engineer to set it up) did not achieve widespread market penetration. By packing the tape and reels inside a plastic case, the Sony U-Matic promised ease of use to corporate users, and the same videocassette approach made video user-friendly enough to move from television studios and institutional settings into homes. The move to cassettes was seen as essential to the creation of a consumer video recorder, but while eager enthusiasts even adapted the professional-grade U-Matic for use in their homes, the machines were still priced well above the reach of most households.

In 1975, Sony introduced the Betamax, a redesigned videocassette recorder based on the U-Matic for home use. It is important to remember that in the early days of home video, there was no supply of prerecorded videocassettes, and the Betamax was fundamentally marketed as a way to record broadcast television for later viewing. Like its ancestor, the first Betamax cassettes could only record an hour of video; Sony designers assumed that the machine would primarily be used for time-shifting broadcast television shows, which generally lasted an hour at most. Tape length was a crucial factor in the early skirmishes between Sony's Betamax and its rival, the VHS ("Video Home System") format invented by JVC and adopted by Matsushita—the story goes that RCA, the major producer of VHS recorders in the United States, refused to start manufacturing its own machine until a tape could record not just an hour-long serial drama, but a full football game. Thanks to an innovative recording head design and thinner tape, the first RCA Selectavision VHS machine arrived in stores with a tape running time of two hours.[5] Though Sony had its own two-hour machine on the market soon thereafter, it never entirely managed to shake off the Betamax stigma of shorter recording time (albeit of higher image quality).[6]

It can be hard to imagine exactly how new and unfamiliar the idea of a video recorder was in the early 1980s. At the time, video storeowners liked to tell a joke about a customer who brought a VCR in for repair, complaining that it had stopped working and would not record any television channels. The storeowner looks the machine over, wondering if there had been some mechanical failure. Finding nothing technically wrong, he ejects the videocassette that was currently in the deck only to see that it had been played and recorded over so many times that the magnetic tape had worn to the point of snapping. Handing the customer the videocassette, the storeowner asks if all of his tapes are this worn, to which the customer replies, "I didn't even know that piece came out!"

As for the cassettes themselves, the first manufacturers of blank cassettes for the Sony Betamax and RCA SelectaVision were in fact Sony and RCA. In 1976, a one-hour Betamax cassette was a piece of high-end technology that would cost you $15 (if you bought in bulk),[7] while retail prices could go as high as $30.[8] Instructions both included with tapes and published in hobbyist magazines outlined the do's and don'ts of videocassette care, including how to store them (vertically, "like they were books"), how to protect them (ideally wrapped in a plastic sleeve or bag when not in use to protect from dust, ashes, and other contaminants), and how to clean them if need be.[9] Moreover, both tapes and recorders carried the same brand label, and manufacturers warned about the potential damage that inferior, imitation tapes might cause to a machine's recording heads.

Initial studies claiming that VCR owners would want to purchase between one and five blank tapes proved to underestimate the desire of early adopters to archive rather than simply time-shift television, and chronic shortages plagued the industry's first two years.[10] The supply problem eventually worked itself out as companies such as 3M and Fuji (and later TDK and Memorex), who were solely involved with magnetic tape fabrication, caught up to the VCR manufacturers.[11] When these competitors introduced their own cassettes to compete with Sony and RCA, they doubled down on the vision of videocassettes as high technology. By emphasizing the technical nature of videotapes, separate from the recorders in which they were played, these newcomers made the case that their "second generation" videocassettes improved on the earlier cassette technology of the VCR manufacturers. One Fuji ad in the summer of 1979 trumpeted: "Finer magnetic particles and more uniform distribution . . . Higher precision and stability in cassette housings, as well as the cassette itself . . . for greater viewing enjoyment and trouble-free recorder performance."[12] Not only would Fuji's improved videocassette technology lead to better images on the TV screen, the ad promised, it would actually improve the functioning of the VCR itself!

Through the late 1970s and early 1980s, time-shifting became increasingly common among American television viewers, but that does not mean that it was uncontroversial. Television networks sold ads based on assumptions of viewers at the time of broadcast, and the studios that supplied the networks with programming considered time-shifting an infringement of their intellectual property. In 1976, Universal Studios sued Sony, arguing that it was liable for manufacturing and selling a device whose primary purpose was piracy. It took almost a decade for the case to make its way through to the Supreme Court, which ultimately ruled in 1984 that home recording of public broadcasts was a legal "fair use", though by then the point was arguably moot, as well over half of American households already owned VCRs.

Prerecorded Tapes

The first attempt to sell videocassettes already recorded with movies and other programming predated Sony's Betamax by over three years. In 1972, Cartridge Television, Inc. announced its Cartrivision videocassette recorder, which played ½-inch magnetic tape from reels stacked one on top of another in square "cartridges". Due to its novel technical design, a Cartrivision cartridge could hold up to two hours of video, and the machines, produced by Avco, had the same basic functionality as the later Betamax and VHS standards.[13]

The real selling point of Cartrivision, however, was the integrated sale of hardware *and* content; the machine was marketed hand-in-hand with prerecorded tapes of mainstream Hollywood movies. The Cartrivision system included two kinds of tapes: black tapes which were for sale only, and which included instructional titles and specially-produced-for-video features as well as blank tapes for television and use with the Cartrivision video camera; and red tapes, which were for rental only, including major Hollywood films such as *Casablanca* (1942) and *Dr. Strangelove* (1964). Hollywood studios, traditionally wary of anything less than complete control over the distribution and exhibition of their films, had been persuaded to license their titles to Cartrivision because the red, rental-only tapes could only be rewound using special in-store equipment (which included a tamper-proof counter for accounting purposes), with the net effect that a customer's rental fee covered exactly one viewing of a given movie, not a second more.[14]

Thanks to a partnership with Sears, Roebuck and Company, Cartrivision went on sale at eighteen Sears stores in the Chicago area in June of 1972, but there was trouble from the beginning. The first recorders were built into a console with a television set (like that early Ampex machine), and salespeople at Sears seemed to be unsure of how to present it to their customers: "They knew how to talk about furniture. They could tell you how to turn a TV set on and off. But they had never run one of these things. They hadn't been trained . . . it was a mess."[15]

Eventually, Cartrivision introduced a standalone deck, but other problems kept cropping up. The prerecorded tapes were sometimes sold on different floors of the store from the recorders, and in some cases renters had to order tapes via UPS and wait days for them to arrive. The no-rewind functionality of the red cassettes proved an irritation to consumers. Moreover, the fabrication process for the cassettes (which involved sandwiching together a strip of blank tape and a master, then passing an electrical current between the two, creating a duplicate of the original) resulted in such poor audio quality that once hand-assembled into the cassette, the audio needed to be erased and re-recorded.[16] Worst of all, the entire stock of prerecorded cassettes began to spontaneously degrade for reasons never fully understood, and there was concern that the use of one of these corrupted tapes could damage the recording heads of a perfectly good Cartrivision recorder.[17]

After all this, the Cartrivision management paused, gathered their resources, and prepared for a do-or-die market test in California. Learning from their mistakes, the company established the rule that "no store could go on line unless we had trained the salespeople in demonstrating the hardware and unless a sufficient supply of cassettes was available and visible with the hardware."[18] Within a few weeks, the test began to show results, but by this point Cartridge Television was hemorrhaging money. Around this time, Avco executives on a trip to Japan saw a demonstration of Sony's next-generation Betamax video recorder and, though Sears was ready to start marketing Cartrivision on a national level, Avco decided to pull out as a manufacturer, leaving Cartridge Television bankrupt.[19]

Another videocassette player would not reach the consumer market until 1975, when the Sony Betamax would be marketed by its manufacturer as a time-shifting device. Akio Morita, the chairman of Sony, did explore potential deals with several Hollywood studios, even announcing a joint venture with Paramount Studios in 1976.[20] At the time, however, none of the Hollywood studios were willing to risk releasing their films on a format that did not offer strict control over users, and their tentative partnerships with Sony fell through.

Soon after the Betamax was unveiled, however, independent suppliers did begin to offer their own tapes with prerecorded content for sale; most of that inventory was either older public domain films or pornography. In the former case, anybody who had a master print of an early film or television show whose copyright had expired could make copies and sell them, royalty-free, and a handful of enterprising suppliers with names like The Nostalgia Merchant who had been offering these "public domain" films on 16 mm and 8 mm film had begun to transfer them onto videocassette, setting up new mail-order businesses.[21] Meanwhile, the pornography industry embraced the new medium of videotape. Pornographers had explored the potential of film from the earliest days of motion pictures, and the "stag film" had been a discreetly tolerated institution in American life since the 1920s.[22] However, it was not until the theatrical success of *Deep Throat* and *The Devil in Miss Jones* in 1973 and the ensuing burst of "porno chic" that adult film broke from the underground into American popular culture.[23] Thanks to their existence on the fringes of mainstream America, those in the adult film business were freer to try new things, among them releasing their films on the new medium of videotape. Mail-order houses began offering videotape copies of recent theatrical sex films (as well as older films by directors such as Russ Meyer and Radley Metzger) as early as 1976, advertising them in the back of film and video hobbyist magazines for prices of up to several hundred dollars apiece. Just a few years later, *Playboy* told its readers in 1979 that "just about every top-quality X-rated movie made in the last several years can be legitimately purchased over the counter."[24]

Surveying the burgeoning video industry in late 1976, a Michigan businessman named Andre Blay who had made a career as a distributor of audio and video equipment saw an opportunity and sent a letter to the heads of the major Hollywood studios asking for permission to license their movies for distribution on videotape. The studio response was less than welcoming. Most did not even bother to write back, and only 20th Century Fox, which was itself preparing to test the waters of the prerecorded videotape market, responded affirmatively. Over the next year, Blay and Fox executives hammered out a deal under which Blay's Magnetic Video would have the nonexclusive right to sell videocassette copies of fifty movies from the 20th Century Fox vault. From today's point of view, the choices were less than ideal; the most recent movie was four years old, and they had all previously been sold for broadcast on network television. On the other hand, these were mainstream Hollywood movies such as *The Seven Year Itch* (1955), *M*A*S*H* (1970), *Patton* (1970), and *The French Connection* (1971), and this was the first time that consumers could actually buy their own copy to watch whenever they wanted.[25]

So, how to get these tapes into the hands of those consumers? Blay tried to package videotapes with VCRs, and also created the mail-order "Video Club of America". In 1977, just before the holiday season, he placed an advertisement in *TV Guide*, offering "original full-length uncut versions of Hollywood's finest movies".[26] At the time, less than one percent of American homes owned a VCR, and Blay's opening words were not addressed to VCR owners at all: "If you've been waiting for the finest in home entertainment before investing in a video-cassette unit, your timing is perfect." Moreover, the first item listed as a "preferred benefit" of joining the Video Club of America is the ability to "save $100s off the list price of the newest Sony Betamax and RCA SelectaVision videocassette recorder/players." Magnetic Video was not just selling tapes to people who already owned VCRs, they were also selling the very *idea* of movies on videotape as a reason to buy a VCR in the first place.

Over the next year, prerecorded video took off like a rocket. Blay expanded his catalog, signing deals with Viacom, Avco Embassy, and the estate of Charlie Chaplin, among others.[27] Within three years, most major studios had opened their own Home Video divisions. Warner Brothers Home Video opened its doors in 1978, followed the next year by Paramount, Columbia, and 20th Century Fox (which, rather than start a new operation from scratch, simply bought Magnetic Video for $7 million). Disney Home Video joined the fray in 1980 along with Metro-Goldwyn-Mayer, which established a short-lived partnership with CBS, and they were soon joined by United Artists, Orion, and even MCA/Universal (who could not deny the growing video market, though their lawsuit against Sony would not be resolved for another three years).[28]

Though it is unclear exactly who did so first, it was around this point that distributors and retailers started referring to prerecorded videocassettes as "software".[29] The choice of language is telling—the manufacturers' reps in particular were used to dealing in electronics such as stereos and television sets, which were traditionally called hardware. "Software" as a category only exists in opposition to hardware; as one salesman put it, "software was something that went into a machine . . . it wasn't what they would consider your normal hard goods."[30] There was something more to this definition, however, than software simply being something that was inserted into a machine. After all, blank tapes in themselves were not software, they were simply accessories. In calling prerecorded tapes "software", these electronics distributors and reps were marking out a specific relationship between prerecorded tapes and video recorders that ignored the materiality of the videocassette, literally naming it in terms of the information it contained rather than its tangible hardware nature.

Video Stores

In 1980, less than three percent of American homes with television sets owned VCRs, but just four years later this figure was close to twenty percent, and as video became increasingly popular among consumers, many would-be entrepreneurs saw an opportunity.[31] In the early days of consumer video, if you watched a prerecorded cassette at home you likely owned it, and you had probably bought it via mail order or at the same electronics store where you had bought your VCR. Within just a few years, however, it was more likely that you had rented a videocassette from one of the thousands of "mom and pop" video stores that had sprung up across the United States seemingly overnight. While retailers of many different stripes continued to add video to their pre-existing stores (with varying degrees of success), by late 1980 there were roughly 900 stores across the United States that put prerecorded videocassettes front and center.

One pioneer of this new industry was George Atkinson, the founder of The Video Station franchise. Atkinson had previously built a business by renting public domain movies on Super 8 mm film (and later U-Matic videotape) to hotels and pizza parlors as a form of free entertainment for customers, and he understood the Betamax not as high technology, but simply as the "perfect movie machine".[32] In 1977, when he heard about the initial batch of movies on videotape released by Magnetic Video, Atkinson looked at his own business and figured that the demand for movie rental would be leaps and bounds above the demand for 8 mm movies and projectors. He placed an ad in the *L.A. Times* that read "Video for rent", along with a coupon for readers to fill out and mail in, and within a week Atkinson had around a thousand inquiries. Unfortunately, Atkinson had no inventory, and without investment capital found himself in a catch-22 situation; in order to make money, he needed a library of videocassettes to rent to customers, but he could not afford to buy cassettes without rental revenues. Soon, he hit on an ingenious solution: renaming his store The Video Station, Atkinson began charging $50 for an annual membership, and quickly raised enough money to build a library of tapes.[33] Once a customer had paid the membership fee, she was entitled to rent tapes for only ten dollars per night, and though Atkinson was happy to sell tapes outright to eager customers, he quickly found that rental comprised the bulk of his business.

In later years, many in the video industry would look at George Atkinson as the "father" (if not the patron saint) of video rental, having reconceived the relationship between his product and his customers. The Magnetic Video approach to videocassette sales framed movies on videocassette as artifacts, tangible objects to be advertised and sold to consumers for their own use. On the other hand, The Video Station offered videocassettes to consumers not as *things* to be bought, but as *experiences* to be rented, and returned within a matter of days to avoid a late fee. By the mid-1980s, the video industry existed as a diverse ecology of national chains like Video Station; smaller regional chains like Washington DC's Erols and the Midwest's Movies-To-Go; and myriad independent "Mom and Pop" stores that had sprouted everywhere. This diverse ecosystem was upended by a Florida businessman named H. Wayne Huizenga who had made his fortune in waste management but saw an opportunity in a more standardized video store experience. Taking over the role of CEO of a small, regional video store franchise called Blockbuster Video, Huizenga proudly declared his intention for the company to become the McDonald's of videocassette rental and embarked on an aggressive campaign of acquisitions and new store openings. Inspired by the fast-food analogy, Portland-based Mark Wattles decided to position his Hollywood Video as the Burger King to Blockbuster Video's McDonald's, and by the early 1990s, the two chains dominated video rental across the United States.

Movies on Videocassette

The videocassette was a key tool in shifting agency from advertisers to viewers. Early video "time-shifters" would edit commercials out of live television broadcasts through expert application of the "Pause" button while recording, and one of the most commonly mentioned uses of the VCR was to fast-forward through commercials when watching time-shifted broadcast programming.[34] One of the adjectives most often applied to prerecorded movies on videocassette was "uninterrupted", a coded way of saying "without commercials". Since movies in the theater did not stop for commercials every fifteen minutes, the commercials that were all too common on network movie broadcasts were understood as artifacts of the particular system of advertising-supported broadcasting that had been dominant in the United States since the 1920s.[35] When commercials ultimately did make their way on to prerecorded videocassettes they were placed before the main program.

The act of translating movies to video raised some tricky issues. For example, since the move to "widescreen" cinematography in the early 1950s, theater screens tended to have different aspect ratios that were much wider than television screens. In the pre-videocassette days, more or less the only option when broadcasting a film on television was to simply center the television camera on the projected film image, chopping off the sides of the film image and hoping for the best. By the 1970s, most movies were encoded onto videocassette for broadcast (and later for rental or sale) using a machine called a Telecine, essentially a modified video camera that converted a filmed image into electrical signals. In addition to tweaking the color and timing of captured video, Telecine operators were able to pan across the widescreen image so that the final product would (theoretically) include the most important elements of a given shot, wherever they might be on the original. Such "pan-and-scan" video transfers were the default for early consumer prerecorded videocassettes, and this particular method of altering the theatrical image naturally went unmentioned in early promotion of movies on video.

Unmentioned, that is, until the mid-1980s. In 1983, Telecine operators at Modern Video Film (one of the pioneers in film-to-video transfers) created a pan-and-scanned video transfer of Woody Allen's *Manhattan* (1979) which was originally filmed in the 2.35 Cinemascope aspect ratio. At the time, virtually all transfers were performed without the permission of directors, many of whom seem to have seen ancillary media such as television or video as irrelevant to their artistic expression.[36] A clause in Allen's contract, however, required that he personally approve any and all versions of *Manhattan*, so Modern Video Film sent him a copy of the cassette. A week later, the company received a faxed message: "Mr. Allen finds this transfer unacceptable. Do it again." Over the next weeks, Modern Video Film created at least three separate transfers, each time sending the product to Woody Allen and receiving the same response.[37] Finally, one technician declared "Maybe he just doesn't like pan and scan" and, with the permission of the now-desperate studio executives, created a transfer that included the *entire* Cinemascope image, with empty space at the top and bottom of the screen to simulate the more oblong shape of the theater screen. As one of the Telecine team remembers, "A week of silence, a fax came back, and it simply said, 'Mr. Allen approves this transfer.' And that was it."[38] Later that year, *Manhattan* shipped on videocassette and RCA videodisc with gray bars on the top and bottom of the screen.[39] This new style of video transfer was dubbed "letterboxing", because the result is rather like looking out through a mail slot—telecine operators, followed by the industry more generally, adopted the British "letterbox" rather than the American "mail slot" because the most popular film transfer devices were manufactured by the British firm Ranks and Tell, who had labeled that particular button with the Anglicism. Since pan-and-scan had been established as the norm for movies on video, letterboxing (though simply an alternate way of transferring from film to tape) was usually cast as an "enhancement", and through the late 1980s letterboxing was identified with higher-end special editions of films (and often more videophile-oriented formats like Laserdisc as opposed to the more common videocassette). This was not always intuitive for viewers who were used to the video image filling their TV screens, leaving video store clerks to explain to frustrated customers that the tape was not broken, and in fact that letterboxing let you see *more* of the film image, not less. It would not be until the advent of the DVD in the late 1990s that letterboxing would achieve a more mainstream identity as a technological practice, ultimately becoming so common that the new High Definition television format moved from the standard Academy 4:3 screen dimensions to a 16:9 widescreen aspect ratio.

The Disappearance of Video

In the early days of the home video industry, blank videocassettes were high technology and prerecorded videocassettes were precious artifacts to be rented over and over to recoup a storeowner's initial investment, but it was a little over 20 years from the initial flood of Betamax and VHS cassettes into the marketplace to their relatively quick decline. Following in the steps of the rapid changeover of the music industry from magnetic cassette tapes to Compact Discs (CDs), the Digital Video Disc (or DVD) offered viewers sharper image quality and a patina of technological sophistication. Less than six years after the debut of the format in 1997, weekly rentals of DVDs outpaced weekly rentals of videocassettes for the first time according to the Video Software Dealers Association,[40] and only a year earlier, Circuit City (the second largest electronics retail chain in the United States) had stopped offering movies on videocassette for sale entirely, turning that shelf space over to DVD. "We're responding to what people are wanting to buy", claimed a company spokesman.[41]

If videocassettes brought movies out of theaters and into the home, allowing everyday consumers to actually hold movies in their own hands for the first time, then DVDs went a step further. A movie on DVD often exists in a context of director commentaries, making-of documentaries, and production stills, all of which direct viewers' attention toward the circumstances of a movie's production and demystify the experience of watching the movie, subtly shifting the nature of movie-watching from immersion in an experience to the abstracted analysis of a text.

Thanks to the unceasing push for higher resolution in consumer television through the 2000s, DVDs came under pressure from higher-definition Blu-Ray discs after that standard won a format war with HD DVD in 2008, and even higher image resolution was promised a few years later by the new 4K Ultra HD Blu-Ray. No matter how high the resolution, however, physical formats have lost ground to digital streams with no physical presence at all; 2016 was the first year that more money was spent in the US on streaming services than on DVD, Blu-Ray, and Ultra Blu-Ray combined.[42] Reed Hastings, founder of Netflix, has said that he always envisioned digital streams as the endgame—his strategy was to "build Netflix first on DVD and then eventually the internet would catch up with the postal system."[43] The trend toward digital distribution of mainstream film and video seems inexorable, though the success of DVD vending machine company Redbox seems to indicate that broadband internet connections are not yet pervasive enough to fully render physical media obsolete. Meanwhile, amid the tangled morass of licensing deals which mean that a movie or television show might be available on a given subscription service one month but another platform the next, it could be that owning a physical copy is the best way to ensure consistent access.

As for time-shifting, disc-based media never really took off as a home recording medium so the videocassette remained the dominant way to record TV for later viewing until the early 2000s, when TiVo introduced the first mainstream digital video recorder (DVR), which used an internal hard drive rather than removable media. Within less than a decade, DVR capability became a standard offering by cable TV providers, and generic cable boxes offer the ability to record multiple channels of live television for later viewing. Per that old video storeowners' joke about videocassettes, that part does not come out anymore.

Notes

1 This section relies heavily on Nmungwun's detailed history of both nonmagnetic and magnetic video recording technologies: Aaron Foisi Nmungwun, *Video Recording Technology: Its Impact on Media and Home Entertainment* (Hillsdale, NJ: L. Erlbaum Associates, 1989).
2 Peter Keane, interview with the author, January 6, 2004.
3 Jonathan Coopersmith, "Pornography, Technology, and Progress," *Icon* 4 (1998): 94–125.
4 Nmungwun, *Video Recording Technology*.
5 Marc Wielage and Rod Woodcock (authors of the "Rise and Fall of Beta"), interview with the author, March 15, 2003.
6 The canonical recounting of the VHS/Beta format war can be found in Cusumano, Mylonadis, and Rosenbloom, "Strategic Maneuvering and Mass-Market Dynamics: The Triumph of VHS over Beta," *The Business History Review*, 66 (1) (Spring, 1992), pp. 51–94.
7 Jim Lowe, *The Videophile's Newsletter*, December 1976, p. 3.
8 Mike Salomon (owner of Video Retreat), interview with the author, May 10, 2003.
9 "How to Care for Your Videocassettes," *Video Magazine*, Summer 1978.
10 James Lardner, *Fast Forward: Hollywood, the Japanese, and the Onslaught of the VCR*, 1st edition (New York: W.W. Norton & Co Inc., 1987), p. 96.
11 "Blank Tape: The Current Market," *Video Magazine*, Summer 1978.
12 "Fuji Videocassette Advertisement," *Video*, Summer 1979.
13 Interestingly, these deals were brokered in part by Sony's Peter Keane, who had left the electronics manufacturer in 1971 and within a year was working for Cartrivision, leveraging his contacts at Columbia Studios and elsewhere to broker licensing deals allowing Cartrivision to distribute tapes of movies.
14 Keane, interview.
15 Lardner, *Fast Forward*, 83–84.
16 Keane, interview.
17 Nmungwun, *Video Recording Technology*.
18 Lardner, *Fast Forward*, 85.
19 Keane, interview.
20 "News and Comment," *Videography*, September 1976.
21 Jeff Tuckman (film collector and Sound Unlimited employee), interview with the author, December 9, 2003. For more, see Daniel Herbert, "Nostalgia Merchants: VHS Distribution in the Era of Digital Delivery," *Journal of Film and Video* 69 (2) (May 3, 2017): 3–19.
22 Eric Schlosser, *Reefer Madness* (Boston, MA: Houghton Mifflin, 2003), 126–128.
23 On the general subject of pornography and early video, see Peter Alilunas, *Smutty Little Movies: The Creation and Regulation of Adult Video* (Oakland, CA: University of California Press, 2016).
24 Howard Polskin, "Tuning into the Videotape Scene," *Playboy*, April 1979.
25 Lardner, *Fast Forward*, 172–173.
26 "Video Club of America Advertisement," *TV Guide*, November 26, 1977.
27 Lardner, *Fast Forward*, 174–175.
28 Wasser, *Veni, Vidi, Video* (Austin: University of Texas Press, 2001), 96.
29 This term was quickly picked up by specialty video magazines; for example, the Winter 1978 *Video Magazine Buyer's Guide* (the first issue of the magazine, which appeared in early 1978) contained the words "Video Software" splashed across two pages, marking the section profiling prerecorded videocassettes.
30 Wayne Mogel, interview with author, April 30, 2003.
31 Frederick Wasser, *Veni, Vidi, Video*, p. 68.
32 Lardner, *Fast Forward*, 176.
33 Ibid., 175–178; David Rowe, "The Real George Atkinson," *Video Store*, June 1983.
34 For a more extensive discussion of VCR users' relationship with their fast-forward buttons, see Bruce C. Klopfenstein, "Audience Measurement in the VCR Environment: An Examination of Ratings Methodologies," in *Social and Cultural Aspects of VCR Use*, ed. Julia R. Dobrow (Hillsdale, NJ: Lawrence Erlbaum Associates, 1990); Carolyn A. Lin, "Audience Activity and VCR Use," in Julia R. Dobrow (ed.) *Social and Cultural Aspects of VCR Use* (Hillsdale, NJ: Lawrence Erlbaum Associates, 1990), pp. 75–92; Barry S. Sapolsky and Edward Forrest, "Measuring VCR 'Ad-Voidance'," in *The VCR Age: Home Video and Mass Communication*, ed. Mark R. Levy (Newbury Park: Sage Publications, 1989).

35 On the early history of advertiser-supported broadcasting, see Susan Smulyan, *Selling Radio: The Commercialization of American Broadcasting, 1920–1934* (Washington, DC: Smithsonian Institution Press, 1994).

36 Lowell Goldman, "Directors on Video," *BoxOffice*, March 1985.

37 Woody Allen (via Allen Eichhorn), personal correspondence (2004).

38 Marc Wielage and Rod Woodcock, interview with author, March 15, 2003.

39 The gray bars (as opposed to black) are an interesting story of boundary-drawing between medium and message: according to Wielage:

> [A United Artists studio executive] said, "You know, I don't think we should put this out with black borders, because that will look wrong. That will look like a mistake. People will think we're covering up some of the picture." I said, "No they won't, it looks fine. It looks like a movie in a theater." He goes "No, no, we should have grey borders. Because that will look more like it's part of the picture and that way they won't think it's a mistake." I tried to tell him as politely as I could, I think that's the stupidest idea I've ever heard in my life. He goes, "No, no this is the way we should do it."

40 David Kaplan, "DVDs Outpace VHS Rentals for First Time," *The Houston Chronicle*, June 20, 2003.

41 Shelley Emling, "Circuit City to Drop VHS Movies," *The Atlanta Journal-Constitution*, June 20, 2002.

42 Andrew Wallenstein, "Home Entertainment 2016 Figures: Streaming Eclipses Disc Sales." Variety.com, January 6, 2017. http://variety.com/2017/digital/news/home-entertainment-2016-figures-streaming-eclipses-disc-sales-for-the-first-time-1201954154/

43 Ashley Rodriguez, "Netflix Was Born out of This Grad-School Math Problem," *Quartz*, https://qz.com/921205/netflix-ceo-reed-hastings-predicted-the-future-of-video-from-considering-this-grad-school-math-problem/.

18

PROJECTING PLAY

The Give-A-Show Projector and Children's Audiovisual Media Toys of the Mid-20th Century

Meredith A. Bak

In a television commercial for Kenner's Play N' Show combination phonograph and picture viewer/projector (ca. 1969) a blonde, pig-tailed girl addresses the camera. "You know what?" she begins, "My brother has a real color TV!" The commercial cuts to her older brother staring into a screen embedded in a plastic console that resembles a cross between a television set and an Apple II computer. He turns to the camera and corrects his sister. "It's not a TV! It's Kenner's Play N' Show." He goes on to explain the toy's picture records, which synchronize audio narration to still-picture slide shows featuring characters such as The Flintstones and Yogi Bear. Astonished by the appearance of Yogi Bear, the little girl repeats again: "Isn't this a nice TV?" and *again* her brother corrects her. The commercial's simultaneous reliance upon and disavowal of television is curious. The Play N' Show is a toy that closely *resembles* a TV and was advertised *on* TV, yet despite the little girl's insistence that it *is* a TV, her brother and the adult voiceover narrator reconfirm that it is a separate medium altogether. The Play N' Show was the latest in a string of media toys sold by Cincinnati-based company Kenner, which boasted a wide product range of toy phonographs including the Close 'N Play, hybrid picture/sound players such as the Play N' Show, toy televisions like the Change-A-Channel TV Set, the Easy-Show movie projector, and the company's most well-known slide projector, the Give-A-Show, on the market from roughly 1959 to 1980.

What follows is a critical discussion of the Give-A-Show, from its earliest iterations, through its heyday, and its eventual decline. Drawing upon material details of the projector, patent records, and promotional materials, I consider the Give-A-Show within the context of several related media toys produced by Kenner and other companies and chart the projector's unique relationship with children's television spectatorship in the 1960s and 1970s. Exploring the Give-A-Show in relation to the television is not a conflation or reduction of the toy's role in the mid-century playscape underpinned by a teleological logic on the basis that it *resembled* a television. Rather, this association was cemented during the first few years of the projector's production and marketing, when the wide-open spectrum of possibilities of a children's projector were more narrowly concentrated in relation to licensed television content, especially the properties of animation studio Hanna-Barbera. Unlike collectors' literature, this is not an exhaustive survey of Kenner's full line of -Show products or rival media toys. Instead, I call upon additional examples as points of particular comparison and contrast of features and to attest to the general ubiquity of media-related toys in the 1960s and 1970s. Likewise, rather than a focus on the Give-A-Show's reception, the focus is placed on the design attributes of Kenner's media toys as they reflect changing relationships with film and television viewing.

Introducing the Give-A-Show

The Give-A-Show projector was a toy projector introduced by Cincinnati-based company Kenner in 1959 that remained a hallmark product for the company for over two decades (Carlisle, 2009: 345). The plastic-bodied projector was most commonly molded in red or blue. Its illumination source was a flashlight bulb powered by three D-cell batteries, which, in concert with the carrying handle incorporated into the body, highlighted the device's portability. Indeed, its central promise was that with it, children could stage vibrant, colorful shows of their favorite cartoon characters whenever they wanted. Kenner was a large licensee of Hanna-Barbera characters, and the Give-A-Show library would include other licensed properties as well. Gradually the allure of this flexible, "on demand" programming would fall out of favor, ultimately contributing to the Give-A-Show's demise as changes in television culture rendered it obsolete. Close examination of the Give-A-Show in conjunction with competing audiovisual toys and Kenner's own expanded line of "-Show" playthings importantly illuminates the role that media content curation and exhibition played in the mid-century American child's playscape.

Simply put, the Give-A-Show functioned as a children's slide projector. The original Give-A-Show most closely resembled earlier models of the magic lantern, but Kenner's subsequent models would incorporate and remediate other audiovisual technologies. As such, the line of toys is perhaps best understood as an amalgamation of several media dispositives—technological configurations that include both material and concrete components, but that also exist on the level of discourse (Albéra and Tortajada, 2010: 11). Understanding this broader context is key to excavating the projector's role in the children's mediascape during the 1960s and 1970s, when domestic media conception and particularly children's media practices were undergoing substantial transformation. Even as it drew upon longer histories of domestic media production and consumption, the Give-A-Show's design (and subsequent iterations), along with Kenner's predominant reliance on licensed content from contemporary animation studios such as Hanna-Barbera, aligned it less with traditions of children's media production and more with children's television spectatorship.

The original Give-A-Show projector was patented by James O. Kuhn (1961) for Kenner. Kuhn's original patent for the toy stressed the simplicity of its construction and the ease of its operation. The projector's main body was two plastic halves joined along a central seam, and a simple switch closed the circuit to turn on the light bulb. The Give-A-Show projected 35mm film slides arranged into 7-image narrative sequences and mounted horizontally in cardboard strips (from the late 1970s, the slides would be mounted in a plastic sleeve with beveled corners, presumably to address issues related to warped cardboard and to facilitate easy insertion). The slides were inserted through a slot behind the lenses, and both the bulb and the batteries could be easily replaced. These features not only contributed to the device's portability, but were also designed to enable children to operate the projector independently of adult oversight. The arrangement of slides into sequences of seven horizontally mounted images gestures to several other formats. View-Master reels, which, by the 1950s had become largely a children's medium, arranged seven stereographic image pairs around the edge of its cardboard reels. The filmstrips' horizontal arrangement (along with the drawn art style of the individual slides and the cartoon subject matter) likewise bears close resemblance to comics aesthetics. Subsequent models of Kenner's "-Show" product range would use different media formats, such as self-contained film cartridges and images mounted around the edge of a phonographic record. Although Kenner's 35mm slides were a standard gauge, the variety of formats its media toys employed during the 1960s and 1970s attests not only to an industrial landscape in which practices of standardization and medium specificity were very much in flux and under contestation, but also specifically highlights how media made for young audiences reflect assumptions about children as consumers and operators of technological apparatus, particularly concerning durability and ease of operation.

The Give-A-Show's function as a projector is fruitfully contextualized within the much broader historical traditions of both domestic media technologies (generally) and media toys for children (more specifically). Scholars have long traced the complex relationships between domestic media technologies and prevailing ideologies of the home, family, and the (often gendered) roles that individuals occupy. Whether in relation to the 19th-century magic lantern, home cinema projector, television, or mobile device, discourses tend to chronicle the uneasy adoption and integration of such technologies into home and family life, variously charting the changing social protocols, design features, and uses of domestic media technologies (Hirsch and Silverstone, 1992). During the 19th century, industrial production and improved lens technology made magic or "optical" lanterns commercially available for middle-class consumers. As Moya Luckett (1995) has explored, while many lanterns and later cinema projectors were likely to be operated by adults, the overall context of home recreation was connected to traditions of children's parlor play with optical devices. Later, home cinema projection would serve as an important practice through which prevailing ideas about family, gender, and ideals about domestic life would be reflected and enacted, as traced in work such as Haidee Wasson's study (2009) of the introduction of 16mm film in 1923 (Singer, 1988). Despite the well-documented practices of home magic lantern and cinema projections as family affairs, comparatively little work has considered the specificity of toy versions of these same media, and children as their attendant users (Basano, 1996; Wells, 2010; Bak, 2015).

By the end of the 19th century, toy magic lanterns, often made of tin, were also widely available at inexpensive prices. Frequently derided for their inferior image quality in comparison to lanterns used in institutional contexts such as churches, schools, and for scientific demonstrations, toy lanterns were nevertheless workable miniatures that allowed children to project shows in their homes, and kids could choose from a wide variety of commercially produced chromolithographed slides covering all subjects. The toy lanterns of the 19th century were often distributed in kits that included show bills and tickets for kids to stage their own shows and charge admission. Toy lanterns were often advertised as investments with claims that children could turn profits with their home shows. Such promotional discourses and packaging demonstrates how purveyors of the toy lantern envisioned play not as simply the act of children projecting images in the domestic context, but instead, the simulation or adaptation of the practices and conditions of commercial exhibition (Bak, 2015). Toy cinema projectors for children were also available from very early on, such as Pathé's Pathé Kid line and a host of others (Schneider, 2008; Museu del cinema, 2009).

This broader historical context demonstrates that when it was introduced, the Give-A-Show would not have been received as a "new" medium from social or technological perspectives. Aside from its plastic construction, its projection capabilities did not represent a radical technological development, nor were the forms of play that it encouraged significantly different from activities associated with earlier kids' projection technologies. By 1959, middle-class children had been "giving shows" to friends and families for nearly a century. What *was* unique about the projector is its positioning in relation to television, and particularly, Kenner's focus on producing slide shows of licensed TV characters, which determined the kinds of shows that kids would give. The Give-A-Show built upon the legacy of earlier modes of home visual entertainment such as the magic lantern and small-gauge cinema to allow kids to watch their favorite cartoon characters wherever and whenever they wanted. Unlike other forms of licensed merchandise that expanded the reach of popular intellectual properties into virtually all realms of children's lived experiences (including toys, games, apparel, and housewares), the Give-A-Show liberated cartoons from their fixed programming blocks, albeit in adapted form. No longer relegated to Saturday mornings (or increasingly, weekday afternoons), characters such as The Flintstones, Yogi Bear, Scooby Doo, and Casper the Friendly Ghost could be conjured up by children at will, projected onto household surfaces in a hazy cone of light. The projector's bright plastic body cast in primary colors, moreover, signified its belonging to the world of childhood, and thus children could stage their shows independently, rather than having potential contestation over the use of a shared family television set or projector.

From Infinite Possibilities to Television Surrogate: The Give-A-Show's Development

When Cincinnati-based Kenner became a sponsor of *Captain Kangaroo* in 1958, it was one of the earliest toy companies to take advantage of nationwide television advertising. The Give-A-Show's close relationship with television, both in the kind of play it facilitated and in the content available for projection is thus unsurprising. Slides featured both licensed and public domain content, such as early sets that included Wild West shows about Wyatt Earp, Annie Oakley, and Wild Bill Hickok. However, the majority of slides produced would be of licensed characters, and the relationship between the slide shows and popular television programs was made prominent on the projector's packaging, which framed the toy as letting kids give shows featuring "TV favorites" (Knutson, 2008).[1] Also, Kuhn's original patent expressly names cartoons (as well as photographs) as possible subjects for projection. The gradual dominance of licensed content for the Give-A-Show, in tandem with the changing design of Kenner's related media toys, helps to chart how the Give-A-Show's association with television contributed both to its ascendance and, eventually, to its decline.

Early iterations of the Give-A-Show and related projectors relied less exclusively on licensed content and instead conceptualized kids' shows within a broader tradition of performance and exhibition. Three years before the Give-A-Show's release, Bill and Sue Severn published a children's home entertainment book called *Let's Give a Show* (1956), which offered instructions and encouragement to children wishing to stage a range of performances in the home, from magician's and ventriloquist's acts, to plays, circuses, and minstrel shows (though race is conspicuously absent from the text). The book's Foreword situates this tradition of performance as more expansive than screen-based media, suggesting "there's more to the world of the theater than watching a movie screen or sitting down before a television set" (Severn and Severn, 1956: viii). However, the book also suggests that children recreate or stage their own movie- and television-themed shows. To adapt the television format for living room performance, the Severns recommend making a stage that resembles "a huge TV screen, or just a dummy camera fashioned out of a cardboard box and wooden legs", then go on to describe the steps in preparing for a variety show. They suggest that "By giving your show a TV background, you can make one of the oldest forms of entertainment seem new and different to your audience" (Severn and Severn, 1956: 93).

Texts like the Severns' home entertainment manual illuminate what "giving a show" might have looked like within the mid-century playscape and echo the assumption that the

Give-A-Show built upon a tradition of live mixed-media performance—a genealogy that might be understood as distinct from television (even as television adapted these same formats for broadcast). Early versions of Kenner's projectors embody this expanded conception of children's "shows". For example, some sets included blank "draw & wipe off frames" so that kids could make their own slides. Similarly, Kenner filed a patent for a combination transparent and opaque projector in 1962, which would be marketed under the name Super Show Projector (Kuhn, 1962). The Super Show came packaged with jointed character puppets in addition to slides, and the text on the box encouraged kids to project a range of materials, such as drawings, photographs, and three-dimensional objects. These accessories and related products link Kenner's toys with 19th-century home lantern shows and early domestic cinema programs that often included both commercially produced and amateur-produced content (Chalke, 2007: 226). Sheila Chalke (2007) argues that the "ephemeral soundtrack" that likely accompanied early home film projection in the form of verbal narration, song, and familial exchange helped render cinema a suitable medium for middle-class family consumption and would serve "as a prelude to televisual experience" (pp. 224, 228). Similarly, Kenner's production and distribution of blank slides, and the release of the Super Show Projector, demonstrate initial interests to broaden the child's play experience to include a range of projection opportunities. However, these possibilities would soon be curtailed as the Give-A-Show's development took shape more clearly around a narrower imperative to project licensed content.

Evidence for this emerges not only in Kenner's production history, such as the development of subsequent "TV-like" toys (discussed below), but also in the scarcity of the Super Show in the vintage and collector's markets, given that it was only released once. Tim Hollis argues that the Super Show's range of accessories and possibilities may have proven "too elaborate" and that children favored the simplicity of the original projector format (Hollis, 2015: 142). Instead, the projector's increased emphasis on licensed content and subsequent versions of "-Show" toys would emphasize children's interaction and play with TV characters over the production and exhibition of original content.

Although the Give-A-Show would prove to be Kenner's most enduring and iconic media toy, the company developed a line of related visual and audio media toys over the course of the 1960s and 1970s. This expanded "-Show" product line (many but not all toys featured the suffix "-Show") would include toys that variously modified and replaced some of the Give-A-Show's core features. Taken as a broader category, these other projectors, such as the Super Show, provide significant insight into how Kenner imagined kids' media play, responded to a rapidly changing mediascape, and envisioned combining audio, still pictures, and moving images in various ways. From the early 1960s, Kenner envisioned products that would add sound to slide shows, such as the Adventure Time Give-A-Show projector, which included sixteen slide shows and two 45 rpm records that offered "Exciting narration!", "Realistic Sound Effects!", and "Music!" Audio cues in the phonograph tracks indicated to children when to advance the slides.[2] The Adventure Time records thus implored children to "hear a show" rather than to give it, instead concentrating their energy on synchronizing the slide changes with the advancing audio track.

In addition to various means of incorporating sound, Kenner also introduced a toy movie projector under the name "Easy-Show" around 1966. The Easy-Show projected short silent films housed in round plastic cartridges and was available in both manually operated and motorized versions. Commercials for the Easy-Show highlighted the projector's "toy-like" qualities—its ability to play cartridges forward, backward, and in "stop-action". Kids watching the films on the Easy-Show were meant to interrupt, arrest, and reverse the reel, suggesting the importance not only of the film narrative itself, but the child's ability to actively manipulate the film in play. In contrast to television or commercial cinema viewing, the Easy-Show afforded its young user a larger degree of control, blending spectatorship with play. In this respect, it might almost be understood as a precursor to other media platforms for kids that allow nonlinear "remixing" possibilities. Kenner's other media toys sustained still other possibilities, particularly reshaping the play experience to more closely resemble TV viewing rather than performing a live show for an audience.

Kenner's Change-A-Channel TV Set, also introduced in the mid-1960s, is exemplary of this shift toward watching TV *as* children's play. The Change-A-Channel was a plastic toy TV set that took "program cartridges" loaded with 16mm film. Each frame of the film was divided in half with two separate cartoons running the length of the strip. A button on the TV allowed children to "change channels" by toggling between these two simultaneously running cartoons. Billed to children as "your own TV set!", the Change-A-Channel's marketing appeal was built upon some of the same features for which the Give-A-Show was known; namely, independent access to television content, as the box said: "see your favorite shows whenever you want—over and over". Kenner's later Play N' Show (Bernard, 1969) is yet another example of this trajectory, its orientation toward TV reflected in its design, likely developed in response to General Electric's Show'N Tell Phonoviewer with Picturesound program discs, introduced around 1964. GE's toy looked like a television set with a phonograph on top of it, which played still-picture slide shows in sync with the records. It also functioned as a standalone record player. Kenner's Play N' Show media format combined audio and still images arranged around a circular record, automating the synchronization of sound and image in contrast to its earlier Give-A-Show's Adventure Time edition, which required manual synchronization. Rather than projecting onto an external (or attached) screen surface, the Play N' Show has a closed body that resembles a TV or even anticipates the design of early home computer consoles, such as the Apple II, and projects its image onto the screen from behind. Given this shift in design that built upon a related product—Kenner's Close 'N Play phonograph, the Play N' Show commercial's insistence that it is *not* a TV is curious. Interestingly, while the words "play" and "show" both function as verbs in the toy's name, the toy itself in fact moves away from practices where children are actively *showing* something, and instead, content is *shown to* them.

The number of other projectors on the market during the Give-A-Show's two-decade heyday also attest to the toy's success. Kenner's projector was preceded on the market by countless film strip and moving-image toys. The British J & L Randall's Merit Ace and SEL Ace, for instance, were sold with slide sets of notable comic characters such as Dan Dare, Archie Andrews, Billy Bunter, and Tiger Tim. Other slide projectors sold under regionally variable names such as the Fortuna, the "Show-A-Show" and the Spectrum, as well as Disney's Magna-slide projector. Kenner also licensed the Give-A-Show internationally, such as to UK toymaker Chad Valley. Later, just as Kenner had attempted to rival View-Master with the introduction of its "See-A-Show" stereo viewer, so too did View-Master venture into projector territory with its Double-View automatic movie viewer. The Louis Marx Toy Company's Flashy Flickers Magic Picture Gun from the mid-1960s was an interesting variation that combined the functionality of a projector with the design of a space-age ray gun. Molded in aqua colored plastic, the Flashy Flickers Magic Picture Gun projected a series of still images from its barrel one at a time when the trigger was pulled, interestingly harkening back to the association between firearms and early photographic technology as imagined by early practitioners such as Étienne-Jules Marey and his photographic gun (1882). The Flashy Flickers' novelty also echoed the portability and everywhereness of the Give-A-Show (the box, for example, suggested projections on walls, ceilings, and the surface of a friend's T-shirt) but also interestingly suggested the suitability of the toy for projecting to a large group. Like the other projectors, many comics subjects were popular for the Flashy Flickers, such as Prince Valiant, Blondie, Hi and Lois, and the Phantom, some well-suited to the gun's design, other titles more incongruous.

Also designed for playback rather than production, Fisher-Price introduced its Movie Viewer in 1973, which was shaped similarly to a camera, grasped by a handle at the bottom, and required the child to peer through a viewfinder while turning a crank. Like Louis Marx's Flashy Flickers gun, which projected while "shooting", the design of Fisher-Price's viewer seemed to blur the distinction between production and viewing, inviting the child to adopt the position of capturing images even as they were watching rather than recording. Despite its camera-like design, the FP Movie Viewer thus forms a curious link with earlier "peep media" like the early mutoscope and kinetoscope, although these toys shed the negative connotations associated with the illicit forms of looking associated with many "peep" devices (Huhtamo, 2006). Kenner followed with a very similar Movie Viewer a few years later that also took film cartridges housed in plastic cases. The shift from toys like the Give-A-Show, which were designed to facilitate play practices wherein the child performed as the show person, to toys like the Change-A-Channel TV Set, the Play N' Show, and the Movie Viewer designed to position children's play as principally spectatorial, echoes the changing design of optical toys a century earlier. André Gaudreault and Nicolas Dulac characterize early pre-cinema devices such as the phenakistoscope (1832) and zoetrope (1833) as placing the user in the position of a "player" who actively manipulates the devices, in contrast to later toys, such as Emile Reynaud's Praxinoscope Theatre (1878), which more fully separated the user from the mechanism of operation, rendering them more of a "viewer" (2006: 234).

Fading Pictures: The Give-A-Show's Decline

By the late 1970s, shifts in both the toy and media industries presented new opportunities for kids to engage with their favorite characters and storylines that would contribute to the eventual decline of the Give-A-Show projector. The importance of the Give-A-Show's unique role delivering colorful "on demand" entertainment would decrease as newer media formats such as VHS videotape became more widely available. Likewise, when the sales of color television receivers began outnumbering the sales of black and white televisions by the early 1970s, the Give-A-Show's vibrant color slides became less of a novelty. Finally, the increased number of households that had one or more television sets alleviated the need for children to stage shows independently because of the demand placed on the set.

According to a report issued by the U.S. Department of Commerce and Bureau of the Census, by January of 1969, 29.4% of U.S. households had more than one television set (U.S. Department of Commerce and Bureau of the Census, 1969). A 1971 poll conducted by Louis Harris and Associates for *Life* magazine found that "nearly half" of U.S. households had two or more television sets and "even among families with incomes of less than $5,000 a year, one in four has a color set" ("Do We Like What We Watch?" 1971: 43). The introduction of second (and even third) sets into households would increase the likelihood that children would be able to watch the programming of their choice without having to contend with competing interests of the adults in their homes. When the VHS format was introduced in the late 1970s, children represented an audience with tremendous potential for development. Although early VCRs were very expensive, entertainment companies were swift to capitalize on the children's home video market. Disney's Home Entertainment Division released the company's first series of ten feature films on VHS in 1980, beginning with *Pete's Dragon* (1977). Hanna-Barbera, which had licensed its character library to Kenner for two decades, would make its cartoons available on VHS from the early 1980s via distributors such as Worldvision Home Video, Goodtimes Home Video, and Kids Klassics (VHS Collector, 2017). The accessibility of children's content on VHS allowed children to watch independently of the time constraints of broadcast television, and although children could rewind and fast forward VHS cassettes at will in order to excerpt, replay, or omit particular scenes, the linear format of the VHS and the continuity of children's programming uninterrupted by commercials likely also encouraged children to watch shows straight through. All of these factors reduced the necessity for children to engage with a standalone toy version of a television or a separate toy like the Give-A-Show that adapted or approximated televisual experience. Instead, kids could watch on a family TV set or a dedicated second or third TV in the home.

In addition to the VCR extending the scope of possibilities for the television set, the introduction of video games presented yet another mode of screen-based interaction that would compete (at least indirectly) with the Give-A-Show. Early game consoles such as the Magnavox Odyssey (1972), Atari's Home PONG Console (1975), and the Atari VCS 2600 (1977) were often framed as family entertainments, some sold with analog components connecting them to board games—a distinct shift from the youth-oriented (and often contested) spaces of the public arcade (Kocurek, 2015). In the wake of these changes, the children's toys and entertainment markets were becoming increasingly segmented by age, and Kenner began targeting ever-younger consumers as the core users of the Give-A-Show. Packaging on later editions of the projector depicted very young children (ages three and up) operating the device and, at times, it was even released under Kenner's line of "Discovery Time" educational toys (1978) and finally under the Kenner Preschool label.

As previously noted, Kenner's various models of "-Show" toys had increasingly moved away from privileging the child's role as live show performer and toward toys that facilitated activities that more closely resembled film and television viewing. This shift toward inviting children to inhabit the position of spectator rather than of show person and performer, should not necessarily be read as an exchange of active, engaged play for "passive" consumption (a dichotomy that has long structured popular discourses of children's media use) (Kinder, 1999). On the contrary, alliances between the film and television industries and toymakers during the 1970s created new opportunities for kids to engage with their favorite characters beyond the screen. Many of these opportunities emerged from contentious discussion over children's advertising on television. Pressure to eliminate children's television advertising by groups such as Action for Children's Television (ACT), founded in 1968, were taken up by the Federal Trade Commission (FTC), but the toy industry successfully lobbied for the continued presence of such commercials and Congress ultimately overruled the FTC ban (Cross, 2009: 185–186). This would pave the way for the program-length commercials (PLCs) of the 1980s; 30-minute cartoons designed to tie directly into toy lines. Though much scholarship discussing this turn laments the loss of a nostalgic playscape connected to intergenerational values replaced by the incoherent, shallow storylines and plastic junk toys associated with PLCs, such sweeping condemnations potentially discount the meanings and values that these toys might have held for their young users (Cross, 2009: 188–227). Kenner, which had actively advertised on television since the late 1950s, capitalized on the overruled FTC ban, and in the 1970s and 1980s, much of its production focus shifted to toys with TV show tie-ins, such as *The Care Bears* and *Strawberry Shortcake*.

Of course, one of the most influential developments within the toy and film industries writ large during the late 1970s was Kenner's acquisition of the license for the *Star Wars* action figures and related merchandise. The popularity of Kenner's 3¾-inch *Star Wars* action figures (a scale that would become the industry standard) dramatically reshaped the relationship between movies and toys. In 1975, after nearly forty years in business, Kenner had reached $100 million in sales. Only three years later (and one year after the first *Star Wars* film was released), that figure had doubled—an unprecedented success attributed to the toy line (Kenner Collector). Kenner's emphasis on the *Star Wars* toys, as well as shifts in television that replaced many of the Give-A-Show's functions, are likely closely related to the projector's decline, although the company's move to figure-based fantasy play was reflected even in products that preceded the *Star Wars* toys, as well. Around 1975, when Kenner introduced its cartridge-based Movie Viewer, it also produced two toys: The Six Million Dollar Man Bionic Video Center and the Snoopy Drive-in Theater. These toys both worked with the same cartridges used in the Movie Viewer, but projected the films onto small screens that could accommodate an audience of a couple of people. Both of these toys, however, were also designed as movie theater toys *for* their respective characters. The Snoopy Drive-In came with a small Snoopy figure in a car, which the child would "drive up" to the screen to show Snoopy the film. The Six Million Dollar Man doll was sold separately from the Video Center, but the Center set came with a doll-sized chair and the box prominently displayed the doll seated in front of the show. Both toys are relatively scarce in collectors' markets today, but they serve as fascinating links between Kenner's investment in media-related toys and dolls and action figures, capturing the moment when the former yielded to the latter. These toys followed the release of Give-A-Show editions of *The Six Million Dollar Man* (1973) and *The Bionic Woman* (1974) and thus represent a transition from the projector to the new emphasis on action figures. By the early 1980s, the Give-A-Show no longer enjoyed the prominence and success that it once did.

Located at the interstices of technology, audiovisual media format, and toy, the Give-A-Show has actually had an influence on a wide range of subsequent toys and technologies. In many respects, the Give-A-Show occupies a unique position as both a media toy that let kids curate content and stage their own performances, but also as a unique delivery mechanism for licensed content—a subset of other licensed merchandise like toys, games, and housewares—that ensured the persistent presence of cartoon characters outside of their televisual appearances. Its influence is thus manifold, connected to contemporary traditions of both media spectatorship more broadly and children's play specifically.

In many ways, the range of products in Kenner's "-Show" line of media toys anticipated the omnipresence of screens and entertainment associated with contemporary children's media culture. The Give-A-Show's portability allowed it to be set up and operated in any dark space, while subsequent models with features such as rear-lit projection surfaces enabled children to watch in a wider variety of lighting conditions. One example is Kenner's "Screen-a-Show" projector, introduced in the early 1970s. Like the Give-A-Show, the Screen-a-Show projected slide shows of still pictures, but its slides were housed in a cassette with a roller apparatus to advance the sequence, and included a small detachable screen onto which the images could be projected. The Screen-a-Show was advertised as "the amazing take-it-anyplace projector". Commercials featured kids watching in predictable locations such as around the kitchen table and in a brightly lit living room, but also in the back of the family station wagon. By contemporary standards, the Screen-a-Show appears unwieldy. It was designed to be set up on a flat surface, thereby maintaining some formal characteristics of traditional projection and exhibition (in contrast to later handheld devices). Nevertheless, its billing as portable and its usability in a wider range of lighting conditions reflect moves toward prioritizing media consumption not only at home, but also on the go. While the original Give-A-Show liberated kids' content from the temporal restrictions of the broadcast television schedule, products like the Screen-a-Show entertained the possibility that kids might enjoy their favorite shows without spatial restrictions, as well.

As discussed, Kenner's media toys would increasingly move in the direction of media viewing rather than focusing on the child's creative or authorial role in staging and narrating the show. However, the original Give-A-Show's reliance upon the ingenuity and activity of the young show person might also be seen borne out in subsequent audio and video recording toys. In 1987, over a decade after Fisher-Price introduced its Movie Viewer, the company released its PXL-2000 PixelVision camera. Molded in gray plastic, the PixelVision more closely resembled camcorders aimed at the adult market and used a standard audiocassette as its recording medium. Its adoption by experimental filmmakers (particularly the work of Sadie Benning) elevated it to the status of not "just" a toy (a familiar pejorative refrain) but also a device that facilitated more sophisticated aesthetic and thematic experimentation (Milliken, 2002). The introduction of digital recording formats in the 1990s flooded the market with children's versions of photo and video recording technologies, as well as a substantial market of audio recording devices. In contrast to the Severns' 1956 book that invites children to engage in TV- and movie-*themed* pretend play, these technologies would turn those possibilities for media production into a reality.

More broadly, the rise and fall of the Give-A-Show provides a powerful example for considering the assumptions that the designers of "child-friendly" media formats and audiovisual toys make about children and their play. As Hollis points out, children's cartoons were available in 8mm and Super 8 formats simultaneously with the distribution of such content for toy projectors (2015: 147). Yet media for children tend to be designed for durability and to be "fool-proof". The Easy-Show Projector's self-contained cartridges were designed to obviate the need for the child to thread the film in a complicated manner, though many collectors remark on the difficulty of loading the cartridges and their tendency to tangle. Fisher-Price's and Kenner's cartridges were housed in even more impenetrable plastic cases, which worked better, but prevented children from taking a closer look at how they worked. Similar preoccupations continue to inform the design of children's tech toys, such as dedicated tablets and mobile devices. Such concerns permeate discussion of both child-friendly hardware and software. Ongoing concerns about durability have resulted in electronic devices being embedded in plush animals or encased by rubber bumpers, while decisions in software and interface design aim to minimize errors and make interactions as easy as possible. Like children's screen-based media of the 19th century and the host of children's projectors in the 20th century, children's devices today still occupy strange territory, both "real" devices with a range of functions, yet also "toy" versions, with limited functionality and constraints in place to address not only the specter of operator error, but now fears over children's safety associated with Internet connectivity. These tensions surface in both product design and software decisions, such as the development of proprietary operating systems, in the case of LeapPad products, or the use of existing platforms, such as Android OS.

Trade-offs, Transformations, and Continuing Legacy

There are both gains and losses associated with the gradual changes in the toy and entertainment industries that rendered the Give-A-Show obsolete. As indicated above, the peculiarity of dedicated children's media formats and toys reflects a series of assumptions about children's capabilities operating and troubleshooting media independently of adult oversight. As children shifted from viewing cartoons via the Give-A-Show, then, they more fully reclaimed and engaged with "adult-grade" audiovisual media, namely the television set and VCR. On one hand, this might be interpreted positively: no longer relegated to play experiences that simulated and adapted television viewing, by the 1970s and 1980s, it was clear that children represented an enormous market and audience, and their increased television viewership might read as a kind of ownership of the technology.

Conversely, the notion that a group becomes invested with power or agency primarily because of their status as a financially viable market is problematic. As a plaything so closely connected to early children's television practices, the Give-A-Show was also a central object through which conceptions of children's spectatorship could be nuanced beyond conventional considerations of the content they saw on TV. For instance, disparaging the commercial imperative that shaped early children's television, Stephen Kline (1995) laments that "children would more or less watch anything that moved sufficiently." Without a financial incentive to produce quality programming (because ads for kids' shows sold for less) producers of children's content—and Kline cites Hanna-Barbera as exemplary of this practice—cut corners in production and recycled themes and plots (p. 125). Like the television shows to which they were connected, the Give-A-Show slides did not offer sophisticated or socially engaged storylines, often reused themes, and exhibited few traditional aesthetic merits. Nevertheless, the popularity and endurance of the Give-A-Show provides a powerful indication of the importance and enjoyment of engaging with television content and characters. Such products provide valuable evidence of the meanings this content might have held for children and provide material touchpoints for researching such issues. Moreover, just as the Give-A-Show's broad spectrum of projecting possibilities eventually narrowed to focus primarily on licensed content, the Give-A-Show's demise heralded the demise of its own distinct slide format, which represented unique extensions of the storylines and characters that were of potential importance to children. The various ways that audio and video were integrated in each of Kenner's media toys, in turn, shaped the way that the stories were told. 3D formats privileged the importance of depth, while film reels could capitalize on motion for humor. While the limited animation style of Hanna-Barbera favored simple actions and especially verbal humor to save money on animation costs, the silent reels of the Easy-Show and Change-A-Channel instead relied on visual humor in static images (Kline, 1995: 126; Hollis, 2015: 145).

Finally, the demise of the Give-A-Show also represented the loss of a "magical" ephemeral viewing environment associated with the impromptu and flexible nature of the original Give-A-Show projector. Just as Ingmar Bergman (1988) remarked upon his first experiences with a toy projector, so too did children playing with the Give-A-Show have an opportunity to darken a room, narrate a storyline, and behold a luminous image on a wall or other makeshift screen. It is hard to measure the importance of such an affective experience, but as children's television viewing replaced the Give-A-Show, it is one practice that became less common or was lost altogether. Of course, subsequent playthings would attempt to cultivate similar atmospheres, for example, by projecting stars or other calming night scenes. However, just as the Give-A-Show experienced its own unique development as a product, so too do contemporary playthings take different shapes in relation to the technological and ideological contexts where they are adopted. One example is the Disney Storytime Theater Projector, currently on the market. The projector works with a companion mobile app. Children collect plastic "Press N' Play" character pieces with embedded RFID chips, which "unlock" story content that is then sent to the projector unit wirelessly. Disney's projector, produced by Tech 4 Kids, aims to be integrated not into the child's established TV viewing or gaming activities, but as a revision of the traditional bedtime story routine.

The projector's historical and cultural significance is also indicated by its inclusion in a range of museum collections. The Give-A-Show and/or its accompanying media can be found in such diverse museum collections as the Science Museum in London, the Bill Douglas Cinema Museum in Exeter, The Please Touch Museum in Philadelphia, and the Strong Museum of Play, in Rochester, to name just a few. Moreover, Kenner co-founder Albert Steiner was inducted into the Toy Industry Association's Toy Hall of Fame in 1993, reflecting the overall significance of the company to the 20th-century toy industry. However, the Give-A-Show's endurance is most saliently visible in private collecting communities, rather than within institutional collections. As a tremendously popular plaything from the Baby-Boomer era, virtually all models of the projector and accompanying media are extant and many are prevalent in the core contexts where vintage toys and related collectors' items are sold, such as flea markets and on-line exchanges such as eBay, for reasonably low prices. The relative prosperity of Baby-Boomer era consumers, and the affective nostalgia associated with an event-based plaything like the Give-A-Show, ensure its popularity.

As is frequently the case with such items, it is the collectors' community rather than institutional records or cultural repositories that offer the most extensive catalogs and organized information about Kenner's media toys. The confluence of both monetary and nostalgic values drives a very close focus on details that aids in scholarly research. Indeed, resources such as Jon Knutson's blog and others were instrumental in the completion of this essay. Despite the often meticulous recordkeeping of collectors, it can still be challenging to answer definitive questions about the Give-A-Show's production, distribution, and eventual decline. Manufacturing processes change, international variants circulate, and Kenner's own internal means for cataloging and sequencing slides within their overall library were inconsistent. Like many "obsolete" media forms, communities of collectors sustain interest in and discussion surrounding the Give-A-Show through on-line activities, and in particular, digitizing images of apparatus, packaging, and in many cases, the slide shows and filmstrips themselves. Many of the original slide programs are not only cataloged on-line, but have been recorded live, hosted on collectors' and hobbyists' blogs and on general sites like YouTube. The largest contributor to this on-line archive is Knutson, though many others participate, including those with crossover interests, such as *Star Wars* fans. Many recorded slide shows have several hundred views, while amateur-produced videos that show the projector apparatus might have several thousand, such as one video of the Easy-Show Projector in operation, which at the time of writing, has over sixteen thousand views, attesting to the ongoing interest in these toys and their media (Preston, 2009).

Conclusion

The Give-A-Show projector's introduction in 1959 situates it squarely within a much broader tradition of home media practices for both families generally and children more specifically. Energized by nearly a century of home lantern shows staged by parents and later by children and their parents and by innumerable toy film projectors for children, Kenner's toy did not represent a "new" media object, nor did it encourage particularly new forms of play. However, what the Give-A-Show *did* offer children was a way to experience TV content independently of network schedules or other demands placed on the family television set. It invited children to reclaim domestic space by rendering blank surfaces spaces for projection. Although Kenner initially imagined possibilities for its Give-A-Show that extended beyond the realm of licensed content, its relationship with television would ultimately become a dominant focus and would sustain the toy for many years. Moreover, Kenner developed a full line of toys that explored distinct media formats and hardware design in an effort to continue imagining and shaping kids' spectatorial practices.

"Toy" versions of media technologies like the Give-A-Show represent a particularly interesting category for study. Their lasting power and the threat of obsolescence is more closely tied to the finite temporality of childhood, as children are often expected to "outgrow" their toys more quickly than more expensive media devices that are considered more family investments. Toys and the proprietary media formats designed for them are especially unstable and make unique assumptions about how their young users will consume and play with them, which are manifest in their design and operation. Also, the affective charge of these toys through their association with childhood is a key factor in their continued visibility within collecting communities, which offer a rich archive of materials to researchers and other interested parties that would likely fall outside the purview of many institutional imperatives. Indeed, the active role that collectors and some museums play in digitizing media from toys such as the Give-A-Show, extensively cataloging and organizing materials, and engaging in sales and trade, challenge the very idea of obsolescence itself. The Give-A-Show reverberates not only within contemporary children's media culture in the form of new digital media and tech toys, but also with those who, through the logics of contemporary new media, distribute the shows to large audiences of unknown sizes, thereby continuing to fulfill the promise of the original toy.

Acknowledgments

Many thanks to my Research Assistant Ryan Bunch for his research support during the preparation of this chapter.

Notes

1 Early boxes, such as one from 1961, noted "TV and all-time favorites", but by the mid-1960s, the phrase had been reduced to "TV favorites", reflecting the increased dominance of licensed content.

2 Most sources, including Jon Knutson's website *Random Acts of Geekery*, date the Adventure Time edition to around 1962; see http://randomactsofgeekery.blogspot.com/2008/06/give-show-projector-list-update.html.

References

Albéra, François; and Maria Tortajada (2010), *Cinema Beyond Film: Media Epistemology in the Modern Era*, Amsterdam: Amsterdam University Press.

Bak, Meredith A. (2015), "'Ten Dollars' Worth of Fun': The Obscured History of the Toy Magic Lantern and Early Children's Media Spectatorship", *Film History* 27 (1): pages 111–134.

Basano, Roberta (1996), "The Magic Lantern as a Toy", *New Magic Lantern Journal* 8: pages 8–10.

Bergman, Ingmar (1988), *The Magic Lantern: An Autobiography*, trans. Joan Tate, New York: Viking.

Bernard, Terry L. (1969), US Patent. USD218713S, Combined Phonograph and Slide Projector.

Carlisle, Rodney P. (2009), *Encyclopedia of Play in Today's Society*, Thousand Oaks, CA: SAGE.

Chalke, Sheila (2007), "Early Home Cinema: The Origins of Alternative Spectatorship", *Convergence: The International Journal of Research into New Media Technologies* 13 (3): pages 223–230. doi:10.1177/1354856507079174.

Cross, Gary (2009), *Kids' Stuff: Toys and the Changing World of American Childhood*, Cambridge, MA: Harvard University Press.

"Do We Like What We Watch?" *Life Magazine*, September 10, 1971, Vol. 71, No. 11, pages 40–44.

Gaudreault, Andre; and Nicolas Dulac (2006), "Circularity and Repetition at the Heart of the Attraction: Optical Toys and the Emergence of a New Cultural Series", in Wanda Strauven, editor, *Cinema of Attractions Reloaded*, Amsterdam: Amsterdam University Press, pages 227–244, 234.

Hirsch, Eric; and Roger Silverstone (1992), *Consuming Technologies: Media and Information in Domestic Spaces*, New York, NY: Routledge.

Hollis, Tim (2015), *Toons in Toyland: The Story of Cartoon Character Merchandise*, Jackson, MS: University Press of Mississippi.

Huhtamo, Erkki (2006), "The Pleasures of the Peephole: An Archaeological Exploration of Peep Media", in Eric Kluitenberg, editor, *Book of Imaginary Media*, Amsterdam and Rotterdam: Debalie and NAi Publishers, pages 75–141.

Kenner Collector, "Kenner History", available at www.kennercollector.com/kenner-history/ (accessed January 21, 2017).

Kinder, Marsha (1999), *Kids' Media Culture*, Durham, NC: Duke University Press.

Kline, Stephen (1995), *Out of the Garden: Toys, TV, and Children's Culture in the Age of Marketing*, New York, NY: Verso.

Knutson, Jon (2008), "Give-A-Show Projector List Update!" *Random Acts of Geekery*, available at http://randomactsofgeekery.blogspot.com/2008/06/give-show-projector-list-update.html (accessed November 10, 2016).

Kocurek, Carly A. (2015), *Coin-Operated Americans: Rebooting Boyhood at the Video Game Arcade*, Minneapolis, MN: University of Minnesota Press.

Kuhn, James O. (1961), US Patent US3100420A, Toy Slide Projector.

Kuhn, James O. (1962), US Patent USD195606S, Combined Transparent and Opaque Projector.

Luckett, Moya (1995), "'Filming the Family': Home Movie Systems and the Domestication of Spectatorship", *Velvet Light Trap*, No. 11, pages 21–32.

Milliken, Christie (2002), "The Pixel Visions of Sadie Benning", in Frances K. Gateward and Murray Pomerance, editors, *Sugar, Spice, and Everything Nice: Cinemas of Girlhood*, Detroit, MI: Wayne State University Press, pages 285–302.

Museu del cinema (2009), *Toy Film Projectors/Projectors de Cinema Infantil. Museu Del Cinema*, available at www.youtube.com/watch?v=0X1GvH6MXXU (accessed August 16, 2018).

Preston, Kevin (ToyKingWonder) (2009), *Kenner Easy-Show Projector*, available at www.youtube.com/watch?v=Egrpr38Ca4w (accessed August 16, 2018).

Schneider, Alexandra (2008), "Time Travel with Pathé Baby: The Small-Gauge Film Collection as Historical Archive", *Film History: An International Journal* 19 (4): pages 353–360.

Severn, Bill; and Sue Severn (1956), *Let's Give a Show*, New York, NY: Knopf.

Singer, Ben (1988), "Early Home Cinema and the Edison Home Projecting Kinetoscope", *Film History*, 2 (1): pages 37–69.

U.S. Department of Commerce and Bureau of the Census (1969), "Households with Television Sets in the United States", Series H-121, Volume 3.

VHS Collector, "VHS/Beta Releases from Worldvision Home Video", available at http://vhscollector.com/distributor-gallery/579 (accessed January 12, 2017).

Wasson, Haidee (2009), "Electric Homes! Automatic Movies! Efficient Entertainment!: 16mm and Cinema's Domestication in the 1920s", *Cinema Journal* 48 (4): pages 1–21.

Wells, Kentwood (2010), "Magic Lanterns: Christmas Toys for Boys", *Magic Lantern Gazette* 22, No. 1 (Spring 2010): pages 10–14.

19

PARAKEETS, MORSE CODE, THE ROAR OF THE CROWD

The Fading Signal of the Modem

Anne C. Deger

Modems have been in use since the 1920s. They are still in use today. This chapter begins with a brief historical overview of modems and their development, from business and military applications to their eventual entrance into the home consumer market. A full discussion of the modem's history in the home would span decades, exceeding the space available here, so this essay will limit itself primarily to the period of time where dial-up modems, along with home computers, transitioned from esoteric hobbyist equipment to ubiquitous consumer device: the 1990s. Each generation of modem users has its own set of memories, its own unique nostalgia, because each generation of modem users has worked under a different set of technological affordances and constraints. This essay examines the ways social spaces of the Internet shifted as larger populations of users became able to access them, and the ways in which that ease of access permanently altered those spaces. As modem technology continued to develop—as broadband access displaced dial-up as the norm, and was itself augmented by cellular data—these spaces shifted again. Eras of modem nostalgia are demarcated by these shifts in technology. The modem has not yet become obsolete, but certain types of modem have obsolesced. This essay explores the habits and practices encouraged by the now-obsolete modems of the 1990s, the 14.4k and 56k dial-up modems which enabled and defined the "Information Superhighway" era of the Internet.

The term "modem" is shorthand for "Modulator-Demodulator" and can refer to a variety of wildly different technologies—anything from a 1920s Teletype machine to a DOCSIS 3.1 standard cable modem able to provide download speeds of 10 gigabits per second. Modems, essentially, allow computers to "talk" to each other, to transmit data: modems take up a periodic waveform (called a "carrier signal" since it serves as a medium for the message being sent) and *modulate* that waveform by acting upon one or more of its properties. Modulation, then, layers a message—digital or analog information—into a carrier signal, one that can be physically transmitted (in this case, sent through a standard telephone line). Modulation can be thought of as "sending mode" for the modem. In "receiving mode", the modem *demodulates*—it extracts the data that has been tucked inside the modulated carrier signal. This technique was initially developed for wireless transmissions—indeed, our terms for commercial radio signals, "AM" and "FM", both refer to this practice, standing respectively for "amplitude modulation" and "frequency modulation".

The ability to reliably and undetectably send data became increasingly desirable during the Second World War, as radar installations came online. The intelligence these installations gathered needed to be sent to commanders to become actionable. It needed to be sent some distance, and it needed to be sent quickly, but wireless transmission held the risk of interception. Early Teletype machines were designed to send written messages. Making use of modulation and demodulation, they could transmit data at a rate of approximately 150 bits per second (bps). A single letter consists of 8 bits, meaning that these machines could send approximately 15 letters per second. In the late 1940s, the United States Air Force adapted Teletype technology for use over the existing national telephone system, allowing radar installations to send imaging data to USAF command centers without installing additional infrastructure. The digital images captured by radar were converted to analog for transmission over copper wire, and converted back to digital images upon arriving at their destination. Using the existing technology of telephone lines as their channels, modems could securely transmit this radar data. The SAGE Air Defense System made use of the first devices referred to by the name of "modem" in 1958: Bell Laboratories' Bell 101 modems, operating at 110bps. These modems were made commercially available in 1959, and transmitted data over ordinary phone lines. The ramping up of the Cold War led the United States military to further pursue the technology, creating the network of military and research sites that would eventually become the Internet.

In 1962, AT&T released the first commercial modem, the Bell 103, which, as a 300bps modem, doubled these transmission rates. Businesses and researchers were the primary purchasers for these devices, given the expense involved and the skill required to use the devices. Through the 1960s and into the 1970s, as bulletin board systems (BBSs) became popular, users needed higher speeds of data transfer to share files and images with each other. Bell developed a 1200bps modem, and manufacturers continued to experiment with protocols, in an effort to send data more efficiently, more quickly, and with fewer errors. The need for speed was counterbalanced with an effort to make modems easier to use, particularly for novices; as computers entered the home market, new users with limited prior experience joined those with computer science degrees in welcoming these strange 8-bit beasts into their homes. In 1977, Dennis Hayes and Dale Heatherington introduced the 80-103A, a modem that connected directly to the user's telephone. Users no longer had to dial the number themselves, or make use of a separate dialer; gone was the handset-and-cradle setup. Their success led them to found the company that would become Hayes Microcomputer Products, and this company would introduce a device that changed home computing: the Smartmodem.

The Smartmodem was not particularly fast—it only transmitted 300bps. Manufacturers had already designed more powerful internal modems, which could be placed directly inside a computer, but these were expensive and fiddly to install and to work with. Heatherington worked out a way to operate the modem in two modes: data mode (where the computer forwarded data for modulation, and that modulated data was sent over the phone line as usual) and command mode (where the computer forwarded data to the modem, which the modem received and took as its operating instructions). Command mode allowed the computer to "talk" to the modem. The Smartmodem would start in command mode and take its orders from the computer, handling the tasks of dialing and hanging up on its own. Living up to its moniker, the Smartmodem could even discern the speed of the computer's serial port. As the first (and, at that time, only) "universal" modem on the market, the Smartmodem fueled its own market demand, as it allowed hobbyists to easily create BBSs. Many of the habits and practices users developed in these highly social spaces came to define interactions on the Internet.

14.4kbps (the kb is for "kilobits"), 28.8kbps and 56kbps modems became available to consumers in 1991, 1994, and 1998, respectively; these were generally internal modems, not standalone units like the Smartmodem. They were hidden within the computer, and often came factory-installed. Consumers did not necessarily select their modems when making their initial purchases, but different brands and models of modem did have different idiosyncrasies, and connection failures required users to become intimately familiar with their personal modem's peculiarities. When the connection process went awry, the user needed to sift through multiple drop-down menus in an effort to find a fix specific to their modem's configuration. Though invisible, the modem always made itself known to the user when dialing up, singing and signaling its presence. Users did not necessarily know how to "speak" the modem's language, and did not necessarily understand the modulation and demodulation process. But users did have to actively participate to initiate the connection, and could always hear that *something* was happening—could hear the modem calling out into the world and seeking its answer.

As always-on cable modems and wireless routers have replaced dial-up modems for home users, however, modems have become silent and invisible. Broadband speeds in the U.S. averaged 11.4 megabits per second in 2015—this is 200 times faster than a 56kbps modem. The connection is constant, with no pause for dialing in. Home users can still purchase their own cable modems, selecting the specifications that suit them best, but many users rent modems directly from their internet service provider (ISP). These devices are usually combination modem-routers, and their users are unlikely to compare technical specifications, or to think much of their presence at all, unless something goes wrong. Consult your own experience: when did you last think consciously of the modem providing your connection while using the Internet? Generally, users notice their presence only in their absence: during a service outage or when a piece of equipment has failed. We only notice the modem when we are suddenly disconnected and, discomfited, attempt to reset it or contact our ISP for guidance. But it was not always so.

"SCREEEEEEEEEEE AAAHHHHHHHHH EEEEEEEEEEE CCCCHHHH HHHHHH CCH CH BERRRRR EEEEEEEEEEEEEEEEEEEEEEEP"

"Beep-beep-beep. Chcck. Eeeerrrhhrr. Bhrrrh. Hccccchhh-ZZzzZZzzz."

"Pshhhkkkkkrrrrakingkakingkakingtshchchchchchchchcch*ding*ding* ding*"

"SS S S K K K K K K R R E E E E E E E E E E E E E E E E E E E E E E didddllllidddddlllllllidddlllllllllllllllleeeeee SSSSKKKKREEEEEEE"

So many habits, practices, feelings, memories are figured under the metonym of the modem. The dial-up noise (perhaps *because* it is a noise?) is not just a thing but a time and a way of life: a world-historical moment, a personal moment, a set of memories, a reminiscence, a longing. It is childhood, coming-of-age, first love, safety, familiarity, novelty, discovery. It is something different for each user, but for every user it exists only in the past tense. It is irretrievable.

Each generation cultivates its own peculiar form of nostalgia; each generation has its own peculiar totemic objects which serve as synecdoche for that generation's lost time. We see these totems percolate up through cultural memory. The rapid accelerative cycles of fad and taste in the contemporary moment of consumer culture often result in these totems becoming popular consumer *objets d'art*: cassette tape tattoos, Lisa Frank branded sneakers, 1970s and 1980s-style ferns-brass-and-glass home décor, fashion trends based around thrift-store finds.

There is a curious phenomenon surrounding this nostalgia, succinctly stated by David Foster Wallace: often, it is the generation just slightly too young to truly remember a popular fad or cultural moment which fixates on the nostalgia for that fad or moment. In Wallace's example, children born after the Cold War's end have a curious fascination with nuclear war games and mutually assured destruction. The current youth fashion for "vaporwave"—an "aesthetic" built around a hodgepodge of 1980s and 1990s cultural markers, is a fine current example of this phenomenon.

Yet despite this current appetite for touchstones of this era, nostalgia for the modem—a defining technology of this time—is confined primarily to those who lived in that cultural moment. The visual aesthetics from the early-to-mid-1990s, the period when home use of modems grew rapidly, are having a cultural moment; television shows from that era are newly available for streaming on platforms like Hulu or getting reboots or remixes on Netflix; even Crystal Pepsi has been returned to supermarket shelves. Brief vogues for obsolete technologies flare up: young hipsters plug corded headsets into their smartphones; high schoolers carry around cassette decks older than they are. It is fashionable to post 1980s or 1990s-style VHS or album covers of modern shows and films to one's Tumblr or Instagram, but the technologies the nostalgic are using to post these images, to access the Internet, do not fall back on old technologies to do so. The young might seek out vinyl records, or Nokia candy-bar-style feature phones, but there is no vogue for dial-up modems. The color palettes and Clip Art images from Windows 95 are remixed and appropriated for the Internet "a e s t h e t i c", but the technology that undergirded Internet access in this time is not reclaimed or fetishized. Nostalgia for dial-up seems confined to those who lived through and with it, and that nostalgia centers on hearing the modem at work. Author Alexis C. Madrigal, who encountered a dial-up recording at the *The Museum of Endangered Sounds* website, describes the effect thusly: "If you grew up at a certain time, these sounds are like technoaural nostalgia whippets." The sounds are fleeting, but rush the hearer into a faraway and half-forgotten world.

> Wow, so many years since I've heard that noise in earnest and I still get a little Pavlovian frisson of anticipation when the lighter static changes to deeper static. Here comes the Internet! All is well! AOL picked up on the other end and life is about to shine like gold, baby.
>
> I readily concede that always-on, fast Internet is something I would not like to do without, but I do sometimes miss that palpable sense of "Oh boy, once the theme music stops playing, tonight's episode of the Internet will start and I bet it's going to be AWESOME" I used to feel back in the days of dial-up.
>
> (Posted by troublesome at 11:26 pm on July 27, 2011)

For those of a certain age, that dial-up noise truly is "the theme music" of the Internet. Access to the Internet, like access to broadcast television, was at this time episodic—confined to discrete chunks, often determined by considerations of access or time of day. Like a television show, the Internet could only be engaged with at specific times. This heightened the sense of occasion felt by its users—the limited quality was part of what made it all feel "special". It is important to note, however, that these users are using the general term "the Internet" for a specific set of experiences bound up with specific technologies of a specific time and place—just as "dial-up noise" is a general term being used to describe the use of particular types of modem by particular kinds of users at a particular point in time.

The method of access has changed, and the speed of operation has increased. The interface technology has changed, but to say the technology has obsolesced might not be quite correct.

The nostalgia for this earlier method of access, however, is undeniable. What is this nostalgia? It is both aural and transitional: aural in the sense that the sound of access, specifically, is missed, and transitional in the sense that users also miss the state change the sound signified. Users miss "going online". They are not (necessarily) nostalgic for a world divided into binary states of "offline" and "online"—they are not willing to sacrifice improved convenience and speed for their nostalgia in the way that audiophiles might sacrifice the convenience of digital recordings for the warmth of vinyl—but they miss the sense of wonder and anticipation, the sense of occasion, which came with "going online".

> I miss the newness, and the complete baffling stumblingness of the Internet. I miss time and disorientation. I miss chatting to strangers without faces online for hours. Those chat rooms still exist, but I do not exist inside of them . . . I miss the time when I visited the Internet rather than lived on it.
>
> (Honor Eastly)

Players of video games will recognize the phenomenon of "time and disorientation" that comes with immersion in virtual worlds of play. Entire days, entire weekends, can disappear; a player may wake up in the morning and sit down in front of a console, still in pajamas . . . then look up, blinking, to discover that it's now dark outside, that an entire afternoon has somehow vanished. Current users of the Internet may still encounter this phenomenon in certain spaces—the "Wikipedia rabbit hole" is a familiar example for many—but it is generally confined to particular webpages. The infinite scroll of social media eats up the user's time, but few users report experiencing lost time, or the sense of being transported. Rather than experiencing this transportation as a trip, both as a journey and as a kind of blackout or altering of consciousness, constant connection tends to produce a low-level state of distraction. This distraction can range from mild to severe, and, in its severe forms, might produce a similar effect of distancing. Certainly the effects on users' awareness of others in the room with them, or of time passing unnoticed, are still in play. But the sense of traveling, of getting lost in a different *place*, not just lost in one's thoughts, seems to have disappeared. It is this sense of magical dislocation that users seem to miss.

Some nostalgic technologies can be emulated. Many arcade, console, and PC games can, with varying levels of success, still be played on original or rebuilt equipment. The *space of play*, however, is trickier to emulate. NES and Atari 2600s are still playable consoles, and their cartridges can still be found in basements, or on eBay, the Internet's collective basement. But playing an original cartridge on an original console—perhaps even the very same console one owned as a child—will not bring back the rush of childhood experience. The shag carpet, the CRT television, the wrappers from snacks, the half-remembered need to do homework, the sound of a parent yelling about a forgotten chore: the pace and pattern of childhood life cannot be recreated by preserved, refurbished, or recreated technologies. The carpet has been ripped out, the snacks have been discontinued, the child has moved out of the house, and perhaps now has children of their own to hassle about chores and homework. Arcades still exist, but a game of *Galaga* (1981) will not feel now as it felt then: the crowds of others pressed up against rows of cabinets, the blinking lights in the narrow dim room, the cacophony of electric beeps and rattling quarters. The knowledge that one is performing a nostalgic act cannot be shaken; the sense of wonder and newness cannot be reclaimed.

As Eastly points out, "those chat rooms still exist"—the technology to maintain them still exists, and versions of these chat rooms can often still be accessed. But the conversations that inhabited and animated those spaces are gone, and efforts to resurrect them are fruitless. Users do not and cannot inhabit those spaces as they once did. In the mid-to-late-1990s, America OnLine's network had many such themed chat rooms available: social spaces for designated topics of conversation (religion, politics, various hobbies), for matchmaking, flirting, and long-distance romance, and for *Dungeons and Dragons*-style themed roleplay. The nostalgia blogger Dinosaur Dracula attempted to revisit one of these spaces in 2013, logging on to AOL and exploring the list of user chat rooms, all of which were at that time still accessible but most of which proved sparsely inhabited:

> The *Red Dragon Inn* was AOL's "roleplaying" chat room, acting as a sort of medieval "bar" in the fictitious land of RhyDin. You'd waltz in, devise some strange character for yourself, and chat with everyone using a phony old world language that came off more like malformed Pig Latin. . . . let's say I went into the room pretending to be a bounty hunter looking for clues. All well and good. I'd interact with people, and they'd play along. If they wanted to engage me, they did. If they didn't, they'd slide out of the conversation, always careful to stay in character (". . . *Lady Ravenclaw moves to the other side of the Inn, distracted by an unseen spectre . . .*").
>
> But man, the moment I got bored or had to leave, the [s**t] hit the fan. Suddenly I was the Terminator. Suddenly I had giant guns and was on a mission to kill everyone, as loudly and with as many caps-locked words as I could type. It took less than a minute to ruin everyone's fun. The good players would do their best to ignore me, but my fellow "noobs" would just as suddenly come at me with flamethrowers.
>
> Other times, we'd try to get everyone to "sing" the then-current pop songs, line by line. Nothing spoiled medieval roleplaying faster than virtual Chumbawamba karaoke.

In 2013, Dinosaur Dracula still had an active AOL account, and discovered that the chat room infrastructure somehow still remained intact and user-accessible. (A screen capture of Dinosaur Dracula's nostalgic adventure is shown in Figure 19.1.)

At first, it seems that all the old functionality is still here. There's still a host to welcome visitors. The interface has changed a bit, with buttons to add emoji or sound effects—in the AOL chat room heyday, users had to enter such things manually, using emoticons or keyboard shortcuts—but the layout remains the same, and the space still works according to the old rules. Dinosaur Dracula still recalls these old rules: employ double-colons to indicate character actions; intersperse narrative and speech; take space into account when playing your role (if one is at an inn, assume there is a bar with a barkeep and patrons), introduce one's character to others and wait to see their reactions, improvising as you go. The space, however, is empty, and since the game only ever consisted of other users, Dinosaur Dracula has nothing to do. Playing "correctly" is impossible, since there is no one to play with; appropriating or misusing the space is equally impossible, since with no players present, there are no rules and norms to break. Some automated actions do remain. Dinosaur Dracula can play dice with the host bot—an innovation, since during the heydays of chat rooms, live human volunteers would play the host's role. But the move to automation has made other actions impossible: the old Internet standby of ASCII art no longer displays as text, but auto-formats or turns into emoji.

The difference perceived between the nostalgic and the current can drive the nostalgic desire to return; ironically, it might be the lived experience of the new technology that leads to a deeper appreciation of what is unique in the old technology. The intangible nature and middling sound quality of music streamed from "the cloud" may add to the romantic luster of vinyl records. But technological advance, in addition to cultural change, forecloses the

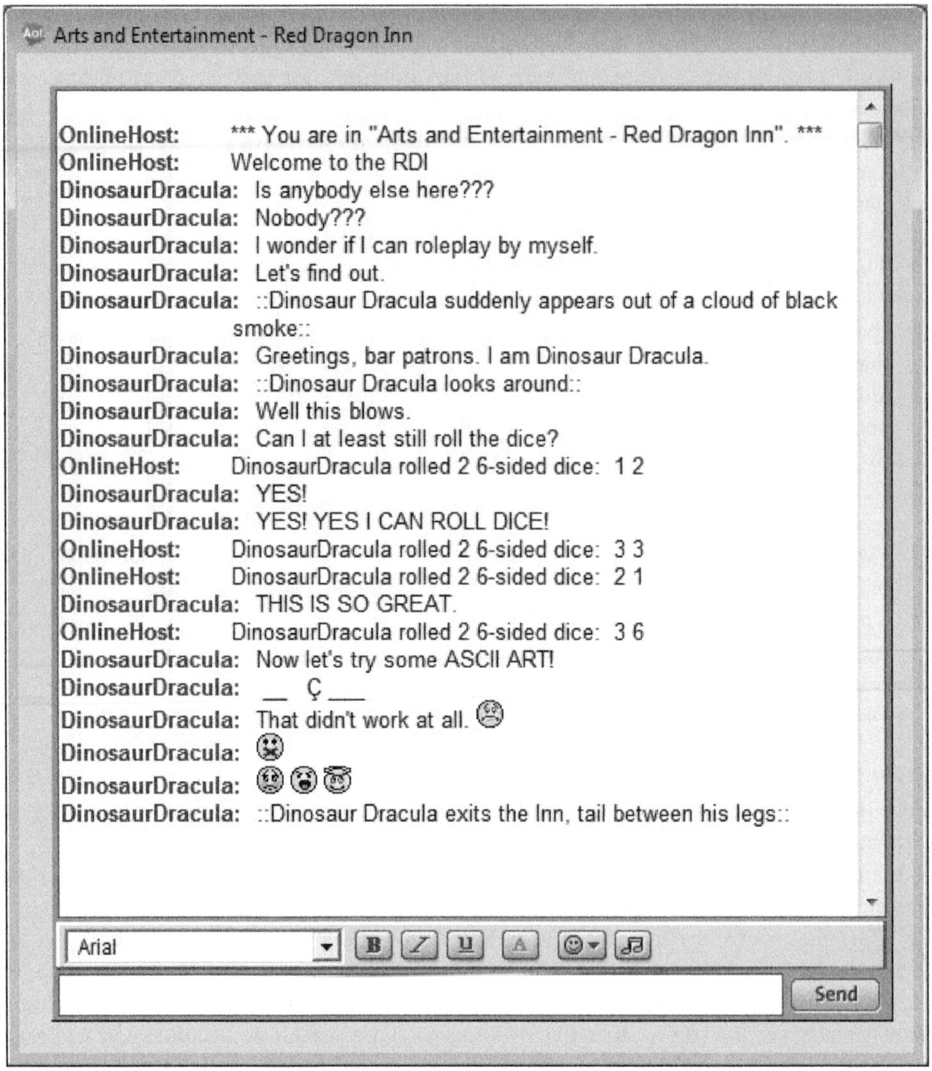

Figure 19.1 A screen capture of Dinosaur Dracula's nostalgic adventure.

possibility of return. Musical artists might wish to satisfy their fans by releasing their work on vinyl, but the master recordings of their work may only exist in CD format. Current interactions on social media might leave users unfulfilled, and the old days of BBSs and chat rooms might hold a similar sort of romance, but the newness and the strangeness, the odd formality and sense of occasion, are gone. Our always-on connections make them impossible to experience authentically.

> I can't believe what a visceral reaction I *still* get during that dial-up sequence. I'm sitting here listening to a freaking youtube video on broadband yet I'm still praying, "pleaseconnect.pleaseconnect,pleaseconnect".

> It also reminds me my stepfather joked that there were gnomes in the modem . . . and even though I knew he was joking, I kinda half believed him as an explanation for all those dropped connections.
>
> (Posted by lesli212 at 11:28 pm on July 27, 2011)

The idea of "magic" or "wonder" is attached to this noise. When we are always online, "online" is no longer a separate, magical, realm. Connection is effortless, immediate. Failure rates are low, and when failure does occur, the cause is often easy to diagnose via software wizard or with the remote assistance of a customer service representative. In the dial-up era, one had to "get online". One had to undertake a quest, and that quest could fail. Getting online required luck (resource availability, a free phone line, a local connection or cheap weekend/evening rates) and skill (knowing how to use the setup, interface, technology, being able to troubleshoot or problem solve or find workarounds if numbers were busy or the ISP was having issues or the modem was not connecting). Using a software wizard felt more like consulting a literal wizard: one had to divine an answer from esoteric and confusing technical language, reading their oracles. Users honed these skills, becoming practiced interpreters of tiny inhuman noises, of the slightest shifts in these unmusical tones. Users learned the quirks of their personal modems—and had to learn an entirely new language every time they upgraded their machines. Modems did not come packaged with dictionaries; users had to learn the language on their own. Few knew the actual protocols involved, the science behind the static and the hiss, but many could still play it by ear.

> I had every bleep in that sequence burned into my brain . . . when it goes from eeee to EEEE and back to eee I know I'm 30% there.
>
> (Posted by nile_red at 11:29 pm on July 27, 2011)

> I used to support dial-up software. It saddens me that I can no longer tell you offhand, from the sound, what the settings on that modem were, or even what the speed was. But I could've, once.
>
> (Posted by darksasami at 11:35 pm on July 27, 2011)

Though the metaphor commonly used to describe the process of dialing and connection is that of the handshake, there is an argument to be made for comparing the process, instead, to birdsong. Like birds negotiating territory or courtship proposals, modems pair and serenade each other, negotiating through song until an outcome is reached. The human waits at the terminal like a matchmaker—perhaps with bated breath and tense shoulders, perhaps with a practiced nonchalance, but always with an ear attuned to the process, alert for a skipped beat, the striking of an off tone. Those with sufficient practice and skill could, by listening, hear what was going awry and intervene, the way a conductor might ask the percussion to play more softly:

> I used to do tech support for an ISP. After a couple of months on the job, if a customer had two phone lines I'd have them hold the phone up to their modem so I could diagnose the problem. I could tell the difference in dial up speeds and the various 56k variants. I had more than one incredulous customer when I tried to do this.
>
> Most of the time the fix was to stick an extra comma or two in their dial-up string so that their modem would skip the 56k handshake and just drop to 33.6. We fixed a lot of "slow connections" that way.
>
> (Posted by mikesch at 12:25 am on July 28, 2011)

Indeed, humans can, with a bit of practice, even sing along:

> I once whistled a 300 baud carrier tone back down the line which put the receiving modem into CONNECT mode . . .
>
> (Posted by benzo8 at 2:58 am on July 28, 2011)

> I once whistled into a mic of my deck connected to my ZX Spectrum and I managed to get the stripes on the screen, although they were rather wobbly.
>
> (Posted by hat_eater at 3:05 am on July 28, 2011)

There is something delightfully right in this: that a technology which connects people should allow people to connect with it in a shared tongue. Users built relationships with these strange singing things. We can interface with these interface devices using nothing more than our own voices, provided we know how to address them in their native language. A bit of thought shows why this should be possible on the technical level—after all, modems were created to make use of an infrastructure initially designed to communicate human speech. Modems had to make use of the tones of the human voice, had to remain within the range of our spoken language while "speaking" with the grammar and syntax a computer could understand. It necessarily follows, then, that humans can, in turn, speak the same pidgin. Few of us learned how, and even fewer of us ever learned more than a phrase or two—no more than a human equivalent of "hello"—but we are able to speak to modems in a way we cannot "speak" to computers, generally. Even when programming in assembly, we require some intermediary: keyboards, or punch cards. This is far more direct.

> Wow. 20 years ago, you couldn't have told me people would be even remotely nostalgic about the dial-up sound. In another 20, will you remember fondly the cell phone feedback noise?
>
> (Posted by crunchland at 12:14 am on July 28, 2011)

In its moment, the dial-up tone was often an anticipatory prelude or an intermezzo to frustration. We would hold our breath, listen for false notes, silently willing the assemblage to function, waiting for the connection to come. We would impatiently drum our fingers on the mouse pad, attempting to regain our composure after getting booted off (and perhaps after yelling at someone else in the household for picking up the phone and interrupting us). We would hope the line was not busy, hope the line was not noisy, hope we could get back to the site or conversation that had vanished when the connection failed. Perhaps this is why the tone has such a hold on those who, for a time, lived and died by it. The uncertain rewards of the tone, freedom or failure, tickles our brains as gambling might. The irregular quality of rewards found in video games hooks players more deeply—this same effect is in play with the dial-up modem.

Other forms of noise-based failure, such as feedback, lack that association. Unlike feedback, which is always an interruption, always a sign of a deteriorating or breaking connection, the dial-up tone also represents connections made. Wait long enough, and, if everything goes right, the harsh noises will signal your success. It is hard for me to think that we will remember that noise with quite the same fondness. Nostalgia is difficult to predict, but the feeling generally described by those who express dial-up nostalgia is: comfort. Anticipation, anxiety, frustration . . . these are appended to recollections of specific instances, but comfort is a constant.

Even as always-on, flat-rate broadband access became available in the early 2000s, many users still kept their dial-up connections, paying for both services instead of transitioning over entirely. Some users kept their dial-up access "just in case" of broadband failure, preferring the relative slowness and frustration of dial-up to the fear of having no access at all, but others seemed to have a sense that different speeds of access were enough to create different forms of access (Anderson *et al.*, 2002: 103). Even in 2015, after 70% of the United States population had gained broadband access, 2.1 million people were still maintaining AOL accounts, many paying $20/month or more for the service. AOL subscriber numbers dropped off sharply in the mid-2000s, as users came to fully trust and embrace broadband, but a committed core hung on (Pagliery, 2015: 2). Some of these users reported that broadband options were unaffordable or unavailable, but others simply did not wish to transition (Horrigan, 2009: 40). AOL makes the transition difficult in some ways—its proprietary e-mail storage system makes transferring old correspondence prohibitively messy—but it is possible. Some users, it seems, simply prefer to live in the Internet as it was, as best they can. But the Internet itself has changed around them. Webpages are far more elaborate, and far more resource-intensive, designed for modern data transfer rates. 56k connections were never meant to handle autoplaying video. Broadband has changed the Internet that we view; even if we can still access it the old-fashioned way, what we can access has been transformed.

Many of the quotes used herein to encapsulate these memories were sourced from MetaFilter, a site which is itself something of a hangover from the age of the modem. A text-heavy, simply-designed site, first launched in 1999 and still reliant on a system that encourages pseudonyms, its norms and traditions seem antiquated in the post Web 2.0 moment. It is possible that these users, above others, are likely to feel nostalgic for this time. Certainly, since the demographics skew older, they are more likely to remember this time than users on newer platforms. Yet they are not alone: users on many social media sites, frequented by a variety of age groups, express a similar wistfulness. Even Buzzfeed, bastion of listicles aimed squarely at the millennials-and-under set, features articles on modem nostalgia, and few "Every 90s Kid Remembers These 20 Things" lists fail to mention "the dial-up noise." They also share in common a conviction that they could be the last generations to know this wistfulness—that digital natives, too young to feel nostalgia for dial-up or dial tones, much less for physical media such as games or records, will not have a generational touchstone, their own Proustian madeleine. A car commercial occasioned the following conversation on a more-solidly Web 2.0 platform, Kinja, regarding obsolescence, technology, and nostalgia. A commenter, going by the name of Vee, responds to Kristin Lee's article on Jalopnik.com regarding this car commercial, a meditation on childhood and loss:

> She's going to grow up with feelings that nothing in life is stable and that things she has come to know and love will suddenly disappear out from under her as soon as they become obsolete.
>
> (Kristen Lee)

> That's true already for most kids. My uh . . . Hmm, guess I should call them niece and nephew? It's not like like [*sic*] my dad's married to their grandmother, but . . . Well anyways, these two kids have gotten used to the idea that everything has a ten year shelf life. To them if it's older than ten years old either it's been hiding from the Upgrade Police or it's an ancient relic from a mystical civilization with puzzles to be unlocked. They're amazed that I have an XBOX. 2001 was sixteen years ago, but come on. It's not like it's medieval technology. I mean sure, it's older

> than both of them, but that's not so strange. They're just stunned that somebody would keep something for so long when you could buy like an XBOX One. It's hard to explain to someone who's lived with disposable purchases their entire lives that sometimes things are irreplaceable. That sometimes the new version lacks what you want, or adds things you don't. That sometimes it's a completely irrational attachment that can never be replicated with the replacement. And trying to do so will leave a dark spot in you [*sic*] life that you'll never get over.

This conversation was sparked, initially, by a car commercial: in a trope that is becoming standard for the auto industry, a family "grows up" with a trusty car, and when the time comes to upgrade, that trusty car is replaced by a new model from the same manufacturer. Many examples in this genre show a grown child inheriting that trusty old car as their own first car, a reliable guardian to shepherd them into responsible adulthood. This commercial chose a rather different path: instead, we see the trusty old car traded in and driven away by a stranger. It is a strange choice to make—and in this discussion, people cast the new car as an unwelcome step-parent, a home-wrecker, an interloper. Does the commercial fail to understand how product nostalgia works? Or are there different types of product nostalgia? How will technostalgia work for technologies that are not tangible, but app-based?

Technology companies are shifting away from the product cycles we have come to consider "traditional". The swift advance of mobile technology has trained many consumers to accept, and even expect, iterative upgrade cycles. It is increasingly likely that new generations of gamers will come of age without a defining console to function as a nostalgia point. Though there is a kind of reverse-novelty appended to old desktops and CRTs, these technologies are, for the most part, not marked by nostalgia in the way game consoles, record players, or even boomboxes have been. Once home computers transitioned from hobbyist equipment to appliances and entered the iterative upgrade cycle, they lost the ability to charm users into close relationships. Perhaps we are witnessing the end of technostalgia; perhaps our relationships with devices are changing, or have already changed. Perhaps coming generations will direct their nostalgia toward nontangible media objects: music, film and television, social media platforms, memes and GIFs; coming generations will then have the peculiar experience of feeling nostalgic for nostalgia itself.

It is too early to know whether this theory will prove true. Time and upgrade cycles will tell. We often associate technostalgia with devices, objects, and physical media; the shift toward converged devices and cloud-based or digital media leads one to speculate that nostalgia will become less common as people have fewer objects about which to become nostalgic.

Bibliography

Anderson, B., C. Gale, M. L. R. Jones, and A. McWilliam. "Domesticating Broadband: what consumers really do with flat-rate, always-on and fast internet access," *BT Technology Journal*, January 2002, Volume 20, Issue 1, pp. 103–114. Available at https://link.springer.com/article/10.1023/A:1014578227619.

Carracappas, Matt. "Old Internet Junk!" *DinosaurDracula.com*, April 9, 2013. Web. Available at http://dinosaurdracula.com/blog/old-internet-junk/.

Chillcut, Brendan. *The Museum of Endangered Sounds*. Web. Available at http://savethesounds.info/.

Davidson, Elizabeth, and Shelia R. Cotten. "Connection Discrepancies: Unmasking further layers of the digital divide." *First Monday*, Volume 8, Number 3, March 3, 2003. Web. Available at http://firstmonday.org/ojs/index.php/fm/article/view/1039/960.

Eastly, Honor. "The internet we miss." Panel presentation, *2016 Digital Writers Festival*, December 6, 2016. Web. Available at https://writersvictoria.org.au/writing-life/featured-writers/the-internet-we-miss.

Gillett, Sharon Eisner, and William Lehr. "Availability of Broadband Internet Access: Empirical evidence." Paper presented at the *Twenty-Seventh Annual Telecommunications Policy Research Conference*, September 25–27, 1999, Alexandria, VA. Web. Available at https://dspace.mit.edu/bitstream/handle/1721.1/1480/LehrGillettTPRC99_0523.pdf.

Horrigan, John. "Home Broadband Adoption 2009: Broadband adoption increases, but monthly prices do too." Study for the Pew Internet & American Life Project, 2009. Web. Available at http://broadband.masstech.org/sites/mbi/files/documents/publications-reports/pew-internet-home-broadband-adoption-2009.pdf.

Kridel, Don, Paul Rappoport, and Lester Taylor. "The Demand for High-speed Access to the Internet: The case of cable modems," in *Forecasting the Internet: Understanding the Explosive Growth of Data Communications* (pp. 11–22), David G. Loomis and Lester D. Taylor, eds. New York: Springer, 2002. Web. Available at https://link.springer.com/chapter/10.1007/978-1-4615-0861-8_3.

Lee, Kristin. "Fuck This Honda Ad." *Jalopnik.com*, April 9, 2017. Web. Available at https://jalopnik.com/fuck-this-honda-ad-1794128517.

Madrigal, Alexis C. "The Mechanics and Meaning of That Ol' Dial-Up Modem Sound." *The Atlantic*. June 1, 2012. Web. Available at www.theatlantic.com/technology/archive/2012/06/the-mechanics-and-meaning-of-that-ol-dial-up-modem-sound/257816/.

MetaFilter. "My God, It's Full of Krchhhhhhhh EEEErrrr EEEErrrrr chhhhhhh." *Metafilter.com*. July 27, 2011. Web. Available at www.metafilter.com/105955/My-God-Its-full-of-krchhhhhhhh-EEEErrrr-EEEErrrrr-chhhhhhh.

Pagliery, Jose. "OMG: 2.1 million people still use AOL dial-up." *CNN.com*, May 8, 2015. Web. Available at http://money.cnn.com/2015/05/08/technology/aol-dial-up/index.html.

van der Heijden, Tim. "Technostalgia of the Present: From technologies of memory to the memory of technologies." *NECSUS, 'Vintage.'* Autumn 2015. Web. Available at www.necsus-ejms.org/technostalgia-present-technologies-memory-memory-technologies/.

Wallace, David Foster. *Infinite Jest: A Novel*. Boston: Little, Brown and Company, 1996.

20

ILLUMINATING OBSOLESCENCE

Eastman Kodak's Carousel Slide Projector and the Work of Ending

Paige Sarlin

> In my end is my beginning.
>
> —T. S. Eliot[1]

Never New Again

On October 22, 2004, the last Eastman Kodak Carousel slide projector rolled off the assembly line in Rochester, New York (see Figure 20.1). There was a crowd watching as four workers completed tasks in sequence, fastening together parts that had been manufactured in China. On this occasion, the projector's final assembly took a little longer than usual—the line was held up because some thirty-five people, ranging from the assemblers to middle management, wanted to sign the inside cover of the machine. As could be expected, digital cameras (both video and still) captured the occasion. Vicky Christakis, one of the employees who had worked on the projectors for over 20 years, planted a kiss on the machine before she rolled it

Figure 20.1 "Rest in Peace, Carousel Slide Projector". Video still from Paige Sarlin, *The Last Slide Projector*, 16mm and digital video transferred to DVD, 2006.

over to her colleague for the final test. A small screen was pulled out across the width of the conveyor belt, the machine was plugged in and turned on. It projected three slides and was deemed suitable for retail. The worker picked up the machine and carried it over to another station for final packaging. Applause followed.

Three weeks later, there was a private Kodak event at the George Eastman House to commemorate the Carousel's significance for Kodak and the Rochester business community that it supported. Neither Vicky Christakis nor her colleagues attended the festivities. Instead, the room was replete with Kodak executives, managers, and dealers as well as small business owners and producers for whom slides were crucial.

Strictly speaking, both of these scenes came many years after slide projectors became outmoded. Neither of these events marked the point at which digital projectors began to replace Carousels, nor the moment in 1979 when the first commercial digital presentation software, BRUNO, was introduced by Hewlett-Packard.[2] But this milestone signaled an endpoint in the slide projector's role in the narrative of technological progress and in the corporate existence of the Carousel as a mass-produced object, though the two cannot be equated. This episode also indicates the non-identity between technological development and corporate strategy. Not simply the result of shifting consumer patterns or the emergence of digital technology, the close-out represents the culmination of a whole host of shifts in the political economy that are less immediate or visible, but no less tangible or powerful. Such a phase-shift in the life of the Carousel needs to be understood in relation to broader trends within American manufacturing and global production.

If obsolescence describes the state in which a technology is no longer sold or used, then the last day of production was just another point in the story of media technology—given that the second-hand market for the Carousel was still going strong and many galleries, museums, and schools still employed these machines in 2004 and continue to employ them in 2017. But this event marked the moment at which employees at Kodak stopped assembling these machines. So the question emerges: *What role does the end of production play in our construction and understanding of obsolescence?*

Dialectics at a Standstill

As a concept, obsolescence is central to the development of media history as a discipline, to narratives of technological progress and to counter-histories that seek to challenge these narratives. Beginning with Walter Benjamin's *The Arcades Project* written in the 1930s, critics have been envisioning the moment when a technology becomes outmoded as a moment to counter the fetish of the new and novelty that dominates corporate and popular accounts of technology.[3] For Benjamin, when a technology becomes outmoded it is released from the tyranny of the market, and its potential as a subject of critical analysis emerges. It becomes possible to produce "dialectical images", to imagine how things might have been, how a certain technology might have been used for emancipatory or even revolutionary purposes.

The slide projector's decline into obsolescence lasted much longer than a moment. No single event or dynamic shaped its trajectory; no solitary scene can be used to encapsulate the economic and cultural significance of its life-cycle. Instead, there is a string of events and turning points that need to be read in order to illuminate the work of ending: what so-called "obsolescence" can mean for this particular machine and for the medium, the industries, and experiences it enabled; for the people who designed and supported it, and who made and transported it, who depended upon its continued production for their livelihoods.

A temporal marker, "obsolescence" indicates that an object has not moved in step with the progressive march of history. This designation refers to something that was once new but will never be new again; it designates that an object, mode of production, notion, or term has a past. Unlike new technologies, obsolete technologies have histories—they are subject to historical forces that extend beyond that of their inventors or developers. Their futures develop at different paces—subject to new timelines, they travel in less regularized circuits of exchange. They are divorced from the realm of production and manufacturing in more than the form of their appearance as commodities and objects of exchange. They are dead labor that is also orphaned. When viewed as inevitable or necessary, a designation of obsolescence naturalizes the rhythms of late capitalism and conflates consumption with production in ways that ignore important economic, political, and social realities. For this reason, it is important that we historicize the category and practice of obsolescence.

The corporate strategy of planned obsolescence emerges at the beginning of the 20th century. Alfred P. Sloan, then director of General Motors, introduced a plan to develop, market, and produce a new model of car every year. This approach to manufacturing transformed the American automobile industry, as it was contrary to Henry Ford's business model. Having invented the assembly line, Ford believed the mass production of high-quality machines would be most efficient if they kept producing the same model, minimizing change and avoiding the need for further investment in new machines for production or research and design. However, within a very short period after the implementation of their plan, GM's bottom line increased and they surpassed Ford in sales. Eventually Ford followed suit and adopted Sloan's approach to manufacturing. Since then, planned obsolescence has become the central corporate strategy within the American auto industry. Many other technology-heavy industries have developed within the milieu in which this ethos was central to growth—and it has become the dominant *modus vivendi* for computer, phone, and camera manufacturing—moving even more quickly with the speed of changes brought about by digital innovation.

Emerging at the same time as Ford, Eastman Kodak subscribed to the Fordist approach to product development and manufacturing. Eastman Kodak prided itself on perfecting and maintaining existing hardware and film stock alongside the development of innovative new models. The company's historic investment in research worked side-by-side with its commitment to supporting and improving hardware and film stock that was meant to last. Building its reputation as an "imaging" company, the market for Kodak's products expanded beyond household consumption to professional arenas—from the film industry to medicine and business. The slide projector's history in this developmental diversification is archetypal, serving a pivotal role at different moments in the growth of the company.

Much that has been written about planned obsolescence focuses on the ways in which this profit-driven strategy has consequences that permeate our society—acclimating consumers to the rampant production of waste, the pace of a market-driven economy, and a culture of disposability.[4] And it is true that obsolescence has ensured that slide projectors have found their way to eBay, museums, galleries, and landfills the world over. But this essay suggests another approach to obsolescence that salvages the category and practice from being colored by an overemphasis on the consumption of new technology. To some extent, obsolescence has been seen heretofore as either a corporate designation or as a cultural or consumer category. But obsolescence can also be a potential political designation when it is understood as a condition of possibility for the construction of a series of "dialectical images", to use Benjamin's concept.

For Benjamin, dialectical images "emerge suddenly, in a flash".[5] Made manifest with the speed of light, they are expelled from the historical continuum and they are redeemable as instants only, not as parts of stories, but rather as configurations which "flash up in the now of their recognizability".[6] Discrete, like slides, they are infinitely re-readable and re-captionable, and as such they require a form and a method to be made legible.[7] For Benjamin, this is "a materialist presentation of history"; a method that makes possible a reading in which "what has been" can "come together in a flash with the now to form a constellation". At the moment of obsolescence, historical movements and forces can be framed by an image, apprehended as "dialectics at standstill". Contradictions and antinomies can be seen and analyzed in their relation, in the dynamic *between* them, not simply as independent factors. Understood not as a singular moment, but as series of turning points, the scenes associated with the end of production allow structural dynamics to become legible in ways that foreground the class character of the processes shaping our social and cultural circumstances and the oppositions between ideology and materiality that define the stories we tell about media technology and practices.

A Machine with a History

The story of the slide projector actually begins in the 1600s with the invention of the magic lantern. The first apparatus organized a lens, a light source, and an image on glass in such a way as to reproduce a drawing in shadows on a wall. The discovery of this arrangement of parts enabled the creation of the medium of the slideshow, whose history has been marked by remarkable versatility. Slideshows have been illuminated by candles, gaslight, and electricity; they have been shown in parlor rooms, classrooms, auditoriums, church basements, and boardrooms; and they have projected millions of images that came to be called "slides" because of the way in which the initial glass plates were shuttled into view on a magic lantern (see Figure 20.2). The medium of the slideshow has outlasted the obsolescence of its technological support a number of times—most notably in 1895, when the magic lantern was declared outmoded and overshadowed by cinema, reduced to a mere light source for the new machines that projected moving images as opposed to still slides. The projector as a category of machine also played a significant role in the history of ideas, providing a figure and a model for ideology and critique.

Figure 20.2 "MacLean Visual Resource Center, Flaxman Library, School of the Art Institute of Chicago", video still from Paige Sarlin, *The Last Slide Projector*, 16mm and digital video transferred to DVD, 2006.

Eastman Kodak's entry into the storyline begins with the development of the Kodaslide projector in 1937.[8] Powered by electricity, this device could be loaded with only one slide at a time and relied on a "douser method" to lower each picture into place. Only two years later, Eastman Kodak released an improved version that utilized a side-to-side slide-changing mechanism rather than the top-down design of the previous model. This year also saw the introduction of the first Kodachrome transparencies fitted with Kodaslide Readymounts, designed especially for this machine. Kodak had already introduced the two-by-two-inch design with the first Kodaslide machine, but the creation of Readymounts established a standard format for all slide film and projectors going forward. This sort of proprietary scheme was characteristic of George Eastman's approach to product development; he consistently sought to dominate the market by providing cameras, film, and the means to develop photographs. In this case, the simultaneous introduction of Kodaslide Readymounts with the new projector ensured the sale of Kodak's new hardware and launched what was to be the primary platform for the circulation and amateur consumption of color photographs until the advent of cheap color prints in the late 1970s (also introduced by Eastman Kodak).

The introduction of the Carousel projector in 1961 was a moment of resurgence. The Carousel's round tray was a vast improvement over other straight-tray models on the market, and this innovation made possible a dramatic increase in use that coincided with and contributed to a whole range of economic transformations. Situated at the nexus of changing modes of production, the Carousel's emergence coincides with the general post-war production boom. The most successful piece of hardware ever made by Eastman Kodak, over 19 million of these vehicles for projection were made during 40 years of continuous production.[9] The machine was a shining example of low-cost mass production and it helped to reconfirm Kodak as a model American company grounded in Ford's paradigmatic approach to manufacturing. Focused on improving existing models, Kodak's investment in scientific research and development is legendary, and the Carousel was a result of this commitment to quality and innovation.

This machine was a peerless vehicle for the projection of images whose myriad uses and popularity far exceeded Kodak's own expectations. More than simply a tool for amateur photography and education, the Carousel became a central instrument for the development of corporate culture. In response to increasing demand, the first professional non-consumer round-tray design, the Kodak Ektagraphic slide projector, was released in 1967. By 1979, the slideshow was the most ubiquitous medium for corporate and educational communications.[10] Production of Carousel slide projectors for European consumers began in Stuttgart in 1964 and these machines were developed alongside the American models, ensuring Kodak's dominance of the global market. This rise in use coincided with the crisis of overproduction that plagued the American economy of the 1970s. Kodak continued to invest heavily in research and development for its booming slide business, introducing significant improvements to existing models until the first major redesign in 1981. But the twin recessions of 1981 and 1983 hit the company hard, and the slide projector department in Rochester suffered layoffs despite the fact that Carousel slide projectors were in constant use in offices, schools, theaters, and conference centers.

The total gross revenues of this industry of producers, photographers, labs, and technicians rivaled the movie business throughout the early 1980s.[11] The slide projector was both a product of the old manufacturing economy and a tool for the "new" information economy. Following trends in manufacturing across the United States, there was a marked increase in worker productivity despite the decline in the number of workers employed in producing the machines. Housed in the huge Elmgrove plant in Rochester, New York, hundreds of workers continued to meet increasing demand through the 1980s into the 1990s, and the slide projector remained profitable even as other aspects of Kodak's business were weakening. In fact, the machine was transferred between different groups during periodic corporate reorganizations in order to maintain the bottom line of failing sectors of the company. Other product lines were gradually outsourced and workers laid off, but the slide projector production and assembly remained in the Elmgrove facility. Different aspects of projector manufacturing slowly began to be moved off site to local manufacturers. It was not until 2001 that the production of projector parts was outsourced to China, which coincided with Kodak's sale of Elmgrove. Only eight of the dozens of workers from the assembly line were transferred to Building 205 in Rochester. They continued the final assembly of these mechanical projectors as the digital revolution swirled all around them. In Stuttgart, Germany, the assembly line had been fully mechanized for many years; by 2004, there were only two workers cranking out thousands of machines in one corner of a large factory (see Figure 20.3).

The Beginning of the End

Near the end of 2003, in the midst of the many discussions of the shift from analog to digital, a group of managers at Eastman Kodak sent out a letter to all Kodak dealers and various slide-related professionals. The letter announced that projector production was going to end

Figure 20.3 "Kodak Factory, Muhlhausen, Germany, 2004", video stills from Paige Sarlin, *The Last Slide Projector*, 16mm and digital video transferred to DVD, 2006.

by spring of the next year. But this was not like the ordinary corporate announcements that herald the introduction of new products and the retiring of old ones. This letter was unique, and read as follows:

> An official statement was issued in September that read something like this: Eastman Kodak Company will discontinue the manufacture and sale of 35mm slide projectors by June of 2004. The company will continue to provide service and support for slide projectors through June 2011.
>
> While those are indeed the facts, those of us who have been intimately involved in the slide projector business for some time have stronger feelings about it. To us, it has had a personality all its own; some would call it indomitable. The character traits of the slide projector industry are comprised of the numerous good qualities of the people who have been our partners for many years – people like you. For such a large, competitive, and, at one time, booming business as this has been, who would have guessed that so much camaraderie would have developed? But, it did. We not only shared the birth, growth and maturity of a business, we shared friendship. Brought together professionally, we connected in personal ways, too.
>
> With that in mind, we who are left to close Kodak's slide projector business offer you our heartfelt gratitude for the personal and professional relationships that have been forged. We greatly appreciate your support for Kodak and its products over the years, and we look forward to your continued cooperation as the trusted slide projector retires. The projector's last day "on the job" here at Kodak will be March 31, 2004. That's when final production orders must be in to the factory. . . . It has been an honor and a pleasure working with you. Sincerely,
>
> Merri-Lou McKeever, Ginger Hunneyman, Joe Paglia, James Auburn, and John Beerse[12]

This document makes clear that the end of projector production was not simply about the paradigm shift from analog to digital or the transformation of a technology for sharing and projecting images. The human consequences of obsolescence are myriad, and this was a milestone in the work lives of many people.

First and second tier management addressed this letter to people who sold, repaired, and used the slide projector. Framing this letter as a thank you, these employees distinguished themselves from the corporate line and acknowledged the social bonds that they had built over years of working with the slide projector and beside each other. By drawing an analogy between the story of the machine and the industry it enabled, this letter called attention to a shift in corporate practices and culture that runs parallel to changes in technology. A touchstone for connection and attachment, the Carousel was a foundation for a production line that represented an approach to manufacturing and business that had become anachronistic itself. Kodak slide projectors were built to last and in order to develop products at that level, there needed to be longevity among the engineers, technicians, and sales team; there had to be a willingness to build relationships with dealers and customers, and to foster trust and confidence in the product within the company as well as on the market. But from the 1990s onwards, that sort of "camaraderie" dissipated across all arenas of manufacturing. Kodak's extended investment in its workers and research supported the image of Kodak as a corporate family and helped to keep unions out of the Rochester factories. But this strategy had become outmoded. The Carousel had persisted past any predictions, having resisted outsourcing and various other forms of restructuring that other manufacturing entities had weathered throughout the 1990s. Even the last day of production had to be extended past the date that this letter announced because of the volume of demand. Clearly not an end to the slideshow as a medium, this moment was the marker of something else that was deeply felt—a transformation at the level of the social, between people.

A few months after the slide projector group sent out their letter, Kodak announced their decision to focus entirely on digital technology, and with that shift they would need to cut at least 10,000 jobs before the end of 2004.[13] This was a huge blow, and it was the first of a wave of layoffs and restructurings that culminated in Kodak filing for bankruptcy in January 2012.[14] There had been waves of layoffs before, but this reorientation was decisive. The stronger feelings of the projector group were echoed and amplified throughout Rochester (see Figure 20.4).

Nostalgic Projections

> This device isn't a space ship—it's a time machine.
>
> —Don Draper

The first season of *Mad Men* ended with an episode entitled "The Wheel" which aired on October 18, 2007. Near the end of the episode, Don Draper pitches an ad campaign for a new slide projector to Eastman Kodak executives. Using the very machine he is trying to promote, Draper explains the ways in which this particular commodity furnishes advertising with an ideal object to sell:

> Technology is the glittering lure. But there is the rare occasion when the public can be engaged on a level beyond the flash: if they have a sentimental bond with the product . . . the most important idea in advertising is new. It creates an itch – you simply put your product in there as a kind of calamine location, but [*there is also*] a deeper bond with the product: Nostalgia. It's delicate but potent.

Figure 20.4 "Workers Leaving the Kodak Factory, Rochester, NY, 2004", video still from Paige Sarlin, *The Last Slide Projector*, 16mm and digital video transferred to DVD, 2006.

Accompanied by the whirring sound of the fan and the regularized mechanical click of the mechanism, the circular tray advances and a series of snapshots of Draper's family including his newborn child fall into view. This scene in the fictional offices of Sterling Cooper Draper Pryce dramatizes the ways in which advertising as an industry has shaped the stories we tell about media technology and the role of the slide projector in cultural and economic development. Simultaneously a family slide show and an advertising pitch, Draper's presentation illustrates the importance of advertising's myth-making to the history of media technology. Nostalgia was crucial to the marketing of the image-making that Eastman Kodak perfected and upon which it capitalized. The company built a multi-billion-dollar business by packaging the Kodak moment, selling the ability to capture experience and preserve memory.[15] But unlike Kodak's other products, this piece of hardware did not capture images. Rather, it enabled the projection and circulation of images; it reproduced and reanimated experiences and memories.

Draper describes the emergence of the Kodak slide projector as a rare moment in industry—and a unique opportunity in advertising—because this product is simultaneously an embodiment of the new and a product that is capable of transporting people to an idealized past, showing them images of themselves and their lives. In this fictionalized account, Draper changes the name of this machine from "The Wheel" to the Carousel—a dramatization that serves to underscore the evocative potential of this piece of hardware and the novelty of its design. This scene demonstrates the ways in which this particular machine was representative of the intersections of commodity culture with the idealization of national and familial relations (see Figure 20.5). A potent object for the production, reproduction, and circulation of ideology, it is no wonder then that this image emerged after the last Carousel projector was made.

On some level, this episode verifies the cultural significance of the Carousel as a machine and the slideshow as a medium. But it is precisely this representation of the projector that I was hoping to supplement by including these other dates in the timeline. In this scene, the slide projector allows for a brilliant celebration and articulation of commodity fetishism.[16] Draper's description personifies this object at the same time that it both erases any reference to the people who made the machine and projects idealized images of family that are detached from the social realities that they document. For Draper, the Carousel "isn't a space ship—it's a time machine". And this scene demonstrates that, from a particular vantage point,

Figure 20.5 "Carousel Multi-Image Slideshow, Germany, 2004", video still from Paige Sarlin, *The Last Slide Projector*, 16mm and digital video transferred to DVD, 2006.

the machine seems only capable of turning toward an idealized past. But the story of this machine extends beyond the parameters of nostalgia and novelty. The Carousel's history precedes and postdates this moment of emergence in ways that are symptomatic of our relation to technology and the transformation of its role in our lives since this fictionalized moment.

For Thomas Elsaesser, a prolific scholar of contemporary cinema, this scene epitomizes the value and power of the Carousel projector as a "solution" to the problems of PowerPoint and boring lectures dominated by linearity and staid juxtapositions.[17] Elsaesser suggests "the Kodak Carousel may be the place to which we need to return."[18] Seen as a gesture toward a reconsideration of history, this proposition is appealing. But as an expression of the type of nostalgia that serves the motor of capitalist accumulation, the logic of commodity fetishism, and the nationalist projects embodied in the promotion of retrograde design and politics,[19] this characterization of the Carousel advances an instrumental view of media technology that ignores the material conditions of technological development and manufacturing. Celebrating the Carousel's beginning as "a place" to which we need to return, Elsaesser's encomium perpetuates a potent fiction—one that encourages reverie rather than critique and values beginnings over endings.[20]

Illuminating Obsolescence

A thoroughly modern category, obsolescence is neither ideologically neutral nor objective. Obsolescence is planned by corporations; it is shaped by technological developers; it is accepted by consumers as an aspect of cultural life under capitalism; it is embraced by artists and collectors; and it is used by scholars to explain the role of media in our lives. Obsolescence is a concept and category, a practice and a designation that derives from material conditions and which impacts many different strains of social and economic relations that extend beyond that of consumption or imagination. Critics are most keen to point out that obsolescence *is* created, a result of market and technological forces. But the playing out of obsolescence from the perspective of manufacturing is an uneven process; it occurs at various paces in assorted places, affecting people in a range of different ways.

For Walter Benjamin, obsolescence allows for a moment in which we can *imagine* a technology freed from the circuits of exchange. But machines like the Carousel do not cease to be commodities after they stop being made new. Just because we can imagine what different *uses* a machine can be put to does not mean that it is entirely free from the circulation of objects—market and commodity fetishism. If anything, since Benjamin, we can see that the fixation on old media has generated an industry in the realm of art and media studies. The retrograde has emerged as a category in the framework of overproduction, an answer to the pervasive anxiety about change in the atmosphere of neoliberal capitalism.[21] At the end of slide projector production, we can recognize the work of ending—the players and processes engaged in the reorganization of labor and technology that the Carousel both enabled and epitomized. At this moment, it becomes possible to reconsider the conditions under which media hardware and software are manufactured. Seen dialectically, these moments invite us to imagine how the social relations of production might have been organized and might yet be organized.

Media technologies are both vehicles of the future and the product and inheritors of the past. Situated in this way, they are machines that both help us to remember and to imagine. These machines offer models for the construction of histories and narratives, of society and technology itself. The story of the Carousel should not be reduced to tales of its domestic or art historical use, to dramatizations of its role in what Deborah Tudor calls "the selling of nostalgia".[22]

This piece of hardware animated other circuits of exchange and value creation over the course of its manufacture. The Carousel was implicated in a historically specific process of outsourcing and the introduction of longer supply chains and more complex logistics. But it was also crucial in the establishment of social relations, of friendships and alliances, of face-to-face contact built on the assembly line as much as in the boardroom or classroom.

When Eastman Kodak stopped the production of Carousel slide projectors, it was the end of an era. The Carousel slide projector had helped to picture the world for over 40 years. It was a vehicle for the projection and illumination of ideas and images that made visible large-scale changes in labor and society. The machine also contributed to transformations in the use and significance of the visual within the very restructuring it made visible. At the end of projector production, the machine was perched on the edge of a certain kind of visibility, it became ready for a kind of critical illumination, one that both inverted and converted an ideological and consumption-focused account of obsolescence into an opportunity to examine the relations between use and production and between things and their apprehension.

In *The German Ideology* (1932), Karl Marx and Frederick Engels liken ideology to a camera obscura in an effort to illustrate how images and ideas of society and the world are related to the conditions they seek to describe. To quote Marx and Engels, "If in all ideology men and their circumstances appear upside-down as in a camera obscura, this phenomenon arises as much from their historical life-processes as the inversion of objects on the retina does from their physical life-process."[23] For Marx, the figure of the camera obscura helps to explain how ideology produces an inverted image; but it also demonstrates the operation by which these inverted images relate to material and historical conditions. The image of the camera obscura provides a representation of both sides of a dialectical relation, the side of production and the side of the image produced. In this way, the camera obscura is a diagram of the very relation that ideology obscures. For me, the significance of this figure is not in its characterization of ideology as a technology of inversion, a notion whose simplicity has been made obsolete, complicated by the critique of many thinkers since Marx. Rather, the image of the technology itself (in this case, the structure of a camera obscura) becomes a tool because it illustrates both sides of a relation. For Marx and Engels, "Life is not determined by consciousness, but consciousness by life."[24] Our thinking is determined by material conditions, the images and descriptions we produce of the world are shaped by this relation. In my reading, this image becomes a demonstration of Benjamin's "dialectics at standstill". The figure of the camera obscura stands in contrast to a one-sided ideological image, just as Benjamin's notion of a dialectical image is meant to stand in contrast to the ideological plot points that make up the narratives of technological progress. In this way, dialectical images do not simply oppose a version of history as progress, but rather they supplement and complicate a one-sided view of history. They show us a plurality of oppositions, contradictions, and antinomies.

The end of production of the Carousel projector marks a moment distinct from that of strict obsolescence. It comes over twenty-five years after what might be termed the "literal" moment of obsolescence, when the first digital presentation software emerged. This distinction helps to make clear that at the moment of the end of production, a complex of forces and dynamics form a nexus that extends beyond the relation between the stories of dis-use and innovation. In contrast, this moment helps to extend a consideration of obsolescence from a focus on the perspective of consumption to one in which changes in modes of production are highlighted. The moment of the end of production shows how obsolescence as a phenomenon explicitly holds together a dialectic between use and production, illustrating not simply a moment in these trajectories but also a change in the very relation between consumption and production—and even more significantly, the distinction between our ideas and images of technology and concrete conditions.

A vehicle for the projection of images, the slide projector was an instrument for business to represent itself to itself and to other constituencies. In schools and boardrooms, the Carousel was a means the slide projector was a vehicle for the projection of images and an instrument for communication. It was a tool that finance capital needed in order to maintain profitability when information became the premier commodity. It is this rise in the importance of immaterial production, the making and distribution of images, advertising and knowledge, that the slide projector illustrates most clearly.[25] To say that the slide projector was an agent of these changes is to mis-diagnose the complications of the ways in which media technology and capital are intertwined. But it is precisely the condition of possibility that this technology helped to organize and structure that contributed to its eventual demise. The story of the slide projector as both a product of the old manufacturing economy and a tool for the "new" information economy is illustrative of the ways that declarations of obsolescence are generally the most visible manifestations of the cycles of hollowing out that have to precede capitalist accumulation. The story of the machine illustrates the transformation from manufacturing to a paradigm of "immaterial" labor[26], it demonstrates the interconnection of cognitive and material labor. In addition, we can see how the organization of social relations changes with the transformation of the means of production and technology. More than a shift in the paradigm of image making and distribution, the end of slide projector production reveals a transformation in our images and understanding of the distinctions between work and leisure, production and consumption, and even our definitions of "the social". From this perspective, the moments associated with the work of ending can be read as dialectics at a standstill—as moments of stillness in the ongoing transformation of forces that will continue to move and effect the forms in which these very changes can be apprehended.

Coda: An Afterlife of Things Made

For Walter Benjamin, the project of history, of writing, and interpreting the past is always in the service of capturing something on the brink of being "irretrievably lost".[27] Benjamin's work diagnoses the difficulties and dangers of nostalgia, of responses to change that forget the present. It is, therefore, instructive to use Benjamin's project as a lens through which to examine the slide projector, whose history is closely tied to the production of nostalgia. The question of the Carousel's relation to "irretrievable loss" is complicated by the fact that the medium it enabled has persisted past these "historic" moments. The mechanisms and dynamics that it illuminates, the forces that determine the image of the slide projector have continued to persist past the fading of this machine. In some respect, this is the point of Benjamin's historiography: to call attention to the persistence of the forces of capitalism, and to suggest that there are moments within its history where this persistence can be interrogated.

This essay began with a description of the last day of production in Rochester, New York, in 2004. Three weeks after that moment, I arrived at Vicky Christakis's house to talk with her about her experience working on the line at Kodak for over twenty years. My cinematographer, Mary Billyou, and I set up our cameras and lights in her living room and she pulled out her Carousel projector and a projection screen (see Figure 20.6). I handed her a round tray with the 35mm slides that I had made from the footage shot on October 22. She had not seen the digital video, so she watched the parade of stills with some disbelief. In stop-motion, the scene was elongated, available for contemplation and inspection. It had a mechanical, methodical rhythm that gave her and I time to talk. We were not swept into the tempo of calendrical time. Instead, the recent past kept coming into view; what-had-been became available to discuss and reconsider. The frozen instants looked different to me sitting next to Vicky; simultaneously abstract and intensely personal. The last day of Vicky's

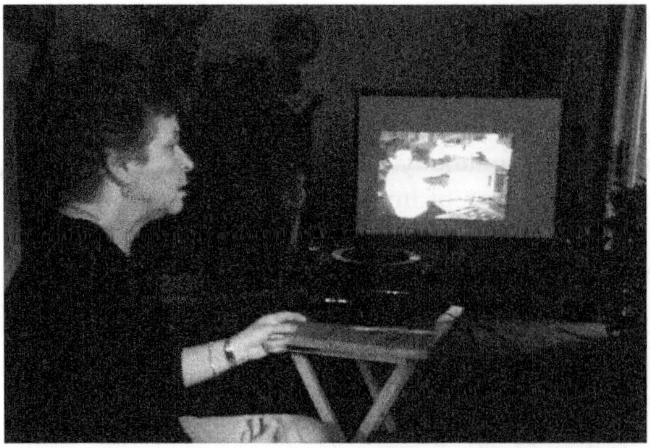

Figure 20.6 "Vicky Christakis, Rochester, NY, November 2004", video still from Paige Sarlin, *The Last Slide Projector*, 16mm and digital video transferred to DVD, 2006.

work-life was represented on that screen. The house we sat in was paid for with paychecks from Eastman Kodak, the company pictured in the photos. Now retired, Vicky sat beside a working Carousel, a machine she had built for years. I sat between a 16mm camera and a digital video camera, the tools I was using to make a documentary film The Last Slide Projector. The finality of events was held in abeyance as we sat in the midst of the contradictions made visible. Confronted with the scene, Vicky giggled when she watched herself kiss the last projector. She turned back to me and said "Someday, you watch, it's going to come back." I smiled. As if on cue, she articulated a desire for an afterlife for the Carousel slide projector, not as an object of consumption or even as the agent for projections, but as the object of production, the organizer of social relations, the basis for a living, a community, and a way of life. Comprised of oppositions and inversions, this scene of my devising was as close as I could get to producing something like a dialectical image for others to see.[28]

To produce a dialectical image of the slide projector at the end of its production is to call attention precisely to the changes in what Marx calls the "general illumination" which bathes everything in its light—the illumination that the dominant mode of production produces. In the 1857 "Introduction to a Critique of Political Economy", Marx writes:

> In all forms of society there is one specific kind of production which predominates over the rest, whose relations thus assign rank and influence to the others. It is a general illumination which bathes all the other colours and modifies their particularity. It is a particular ether which determines the specific gravity of every being which has materialized within it.[29]

To some extent, it is the role of projection in the analysis of the transition from one "dominant" mode of production to another that the slide projector's history helps chart. From the vantage point of decline and the work of ending, it becomes possible to articulate and image the contradictions inherent in various changes in the mode of production and to begin to apprehend the significance of technological change in a way that is much broader than the singular attention on use and consumption can reveal. By offering this final figure of Marx's "general illumination" and extending Benjamin to consider the moments of at the end of production, it becomes possible to reconsider the role of manufacturing and labor in the imaging and apprehension of the present (see Figure 20.7).

Figure 20.7 "Last Projector", video still from Paige Sarlin, *The Last Slide Projector*, 16mm and digital video transferred to DVD, 2006.

Obsolescence is made; it is contingent. The figures and technologies through which this can be made visible are involved in the very processes of that production. Considered "in light" of Marx's model of ideology, today's dominant mode of production can and should be read in relation to the historical processes by which this mode of production came to dominate and determine the appearance of all the objects in its light. It must be read in relation to the "other" side of the figure, the material conditions and labor that make this illumination possible. The mechanical models of the camera obscura and slide projector remind us to think spatially, and to hold the dynamics and the antinomies in relation to one another, dialectics that are often obscured within the digital ether, relations and oppositions that might be changing but which are not collapsible. Among these dialectics are the relations between users and producers, activity and analysis, production and consumption, description and diagnosis, physical and intellectual labor, and representation and enactment. In this way, a dialectical image of the slide projector, with its histories and legacies, can make visible the dynamics and contradictions of capitalism, but can remind us that the point is to imagine how to render that technology and those forces obsolete.

Notes

An earlier version of this chapter appeared as "The Work of Ending: Eastman Kodak's Carousel Slide Projector", *PhotoResearcher*, No. 24, October 2015.

1 T. S. Eliot, "East Coker" in *The Four Quartets*, San Diego: Harcourt Brace & Company, 1988.
2 Another important event in this timeline of "strict" obsolescence occurred in 1987 when Microsoft first shipped PowerPoint to customers. See Robert Gaskins, *Sweating Bullets: Notes about Inventing PowerPoint*, San Francisco: Vinland Books, 2012; and https://en.wikipedia.org/wiki/Presentation_program.
3 Walter Benjamin, *The Arcades Project*, edited by Rolf Tiedemann, translated by Howard Eiland and Kevin McLaughlin, Cambridge, MA: Belknap Press, 2002. See "Convolute N", pages 469–481.
4 See Vance Packard, *The Wastemakers*, New York: Van Rees Press, 1960; Jonathan Sterne, "Out with the Trash: On the Future of New Media" in Charles R. Acland, editor, *Residual Media*, Minneapolis: University of Minnesota Press, 2007, pages 16–31; and Lisa Parks, "Falling Apart: Electronics Salvaging and the Global Media Economy" in Charles R. Acland, editor, *Residual Media*, Minneapolis: University of Minnesota Press, 2007, pages 32–47.

5 Benjamin, *The Arcades Project*, [N9,7], page 473.
6 Benjamin, *The Arcades Project*, [N9,7], page 473.
7 Walter Benjamin writes: "The object constructed in the materialist presentation of history is itself the dialectical image. The latter is identical with the historical object; it justifies its violent expulsion from the continuum of historical process." Benjamin, *The Arcades Project*, [N10a,3], page 475.
8 Merri-Lou McKeever, "A Brief History of Slide Projectors", 2004, available at http://resources.kodak.com/support/pdf/en/manuals/slideProj/history.pdf (24.08.15).
9 Information concerning the history of the audiovisual industry and Carousel projector production was compiled from interviews with Tom Hope, publisher of *Hope Reports*, the leading industry report on slide projector and AV industry sales from 1962 until 2004, and Merri-Lou McKeever, Manager of the Presentation Group at Eastman Kodak from 1994 to 2004.
10 Thomas W. Hope, "Large Screen Presentation Systems", Rochester: Hope Reports, 2000.
11 Thomas W. Hope, "Presentation Slides V: Electronic and Film – Update", Rochester: Hope Reports, October 1999, page 7.
12 James Auburn, John Beerse, Ginger Hunneyman, Merri-Lou McKeever, and Joe Paglia, "Letter to Kodak Dealers", Rochester: Eastman Kodak, November 2003.
13 Ben Rand, "Worldwide Kodak Layoffs", *Democrat and Chronicle – Rochester, N.Y.*, January 23, 2004.
14 Steve Sink, "Kodak Files Chapter 11 Bankruptcy", *Democrat and Chronicle – Rochester, N.Y.*, January 19, 2012.
15 Kamal A. Munir and Nelson Phillips, "The Birth of the 'Kodak Moment': Institutional Entrepreneurship and the Adoption of New Technologies", *Organization Studies*, November 26, 2005, pages 1665–1687; Reese V. Jenkins, "Technology and the Market: George Eastman and the Origins of Mass Amateur Photography", *Technology and Culture* Vol. 16, No. 1 (Jan., 1975), pages 1–19.
16 Marx explains: "The commodity-form . . . is nothing but the determined social relation between humans themselves which assumes here, for them, the phantasmagoric form of a relation between things." Karl Marx, *Capital*, vol. 1, Frederick Engels, editor, Samuel Moore and Edward Aveling, translators, 1887, New York: International, 1967, page 165.
17 Thomas Elsaesser, "Kodak Carousel" in *Objects of Knowledge, of Art and of Friendship: A small technical encyclopaedia for Siegfried Zielinski*, David Link and Nils Röller, editors, Leipzig: Institut für Buchkunst, 2011, page 70.
18 Ibid., page 71.
19 Christian Thorne, "The Revolutionary Energy of the Outmoded", *October*, Vol. 104, Spring 2003, pages 97–114.
20 In this instance, Elsaesser reveals himself to be an archaeologist who follows in the line of Siegfried Zielinski—celebrating the potential of an historical approach to emphasize both the humanistic and artistic possibilities inherent in media technologies. In contrast to Elsaesser, Jason Mittell reads against the grain of the scene, locating in it a "criticism of the illusion of nostalgia." I am not certain I believe that this scene serves as critique so much as reproduction—a reproduction of the false paradox between spectacle and reality. As Irene Small has argued, the dialectic between depth and surface that *Mad Men* perpetuates needs to be countered because it serves to perpetuate the investment in image-making that places profit and innovation as the high water marks for technology. Jason Mittell, "On Disliking *Mad Men*", *Just TV* 29 (2010). Irene Small, "Against Depth" in *Mad Men, Mad World: Sex, Politics, Style, and the 1960s*, Lauren M. E. Goodlad, Lilya Kaganovsky, and Robert A. Rushing, editors, Durham: Duke University Press, 2013, pages 181–194.
21 Christian Thorne, "The Revolutionary Energy of the Outmoded."
22 Deborah Tudor, "Selling Nostalgia: *Mad Men*, Postmodernism and Neoliberalism", *Society*, Vol. 49, No. 4, 2012, pages 333–338.
23 Karl Marx and Frederick Engels, *The German Ideology: Part One*, edited by C. J. Arthur, New York: International Publishers, 2004, page 47.
24 Ibid.
25 Information concerning the history of the audio-visual industry and Carousel projector production was compiled from interviews with Tom Hope, publisher of *Hope Reports*, the leading industry report on slide projector and AV industry sales from 1962 until 2004, and Merri-Lou McKeever, Manager of the Presentation Group at Eastman Kodak from 1994 to 2004.
26 Roggero, Gigi, "The General Illumination Which Bathes All Colours: Class Composition and Cognitive Capitalism for Dummies", *Culture Unbound: Journal of Current Cultural Research*, Vol. 6, 2014, pages 125–135.

27 A different version of this complete passage reads: "The dialectical image is one flashing up momentarily. It is thus, as an image flashing up in the *now* of its recognisability, that the past . . . can be captured. The redemption which can be carried out in this way and in no other is always only to be won out of the perception of that which is being lost irretrievably." Walter Benjamin, "Central Park" in *Selected Writings Volume 4*, Cambridge, MA: Harvard University Press, 1996–2003, page 190.

28 Max Pensky, "Method and Time: Benjamin's Dialectical Images" in *The Cambridge Companion to Walter Benjamin*, David S. Ferris, editor, Cambridge University Press, 2004, pages 177–198.

29 Karl Marx, *Grundrisse*, New York: Penguin Books, 1993, page 107.

21

"POOR BLACK SQUARES"

Afterimages of the Floppy Disk

Matthew Kirschenbaum

Residual Media

It was an obscure government report, but it contained a tidbit that went viral immediately: in May, 2016, a white paper released by the U.S. Government Accountability Office on the need for federal agencies to address aging legacy systems noted (more or less in passing) that elements of the nuclear command and control system for Minuteman missiles still relied on a forty-year-old IBM Series/1 computer serviced by 8-inch floppy disks. "Introduced in the 1970s," the author explained, "the 8-inch floppy disk is a disk-based storage medium that holds 80 kilobytes of data. In comparison, a single modern flash drive can contain the equivalent of more than 3.2 million floppy disks." The report also included a helpful illustration depicting one of the matte black square-shaped disks with a donut hole in the middle alongside its paper slip case.[1]

The Internet loved it. The juxtaposition of floppy (*floppy!*) disks with nuclear-tipped ICBMs seemed to encompass everything that was absurd about both government bureaucracy and Cold War strategic thinking—*WarGames* (1983), *Dr. Strangelove* (1964), and *Brazil* (1985), all at once.[2] It was left to the *Washington Post* to explain that the military logic (such as it was) was that the IBM Series/1 was isolated from any networked threat and the air-gap bridged by floppies was a security feature, not a bug. (The *Post* also reported that the system was due to be overhauled and replaced in 2017.)[3] Regardless, the entertainment value of this news item—feeding the Internet's insatiable appetite for irony and oddity—underscores the extent to which the floppy disk has lodged itself in the memory of a generation weaned on the totems and paraphernalia of early home computing. It also suggests other contradictions worth our notice: that floppies are self-evidently obsolete but still stubbornly useful and useable, much like a typewriter or landline telephone in certain circumstances; that they are ancient and obscure technology—such that the GAO report's author felt obliged to explain the most basic facts about them—but yet everyone devouring the story online knew exactly what a floppy disk was, and so the delighted (if slightly queasy) reaction.

The floppy disk is thus an exemplar of what Charles R. Acland and others, after Raymond Williams, have asked us to call residual media, a formulation they employ as an alternative to the relentless presentism of so-called "new" media. Residual media, by contrast, reveal the persistent medial continuities between and across different historical moments, and they trouble our standard progressive narratives of technological advancement.[4] My locution of the afterimage in the title of this essay is meant to evoke this same lingering quality, but it has two additional valences as well: the first is to account for the iconography of the floppy in popular visual culture; the second is to direct us to a specific set of technical preservation practices for floppy disks, a topic which this essay treats in some detail. The floppy's afterimage is not just the fossilized remains of a dead medium from the Jurassic era of computing; it is—quite literally as we will see—the means by which the programs and data once stored on floppies are reanimated.

A floppy's virtual afterimage presides over the very window in which I write this, a tiny thumbnail icon in the privileged top left corner of the screen that initiates the Save command—itself an increasingly residual function, as most productivity software nowadays AutoSaves by default.[5] Yet given their status as residual media, floppies are also non-virtual remainders taking up space—a lot of us still have a shoebox or two of them tucked away somewhere. As they pile up, the consequences can be deadpan amusing: "Give us a sec. We goin through floppy disks", tweeted the Roots's Questlove, with an accompanying Instagram photo.[6] In this they can function as what museum professionals term *numinous objects*, seemingly ordinary everyday things that take on an aura because of the personage they have been associated with. Derrida anticipated as much: "Some particular draft that was prepared or printed on some particular software, or some particular disk that stores a stage of a work in progress—these are the kinds of things that will be fetishized in the future."[7] Sure enough, a couple of years before his death, a fan fished a dozen or so floppy disks belonging to John Updike from the author's curbside trash and posted pictures on the Internet.[8] When Prince died, Anil Dash tweeted a photo of a 3½-inch diskette emblazoned with the famous "Love Symbol" the artist had begun using in 1993; the diskette contained the font for the glyph, one of many that had been prepared and mailed to publishers and news outlets so it could properly be rendered in print.[9] The disk becomes a totem of a particular individual.

Floppies have proven generative in other ways too. Writers are often given to reminiscing about their early computer experiences; for example Karl Ove Knausgård who, in Book 5 of his autobiographical serial *My Struggle* recalls the "green luminous futuristic letters" on his monitor that are then "saved onto the thin little disk and . . . brought back to life with one tap of a finger, like the seeds that had been trapped in ice for hundreds of years and then, under certain conditions, could suddenly reveal what they had contained all this time, and germinate and blossom."[10] Other kinds of artists have also found uses for the disks' distinctive square afterimage. The photographer Jim Golden, for example, uses the regular geometry and enamel colors of floppies to compose striking visual collages.[11] London-based artist Nick Gentry has similarly used them as a patterning element in his work, employing mosaic-like arrangements of diskettes to create portraits.[12] YouTube can be mined for homegrown videos wherein the grinding and gnashing of old floppy disk drives is orchestrated to reproduce the *Star Wars* theme or similar.[13] On the Cartoon Network's popular *The Regular Show* (2010–2017), Floppy Disk joins 8-Track, Betamax, and Reel-to-Reel to form an animated quartet known as the Guardians of Obsolete Formats. Floppies have inspired pillows, coasters, notebooks, and coffee tables (complete with drawers, thus turning the table back into a literal storage device (see Figure 21.1)).[14] There is a floppy emoji and there are floppy tattoos (see Figure 21.2). It might be harmless hipster nostalgia, but such persistence should not surprise us: floppy disks have been willfully defying their own obsolescence ever since that wagging, flexible descriptor was applied to their most widely distributed form factor, the patently inflexible hard plastic 3½-inch disks that had begun appearing by the mid-1980s, only to be superseded themselves by what were termed "hard" drives in contrast.

Hard drives, in fact, historically predated floppies, which were once regarded as the more advanced medium. Hard drive or hard disk technology—itself the successor to a wide variety of exotic storage media, ranging from punched paper tape to mercury delay lines and so-called

Figure 21.1 The Floppy Table. (Source: www.floppytable.com/)

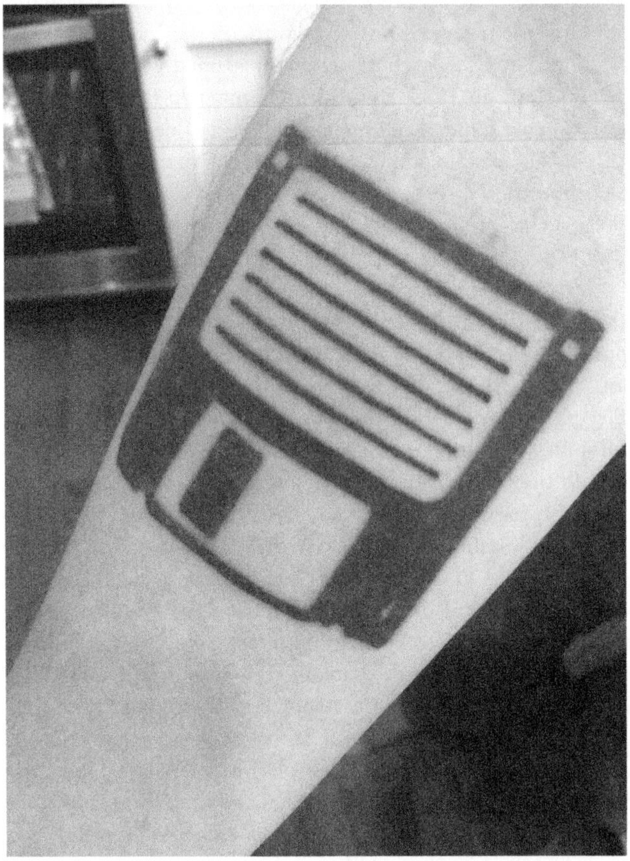

Figure 21.2 Floppy tattooed on the right forearm of Trevor Owens, a digital preservationist. The tattoo is based on an image included in the Noun Project. Photo by the author.

Williams tubes or cathode-ray tubes to magnetic drums and ringlets, as well as, eventually, magnetic tape—was initially developed by IBM in the 1950s, with a system called the RAMAC 305. Displayed at the 1958 World's Fair, the 1960 Olympic Games, and soon thereafter in glass-walled showrooms belonging to corporate clients including United Airlines, the RAMAC was an industrial refrigerator-sized unit moved with a forklift. Its storage capacity? About 5 megabytes.[15] Floppies only became commercially available in 1971 (again from IBM) in the original 8-inch format; by 1976 Wang Labs had introduced the so-called "minifloppy", smaller at 5¼-inches, less cumbersome, and less expensive. Though the storage capacities of both early formats were dwarfed by a hard disk, their portability and affordability were understood as key advantages in the burgeoning market for home and office computers.

Floppies were media that wore their material particulars on their (paper) sleeve, literally: single-sided, double-sided, single density, double density, high density, hard-sectored, soft-sectored—incantatory speech brushing the particulars of magnetic recording technology across the tongue. This was guild speech, and further initiation yielded further mysteries: tracks, sectors, nibbles. (Floppies, in fact, demanded a literal act of initialization in order to be ready for use, one of the first rituals home computer users mastered.) Jerry Pournelle, a science fiction author and correspondent for *BYTE* magazine, devoted continual column ink to tracking ongoing changes in the industry: "Shirt-pocket floppies haven't been out long enough; I don't recommend them just yet", he wrote in 1984, meaning the then new-fangled 3½-inch disks. "I believe they (along with hard disks) will eventually replace both 8-inch and 5¼-inch disk drives; but not yet."[16] Pournelle's prognostication proved right: by the time the first integrated systems such as the Apple, Commodore, Kaypro, and Osborne appeared, the 5¼-inch disk had emerged as the preferred option, appreciated for its convenience and relatively capacious 140 KB of storage. But the rigid 3½-inch disks eventually supplanted them as the industry standard. Smaller and more durable, sheathed in a hard plastic carapace and with significantly larger storage potential, it was inevitable.[17] Even then though they were still universally known as floppies, right up until when Sony (the last company to commercially manufacture 3½-inch disks) finally ceased production in 2010.[18]

The floppy disk—with its seemingly self-contradictory geometry, rectilinear rather than circular—furnished the unlikely platform for the first popular digital culture the world had ever known. Floppies were to personal computing what the paperback book was to publishing and mixtapes were to pop music. They made it trivially easy to move material from one computer to another, the classic incarnation of the so-called "sneakernet". Unlike a hard drive, sequestered deep inside the recesses of the machine, a floppy was something you held in your hand. They were visible *and* tangible. (Indeed, the instructions that came with floppies were replete with warnings about which areas of the diskette you could and could not touch.) Even the write-protect mechanisms yielded haptic engagements that quickly became muscle memory, whether applying (and peeling to remove) an adhesive bandage from the diskette's outer envelope or manually thumbing a plastic switch up and down. Meanwhile, a hole-puncher let you double-side the diskette, a second write-protect notch transforming the floppy into a "flippy" disk and thereby doubling your storage capacity (useful when you were eleven years old and a fresh box of floppies was something you had to save your allowance for). And like some talisman out of a Borges story, these strange new objects (suddenly seemingly everywhere) spawned additional new artifacts in their wake—an enormous array of purpose-built containers, boxes, bins, travel packs, albums, flip-books, and carousels were available for home and office, made of plastic, metal, pressboard, and occasionally hand-tooled from real wood. Storage, it turned out, that demanded a goodly amount of storage for itself.

Floppies also quickly came to cohabitate with other media, notably books and magazines. When Microsoft Word was released, a copy of the program was included on a floppy slipped into the November 1983 issue of *PC World*, the first time that particular gimmick—soon to become a staple of companies such as AOL—had been seen in the industry. A year later Robert Pinsky, later to become Poet Laureate of the United States, co-wrote an interactive text adventure called *Mindwheel* (1984) that began as prose fiction you read on the printed page and continued on a floppy disk you plucked from the sleeve at the back of the book and slotted into your Commodore, IBM, or Apple computer.[19] Anticipating an entirely new literary form, the publisher, Brøderbund, trademarked the phrase "Electronic Novel". Libraries, for their part, had to develop new lending policies for the floppies that found their way (Trojan horse-style, inside of the books) into their stacks. Floppies were thus mass media that could be used for promotions and publications, but they were also personal media—you could outfit them with labels and annotations and decorated sleeves (see Figure 21.3). And like the mixtapes that were ubiquitous at about the same time, you could lovingly curate them, loading up an individual diskette with programs, text files, clip art, fonts, or whatever else you wanted to share. We might even say floppies were social (or at least sociable) media.

Because the disks could only hold so much, floppies made the weight and heft of software something felt by nearly every user. It was not uncommon for programs to sprawl across a dozen or more diskettes which, to install, one had to sit and assiduously attend to, swapping them in and out of the drive to load the entire thing (and perish the thought of a read error on the final disk). Even once installed, many programs demanded the regular swapping of floppies, as between a program disk and a data disk, back and forth, again and again. (Two drives on the same system were thus desirable to reduce this need for fumbling.) But perhaps nothing served to demonstrate the curious physicality of software as convincingly as a gag built into Jordan Mechner's first published game, *Karateka*. Released by Brøderbund for the Apple II in 1984, *Karateka* was a side-scrolling martial arts fighting game with unusually good

Figure 21.3 5¼-inch floppy disks labeled and decorated by the writer and artist Judy Malloy. Photos courtesy of Lori Emerson, used with permission.

graphics and its own musical score, as well as dramatic, cinematic cut scenes—all of it packed onto a single 140 KB floppy, a truly formidable piece of software engineering. The diskette, however, also contained a second copy of the game on its reverse side; and if a player were to accidentally insert it into the drive upside down, the game would obligingly appear on the screen but *also* upside down! In retrospect, it is exactly the sort of thing a puckish young designer (Mechner was just twenty at the time) would do—and it was a programming trick, not an actual consequence of the disk's being inverted. Nonetheless, the effect speaks powerfully to the way in which the medium of the floppy created a tangible sense of connection between how the user handled the disk and what went on inside of the computer, much as playing a record backwards could be exploited to similar ends.

The limited storage capacity of the disks—ranging from 80 KB for 8-inch disks in the early 1970s to 140 KB for the 5¼-inch to (eventually) 800 KB and then more than a megabyte with 3½-disks—their capacity *increasing* even as the physical dimensions shrank owing to advances in magnetic recording—was a palpable constraint, but it also spurred innovation (and a healthy respect for efficiency) in programming. Perhaps the greatest testament to this limitation is the digital artistic practice known as the demoscene. Though seen on a variety of platforms today, the demoscene originated with software pirates who, when cracking a disk (more about this in the next section), took to adding ever more elaborate ruffles and flourishes to the intro screens of hijacked programs as signatures of their work. Because there was often very little space on the disk to spare, these additions demanded complex bags of tricks that evolved into their own form of competition between rival hackers and crackers. The results were virtuoso demonstrations ("demos") of how to maximize dramatic visual and audio effects while minimizing resources such as processor speed, memory, and disk storage through creative coding.[20]

Floppies even had a distinctive aural dimension. Indeed, unlike the smaller sizes that came after them, 8-inch diskettes were kept in continual motion by the drive, creating a background noise that some users found distracting; it was regarded as a drawback. But that familiar grinding crunch could also be expressive: an experienced practitioner could tell a lot from the sound of the drive's churning as it tried to read an unreadable sector. Maybe the disk was caught in a copy protection loop, or maybe it was a bad sector (the underlying media physically damaged); those were both recognizable sounds. The noise made by the drive could also function as an important cue in games, when the sudden sound of the floppy being accessed meant a new level or a new piece of the puzzle was being loaded into RAM memory. Even the time required to read the floppy into memory could prove expressive, the lag contributing to the sense of a journey in a game such as *Oregon Trail* (1985). Less poetically, loading a new level from the floppy gave harried players a chance to flex their fingers and breathe.

The humble, seemingly unadorned floppy disk offers a rich index to the material culture, technical practices, and creative constraints of the personal computing era. Floppies have been continually defined and redefined by their inherent contradictions, ranging from the superficial—that their outer shape is in fact square, or that (in time) they grew rigidly inflexible—to the more profound: limited in size and capacity but seemingly endlessly expressive and generative, a storage solution that is undeniably obsolete even as it is still visibly and palpably present. As we will see in the next section, the afterimages of floppies endure in unexpected and even remarkable ways.

Shoebox Stories

Floppies might have been all that, but they could also be the wellspring of considerable frustration and woe. Any disk could fail at any time, and the most basic lesson every home computer user internalized was backing up their work. Backups had to be made early, and often. This applied to your data—your spreadsheet files, say—but it also applied to the programs themselves. Shrink-wrapped software was expensive, and a customer who bought a game or a word processor expected to be able to make backup copies of the program for the day when the original disks failed. (Keep in mind that for systems without a hard drive, the program disks would be in continual use.) There was one problem: if the owner could copy a program disk for backup purposes they could also copy it for other purposes, including passing it along to friends or even selling bootlegs at a discount. Piracy became commonplace, and a generation of users who otherwise might not have imagined themselves especially technically inclined learned their way around disk operating systems and low-level data structures just by trying to dupe a copy of *Miner 2049er* (1982) or *Choplifter* (1982) from a friend.

As a result, software developers waged an ever-intensifying battle with software pirates, employing ever more recondite techniques which sometimes seemed to bend the laws of magnetics to encode data in novel and unpredictable ways. For example, one form of copy protection perfected by a programmer named Roland Gustafsson involved tinkering with the divisions between the different sectors (or cells) of the diskette's internal formatting, so that a standard 140 KB floppy could actually be made to hold a couple of dozen extra kilobytes; when some hapless individual attempted to "crack" such a floppy they would find that it—somehow—contained more data than they could save on to a duplicate disk of supposedly the same size.[21] Pirates, for their part, matched each anti-copy scheme with another generation of cracking tools designed to defeat it. Some companies went low-tech, selling software for pennies on the dollar and simply imploring users to buy it rather than copy it; others made programs dependent on access to a manual and (especially for games) other kinds of paraphernalia, so-called "feelies" packaged with the disk itself. (Infocom set the gold standard for this approach with their best-selling interactive fiction titles and their feelies are now collector's items.) By 1992, the problem had become so acute that the Software Publishers Association produced a 9½-minute music video entitled "Don't Copy That Floppy" featuring an extended rap routine of the titular jingle intercut with earnest dialogue between two precocious teens. The performance was unquestionably excruciating; the message, however, was uncompromising: if you keep on copying those floppies then software companies will not make money and programmers (maybe you want to be one yourself some day?) will not get paid; and the industry, like the screen, will go dark; game over.[22] (And if that was not enough, the lyrics relentlessly hammered the contention that copying a floppy was not just morally wrong, it was illegal.)

If the battle against illicit copies was once seen as life and death for the software industry, today the ability to successfully copy (migrate) data from obsolescent media is vital to preserving the content stored on them. Despite expectations to the contrary, data on disks from the 1970s and 1980s is often still recoverable by digital conservationists. Nonetheless, the realities of magnetic recording and decomposition can only be staved off for so long; with each year that passes more and more data on those disks decays. As the computer historian and archivist Jason Scott warned on his blog: "Someone has to break it to you, and that person is me. It's over. You waited too long. . . . With some perseverance and faced against all the odds stacked against you, something might get out of these poor black squares, but I would not count on it."[23]

Transferring or migrating data from floppies is a relatively straightforward procedure for the digital conservationist, at least with 5¼- and 3½-inch formats (8-inch and other sized disks are less common and present greater challenges).[24] Tools and hacks exist for connecting old drives (whether scrounged from eBay or dusted off from the basement) to contemporary systems with hardware bridges; at that point, if the diskette itself is still physically intact and if the data stored on the media has not been compromised, the bits can usually be read from the floppy, yielding what is known as an "image"—a complete virtual surrogate of all the information once stored on the disk. A disk image is not a photograph of a floppy but rather a bit-by-bit copy (sometimes also known as a bitstream) of the source, including partially deleted fragments of old files and other extraneous data (a tool called a hex editor can be used to examine the raw bytes of the image and extract strings of text or other data). Disk image files are thus self-contained digital objects of the same size (80 KB, 140 KB, 800 KB, etc.) as the original media that can circulate (and multiply) as content on our own contemporary systems and networks. Minus the actual material qualities of the diskette—its label and outward appearance—they are true simulacra of those "poor black squares", such that a disk image created under proper circumstances is legally admissible as evidence in a court of law. A more common use to which disk images can be put is to be run with emulation software in order to recreate the experience of some now-vanished program. They are talismanic in that sense, or uncanny, but the process of obtaining one also inculcates an extreme awareness of the mundane array of connections, cables, controllers, hacks, patches, and workarounds required to circumvent the fundamental incompatibility between a media format now some thirty or forty years old and the hardware coming off assembly lines today. As Jonathan Sterne reminds us, the reality of computing is often characterized by precisely such juxtapositions of "old and new" media, with obsolesced devices, defunct and dysfunctional cables and connectors, and spare parts accumulating and commingling.[25]

Some remarkable recoveries from floppies have now come to light. For example, in April 2014, a team of scholars, artists, and engineers based at Carnegie Mellon identified and retrieved digital paintings created by Andy Warhol on an Amiga 1000 computer in 1985, stored on 3½-inch floppy disks at the Andy Warhol Museum. After some elaborate intermediary steps, including reverse engineering the proprietary format in which the files were originally created, the previously unseen artworks were released to the public.[26] As befits a find of this magnitude—a dozen new Warhols!—press coverage was extensive. Some six months earlier a similar recovery had taken place at the New York Public Library, which owns Timothy Leary's papers. As one press outlet noted, the "papers" include some 300 floppies, "containing notes on everything from cybersex to cryogenics [and] letters to famous actors and artists."[27] Also among the digital detritus were prototypes of several incomplete video games designed by Leary (one of them based on William Gibson's novel *Neuromancer* (1984)); the games, which had not been accessible for decades, are described as their own quirky, whimsical expression of Leary's worldview.

Both the Warhol and the Leary stories involve (not altogether coincidentally, one might argue) two members of the counterculture born in the same decade who both began using personal computers for creative experimentation at about the same time, and whose efforts were retrieved from the floppy disks containing their frozen "seeds" within mere months of one another. And yet, with all those shoeboxes tucked under beds and in basements, much of the most important preservation and recovery work around floppies is being done by individuals, fan communities, and institutions that do not necessarily have big white pillars out in front. Jason Scott (who now works for the Internet Archive, which actually *does* have big, white pillars out in front) was for many years a free-agent, initiating and pursuing his own projects. One person who sought him out was the aforementioned game designer and screenwriter Jordan Mechner. Though Mechner's career had been launched by *Karateka* he would become best known for his *Prince of Persia* franchise, another casually Orientalist series of titles growing ever more sophisticated as they migrated to different platforms and publishers, culminating in a feature-length Disney film in 2010. Mechner is a self-described packrat, having saved arcana from every stage of his career. Fans, for example, can go online to peruse the original sketches and storyboards done for *Karateka*. And yet, like some Derridean parable, Mechner's personal archive contained an absence at its very center: the original source code for the first Apple release of *Prince of Persia* (1989), maddeningly missing for many years. Enter Mechner's father, with shoebox: cleaning house he had found a cache of floppy diskettes, which he shipped to his son. Among them was one whose label suggested it might contain the long-lost code.

The rest of the story is quickly told. Mechner enlisted Jason Scott, who in turn brought in a retrocomputing expert named Tony Diaz. Attended by a reporter and photographer from *WIRED*, the trio convened at Mechner's homestead in Hollywood. The diskettes Mechner's father had found were literally covered with dust, and there was no guarantee a successful recovery would be possible; nonetheless the stage had clearly been set for an event. Diaz had his methods, including gently bathing the diskettes with detergent if need be. He also brought with him an impressive array of vintage Apple II computers that he has hot-rodded with enhancements such as Ethernet ports. A number of bits and pieces from Mechner's career were salvaged that day, the *Prince of Persia* code included; also several prototypes of unreleased games. Some of the disks, true palimpsests, contained multiple layers of data, thereby reminding us of another critical affordance of the floppy—that unlike the CD-ROMS that followed them, they were reusable. As narrated by *WIRED*:

> Mechner keeps handing disks to Diaz, and more golden idols keep spilling out. Like many computer kids of his day, Mechner used old floppies and rewrote them with new data. One of them used to hold part of Roberta Williams' early Sierra On-Line adventure game *Time Zone*. Now, Mechner says, it holds the only copy of one of his earliest games.[28]

Scott, meanwhile, was cloning additional diskettes and uploading the salvaged code to GitHub, where it was instantly accessible to onlookers from all over the world—word of the event had leaked out, and people were logged on and waiting. The *Prince of Persia* source code has since been pored over by enthusiasts, shared, commented on, and generally cherished.

WIRED's prose here is worth quoting not only for the color it adds to the description of the event, but also for several other qualities of what we might think of as the shoebox story, an emerging genre of computer folklore. As the headlines read—invoking one of the franchise's most successful installments—the *Prince of Persia* had been rescued from the sands of time. The rescue, however, is (knowingly, one assumes) staged by the journalistic coverage as just such an exotic escapade as is portrayed in the games themselves, replete as they are with acts of plunder and abduction. The tomb raiding is depicted here complete with middle-aged computer geeks in the swashbuckling role of media archaeologists. Complementing this stock set-up is the aura of the lost original, the "only copy" of the isolate object of desire, that oxymoronic phrase embodying all of the contradictions of memory and materiality when that which is unique and precious is preserved through proliferation (digital archivists are fond of the acronym LOCKSS: Lots Of Copies Keeps Stuff Safe).

The technical achievement of this and other data recoveries (like the Leary or the Warhol) is genuinely great; their importance to their constituencies is undeniable; but when the after-images of these lost floppies come in to contact with cultural imagery as casually well-worn as Indy's leather jacket, they leave the shoebox story at risk of replicating a narrative tradition vexed by acts of appropriation and exploitation. Many of the programs now dormant on quiescent floppies belong to a category known as abandonware, their copyright owners impossible to contact or even identify owing to mergers and takeovers and buyouts in the industry. This is to say nothing of the personal data that inevitably turns up on diskettes, raising issues of ethics and privacy. Who has the rights to the collective heritage of a moment described by one of its participants, a member of the computer underground, as follows: "'People' who never existed did things that never took place, upon a stage of fragmented software that currently sits on a hundred thousand disks in dusty boxes, chronicling events that happened only by mutual wish-fulfillment."[29] What happens, in other words, when the owner and originator of the source code is not there standing over the shoulder of the experts seeking to retrieve it?

A corresponding frame of reference can be found in the assiduous work of the Software Preservation Society (SPS). The imposing name notwithstanding, they are a tiny group of technologists based in the United Kingdom who have dedicated themselves to creating solutions for locating, duplicating, documenting, and preserving disk images of classic computer games, especially those released for the Commodore Amiga. They are also an advocacy group who are vocal about the need for game preservation, and they are distinguished by a radical commitment to authenticity, meaning that they will only acknowledge as legitimate those disk images that can be proven to have originated from unaltered factory-written exemplars of the original title. "I don't count games as preserved if they have been modified in any way by a third party, not only that, I do not count games as preserved if they cannot be proven to be unmodified", they write on their Web pages. "Museums", they go on to note, "expend a large amount of effort and money in determining the originality of the artifacts they hold."[30]

As the willful and repeated comparison to museum curation and traditional artwork makes plain, the SPS sees itself as part of a cultural heritage continuum that seeks to employ the same stringent standards of authenticity and verification commonplace in traditional conservation to the arena of computer games and software. Software and games, however, were especially vulnerable to corruption as unauthorized (pirated) copies were passed from hand to hand. Sometimes (as happened with Mechner's own diskettes above) fragments and traces of past programs remained readable on the media; moreover (as we have also seen) because of the competitiveness between rivals, so-called "crack screens" (soon to become demoscenes) were frequently grafted onto the opening credits of a pirated game so a cracker could enhance their reputation. Finally, most home computer games shipped without the write-protect feature on the diskette enabled, which was necessary in order to accommodate many seemingly mundane features of gameplay, including saving high scores back to the disk and saving game sessions in progress. No disk that cannot be proven to have been unaltered from a factory-level master can therefore be wholly reliable as a preservation copy because there is no guarantee that the actual software's functionality has not been compromised at some point in its past life; in other words, any written intervention in the disk's data—a saved game, a crack screen—could, in theory, introduce bugs that would crash the program or otherwise corrupt its behaviors. Given the radically constrained storage space on a disk, this was more common than people realize; crack screens and demoscenes, for example, were sometimes added at the expense of actual program data.

The technology with which the SPS is closely associated is a hardware device called the Kryoflux, one of several controller cards available for connecting legacy disk drives to modern systems for purposes of data migration (see Figure 21.4). Unlike its competitors, however, the Kryoflux operates at the symbolic register of the actual magnetic encoding on

Figure 21.4 A Kryoflux (at right) in use with a 3½-inch Sony drive. Source: http://amigax1000.blogspot.com/2014/03/kryoflux-on-x1000-and-a4000t.html.

the disk, the "fluxes" whose sequences are mathematically reconstituted as the zeroes and ones of binary computing. For floppy disks do not store anything as simple as actual bits; rather, they record sequences of magnetic flux reversals that are themselves encoded using recondite schemas bearing names like Run Length Limited and Group Code Recording. Magnetic Force Microscopy, a variation on electron microscope technology, is able to visualize the fluxes on the surface of the media, yielding startling, alien imagery which—though it encourages us to resort to inscriptive metaphors like "read" and "write"—is properly apprehended as a signal the firmware seeks to interpret from the media. Not magnification then, but stochastic prediction and rendering.[31]

The archivist-turned-media-theorist Wolfgang Ernst anticipates what is at stake in these obscure details and terms: "Technological media that operate on the symbolic level (i.e., computing)," he writes, "differ from traditional symbolic tools of cultural engineering (like writing in the alphabet) by their registering and processing not just semiotic signs but physically real signals."[32] These signals (or fluxes) are temporal rather than semiotic markers, very much manifestations of the time-criticality and micro-temporalities Ernst understands as definitive of modern media; or, as the SPS put it with regard to the Kryoflux, "Reading . . . can be done by timing, at a very fine resolution, how far the flux transitions are apart and then deriving the intended bit cells."[33] In other words, the appropriate frame of reference here is sampling as much or more than inscription; and within this frame, we enter into wholly new registers of media perception and individuation; what others have termed *variantology*: "We can see when this happens as each mechanical drive has its own 'fingerprint,' made by the 'crackles and pops' unique to each one", claims the SPS. "Disk images made from commercially mastered disks have far less of these mechanical stutters due to the high quality equipment used."[34]

And yet, the higher the tolerances of the hardware, the more chimerical the quest for absolute fidelity becomes. Practitioners working with the Kryoflux report that even routine fluctuations in the electrical current being used to power the drive can impact the rendering of the flux stream as captured by the device. "Computing", as Jean-François Blanchette observes, "is material through and through. But this materiality is diffuse, parceled out and distributed throughout the entire computing ecosystem."[35] If that ecosystem includes not just magnetic flux transitions but the very electrical current—the deep core of the power grid itself—then we should ask what media archaeology ultimately seeks to excavate, or more precisely how far down those excavations go, and where the fault-lines finally tremble. Get close enough to the metal and you just might find a wormhole waiting there, one that leads back outward, to the planetary-sized ecosystems on the other side that are necessary to still spin these diminutive disks containing worlds.

Flash Back

There are other shoebox stories to tell. One might be the restoration of Canadian writer bpNichol's "First Screening", a suite of text experiments for the Apple II, from a twenty-two-year-old floppy disk.[36] One might be the capture and release of a disk image of William Gibson's elusive electronic poem "Agrippa", famously designed to (appear to) encrypt itself after a single reading; with the disk image now in hand the text can be read any number of times in an emulator, thus preserving the poem's existence while, arguably, destroying its very essence.[37] And then there is the shoebox story that centers around widely publicized efforts to locate and restore copies of the earliest HTML "pages" ever published on the Web. This one also made headlines: "The first draft of the World Wide Web has gone missing", technology news outlet CNET reported in May 2013, "with perhaps one of the only copies of the very first Web site floating around the world's drawers or attics on a floppy disk somewhere."[38]

The story is actually more complex than that: as National Public Radio in the United States was the first to relate, in 1992 Tim Berners-Lee had saved a complete copy of the early CERN site to a 3½-inch floppy for use in demos while traveling. (A first generation "Web site", after all, was just a collection of non-dynamic text and image files that could be accessed locally from removable media as easily as from a Web server.) A version of the 1992 CERN site in this state had, in fact, been intermittently available online for some time.[39] CERN's very first Webpage, however, was created in 1990; the speculation, therefore, is that a version of it from as much as two years earlier than Berners-Lee's demo disk might also still exist, most likely not on a floppy but a large-format optical disk. "It was such a beautiful object, that optical disk, that someone maybe has it on their coffee table or their bookshelf", NPR quoted CERN's current Web historian Dan Noyes.[40] There are, then, two separate disks in question, the 3½-inch demo floppy and the absentee optical disk. Several days after the initial NPR story, the University of North Carolina's Paul Jones reported having a copy of the CERN site—retrieved from yet a third disk—dating from 1991. Likely other copies are extant as well. The story illuminates a complex, non-progressive historiography in which portions of the early Web were routinely downloaded and consigned to floppies and other removable media, as backups, demos, and the like. The floppy disk, universal iconic token of obsolescent media, thus becomes the conveyor and purveyor of the very earliest incarnations of a contemporary medium that we are routinely asked to accept as having now assumed its apotheosis in "the cloud".

If, as Erkki Huhtamo claims, media archaeology is the study of "cyclical phenomena which (re)appear and disappear and reappear over and over again in media history and somehow seem to transcend specific historical contexts", then floppy disks ought to be a primary object for such attention.[41] The "poor black squares" preemptively eulogized by Jason Scott are easily mistaken for transitional technologies, inserted into computing history between the industrial regimen of tape and the silent, invisible, always-increasingly capacious hard drives that replaced them, along with other near-forgotten disk formats such as Zip and Jaz, and optical media such as CD-ROMs. None of these, however, has instilled the affection in their users, and none have proven so iconic of what it means to "save" something digitally. The stubborn resilience of the floppy's reappearance in media histories and the persistence of its afterimages perhaps suggests why neither the raiding of lost antiquities nor the kind of curatorial fundamentalism espoused by the SPS is the most appropriate range of reference for their continued significance today.

On the contrary, floppies are not rarefied or precious; they were and are ordinary everyday media, ragged and gummy, often dust coated, maybe a touch ridiculous in retrospect. (One much loved bit of lore has it that the 5¼-inch square was derived from the size of the cocktail napkins at the bar frequented by technicians from Shugart Associates, the firm that did the engineering work.)[42] They were not black-boxed like a hard drive or shiny and cyber-slick like a CD-ROM. Floppies, whose physical dimensions became common synecdoche for the objects they delimited, were as much a product of their moment as monochrome screens and dot matrix printers. But they were somehow more intimate than either, the object that passed from the hand of the user to the mouth of the drive and thereby brought the machine to life with a game that had been saved in *medias res* the previous afternoon or a letter or a manuscript that was slowly taking form. Floppies were where the data lived, and where people's digital lives slowly accumulated. They are perhaps, as William Gibson indeed recognized in "Agrippa", most closely analogous to snapshots, something else we tend to keep in shoeboxes and something else whose afterimages remain with us long after the original has passed from view.

The not yet obsolesced media technology that comes closest to the floppy in spirit is surely the "flash" USB stick, which is also a semi-disposable readily portable form of storage suitable for accessorizing and casual exchange. "These days, the things that seem to turn up all over the place—lurking in pockets of different bags, filling drawers, and junk boxes, dropped down the back of desks . . . There is almost certainly one within ten feet of you right now", writes information scientist Paul Dourish in an article in *The Atlantic*.[43] He notes that despite the pervasively networked world we inhabit, flash drives are still among the most reliable ways of moving data from one computer to another, especially on the spur of the moment—across the table in a coffee shop for example, or in a meeting.[44] For Dourish, "The flash drive exposes the great lie of technological progress, which is the idea that things are ever really left behind." This is a sentiment similar to that which we have already seen in Acland, Sterne, Ernst, and Huhtamo, but Dourish here has something quite specific in mind. Flash memory sticks, it turns out, employ a technology known as the File Allocation Table (FAT) to manage the data we store on them. First introduced in 1977, the FAT was originally developed at Microsoft for use with—you guessed it—the floppy disk.

Notes

Dedicated to Jason Scott, aka @textfiles. A handful of non-contiguous sentences (and associated notes) were originally published in a different context in my *Track Changes: A Literary History of Word Processing* (Cambridge: Harvard University Press, 2016), pages 219–220.

1 For the complete GAO report, see: www.gao.gov/products/GAO-16-696T.
2 There have been similar nexuses of humor in the past; for example in the wake of the complications around the initial online rollout of the Affordable Care Act in 2013, a satirical *Onion* story about a "new and improved" Obamacare website to be released on 35 floppy disks. See www.theonion.com/article/new-improved-obamacare-program-released-on-35-flop-34294. *The New Yorker*, meanwhile, ran a Barry Blitt drawing entitled "Reboot" for its November 11, 2013 cover,

depicting President Obama—oversized cordless phone in hand—standing next to an anxious Katherine Sebelius and an aide inserting a floppy disk into a hulking desktop computer.

3 Brian Fung, "The Real Reason America Controls its Nukes with Ancient Floppy Disks," *Washington Post* (May 26, 2016). www.washingtonpost.com/news/the-switch/wp/2016/05/26/the-real-reason-america-controls-its-nukes-with-ancient-floppy-disks/?utm_term=.a067ed37c1a6.

4 See Charles R. Acland, "Introduction," *Residual Media* (Minneapolis: University of Minnesota Press, 2007), pages xix–xx.

5 Debates over whether to replace the floppy icon are a staple on tech blogs and discussion forums. Connor (Tomas) O'Brien considers the issue from the standpoint of graphic design and notes that the floppy's distinctive shape, with one beveled corner, contributes toward the icon's being uniquely recognizable: http://connortomas.com/2013/04/in-defence-of-the-floppy-disk-save-symbol/.

6 http://instagram.com/p/WdgkYcwaxd/

7 Jacques Derrida, "The Word Processor," *Paper Machine*, trans. Rachel Bowlby (Stanford: Stanford University Press, 2005), 29.

8 See www.theatlantic.com/entertainment/archive/2014/08/the-man-who-made-off-with-john-updikes-trash/379213/.

9 See http://nymag.com/selectall/2016/04/princes-legendary-floppy-disk-symbol-font.html.

10 Karl Ove Knausgård, *My Struggle: Book Five*, trans. Don Bartlett (New York: Archipelago, 2016), 396. I am grateful to Justin Tonra for pointing out this passage.

11 See www.jimgoldenstudio.com/Portfolio/Relics-of-Technology/thumbs.

12 See www.miaminewtimes.com/arts/nick-gentry-brings-floppy-disks-back-to-life-at-robert-fontaine-gallery-6501192.

13 See www.popularmechanics.com/technology/audio/a21740/floppy-disk-orchestra-star-wars/.

14 See www.thisiswhyimbroke.com/floppy-disk-table/.

15 For more on IBM's RAMAC and the history of early hard drive technology see Kirschenbaum, *Mechanisms: New Media and the Forensic Imagination* (Cambridge: MIT Press, 2008), especially chapter 3.

16 Jerry Pournelle, *The User's Guide to Small Computers* (New York: Simon and Schuster, 1984), 61.

17 In fact the industry experimented with a number of different form factors, most notably the 3-inch disk developed for use with the popular Amstrad computer line in the UK.

18 According to www.dailymail.co.uk/sciencetech/article-1269142/Floppy-disks-terminated-Sony-stops-production.html.

19 See www.newyorker.com/books/page-turner/when-robert-pinsky-wrote-a-video-game.

20 For an introduction and overview of the demoscene, see *Demoscene: The Art of Real-Time*, ed. Lassi Tasajärvi (Helsinki, Finland: Even Lake Studios, 2004).

21 For more on this technique, see: http://fabiensanglard.net/prince_of_persia/pop_boot.php.

22 Watch it (or don't) at: www.youtube.com/watch?v=up863eQKGUI.

23 See http://ascii.textfiles.com/archives/3191.

24 For specifics of this process, see Kirschenbaum, "Ancient Evenings: Retrocomputing and the Digital Humanities." In *A New Companion to Digital Humanities*, eds. Susan Schreibman, Ray Siemens, and John Unsworth (Hoboken, NJ: John Wiley: 2016): 185–198.

25 Jonathan Sterne, "Out with the Trash: On the Future of New Media." In *Residual Media*, ed. C.R. Acland (Minneapolis: University of Minnesota Press): 16–31.

26 Read the press release here: www.cmu.edu/news/stories/archives/2014/april/april24_warholworks discovered.html. And the complete technical report here: http://studioforcreativeinquiry.org/public/warhol_amiga_report_v10.pdf.

27 www.wired.com/2013/10/timothy-leary-video-games/. See also James A. Hodges's technical report on *Mind Mirror*, Leary's only commercially published game: http://mediaarchaeologylab.com/wp-content/uploads/2016/10/JamesHodges_MindMirror_MALwareTechnicalReport_10-2016-1.pdf.

28 Gus Mastrapa, "The Geeks Who Saved Prince of Persia's Source Code from Digital Death," *WIRED* (April 20, 2012): www.wired.com/2012/04/prince-of-persia-source-code/.

29 The quotation is from Patrick K. Kroupa, "The Akashic Records of Cyberspace" (1993): http://exciteddelirium.net/the-akashic-records-of-cyberspace/. Kroupa was the founder and proprietor of the quasi-legendary computer bulletin board MindVox.

30 See www.softpres.org/article:importance_of_digital_preservation and www.softpres.org/article:importance_of_data_authenticity.

31 For more on MFM and these other details, see Kirschenbaum, *Mechanisms*, especially chapter 1.

32 Ernst, "Media Archaeography: Method and Machine versus History and Narrative of Media." In *Media Archaeology: Approaches, Applications, and Implications*, eds. E. Huhtamo and J. Parikka (Berkeley: University of California Press, 2011): 242.

33 See www.softpres.org/glossary:bit_cell.

34 See www.softpres.org/faq:imaging_disks:disk_modification.

35 Jean-François Blanchette, "A Material History of Bits," *Journal of the Association for Information Science and Technology* 62.6 (June 2011): 1042–1057.

36 For details see https://vispo.com/bp/introduction.htm.

37 See Kirschenbaum et al., "'No Round Trip': Two New Sources for *Agrippa*," *The Agrippa Files* (December 2008): http://agrippa.english.ucsb.edu/kirschenbaum-matthew-g-with-doug-reside-and-alan-liu-no-round-trip-two-new-primary-sources-for-agrippa.

38 CNET reported the story on May 23, 2013: http://news.cnet.com/8301-17938_105-57585922-1/search-is-on-for-lost-first-draft-of-first-web-page/. For NPR's original May 22 piece, see: www.npr.org/2013/05/22/185788651/the-first-web-page-amazingly-is-lost. A May 30 follow-up in *The Atlantic* details Paul Jones's discovery: www.theatlantic.com/technology/archive/2013/05/world-we-have-lost-the-first-webpage-professor-oh-i-have-a-copy-of-it-right-here/276387/.

39 Currently available here, at the original URL: http://info.cern.ch/hypertext/WWW/TheProject.html.

40 The "beautiful object" in question is a Canon optical disk manufactured for the NeXT computer: "Steve Jobs' vision for the future was simple: without any other kind of permanent storage, users would keep their entire universe of files and operating system on a disk. . . . They could move from machine to machine, taking with them hundreds of megabytes of digital files." See www.thegogglesdonothing.com/archives/2010/03/next_optical_discs.shtml for more.

41 Huhtamo, "From Kaleidoscomaniac to Cybernerd: Towards an Archeology of the Media": http://web.stanford.edu/class/history34q/readings/MediaArchaeology/HuhtamoArchaeologyOfMedia.html.

42 As narrated here, for example: http://archive.computerhistory.org/resources/text/Oral_History/5.25_3.5_Floppy_Drive/5.25_and_3.5_Floppy_Panel.oral_history.2005.102657925.pdf.

43 Paul Dourish, "Why Flash Drives Are Still Everywhere," *The Atlantic* (June 30, 2016): www.theatlantic.com/technology/archive/2016/06/why-flash-drives-are-still-everywhere/489458/.

44 Flash drives are also the backbone of samizdat information exchange in places where networking is either not ubiquitous, or subject to authoritarian control. In Cuba, for example, this system is known as *El Paquete*. See www.fastcompany.com/3048163/in-cuba-an-underground-network-armed-with-usb-drives-does-the-work-of-google-and-youtube.

22

VIDEO GAME CARTRIDGES

The History of Durable, Removable, and Portable Software

Michael Thomasson

The word "cartridge" is derived from the French word "cartouche", and its origins stem from military applications dating back to the 1500s. The initial ammunition cartridges were nothing more than charges of gunpowder bundled for quick access, usually housed in waxed paper, linen, or even animal tissue. By the turn of the 17th century, ball and powder wrapped together were commonplace. New chemical concoctions improved the powder formula, while the idea of replacing the conventional spark with a hammer blow necessitated the introduction of metal tubes. This all led to the development of self-contained fire cartridges consisting of a metal case, a propellant charge, a projectile, and a primer—and forever changed the world. The cartridge concept was eventually employed into other facets of life such as camping stoves, washing machines, and ink and toner cartridges used for printing. It seemed to lend itself especially well to media applications, with radio stations using Fidelipac (a.k.a. NAB) cartridges for broadcasting in the 1950s. The concept stuck, and was adopted by magnetic tape media such as the Muntz Stereo-Pak (1954), 8-track tape (1964), compact cassette (1968), and dozens more . . . ultimately leading up to the video game cartridge.

From its inception, and for almost three decades the video game cartridge was the primary means to distribute video game software on home gaming consoles. Simple in design, it was perhaps the most user-friendly process to launch computer software; a sturdy plastic or metal casing that could produce a game within seconds of plugging it into a console. It required no typing of command lines or other complex maneuvers by the user; in fact, it could be inserted and launched with only one hand.

The first commercial home video game system was the Magnavox Odyssey (1972), referred to as the Philips Videopac G7000 in most markets outside of the United States. It was engineered by Ralph Baer and first introduced a cartridge of sorts to the video game industry. However, they are not considered to be true cartridges, since they did not contain any read-only memory (ROM). These "game cards", as they were called, were instead simply plug-in jumper cards with dedicated logic circuits. They did not house a game internally, but only modified the internal circuitry of the console, allowing instructions residing within the Odyssey console itself to be accessed for interaction.

Other early video game consoles were standalone in design. Each unit featured a single game, or a variation on a game, such as Sears's *Tele-Games* (1975) or Coleco's *Telstar* (1976) which played *PONG* (1972) or similar ball-and-paddle games. They essentially housed General Instruments (GI) AY-3-8500 "Ball & Paddle" integrated circuit (IC) or similar chip, which output each game on a television screen via RF modulation. The AY-3-8500 was often referred to as "Pong on a chip" by nontechnical laymen. Later gameplay evolved, offering Evil Knievel-style motorcycle stunts via Atari's *Stunt Cycle* (1976) or the ball-and-brick-breaking gameplay of *Breakout* (1976) or *Video Pinball* (1978). The games were fun, but often it was only so long before the novelty wore off and the games themselves began to seem routine. When that happened, there was no way for the publisher to upgrade the console to make it seem "fresh" again.

Development of the ROM Cartridge

Before the development of the cartridge, inserting and removing socketed electronic assemblies had been an activity reserved for trained technicians, engineers, and military personnel. Furthermore, other forms of interchangeable software required costly hardware such as spinning magnetic disks, or called for special equipment to read media such as paper tape or Hollerith punch cards. All of these options were impractical and cost-prohibitive to be used for consumer goods.

When Ted Hoff and Intel revealed the 4004 microprocessor, multiple circuit boards full of electronics were replaced by a single chip smaller than a fingernail. A few revisions later, Alpex Computer Corporation engineers Lawrence Haskel and Wallace Kirshner chose Intel's third chip, the Intel 8080, and stuffed it into a 16 × 6 × 5-inch metal box. They accompanied it with an excessively convoluted keyboard for user access, and created what they called the Remote Access Video Entertainment project, referred to as RAVEN (1974).

After playing video ping-pong on the Odyssey in a department store, Haskel left inspired and impressed, and proceeded to program a hockey game for the RAVEN console. He followed this up with *Tic-Tac-Toe, Shooting Gallery*, and *Doodle*, a primitive art program. As the game library grew, it was evident that a more efficient means of loading software was needed. Kirshner's solution was to take an erasable-programmable read-only memory (EPROM) chip, mount it to a circuit board, and then bond the chip's pins to a durable 25-pin connector that could bear frequent insertion and removal. Following a trip to the local Radio Shack to purchase a 3 × 5-inch black plastic box, the first video game cartridge was fashioned!

Alpex did not have the funding to go it alone, and since the RAVEN system was already television-based, they looked for partners within the TV industry such as Motorola, RCA, Sylvania, and Zenith. Finding no interested parties, Kirshner and Wallace then approached semiconductor manufacturers, who were launching consumer electronic appliances such as pocket calculators, digital watches, and electronic clocks.

Fairchild Semiconductor signed on to license the technology and renamed it STRATOS. The chief hardware engineer, Gerald "Jerry" Lawson, was part of the infamous Silicon Valley Homebrew Computer Club that produced such industry notables as Apple co-founders Steve Wozniak and Steve Jobs. Jobs had worked for Atari and introduced Jerry to Atari's co-founders Ted Dabney and Nolan Bushnell, who in turn demonstrated *Computer Space* (1971) to Jerry, inspiring him to build his own game. Implementing microprocessors obtained via his employer Fairchild Semiconductor, he engineered the game *Demolition Derby* (1973) in his garage. (Lawson's original *Demolition Derby* was worked on during 1972–1973, and believed to be completed in 1973; but the game itself was not commercially made available until 1977 by Chicago Coin.)

Jerry Lawson's first step toward production was to switch the competitor's Intel 8080 CPU with Fairchild's own F8 chip. The archaic keyboard was replaced with a new type of hand controller. The system was renamed again as the Fairchild Video Entertainment System (VES) and was ultimately released in August of 1976 as the first video game system to employ its own microchip. When Atari released its first cartridge-based console, branded the Video Computer System (VCS), Fairchild changed the name of the VES again to Channel F (short for Channel Fun) to avoid confusion between the two competing units. When Fairchild's video game cartridges were first introduced, they were referred to as "videocarts". What made ROM game cartridges revolutionary was that game systems were no longer limited to what was manufactured in the factory during production. The lifespan could be prolonged via cartridges that housed their own program or game, essentially giving a new lease on life for the product. The videocarts were yellow in color and generally mimicked the design of an 8-track cartridge, the trending music delivery system of the time. They had an ingenious spring-loaded mechanism that protected the 22 gold-plated contacts hidden and protected within, which allowed the cartridge to be inserted into the console without fear of electrostatic discharge, or physical destruction of the semiconductors.

The debut videocart for the Fairchild VES was *Videocart-1* (1976) which featured four distinctly different games. The cartridge offered the traditional pen-and-paper game *Tic-Tac-Toe*, a simple *Shooting Gallery*, an art program that allowed the user to *Doodle* computer graphics on screen, and finally an interactive kaleidoscope referred to as *Quadradoodle* (note that two of the four games were salvaged from Lawrence Haskel's work on the aforementioned RAVEN project). Twenty-six different videocarts were published for the Channel F in most markets.

In January of 1977, RCA entered into the home video game market with the release of the RCA Studio II. The Studio II cartridge resembled the original Magnavox Odyssey circuit board game cards, except that they contained ROM chips. Approximately a dozen (11 in the US, 14 in France, 14 in Australia) game cartridges were released for RCA's gaming console.

Unlike the lesser-known systems that preceded it, Atari's Video Computer System (VCS), released in July 1977, is the console that really brought video game cartridges to the masses. While Atari struggled to gain market share, as did Fairchild and RCA, traction slowly set in as Atari published *Breakout* (1978), *Night Driver* (1978), *Adventure* (1979), and others until the industry exploded in January of 1980 with the release of *Space Invaders* (1978).

In late 1977, the pinball and slot machine manufacturer Bally released the Bally Professional Arcade, later renamed as the Astrocade. The cartridges for Bally's home console resembled the cassette tape, which surpassed the 8-track as the music industry standard.

In December of 1978, Philips and Magnavox released another sequel to the Odyssey, named the Odyssey^2. Unlike its predecessors, which used discrete logic circuit cards, the Odyssey^2 featured a microprocessor and used ROM cartridges for storing programs, allowing a variety of games styles to be played on the console.

When four disgruntled programmers left Atari in 1979 to become the first third-party publisher, Activision, they feared legal retribution. To protect and distance themselves, Activision purposely altered the design of the Atari-compatible cartridge. The physical dimensions of the cartridge were altered by adding molded finger grips, the face and spine labels were combined into one, and the hinged doors were removed and replaced with a sponge to protect the circuit board. Internally though, Activision's cartridges were still the standard ROM cartridge.

The use of ROM cartridges promoted massive growth within the industry. As the market flourished, conglomerates such as Mattel, the world's largest multinational toy manufacturer, took notice. In 1979, the toy giant released the Intellivision, a portmanteau of "intelligent television". Mattel introduced the "fill line" concept to gaming cartridges—a simple solution to a problem no one had. Basically, a line was molded on the side of each cartridge that showed how far to press the cartridge into the Intellivision unit.

While the American market was at the forefront, Japan produced a few exclusive consoles of its own. The Japanese toy maker Bandai released a series of gaming machines under the TV Jack line. The initial models, premiering in 1977, were ordered sequentially (1000, 1200, 1500, 2500, 3000) and were standalone units. However, by 1978, home video games released as cartridges became the norm, and the TV Jack 5000 debuted with a cartridge slot. Similar to GI's AY-3-8500 "Pong on a chip", the game cartridges themselves did not use a ROM chip and were not programmable. Instead, they used single dedicated chips manufactured by General Instruments (GI) and were referred to as PC-50X cartridges. The dedicated chip contained most of the game circuits and, when combined with a television's non-modulated video signal, would construct a game.

The following year, Bandai released the Super Vision 8000 (1979). Three years behind the release of Fairchild's Channel F, the Super Vision 8000 contained a central CPU and became the first console in Japan to use programmable ROM cartridges.

The first handheld gaming console was the MicroVision (1979), released by the long-established board game manufacturer Milton Bradley. The MicroVision's cartridges were unique in that they were larger than the standard console cartridges and almost as large as the handheld device itself. Interestingly, and unlike nearly every other gaming device, the CPU for each game was included on the cartridge as opposed to being housed within the console, along with the game code, which was also included on that very same chip. Each cartridge also featured a keypad which revealed up to twelve plastic membrane buttons unique to each individual title.

Also in 1979, APF Electronics released the APF-MP1000. It was primarily an 8-bit cartridge-based home video game console, but it could be expanded to a full-fledged home computer by docking it within the IM-1 computer component. Combined with the full-sized keyboard and cassette tape drive, it became the APF Imagination Machine.

The home video game market exploded as a parade of consoles emerged in 1982, including Emerson's Arcadia 2001, Coleco's Colecovision, Milton Bradley's vector-based Vectrex, and Atari's follow-up to the VCS, the 5200 SuperSystem.

Cartridges for Computers

The late 1970s and early 1980s brought the computer revolution from military and educational institutions into the home *en masse*. In addition to using magnetic media, many early home computers by Commodore, Texas Instruments, and even the Atari line of computers, used ROM cartridges.

In a trend that familiarized the general public with new computer technology, the video game system was a Trojan horse of sorts. A home video game system was often the first computer within many homes, and the simplicity of cartridge-based media made the transition simpler. Most of the major players within the industry attempted to upgrade their game systems into full-fledged home computers; Atari with its cancelled *My First Computer* and *Graduate* plans, Mattel with its Intellivision Keyboard Component, and Coleco with the ADAM. With the success of the Intellivision, Mattel followed suit to compete with Atari in the home computer market by releasing the Aquarius Computer (1983). It also played games via cartridge, although the Aquarius's cartridges were unique in design and not compatible with Intellivision cartridges.

On July 15, 1983, Nintendo released the Family Computer (Famicom) in Japan, a dedicated home video game console, despite the term "computer" in its name. That very same day, SEGA released its first home game console, the SG-1000, and its computer counterparts, the SC-3000 and SC-3000H. Game software for all the aforementioned used traditional cartridges.

The Atari XE Game System (XEGS), released in 1987, was a hybrid computer/console. While it was fully backwards compatible with Atari's 8-bit computer cartridges, it had its own library of XE games on cartridge, many of which were titles originally released on 8-bit cassette or 5¼-inch floppy disk, such as *Blue Max* (1983) and *Archon* (1983).

Card Cartridges

Hudson Soft, a prominent Tokyo-based software publisher, was the first company to release video games on credit-card-sized cartridges; electronically erasable, programmable read-only memory (EEPROM) devices referred to as Bee Cards. They were a bit thicker in depth than a credit card, but very small and compact in comparison to other software distribution methods in 1984. Hudson Soft developed them for use with MSX-compatible computers, but they were also adopted by Atari Corporation for use with Atari's PC Folio palmtop computer and other handheld PCs. In 1983, Control Video Corporation (CVC) introduced the GameLine cartridge. While it was not a card cartridge per se, it was the premiere technology-bridging cartridge device which paved the way for SEGA's impending Card Catcher product. The GameLine included an internal 1200 baud modem which utilized a phone jack mounted on the side, and is historically important since it marked the first time that video games could be downloaded through a telephone line.

SEGA's second home system, the SEGA-1000 Mark II (1984), used traditional game cartridges, but also implemented the SEGA Card, allowing insertion through an additional input slot. Games requiring less memory were placed on cards as a price-saving maneuver. A cartridge called the Card Catcher was released in 1985 and served as a pass-through device allowing SEGA Cards to be used with devices that did not have card slots, such as the original SG-1000 and SC-3000 computer line. This concept of using a cartridge to play another type of cartridge was the first of many such devices used by SEGA to allow backwards compatibility.

SEGA's third entry into the home market was the SEGA Mark III (1985), rebranded as the SEGA Master System (SMS) in many regions. Like its precursor, it also contained input slots for conventional cartridges as well as games on cards. When SEGA later released the Genesis (also known as the Mega Drive) it allowed compatibility with its predecessor, the SMS, via the Power Base Converter. This device, which plugged directly into the Genesis cartridge slot, recognized Master System game cards in addition to cartridges.

Hudson Soft collaborated with the Nippon Electric Company (NEC) to design the PC Engine (1987). The Bee Card was renamed the HuCard, abbreviated from the Japanese word 'HyūKādo'. In the United States, where the PC Engine was rebranded as the TurboGrafx-16, HuCards were referred to as TurboChips to coincide with the name change. The successor to the PC Engine, the SuperGrafx (1989) also used an enhanced version of the HuCard, renamed the Super HuCard.

It should be noted that the SEGA Card evolved in 1985, with the newest rendition being rewritable, allowing data entered during the game such as custom character names, high scores, and game progression, to be stored on the cards themselves for future retrieval once the device had been powered off. NEC followed suit with a HuCard upgrade in 1991.

Bank Switching

As video games became more popular, players expected and demanded more and more from the gaming hardware. The engineers behind Atari's VCS never expected the seasonal console to be on the market for a decade and a half. Over time, this forced the programmers to jump through all sorts of hoops to exploit the hardware.

One such hardware trick was bank switching, first implemented in 1981 by Brad Stewart, when Atari refused to allow him to use more than 4 KB for the home console version of *Asteroids* (1979). To work around the memory problem, he divided portions of the program into disjointed sections. The Atari's CPU could only recognize one bank at a time, but it allowed the VCS to exceed the 4 KB limit. In time, the VCS more than quadrupled the capabilities originally calculated by its designers.

In 1985, Nintendo introduced the Famicom in America as the Nintendo Entertainment System (NES) and revived the dormant gaming industry in North America. The method of adding more capacity to video game cartridges via bank switching continued on the NES. Nintendo actually included a special chip (Multi-Memory Controller) just for the explicit purpose of adding bank switching capabilities to the unit. Nintendo Game Boy (1989) cartridges improved on the controller chip with the Memory Bank Controller (MBC), which expanded both ROM and static RAM. SEGA Genesis cartridges using more than 4 MB also used this modus operandi.

Battery Back-Up in Cartridges

Of particular interest is the inclusion of a battery-backed static RAM semiconductor within cartridges to save game progression or other data. The concept, originated by the Philips-owned company Probe 2000, was to be implemented with the release of *Lord of the Dungeon* (1984) for the Colecovision. However, the North American video game industry crash prevented it from reaching wide-scale production.

In Japan, Nintendo released the Famicom Disk System in 1986. The peripheral device attached to the Famicom allowing the use of games stored on a "Disk Card", a proprietary form of floppy disk. A feature of the Disk Card allowed game progression to be saved, allowing for longer stories to be told over multiple gameplay sessions. Seven games premiered with the Famicom Disk System, including the immensely popular game *The Legend of Zelda* (1986). Since the Famicom Disk System was not released outside of Japan, Nintendo needed another solution to publish the game overseas. Eighteen months later, Nintendo officially introduced cartridges with internal battery back-up when it published *The Legend of Zelda* (1986) on August 22, 1987 in North America.

The Game Cartridge Surge

When Atari released the VCS in 1977, it never occurred to them that other companies would release software on their hardware platform. That all changed when several disgruntled Atari employees formed the first third-party software company, Activision, in October of 1979. After a banner year, other Atari personnel defected to create Imagic. By 1983, there were more than 30 companies competing against Atari for shelf space. Nearly 400 games were released by third-party publishers for the Atari VCS alone. When Mattel, Coleco, and other companies released their own platforms, the amount of software being produced was even more staggering! And then Nintendo entered the fray . . .

An avalanche of cartridge software dropped for the NES, releasing well over 700 titles for the platform over the system's first decade. Nintendo wanted to prevent what happened to Atari from occurring on their console. To combat and deter such competition, they implemented a very clever system. Embedded in each cartridge was a custom Checking Integrated

Circuit (CIC), commonly referred to as the "10NES" or the infamous "lockout chip". The chip communicated with the microcontroller within the NES unit to approve licensed software or deny illegal software. The CIC was the key to unlocking the NES, and if authorization was not received, the CPU reset the system, preventing the game from operating.

With the renewed interest in video games, Atari finally released the delayed Atari 7800 (1986) to try to capitalize on the craze. SEGA at long last found success with the Mega Drive (1988), also known as the Genesis in many regions. In 1990, Nintendo launched the Super Famicom (also known as the Super Nintendo). Combined, SEGA and Nintendo manufactured billions of video game cartridges during their heyday.

A unique cartridge released for the SEGA Genesis was *Sonic & Knuckles* (1994). During development, the game's scope grew exponentially, delaying the title's release. As a result, SEGA chose to split Sonic's third release into two separate components. *Sonic 3* was released in February, followed up in October with the release of *Sonic & Knuckles*, a cartridge that included "lock-on" technology. The top of the newly styled cartridge featured an opening hatch that allowed the insertion of *Sonic 3*, as well as *Sonic the Hedgehog 2* (1992). Combining either of the Sonic games with the *Sonic & Knuckles* cartridge altered game levels, added extra characters and character abilities, and unlocked special stages of gameplay.

Cartridges Inside Arcade Games

Arcade operators went to much trouble and often great expense to change games in their facilities. Since each coin-operated arcade game that was released had customized hardware designed uniquely to meet the technical requirements of a specific game, it was not cost effective. In order to address this problem, some manufacturers offered some rather innovative solutions.

In 1980, Data East introduced the DECO Cassette System. In place of a standard motherboard, the norm at the time, the game code was delivered via an ordinary audio cassette. When paired with a small dongle installed in a base cabinet, the game software would be copied to the coin-op's RAM during start-up in order to be played. Nearly 50 DECO-compatible games were distributed by Data East.

In time, some cartridge-based arcade system boards were developed. SNK's 1990 Neo•Geo Multi Video System (MVS) and Sammy's Atomiswave (2003) are examples that implemented cartridges to help arcade operators swap games more easily, saving time and financial resources.

SNK also released a home version of its arcade platform, referred to as the Advanced Entertainment System (AES), in 1990. It naturally used cartridges over motherboards, although these cartridges were not compatible with the arcade MVS. Since home Neo•Geo games were identical with those used in the arcade, as opposed to subpar, scaled-down, and reprogrammed adaptations of their arcade titles, they required more memory than typical game cartridges of the time. As a result, the games retailed at $200 or higher per title; almost quadruple the price of cartridge games released for competing gaming platforms. SNK supported the AES for fourteen years, with *Samurai Showdown V Special* (2004) being the final official cartridge release. While the MVS was a grand success, the AES sputtered in the marketplace and over time became more of a cult hit.

SNK's second arcade outing, the Hyper Neo•Geo 64 (1997), was an outright failure. Its intent was to replace the aging MVS hardware by hosting a new 64-bit RISC processor allowing 3-D graphics as well as other bells and whistles. Instead, after several lengthy delays, it was too little too late and offered hardware grossly underpowered compared to the competition. Only seven game cartridges were released for the platform before it faded away 21 months after its introduction.

The Last Wave of Console Cartridges

The Compact Disc (CD) was co-developed by Sony and Philips in 1982 to store audio recordings. In 1994, Sony and Denon, the brand name of Nippon-Columbia record company, adapted it to store data, engineering the CD-ROM. The new medium was used by many large computer companies of the time, including Philips, Apple, NEC, and Microsoft—all of which ultimately released CD-ROM-based gaming consoles. There are many reasons why the CD-ROM replaced the cartridge (see the next section), but it all started with NEC's PC Engine add-on in 1988. While the CD-ROM was first introduced to gaming consoles as an additional accessory, the first dedicated CD-ROM game consoles were the FM Towns Marty (1991) in Japan, the Philips CD-i (1991), and the Amiga CD32 (1993). None of these systems included a cartridge slot.

Atari, once the industry leader, hoped to return in force with the release of the Jaguar (1993) console. Ultimately a commercial failure, only 67 licensed titles were published on cartridge for the system. Atari eventually released a CD-ROM add-on to try to reduce costs, maximize profits, and become relevant in the age of the CD-ROM. Atari failed on that front as well, with the Jaguar becoming the final console released by the once-proud company.

SEGA had mostly abandoned traditional cartridges with the introduction of the SEGA CD (1991). This add-on allowed CD-ROM software to play on the SEGA Genesis, adding new capabilities such as better audio and full-motion video that came with the new storage medium. However, SEGA did release a secondary cartridge-based Genesis peripheral known as the 32X in 1994, which doubled the 16-bit processor to a full 32 bits.

In 1994, SEGA also released the Saturn, which primarily used CD-ROM software. The Saturn's hardware did incorporate a cartridge slot, although it was not used to launch software. Instead, it served as an expansion bay of sorts, allowing easy introduction to extended RAM (housed within a cartridge), permitting the use of cheat devices such as the Game Shark or Action Replay, and allowing Saturn users to access the Internet via the Net Link device.

The Nintendo Virtual Boy (1995) introduced a head-mounted display to console gaming. While it had more in common with table-top gaming devices, it was a console just the same. The Virtual Boy was the second and final 32-bit console to use cartridges. Harshly condemned by critics and consumers alike, Nintendo's hybrid experiment became the company's only significant commercial failure. As a result, only a mere 22 titles were published for the Virtual Boy.

Nintendo initially considered using the CD-ROM format for its Nintendo 64 (N64) console, released in 1996; however, after failed attempts to work with CD inventors Sony and Philips, Nintendo chose to continue with the cartridge format, much to the disappointment of consumers and against the direction of the video game industry. Pushing against the trend, Nintendo forged ahead with cartridges as its competitors SEGA, NEC, Sony, and others embraced the CD-ROM surge. Remaining within the company's comfort zone diminished Nintendo, as the company lost the support of many key developers such as Shiny Entertainment. Furthermore, some publishers snubbed the N64 system, pulling all support for the console. Powerhouse publisher Square even migrated the popular *Final Fantasy* franchise to the PlayStation to take advantage of the CD-ROM's ability to store and play back large amounts of data including full-motion video. While the average N64 cartridge was 8–16 MB in size, much larger than carts released for previous gaming systems, it still paled in comparison to the CD-ROM, which offered almost ten times the maximum capacity of an N64 cartridge.

The N64 is regularly, and incorrectly, reported as being the last game system to make use of video game cartridges. It was, however, the final system to use cartridges released by a leading player in the industry. Nintendo transitioned primarily to CD-ROMs with the N64's successor, the Nintendo GameCube in 2001, while still manufacturing cartridge-based games for the N64 well into 2002. Regardless, there were still other smaller companies releasing cartridge-based video game equipment beyond Nintendo's abandonment of the medium.

One such cartridge-based system was SSD Company Limited's XaviXPORT, released in 2004. The XaviXPORT used thin rectangular cartridges that actually pressed into the system in a unique manner. Instead of a traditional game console cartridge slot which accepted the smallest face of a parallelepiped-shaped cartridge, the XaviXPORT cartridge was inserted face down into the console, which collapsed into itself to accept the game cartridge, almost like an elevator. The system itself was simple by design and contained little memory and minimal processing power. To compensate, the cartridges themselves included a custom proprietary multiprocessor.

The XaviXPORT did not use conventional controllers, but instead used wireless controllers shaped like sports equipment, such as tennis rackets, baseball bats, golf clubs, and fishing rods. These real-world accessories used motion sensors to represent user actions on screen several years before Nintendo launched its GameCube successor, the Nintendo Wii. Some cartridges, such as *Bowling* (2004), had a small motion-sensing camera mounted catty-cornered atop the cartridge itself.

Video game cartridge use did not cease with the demise of SSD's XaviXPORT, although it was relegated to mostly handheld and educational devices, discussed later in this essay.

The Pros and Cons of Game Cartridges

Cartridges have many advantages over other game storage and delivery systems. The primary advantage is speed. Since cartridge memory is mapped into the normal address space of the system, it is read like normal memory. Thus, it is not impeded by the process of shifting data and typically uses less RAM, leaving memory open to address other processes. Programs loaded much faster off a cartridge than from cassette tape and other magnetic media during the initial home computer era. Even today, cartridges load data more rapidly than multi-speed CD-ROM drives or even the digital download of streaming content.

The cartridge format can be more reliable than its competitors. While it can be damaged, the exterior outer shell of the cartridge works as a shield, making it more robust. In 1994, hobbyist John Earney documented abusing standard Atari cartridges in a direct attempt to destroy them. After exposing them to extreme conditions such as boiling them in water, storing them in a freezer, burning them in direct flame, subjecting them to high-powered magnets, bombarding them with electrical current, and even driving over them repeatedly in his vehicle, Earney discovered that the cartridges still functioned perfectly. In most cases, a cartridge cannot be made unreadable by damage the way a CD-ROM becomes unreadable when scratched. In addition, it cannot be stretched, broken or unreeled like a cassette tape, or accidently erased like magnetic storage. Dust and dirt can accumulate on cartridge contacts, but are easily removed with rubbing alcohol.

As the amount of computer memory expanded and at the same time became cheaper, computer companies began incorporating disk drives, removing cartridge compatibility entirely. The cycle repeated itself decades later when CD-ROM drives began replacing cartridge ports, since CD-ROMs offered substantially more space and exceeded the practical limits of the traditional video game cartridge.

Another benefit of the cartridge is that it deterred piracy. Floppy disks and CD-ROMs are inexpensive and easy to replicate. Cartridges require a more demanding level of manufacturing, which discourages illegal copying. At the same time, such production is more expensive than other media to produce, placing cartridges at a disadvantage. Cartridge manufacturing typically required larger production runs to even be competitive in price. It is a slower process than pressing CDs, and if quantities manufactured do not meet market expectations, then surplus issues can arise. If publishers print too many, they are left holding expensive unsold inventory. If they underestimate demand, then a shortage of product results in delays of replenishment. In the end, using CD-ROMs usually proves to be a less risky endeavor.

Cartridges Within Portable Gaming Devices

Ten years after Milton Bradley's revolutionary MicroVision (1979), Nintendo released the Game Boy in 1989. It premiered with much fanfare, as it was bundled with the popular puzzle game *Tetris* (1984). The cartridges for the Game Boy were standard ROM cartridges, but referred to as "Game Paks", and the system's LCD screen could display only four shades of greenish-gray.

The popularity and success of Nintendo's Game Boy spurred many challengers into the handheld gaming arena. Soon, cartridge-based handhelds such as Atari's Lynx (1989), SEGA's Game Gear (1990), Bitcorp's Gamate (1990), and Watara's Supervision (1992) were competing for a foothold in the marketplace. Despite the crowd, the Game Boy dominated and saw many renditions of its design over the years. The Game Boy Color (1998) added a color display, while the Game Boy Advance (2001) transitioned the series from 8-bit to 32-bit. In total, the Game Boy series sold over 200 million units . . . and just over 878 million cartridges!

During this time, NEC released the NEC PC Engine GT (1990) in Japan. It was a portable version of the PC Engine and used the same HuCards. This marked the first time that a portable unit could play the same media used by its "big brother"! It reached North American shores in 1992 as the TurboExpress.

The Super Nintendo (SNES) even received a *Super Game Boy* (1994) cartridge that allowed Game Boy and Game Boy Color game cartridges to be played via the SNES. Not a traditional cartridge, it housed a CPU consisting of the same hardware found inside the Game Boy.

Other portable gaming devices that used cartridges were Tiger's Game.Com (1997), SNK's Neo•Geo Pocket (1998) and Pocket Color (1999), and Bandai's WonderSwan (1999) and WonderSwan Color (2000). With the Pokémon franchise in full swing, Nintendo released the Pokémon Mini (2001) along with ten games on cartridge. Approximately 3 × 2 inches in size, it was the smallest Nintendo handheld to use cartridges. Of course, its cartridges were even smaller, approximately a quarter of the size of the unit. While this seems tiny, the world's smallest cartridges probably belong to the AGP X-System. They measure a mere $^{11}\!/_{16}$ by $^{13}\!/_{16}$ inches and are barely bigger than a thumbnail.

Gamepark's GP-32 (2001) introduced smart media cards (SMC) to handheld gaming. SMCs used electronically programmable non-volatile flash memory, allowing data to be stored permanently without an extra power source such as an internal battery. Flash memory became the standard as other portable gaming devices followed suit: Nokia's N-Gage (2003) used MultiMedia Cards (MMC), while Tapwave's Zodiac (2003) employed Secure Digital (SD) cards. Tiger's Gizmondo (2005) made use of both MMC and SD flash cards. When Nintendo adopted flash technology for its DS Game Cards in 2004, it enclosed the

EEPROM chip in hard plastic, essentially making cartridges out of the cards. This procedure was also adopted for its Nintendo 3DS (2011) platform, still in regular use today.

Sony bypassed cartridges with the PlayStation Portable (PSP) in 2004, choosing the Universal Media Disc (UMD) instead. UMDs were miniature 2½-inch optical discs designed specifically for the PSP, although Sony hoped other products would adopt the technology. Poor performance in the marketplace ended such prospects. The format was eventually abandoned in favor of the PlayStation Vita Game Card, a proprietary flash-based memory card, when the company released the PlayStation Vita in 2011. Vita cartridges are also compatible with PlayStation TV (2013), a media streaming set-top box.

Learning System Cartridges

While the mainstream entertainment game system manufacturers such as Sony and Microsoft went full-throttle with CD-ROMs, there was an often overlooked educational niche making good use of the video game cartridge format. In fact, it is still the de facto standard for learning systems. The durability of the cartridge warranted the extra expense over CD-ROM technology in order to withstand the often brutal handling of the games by young children.

LeapFrog leapt first after the co-founder of the company could not find a satisfactory solution to help teach his son to read, and designed what ultimately became the LeapPad in 1999. A year later, Fisher-Price followed suit with their Pixter line (2000–2003). Interestingly, the video game pioneer SEGA took an interest in LeapFrog products and released them to the Japanese market in 2002. However, the educational cartridge did not really take off until LeapFrog launched the Leapster Learning Game System in 2003, followed by V.Tech's V.Smile line in 2004, making use of its branded "smartridges".

For several years, VTech and LeapFrog competed against one another, launching several more cartridge-based systems and handhelds, until April 2016, when VTech took control of LeapFrog. Despite the acquisition, the LeapFrog brand still continues today. New games for the LeapPad3, the latest creation from LeapFrog (as of 2017), are still being made available on cartridge media and digital download via the LeapFrog App Center. One such game is *Finding Dory: Mathematical Memories* (2016).

Cartridge Revival

Hoping to make physical media relevant again, Nintendo's tablet-like hybrid video game console, the Switch (2017), uses Game Cards. When the home console industry moved to CD-ROM storage to gain more storage space during the 1990s, it eventually capped out at 50 GB with Sony's Blu-ray disc. Meanwhile, solid state drive technology has been growing exponentially and currently can hold up to 1000GB, or one terabyte. Since the largest games currently available on the market do not even surpass 80 GB, a future involving cartridges as a viable medium might once again be possible.

Bibliography

Edwards, Ben J. (2015, January 22), *The Untold Story of the Invention of the Game Cartridge*, retrieved from www.fastcompany.com/3040889/the-untold-story-of-the-invention-of-the-game-cartridge.

Herman, Leonard (2016), *Phoenix IV: The History of the Videogame Industry*, Springfield, NJ: Rolenta Press Publishing.

Wolf, Mark J. P., editor, (2012), *Encyclopedia of Video Games: The Culture, Technology, and Art of Gaming*, Santa Barbara, CA: Greenwood Press.

23

DIGITAL DATA DEMISE

Obsolete Digital Data Formats

Gary Locklair

You sense the quest is almost over. You can see the object with its unnatural shape and strange gleam. As you carefully examine the object, you conclude it did not have a terrestrial origin, yet it is not a naturally occurring object either. Close examination reveals what must be information content on the object. This object must have had an intelligent creator. Fascinating! But, what does the information mean? You can discern lines, shapes, and patterns, but how will you decode the information content the object possesses? Are the

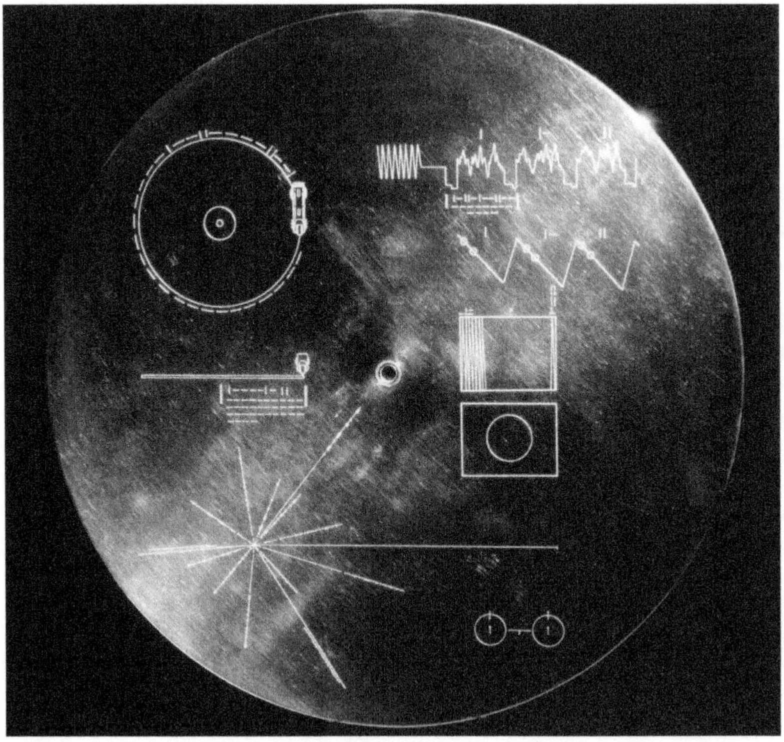

Figure 23.1 The object. (http://voyager.jpl.nasa.gov/spacecraft/goldenrec.html). Courtesy NASA/JPL-Caltech

patterns letters, numbers, sounds, or images? Is it meant to convey a greeting, a warning, or a map? Where did this strange object come from and what was its purpose?

Let us dispense with reality, put on our science fiction hats, and imagine there are intelligent beings outside of Earth who find the golden record aboard the Voyager 1 space probe launched from Earth in 1977. Would they be able to decode and understand its message?

A similar scenario is played out here on Earth almost every day as people and companies struggle to decode obsolete digital data file formats. Consider the BBC's updated Domesday Project of the 2010s.[1] In the early 1980s, the British Broadcasting Corporation embarked on a project to create a modern version of William the Conqueror's Domesday Book (completed in 1086). The Domesday Project was completed in 1986 on the 900th anniversary of the original work. The BBC's Domesday Project was meant to chronicle ordinary life in the 1980s and was a huge success with more than a million contributions from citizens. However, the Domesday Project information was originally released on laserdisc. Accessing laserdiscs required expensive technology in the 1980s and the information did not reach a wide audience. Twenty-five years later when the BBC decided to make the 1986 version available online, extracting the data from the original laserdiscs proved difficult. Software for the original project was created using Basic Combined Programming Language (BCPL), an obsolete language. The images were not digital but were represented as analog, single-frame images encoded on the laserdisc. The BBC expended tremendous effort and money to recover the original data which had been created in a soon-to-be-obsolete format.

Recently the father of my daughter's friend died unexpectedly. He had created a message for his children a number of years ago. He wanted to leave them a "time capsule" of his thoughts at that time. Unfortunately, the family found the message was on a floppy disc. Even if they could find a floppy disc drive, the odds of them being able to read a document created in XyWrite is low. The use of an obsolete data file format might mean the data itself is unreadable.

Anyone who has used computer systems for a significant amount of time has likely encountered the problem of an obsolete digital data file format. After 30 years of teaching, I was interested in looking at assignments from my first year of teaching. The fact that my files were on a floppy disc was not particularly problematic. The fact that the files were created on a Hewlett-Packard HP 150 touchscreen PC using a word processor known as Executive MemoMaker was a problem! I could not find an easy way to convert and read the digital data files from the 1980s. It is amazing to consider that even with the power of modern technology we might not be able to access data created in the recent past. Some practitioners have coined the term "digital dark ages" to refer to the massive amount of digital data that is stored, but not currently accessible since it is in an obsolete file format or accessible only via obsolete hardware.[2]

The Problem

A computer system is hardware, software, and people working together to solve problems. It is usually clear and obvious when hardware becomes obsolete. Even if the hardware functions correctly, technology advances swiftly, resulting in devices that are smaller, faster, and cheaper than the year before. Apple Computer has demonstrated this repeatedly as a new iPhone has been released almost every year since 2007 (and some might refer to this as planned obsolescence). It is not so clear and obvious when software becomes obsolete. If software cannot be executed on current hardware, then it might be considered obsolete. However, even if a software application runs on current hardware, it might have been superseded by later versions. A later version could have additional features and enhancements (for example, it might execute faster). These updates are visible and obvious. Not so visible and obvious are the changes made to the internal data file format. Software is malleable; that is, easily changeable. The format the software application uses to create, open, modify, save, and access its data is deeply hidden from the user. The user would be unaware if the software were using an obsolete data file format.

Software creates data files. Data are the "raw material" of computerized processing. Information technology takes data as input, processes it via algorithms, and produces information as output. People refer to the information and then make problem-solving decisions.

It is the combination of hardware and software that determines how data are stored. Analog computers existed at one time, but since the 1960s digital systems rule the universe of computation. Digital computer systems represent everything in binary form. How can data formats be a problem since computers only deal with data in binary form? Conversion should be easy and simple, right? Unfortunately, the answer is "no".

Data file formatting is almost completely transparent to the user. The vast majority of people have no idea how data are organized and stored on their computer system. The organization of data in memory and on storage media is vastly different from how we see the data on our monitors. In order to see the data in human understandable form, the computer must translate from the data file format into words, numbers, images, and sounds that we recognize and understand. This is by design. Software exists to abstract hardware. In other words, software makes using the hardware easier. A typical user does not need to understand exactly how computers work (that is, what is "under the hood"). The software abstracts (simplifies) the complexities of the hardware for the user. The disadvantage of this abstraction is that users are unaware of how the data are actually organized for access. We rely on the software to understand the digital data file format. If the software becomes obsolete or makes changes to the format, we might be unable to retrieve the data at a later time.

Files

A computer system is a combination of software, hardware, and people. No single component functions independently. The term "file" is a generic name given to any logical entity on storage. A file could be an application package, user data, set of operating system commands, recently created information, or the like. In most operating system environments, a file's complete name consists of two parts: file name and extension. The extension is a key indicating the type of the file; often, the complete file name is written with a "." between the file name and its extension. On a Windows-based personal computer, the extension "exe" indicates that the file is an "executable"; that is, a machine language application package. Other extensions indicate data files created by various application packages, such as "docx", "xlsx", or "pdf". In most operating systems, specific data file extensions are associated with application packages that can access the particular file format. For example, the extension "pdf" is associated with the Adobe Acrobat™ application package. In this manner, users do not need to know which application package can access a particular data file. Instead, the operating system association ensures that when a user requests access to a particular file, the appropriate application package will be run to load the data file. File associations are created when application packages are installed and can usually be changed by the user if necessary. Many graphical operating systems also use the file's extension to display an appropriate icon with the file. The chosen icon is based upon the pre-existing file association. For example, an icon of a pie chart might appear for a file whose extension is associated with a presentation graphics application package.

Digital data file storage has advantages and disadvantages. There are concerns associated with the digital nature of computer data files. On the one hand, the power of storage is related to the digital nature of data files and especially the medium used to represent data. Both electronic and magnetic media can be easily changed or modified. This ease of manipulation allows a user to correct a misspelled word in a data file without a trace of the original misspelling. Modifying hardcopy usually leaves a mark, such as white-out, while changing a data file can be done easily with no evidence of the original data left behind. On the other hand, the digital nature of data files is a double-edged sword as it is often not possible to detect if someone else has maliciously altered the contents of a file. Data files show no visible signs of tampering. Even the date of modification or access is just digital data stored within an operating system's file table; these dates can also be easily modified. If you receive a letter in the mail and the enveloped is ripped and taped, you might assume that someone had seen the contents. Reading a digital data file leaves no evidence. An e-mail message could have been read by others, or even altered, without the recipient's knowledge.

Binary Form Representation

Everything within a computer system (the processor unit, memory, and storage) is represented using binary form. If the software algorithms and user's data could be seen, they would appear as sequences of 0s and 1s. In fact, neither can be visually seen since current digital computers are electronic in nature. The term "binary" refers to two states such as 0 and 1. Everything in the processor unit is represented in binary form, rather than in a familiar human form of letters and digits, because the processor unit is fundamentally constructed of switches. A switch is in one of two states: off or on. This realization is abstracted to the binary form concept of 0 or 1. In order to understand the process activity, one needs to understand binary form as the "native tongue" of the processor unit.

Although internally everything is represented using binary form, there are different codes for interpreting binary form representation. For example, there is one representation for instructions and a different representation for data. Data might either be numeric or characters and each has a different internal representation. It is a mistake to refer to binary form as "numbers". Binary form consists of two symbols, and combinations of those symbols can be interpreted in many ways depending upon the file format. In other words, meaning is imposed on the symbols based upon the code used to store them. A computer system does not assume that binary form represents numbers (or anything else). The computer interprets the symbols based upon the digital data file format.

If you found a document and it contained the symbols "MIX", it would be easy to assume those symbols were letters which stood for the English word meaning "to combine together". It turns out, however, that the MIX in this document actually represented 1,009 as the symbols were Roman numerals. In a similar fashion, computers do not work with binary numbers; they work with symbols; those symbols might be interpreted as numbers, but they do not have to be. Understanding a language is foundational to understanding in general. Understanding data file formats is the key to interpreting their content.

Base-2 Binary Form

Numbers are represented within a computer system using base-2 binary form. In base-2 binary form, 01001101 represents the number 77. A computer would represent the number 61 as 00111101. As used today, base-2 binary form was formulated by Gottfried Leibniz in the late 17th century, although binary numbers were likely used in antiquity. In 1703, Leibniz published "Explanation of Binary Arithmetic, which uses only the characters 1 and 0". Interestingly, Leibniz saw base-2 binary form numbers as a Christian symbol. He wrote to the Duke of Brunswick, "[referring to the concept of] the creation ex nihilo through God's almighty power. Now one can say that nothing in the world can better present and demonstrate this power than the origin of numbers, as it is presented here through the simple and unadorned presentation of One and Zero or Nothing."[3]

George Boole developed an algebra of logic in the 19th century that forms the basis of computer hardware today. Boolean algebra deals with two states. While Boole thought in terms of true/false or yes/no, the two states can also be represented by computer hardware switches being on/off, and therefore bits being 1/0. In the 20th century, Claude Shannon and George Stibitz demonstrated the computational power of combining Boolean algebra with base-2 binary form numbers.

Base-2 binary form is interpreted in the same manner as any other place value system using a decoding algorithm. The value of each place is determined by raising the base to a successive power, beginning with 0 in the least significant place. Comparing base-2 with the familiar base-10 system yields,

Base-2	Place values are powers of 2				
	. . . 2^4	2^3	2^2	2^1	2^0
	. . . 16	8	4	2	1

Base-10	Place values are powers of 10:				
	. . . 10^4	10^3	10^2	10^1	10^0
	. . . 10,000	1000	100	10	1

Because there are only two digits, base-2 numbers will grow in length quickly. The first ten values in base 2 are represented as: 0, 1, 10, 11, 100, 101, 110, 111, 1000, and 1001. The last number represents the value 9 in base 10. It requires four base-2 digits to represent a value which requires only a single base 10 digit.

To understand the value of a base-2 number, the place value "decoding" algorithm can be employed. The value of the base-2 number 10110 is calculated as follows: (1 * 16) + (0 * 8) + (1 * 4) + (1 * 2) + (0 * 1). Thus, 10110 has the value of 22. Actually, base-2 can be easier to decode than base-10 since there are only two digits. It is easy to "shorthand" the decoding algorithm by remembering that the only possible digits are 0 or 1. A 0 in the place value is ignored, whereas a 1 in the place value results in the place value itself being added to the running total. The number 1010 results in a value of 10 since there is a 1 in the 8s place and a 1 in the 2s place.

Typically, base-2 numbers are written in groups of eight binary digits (bits). Eight bits form 1 byte which is the fundamental unit of memory and storage in a computer system. If the base-2 number has fewer than eight digits, leading zeros can be added to the representation without changing its value. Thus, 10110 would be represented in 8 bits as 00010110. In fact, memory represents this base-2 number using a series of eight hardware switches. A switch that is "on" might represent a "1" while a switch that is "off" might represent a "0". The base-2 number 00010110 is an abstraction of the physical implementation using eight switches (three of which are "on").

The largest value that can be represented in one byte is 11111111, or 255 in base-10. In order to represent bigger values, multiple bytes are logically connected and the place value system is extended. Representing a number with two bytes means that the "higher order" byte will extend the place value system of the "lower order" byte. The first bit position in the higher order byte will be the 2^8 or 256s position. The second bit position will be the 2^9 or 512s position and so on until the final bit position which represents the 2^{15} or 32,768s place. Using two bytes, values of up to 65,535 can be represented. If larger values are required, additional bytes are used and the place value system is extended further.

ASCII Binary Form

Characters are represented within a computer system using American Standard Code for Information Interchange (ASCII) binary form. Technically, most computer systems employ Unicode for characters. But as a subset of Unicode, ASCII is easier to investigate. ASCII is a standard method of representing characters using binary form. Certain combinations of 0s and 1s are used to stand for each character. For example, the letter "A" is represented in ASCII as 01000001.

ASCII and all other encoding techniques work due to abstraction. Abstraction is an agreed-upon method of simplification. Rather than carrying around large amounts of cash, most people carry a checkbook or credit card. A check is an abstraction of money. It is more convenient to write a check and the system works because of an agreement between you, the merchant, and your bank. In a similar fashion, abstraction allows us to represent characters using binary form. An "agreement" or standard allows specified sequences of 0s and 1s to represent certain characters.

At this point a problem might arise. What, exactly, do the contents of a byte represent? The binary form contents of a byte, say 01000010, could be interpreted in at least two distinct ways. The contents might represent a number. If the binary form represents a number, then the base-2 decoding algorithm is used to determine its value (which is 66). On the other hand, the contents could represent a character; then the ASCII scheme is used to determine its value (in this case, "B"). How does the CPU determine which representation is used—how is it interpreted or decoded? The answer involves software. The hardware alone is not able to determine which binary form representation is present; the combination of hardware and software working together is necessary to make the determination. The software field must indicate what the contents of a byte represents. A field is a placeholder and descriptor of data and therefore must describe what representation (e.g., base-2 or ASCII) is being used. Recall the convention for entering data into a spreadsheet. If digits are entered directly, the spreadsheet cell interprets the data as a number and it is represented internally using base-2 binary form. If digits are entered after an apostrophe ('), the spreadsheet cell interprets the data as a label indicating characters, and it is represented internally using ASCII binary form. In this case, a user command is given to signal which representation is to be followed.

It should be noted that there is a difference between the *number* "5" and the *character* "5". The symbol "5" is a digit and it could be interpreted (and therefore represented) differently. Which representation is used depends upon the application, and it certainly makes a difference in context! When an hourly pay rate is needed, "5" indicates a number and should be represented internally in base-2 binary form. When an apartment "number" is required, "5" indicates a character and should be represented internally in ASCII binary form. Software fields will provide the context for the representation and ultimate decoding of the input data.

So far we have investigated two types of binary form: base-2 and ASCII. The term "binary" merely refers to strings composed of bits. Thus, to proclaim that something is in binary does not provide enough information for its decoding. Base-2 applies that particular decoding scheme to a bit string while ASCII applies a different decoding scheme to a bit string. A string of bits could represent many different things besides numbers and characters. Sounds and pictures are also represented using binary form and specialized decoding schemes, often called "file formats", are used to represent them.

Machine Language Binary Form

In addition to data, instructions are also represented internally in binary form. Since neither words (add) nor symbols (+) can be remembered or stored directly, software algorithms are also represented using 0s and 1s. The creators of the CPU designed it to execute a certain machine language. Various combinations of 0s and 1s are recognized by the CPU as valid instructions that can be performed. For example, 01010001 might mean "add". Machine language is the lowest level instructional language for a computer; few people program computers in machine language as there are better, and less error-prone, languages available.

The "native tongue" of the processor unit is binary form. All processing is carried out on data and by instructions that are composed of 0s and 1s. This convention is due to the fact that the processor unit is fundamentally composed of on-off switches.

Data File Obsolescence

What does it mean for a digital data file to become obsolete? It means the organization of data is not understood by any current system or software attempting to access it. Therefore, the data itself is unreadable. It is not possible for the current software to translate the binary form into intelligible form since the format of the data is unknown. Barring physical damage, we can directly recover the underlying 1s and 0s, but what does the binary form mean and what is its context?

If there are well-understood codes for binary form (base-2 for numbers, ASCII for characters, and so on), how can data files become obsolete? One contributing factor we examined earlier is that a binary string might be interpreted as either a number or a character. Formatting is necessary to determine which it is. Formatting is absolutely vital for file integrity. While computers cannot understand the difference between characters and numbers, humans can. We expect our names to contain characters and our paychecks to contain numbers.

As an example, consider translating from ancient Hebrew, as found in Old Testament manuscripts, into English. The Hebrew letters are known and identifiable. However, the Hebrew text contains no spacing and no punctuation. In addition, there are no written vowels in Hebrew (vowel pointing was added later as an aid to pronunciation and understanding). Thus, translating from ancient Hebrew into English is a difficult process, especially without the formatting supplied by spaces, punctuation, and vowels. But at least storing numbers should be easy and there should be no obsolescence when it comes to them, right? Unfortunately, that is not the case.

While binary form is the foundation for numeric representation, there are many kinds of numbers. Earlier we saw how binary form can represent small integer values. A small integer of up to the value 255 will fit in a single byte, but what if we have longer integers or what if we need to store real values (numbers with fractional portions)? In both of these cases the formatting must be known in order to correctly decode the number.

Let us consider some additional examples involving numbers. When multiple bytes are needed to represent large values, an obvious formatting question becomes "which byte is the first byte?" While humans normally read digits left to right with the most significant digit being in the leftmost position, computers are under no such restriction. A computer could assume that the most significant byte was the first one, or the computer could operate assuming the most significant byte was the last. So, the digits "1 9 7 7" might represent the value 1,977 or 7,791. This concept is referred to as *endianness*. Computer systems may treat numbers in a "big endian" manner with the first byte representing the most significant digit (1977) or they may treat numbers in a "little endian" manner with the first byte representing the least significant digit (7791). With multiple bytes, the formatting must declare where the "front" of the number is located. Without this knowledge, the number cannot be decoded correctly.

When a real or floating point value needs to be represented, the most common format is IEEE 754. This standard specifies how to represent a number using binary form by breaking the value into three distinct pieces: a sign, a significand (mantissa), and an exponent. Imagine representing floating point values with 16 binary digits. The most significant bit would be the "sign" bit. A 0 in the sign bit indicates a positive value while a 1 represents negative. The next five bits would be reserved for the exponent. The final ten bits would represent the significand. We can think of the floating point value in this format: sign, significant * 2^exponent. There are several "tricks" performed in floating point representation. The first is biasing (or indexing) of the exponent. In order to represent both positive and negative exponents, a "middle" value is used to represent 0. Thus 01111 represents the exponent value 0, while 10000 represents the exponent value 1. Additionally, the significand is normalized. When a floating point value is written, we begin with the most significant non-zero value. Thus, the first bit of the significand will always be a 1. Because of this, we can "assume" the 1 and have the significand begin with the next digit in the value. Thus, the 16 bit value "010000100000000" would be + 1.1 * 2^1.

If the computer encountered a string of 1s and 0s, the formatting provided by the IEEE 754 standard would indicate where the three pieces of the value (sign, significand, exponent) were located. In other words, the format determines how many total bits are in the value (32, 64, 128, or more) and how many bits represent the exponent and significand, respectively. Without this information, the value would be unreadable by the software and thus not displayable to the user.

Storing and retrieving characters seems less straightforward than numbers. How can characters, such as this sentence, be represented in binary form? As we saw before, the answer is found in using a convention or standard. The complexities of symbols in an alphabet are abstracted (simplified) into bit strings. This idea is not unique to computer systems. Morse code used the same principle of abstraction to represent every letter in the English alphabet with a sequence of dots and dashes. The representation of the letter "S" by three dots was part of the formatting of Morse code. In addition to the dots and dashes of the letters, the time interval between letters, the space, was also part of the formatting.

At one time, there was no standard for characters in computer systems. Each manufacturer used a different convention for representing characters. This, of course, limited communication and interoperability between computer systems. In the 1960s the American Standards Association (currently the American National Standards Institute) developed the ASCII code as a uniform way to represent characters in binary form. The ASCII code can represent 128 unique entities, including the 26 upper case, 26 lower case, and special symbols as found on computer keyboards. Today, ASCII is a subset of a larger standard used to represent characters in multiple languages, Unicode. Unicode represents characters in more than 100 different languages. The same principle of abstraction and representation as found in ASCII is utilized in Unicode.

Digital data represents more than just numbers and characters. Modern digital data files contain binary form that represents images, video, and sound, among other entities. There are dozens of widely used formats for each of these entities. For example, a graphical image might be represented using either a raster (bitmap) or vector (mathematical) file format. Among raster formats there are JPG, PNG, GIF, TIFF, Exif, BMP, and PSD file organization methods, along with many others.

Example Data File Formats

Digital data file formats are complex. Consider all of the elements within a typical word processing document. A document contains not just text but hypertext including images. In addition to the data, there is an abundance of formatting including tables, bold text, links, and the like. How do you specify that a character string should be bold, and how do you distinguish the command "bold" from the data for characters or numbers? These questions are answered via the file formatting. If the software becomes obsolete, or if newer versions of the software change the data file formatting, these questions become unanswerable and the data is lost. At one time, WordPerfect was the most widely used word processor for personal computers. Over time, newer versions of WordPerfect were introduced with additional features. The additional features necessitated changes to the underlying data file format. Today, WordPerfect has been eclipsed by Microsoft Word as the dominant word processing platform. These factors have led people to speculate that the older WordPerfect file formats are either obsolete now, or will be shortly.

To see the scope and magnitude of the issue, let us investigate the Microsoft Office Word digital data file format.[4] Versions of Microsoft Office Word from Word 97 through Word 2003 used the .doc data file format as their standard. In the .doc file format, the "character" serves as the basic unit of data. A "character" can be an actual character represented in ASCII (or Unicode), but it can also refer to formatting and other descriptive items of metadata. Everything in the document resides in what Microsoft calls the Word Document Stream, which is basically the logical linear collection of bytes that comprises the document within the file. The stream begins with a header that helps the software navigate the characters within the file. The header is referred to as the File Information Block. The File Information Block is a "master" piece of formatting information that points to all of the other data in the file. Each entry in the File Information Block is a pair of integer values. The first value identifies the location within the file for the data while the second value identifies the length of the data. Because of this pointer mechanism, sequential characters in the user's document are not necessarily adjacent in the data file.

As with most file formats, the .doc format includes information that describes the actual data within the file (for example, this character is in italics). Data that describes the characteristics of other data is termed metadata. Each data stream and each record in the file format contains both data and metadata. Being able to distinguish between the data and metadata is crucial for the software to correctly display the binary form data in human readable form. The interested reader who would like to learn more about the .doc file format should use the Office Visualizer Tool, offvis.exe, provided by Microsoft. This software allows the user to glimpse the structure and organization of the .doc data file.

Microsoft has updated and expanded their Microsoft Office Word data file format over the years. Version 2007 of Office introduced a new XML-based data file format. And, beginning with version 2010 of Office, Microsoft updated the binary file format again. File formats are updated to allow new capabilities and additional features. While this is good and useful, it is a double-edged sword. New file formats cause compatibility issues with data stored in previous file formats. Microsoft Office Word can access files stored in a number of different formats, yet this results in larger software programs and the probability of errors accessing the different formats increases.

XML, eXtensible Markup Language, is metadata designed to explain data. XML uses "tags" that describe data. Because the tags can be defined by the author and the software system, XML is extensible and not limited to pre-defined attributes. XML is considered an "open standard" as the internal architecture is freely available and documented. The roots of XML are found in SGML (Standard Generalized Markup Language). XML was formalized in the late 1990s. Today, XML is widely used for both online data and software applications. XML is different from HTML, however. Whereas HTML focuses on the formatting of the data, XML also includes information describing the meaning of the data. XML provides rules for not only digital data file formats but also for easily allowing humans to read and understand the data. For example, while HTML uses tags to specify document formatting, <bold> important words </bold>, XML uses tags to specify data meaning, <sales data> $3000 </sales data>.

How Do Data File Formats Become Obsolete?

Digital data file formats can become obsolete for several reasons. If the software that created the data file becomes obsolete, then often the data file format becomes obsolete also. WordStar was a dominant word processing application for personal computers in the late 1970s and through the 1980s. It was eventually displaced by Word Perfect. The data file format for WordStar was relatively simple by today's standards. It did have one unusual feature, however; the most significant bit in the byte was reserved to indicate that the character was the last one in the current word. Because WordStar is no longer produced and supported, users with WordStar data files must attempt to use other applications to access their data files. At one time, Microsoft offered "converters" for Office Word which allowed the software to access WordStar documents. However, finding any word processing software that can read WordStar data files directly is difficult.

Even if current software claims it can read an obsolete data file format, the results are often less than perfect. Because data file formats are not always well documented (and due to other factors), the conversion from the old data file format to the current format often results in missing data, extraneous data, or failed formatting. In other words, the conversion procession does not always work well.

Many people have experienced this scenario; you open an old data file using your current word processor and gibberish (garbage characters) is displayed, such as:

ÐÏà¡±á >______________________ þÿ ? A þÿÿÿ >
ÿÿ
ÿÿÿÿì¥Á Y

ð¿ b'

bjbj:&:&

-6 XLbXLb K ÿÿ ÿÿ ÿÿ · Ž Ž Ž Ž Ž ÿÿÿÿ ¢ ¢ ¢ 8 Ú ö $ ¢ Îø õ õ õM O O O O O O$
Æ- ¶|! È s Ž õ õ õ õ õsŽ Ž Û ˆ ===õ² Ž Ž M= õM== T h e F o u n d a t i o n s
o f C o m p u t e r S c I e n c e = ÿÿÿÿ

FÊ. . . ______________________ iÒ ÿÿÿÿ §v =9 ž0 Î= D" ® D" =D" Ž =ü õ õ=õ
õ õ õ õs sËr õ õ õÎõ õ õ õ ÿÿÿÿ ÿÿÿÿ ÿÿÿÿ ÿÿÿÿ ÿÿÿÿ ÿÿÿÿ ÿÿÿÿ ÿÿÿÿ õ ÿÿÿÿ ÿÿÿÿ
õ ÿÿÿÿ ÿÿÿÿ ÿÿÿÿ ÿÿÿÿ D" õ õ õ õ õ õ õ B

Within the gibberish you can make out some text since those characters were originally saved in ASCII. What are all the other "garbage" characters? In the original data file, these characters represented formatting information. In other words, they were metadata and not data. Because the current software does not understand the old data format, the characters are not interpreted correctly as formatting. Instead, the software attempts to display everything as ASCII text, resulting in gibberish. The characters that show up as garbage were originally binary strings that represented formatting metadata. Since the metadata formatting is not understood or recognized, the software converts it to ASCII text which is meaningless to the user.

Software application packages are often updated to support new features and functions. When this happens, there might be changes to the data file format. When word processors first accommodated text formatting, the data file needed to change from a stream of just ASCII characters to a mixed file containing both text and formatting information. The file needed to specify that certain text was bold, for example. When word processors began accommodating images within the file, another data file format change was required. The file now must handle both ASCII for the text and an image file format.

As new technology emerges and more features are added, it is often useful to change the data file format. While the new format provides increased capability, it is now incompatible with the previous file format. While it is possible for the software to work with both file formats, the software would require that instructions for both file formats be maintained within the application. This results in larger software which becomes more difficult to create and maintain. Some applications packages do support older data file formats; for example, Microsoft Office supports both the older .doc file format and the newer .docx file format.

Microsoft Works was a low-cost alternative offered by Microsoft for users who did not need the capabilities of their flagship products, such as Office Word. Works was supported by Microsoft from 1987 through 2010. Data files saved in later versions of Works (from about version 6 and onward) can be opened in Microsoft Office applications using converters; however, data files from previous versions of Works are not accessible as no commercial converter exists. In this case, the very legitimate benefit of a lower-cost application with a restricted feature set ended up rendering millions of data files as orphans, unable to be accessed by the next generation of applications.

Why Are Obsolete Data Formats a Problem?

Preservation of data, long-term storage for record keeping, and historical archives all require access to data. Anyone can visit their community's library and locate a book written one hundred years ago. Upon opening the book, the contents can be read and comprehended (as long as the book is written in a language the reader understands). By contrast, it is extremely difficult to access data stored on a computer from the 1940s. Due to obsolete hardware and unknown data file formats, the retrieval process would be a nightmare.

The Y2K issue can offer some insight and guidance. The "Y2K bug" or Y2K issue refers to the potential problems computer systems were expected to experience when the date changed from 1999 to 2000—the turn of the new millennium (Y2K stood for the Year 2000). There were predictions of catastrophic system failures after midnight 31 December

1999. While there were problems, many of the issues were resolved beforehand or corrected afterwards. It was feared that date calculations would be processed incorrectly, especially those dates that spanned the turn of the millennium.

In the mid-20th century, many computer systems specified a two-digit field to hold the data for the year. Many people still do this when they write out a date, such as 8/13/21, where the "21" stands for the year, 2021. Storing only two digits for the date field saved memory and storage space in the computer system. At one time, memory and storage were expensive and saving space was an important task of a programmer.

If you want to calculate a person's current age, you would subtract their year of birth from the present year (ignoring months and days, obviously). For example, a person born in 1956 would be 65 years old in 2021. The algorithm to compute the person's age is {current age} = {current year} minus {year of birth}. Notice what happens if only two digits are used for the year, rather than all four digits; using the same correct algorithm produces the incorrect, and nonsensical, value –35 (21 – 56).

The lesson from the Y2K issue was that formatting *was* the root of the problem. Before a system can process data into useful information, it must be organized well and correctly. Data formatting can be just as much a problem as incorrect algorithms within a system. When programmers create software systems, they consider both algorithms (the instructions specifying how to process) and data structures (the organization of the data). The digital data file format is just as important as the processing.

Various laws and governmental agencies require recordkeeping and access to digital data for specified periods of time. For example, the U.S. Internal Revenue Service requires taxpayers to keep certain records for three years. However, the IRS deems that other records need to be kept "indefinitely".[5] The U.S. Equal Employment Opportunity Commission requires that certain records be kept for at least one year beyond an employee's termination.[6] If the employee has served the company for 40 years, the likelihood that the original data is being stored in an obsolete data file format is high. I was originally planning to submit this essay to the editor in an obsolete data file format, but I realized the chances of my essay being including in this volume would be minuscule if I did!

Even if there were no rules or regulations, people would want to have access to historic data. We are all interested in our roots and origins, and historic insight provided by data from the past shapes us in the present. The meaning of anything is connected to its origin. Without understanding the past, the future is perilous. Scholar C. S. Lewis put the past into context well when he stated,

> We need intimate knowledge of the past. Not that the past has any magic about it, but because we cannot study the future, and yet need something to set against the present, to remind us that the basic assumptions have been quite different in different periods and that much which seems certain to the uneducated is merely temporary fashion. A man who has lived in many places is not likely to be deceived by the local errors of his native village: the scholar has lived in many times and is therefore in some degree immune from the great cataract of nonsense that pours from the press and the microphone of his own age.[7]

Egyptian hieroglyphics were undecipherable until the discovery of the Rosetta stone. Created in about 200 AD, the Rosetta stone recorded Egyptian priests honoring the pharaoh by listing his accomplishments. The text on the Rosetta stone was written in Egyptian (both hieroglyphics and demotic script) and Greek. When the Rosetta stone was re-discovered in 1799, it provided the key to unlocking the meaning of hieroglyphics. After a twenty-year study, Jean-François Champollion deciphered the hieroglyphs by comparing the Greek text to the symbols written in Egyptian. Unlocking the meaning of hieroglyphics ushered in a new era of understanding the peoples of Northern Africa. This understanding provided insight into our cultural origins. The meaning of anything is connected to its origin. Historical data is key to human understanding.

Possible Solutions

The use of standards might be one possible solution to prevent obsolete data file formats. If all software applications used the same, standardized format, data would be accessible across time.

For reference, let us consider the history of multimedia electronic mail. Originally, e-mail was limited to text characters. If the data stream is limited to characters, encoding and decoding are straightforward. Each byte represents a single character. As long as the scheme is known, such as ASCII, it is a simple matter to both change the text into binary form for transmission and change the binary form into human form for display. By the 1980s computer scientists were working on systems to send more than just text via e-mail. During that time, a number of systems were developed and each used its own digital data format. Two of the most popular were Diamond from BBN and Andrew from Carnegie Mellon University. Each system allowed messages that included elements beyond plain text, such as images. Because each system employed a unique data file format, the systems could not exchange multimedia e-mail with each other. In other words, the systems could not interoperate. The reason for the lack of interoperability was due to incompatible digital data formats used in the systems. In the following years, companies such as NeXT (founded by Steve Jobs) and Microsoft introduced their own proprietary multimedia data file formats. Without interoperability, multimedia e-mail was seen as interesting but not useful. If the typical user cannot be sure a multimedia e-mail message will be readable and understood by the recipient, plain text will be sent rather than multimedia content.

It would seem reasonable that a standard for multimedia e-mail was needed. Creating a standardized digital data file format would solve the interoperability problem; as long as all systems used the same file format, multimedia messages could be sent among every system that adhered to the standard. There are several standards for Internet e-mail. For example, Simple Mail Transport Protocol (SMTP) specifies how to send and receive ASCII characters. In the early 1990s, Multipurpose Internet Mail Extensions (MIME) was developed to handle multimedia content in e-mail. Rather than explicitly specifying a data file format, MIME specifies a way to identify different sections of an e-mail message. Thus, an e-mail message could have a standard text section (with characters in ASCII, for example) and another section that contains a graphic image. The format of the graphic image is not specified as part of the standard; thus, the image could be in any image data file format such as .gif or .jpg.

There are a number of digital data file formats that might be considered standards today. Adobe's portable document format (.pdf), Microsoft's Word format (.docx), and Open XML are three possibilities. Adobe's portable document format originated in the Camelot project begun in 1991, with an initial release on June 15, 1993. The idea was to allow any device to capture and share digital data regardless of the underlying system hardware and software. The resulting digital data file could be exchanged across diverse systems and still be read and accessed. The pdf specification was released as an open standard in 2008 through ISO 32000, an international standard. It should be noted that there is an ISO standard using portable document format for archival purposes, PDF/A. By accurately and completely capturing the "as printed" view of data, PDF/A is designed to preserve digital data files for long-term storage and future access.

Why aren't standards the universal solution to prevent obsolete digital data file formats? It would seem sensible to adopt a standard and then have that standard carried forth in perpetuity. The ISO 32000 PDF specification could serve as a current and future standard, for example. As with any technology, there are advantages and disadvantages. One disadvantage to the PDF is the size of the file it creates. If storage space is at a premium, retaining a large number of PDF files is problematic. Companies and individuals with numerous files might be reluctant to devote significant storage space to PDF files. While a number of word processing applications can read and write PDF files, no mainstream word processor uses PDF as its native file format. For example, Microsoft Office Word can read and write PDF, but it does so via conversion. Many users find it easier to employ their word processor's native file format which does not require extra steps to access.

As technology changes, it becomes more difficult to offer a universal solution applicable for all time. Even the PDF/A standard has changed over the years. The original PDF/A specification (version 1) did not allow "embedding" other files within the document. Version 2 of PDF/A allowed embedding files as long as they were in the PDF/A specified format. Version 3 of PDF/A permits the embedding of almost any file format.

There is, of course, no law mandating these standards. Individuals and companies are free to choose any digital data file format. Without considering the long-term consequences, it is easy and simple for users to save data files in the application's default file format. Computer technology has overwhelmed us with the "here and now". Users are enthralled by the power of technology and believe the present is the future. The "tyranny of the immediate" imposed on us by technology means that we oftentimes fail to pause and learn from history, or to imagine a future that is different from the present.

Even with standards, sometimes the specific implementation is inaccurate. The web gives us an example of this issue. Webpages are supposed to conform to standards, HTML being one. Even with a published standard, different browsers interpret the standard slightly differently. In other words, the standard can be ambiguous and different systems might interpret the standard differently. Most users are familiar with this if they employ multiple browsers to access the web. A webpage that "works" using one browser might appear "broken" when using a different browser.

While there is no absolute solution to the problem of digital data file formats becoming obsolete, an understanding of the concepts and issues underlying data files will equip individuals and companies with the intellectual tools to prepare their data for future access. Here are some recommended best practices when dealing with digital data files. Save important documents in several data file formats when considering long-term storage and archiving. In addition to saving the data in the application's native format, choose an additional format that is a current standard. It is wise to save a copy in a plain-text format also. Although all formatting and non-text items will be lost, the data itself will be preserved. The probability of future systems being able to access plain text in ASCII is very high, so the data will be available and understood for years to come.

Conclusion

Computers are powerful tools, due in no small part to digital data files. The ability to easily access, manipulate, and transfer data digitally has revolutionized civilization. The amount of data currently stored in digital files is astronomical. Unfortunately, this data is in danger of becoming unreadable in a rapidly advancing technological world. As new hardware and software are developed, it is possible that the format used to save data in the past will not be recognized by future systems. To prevent a future "digital dark age", understanding and planning are required in the present. The specter of obsolete digital data file formats can be prevented when armed with the concepts related to digital file formats and when proper foresight is applied. Human intelligence and creativity are still the most powerful aspects of a computer system.

Notes

1 See www.bbc.com/news/technology-13367398
2 www.bbc.com/news/science-environment-31450389
3 J.E.H. Smith (2008). *Leibniz: What Kind of Rationalist?* Springer. p. 415. ISBN 978-1-4020-8668-7.
4 https://msdn.microsoft.com/en-us/library/office/gg615596(v=office.14).aspx
5 www.irs.gov/businesses/small-businesses-self-employed/how-long-should-i-keep-records
6 www.eeoc.gov/employers/recordkeeping.cfm
7 C.S. Lewis, "Learning in Wartime", 1939, www.calvin.edu/~pribeiro/DCM-Lewis-2009/Lewis/Learning%20in%20War.doc

24

LASERDISCS

On the Way to a Digital Video Future

Stephen Mamber

The story of LaserDiscs is inextricably bound with the rise of cinephilia and a huge public interest in viewing movies, mostly at home. They can now be looked at as taking an important role as one of the steps in a longer process, the shift in media from analog to digital. As for a time LaserDiscs were available with digital audio tracks, they can actually be seen as a hybrid medium, although other advantages of LaserDiscs over VHS tapes were more important; widescreen letterboxed transfers, many of newly restored prints, individual frame access on CAV (constant angular velocity) discs and rapid access on all of them, and innovative special features are among the things we can remember and celebrate.

Our view of "obsolete technology" might be more complex than it first appears, particularly in the age of the Internet and eBay. Initially, it can seem like we are feeling nostalgic for earlier periods in our lives, our histories caught up with our use and enjoyment of the devices themselves, much as the association with music or movies from our younger days has an inevitable connection with the times when we experienced them. We want to think that there is still some life in old technologies, in whatever form, because it means that they are no more obsolete than our memories of times past (and not to mention ourselves). Whether we consider them truly dead, or transitions to newer technologies, can also affect our sense of obsolescence. Radio is still alive and well, but changed in terms of the devices themselves, and has left many casualties in its wake. We can love the wooden consoles and beautiful dials of old radios, or even feel nostalgic about early transistor radios, but those feelings are for the lost objects themselves, rather than what we used them for. Maybe old radios with large booming speakers sound better to us, just as other analog experiences may allow for the same feelings, but this has little to do with our love for a particular medium, which may be marvelously thriving. With radio, one has only to look at (and listen to) radio.garden on the Internet to see the tremendous opportunities now afforded to us. And since everything seems to be available on eBay for under $50, no matter how expensive originally, there is a false sense that no technology is ever obsolete.

It can be difficult to acknowledge that certain technologies can become obsolete for legitimate reasons, especially as digital technologies have offered tremendously improved alternatives. Film editors can still love Steenbecks and splicing tape, but it is hard to argue that Avids or Premiere cannot do the same job better. We can chart the pros and cons of the old and the new, and if there are enough pros to the earlier stuff, we can certainly miss what we once had. Especially with technology and media, though, it is difficult not to acknowledge that things have greatly improved—that televisions are better, that cable is better than our previously limited over-the-air broadcasts, and yes, DVDs, Blu-ray discs, and streaming offer much that VHS tapes and early LaserDiscs could not. So we can look back to the special pleasures of old devices, but we can also consider how they sowed the seeds of what was to come.

So, we can try, if we are able, to separate our love for the physical objects from what we actually used them for, as difficult as that might be. In the case of LaserDiscs, we would probably want to start with the desire to have prerecorded content available when we wish. This would go back to watching movies in theaters, where we could only go at specified times (and places), and where repeatability meant staying for a second showing or coming back another day. As much as we can love 35mm film in theaters, we have to acknowledge that they offer a certain kind of fixed experience which was limited in relation to what we have available for us in movie watching today. (Now, when a new movie is released in theaters, it can feel as if it is not really available until there is a DVD out.)

Technological obsolescence in the film-video arena happens at both of what we now more commonly call the hardware and software levels. Pioneer LaserDisc players currently appear on eBay in the category of "Vintage Electronics", the same place where they stick pushbutton telephones and vacuum tubes.[1] While some can be had as cheaply as $40, "AS IS" is affixed to most listings of them, so you can guess what that is likely to mean. When the players go, the discs either remain as collectibles or curiosities. (The same has happened with VHS, surprisingly quickly given how ubiquitous the format was, although in a way it is enjoyable to see a poor quality video delivery system die a death it has long deserved. So much for obsolescent technology nostalgia.)

After theaters, television, of course, brought films into the home, and while it was also a technology that first could be decried for lesser picture quality, the greater availability of product as well as their being free (just all those pesky commercials) made serious and extensive movie watching possible in ways that movies in theaters could not. It is wonderful, for example, in *A Personal Journey with Martin Scorsese Through American Movies* (1995) to hear him recount the times and circumstances when he saw certain films, often beginning with or including viewings on an "Afternoon Movie" program, which offered its own pleasures and opportunities not supplanted for him by later screenings under better circumstances.[2] Serious cinephilia begins then, even if it was still possible before, and has certainly expanded since.

The next major step in the ability to see movies (both in terms of availability and cost) was videotape, which started, as many of these technologies do, as too expensive for middle-class consumers until it eventually trickled down. Three-quarter-inch tape recorders and players (especially what Sony called their U-matic devices) first appeared on the market in the early 1970s, and while still somewhat expensive, offered the first home possibilities for off-the-air recording and the opportunity to build a collection. (U-matic machines can still be found on eBay in the $50–$100 range, and some of them might actually still work.) As this was also the time of the first movies available without commercials through cable, the timing is hardly accidental. Machines to record movies and better quality versions of the movies themselves went hand in hand. Also, boutique cable offerings, such as the "Z" Channel in Los Angeles, were the first place to see movies on television in letterboxed format and also in "director's cut" versions, two attractions that were advanced by LaserDiscs but not started there.

We might see the rise of VHS tapes, which began in 1976 in Japan (and a year later in the U.S.) as a step backward, as it was rare to find a widescreen film ever letterboxed (the infamous pan-and-scan was the typical VHS treatment), but one cannot deny the huge impact of an inexpensive technology which offered both the ability to record off the air and to view commercial tapes. (And put a couple of machines together, and you could even make degraded copies.) The late 1970s, again not coincidentally, saw the rise of video rental stores, where films could be rented for a few days (and for a few dollars), demonstrating a huge public appetite for viewing films at home. (The aptly named Blockbuster video rental chain, taking a cinema term into the video world, was the most popular, having at its height in 2010 over 6,500 worldwide stores.)[3] This led to the availability of titles which were not sufficiently commercial for broadcast TV, or had content deemed unsuitable for over-the-air viewing. The widest manifestation of this impulse was the great popularity of pornographic adult content on VHS (well documented in many histories of television).[4] VHS was also part of the great VHS–Betamax debate, where the latter (and superior) Sony tape cassette technology quickly lost out to the cheaper and RCA-supported VHS. (We will see a similar battle in LaserDisc vs. VHD, in Blu-ray vs. HD DVD, and even in the great and continuing discussion of Apple vs. Windows). These competitions were both commercial and cultural, forcing the user to determine where cost considerations and sometimes minor technological differences enter into one's preferences and purchase decisions. That VHS demonstrated conclusively that lots of people wanted to see a wide variety of films had an impact on all subsequent video-viewing technologies, including LaserDiscs. Whether enough people would be attracted by various forms of superior cinematic experiences became the question that LaserDiscs, as with subsequent technologies, looked to answer.

A Little More on the LaserDisc and Film Studies

If we are looking at parallels, we can see another overlapping history between LaserDiscs and film studies, and most particularly, the desire to engage in close visual analysis (sometimes called textual analysis). The rise of auteur criticism and a strong interest in *mise-en-scène*, carrying over from French film criticism to the American academy, showed a hunger for freezing the (high-quality) image, playing it backwards and forwards, jumping around from example to example. That this preceded (just barely) the popularity of the LaserDisc among film scholars can be demonstrated by a rare and almost forgotten (and expensive at the time) piece of equipment, the 16mm analysis projector. A popular model was offered in the mid-1970s by a company called NAC Image Systems, who billed their products as "Videographic Motion Analysis Systems", and proudly advertised the ability of their projector to present 16mm films "FLICKER-FREE at all projection speeds – forward and reverse".[5] Kodak also offered a similar product.[6] In order to project film slowly and not see it burn up (as in Bergman's *Persona* (1966)), a high-volume blower kept cool air aimed at the hot projection lamp. It was thus possible to engage in slow and repeated analysis of a chosen sequence, although, of course, if one wanted to do the same with a sequence elsewhere in the movie, it was necessary to unthread and then rethread the film (a quite cumbersome task), possibly with pieces of white tape to mark where one wanted to go. Analysis projectors are interesting not just as historical curiosities, but as demonstrating what film scholars wanted to do before there was really a suitable technology to do it. Equivalent technologies might be 19th-century devices such as the zoopraxiscope showing so clearly the hunger for moving images satisfied soon after by the Lumière brothers.

When LaserDiscs fulfilled this desire for analysis, without any damage whatsoever to the disc, it also became possible to capture still images for the purpose of writing about a film. (There were also cumbersome devices for this which could print images from individual movie still frames; again, a device whose presence demonstrated the desire to use existing technologies in ways that were about to become much easier.) No wonder popular film critics Gene Siskel and Roger Ebert in the 1980s extolled the superiority of LaserDiscs as a medium to study film, also anticipating the coming of higher-definition television as an outgrowth of what LaserDiscs demonstrated.[7] As this was also the time of the so-called film school generation of directors, there was a ready group of creative artists who could see the link between the films they made and the public interest in studying them further. Once again, Martin Scorsese, a dedicated preservationist as well as great film director, a graduate of the NYU film school, is but one example among many who showed a serious concern for (and involvement with) how their films were presented on LaserDisc, clearly knowing that they would be given a different kind of attention there than in theaters, and could be presented in a scholarly context.

Not to get caught up in LaserDisc obscurity, we should at least briefly note the quite bizarre competing system called Selectavision, on which RCA reportedly spent over $600 million in 1970s dollars to develop and market, which competed briefly with LaserDiscs before dying a well-deserved death. Using what were called CED discs (for Capacitance Electronic Disc, in case anyone cares), they resembled vinyl records and were made of the same material. Similarly, they used a needle at the end of an arm which physically read the surface of the disc as it turned. A rather large problem with this was that after as few as 10 or 12 plays, the discs would then show signs of wear on the screen when played. (A nice review of this format is on one of the Retro Tech programs on YouTube, where it is decided that this format was never very good in the first place and never worked very well.)[8] But, this is a moment of transitional obsolescence worth noting as well, as the last grooved medium, which started with the Edison Cylinder Phonograph, patented in 1877. While vinyl audio has enjoyed a recent resurgence, this is the same technology as it has always been. Indeed, its retro nature has been part of its recent appeal. No nostalgia for grooved video vinyl will likely ever occur, even if one were to locate a player, as this is one of the rare cases where a more expensive superior technology (the LaserDisc) beat out the cheaper inferior alternative.

One can see where libraries and rental outlets would prefer a more permanent medium like LaserDiscs, which did win out despite rampant tales of potential "laser rot", a not entirely apocryphal phenomenon which certain individual LaserDiscs were subject to, owing to manufacturing defects. By the end of 1990, there were reportedly over a million players sold. As the VHS market was in the $5 billion dollar range during this period, it is clear that LaserDiscs were always a niche market, but that market had an impact on movie fans and scholars that was arguably far stronger than its lesser numbers would suggest.

LaserDiscs had unique capabilities which strongly affected the form that DVDs and even streaming media have taken today, so a look at some of these is useful, as well as paying homage to an innovative company whose existence alone made LaserDiscs so important.

Deleted Scenes and Director's Cuts

These two features go hand in hand, and were both major bonuses that LaserDiscs first offered. These are among the notable "Supplementary Features" which now abound on DVD—some titles have special features which, together, are longer than the films they discuss. From the LaserDisc era on, this was one of the reasons to buy the title, for either those fans who had seen the film many times before or who already owned a VHS tape version (which rarely had special features because they had to be stuck on the end of the tape and could not be easily located or accessed). These added elements are also part of a completist impulse well served by multiple versions and alternative abandoned segments; the database-like and archival underpinnings that began with LaserDiscs. The CD box sets of today, with every version of a song now collected, or every work by the artist in question, is one form of this combination of scholarly attention and marketing savvy which starts with LaserDiscs, as do the box sets of television and movie series, and collections by actor or genre. (While there were tape versions attempting similar things, it was on shiny discs that sufficiently large quantities could be collected and easily accessed.) One fine example, that still has never appeared on DVD, is Francis Ford Coppola's *The Godfather Trilogy 1901–1980* (1992), a chronologically recut version of his trilogy with additional scenes. The LaserDisc was labelled and came out in 1992.[9] The movie in this form is a whopping 583 minutes and took up no less than 11 LaserDiscs. (There also were two additional LaserDiscs, including a 73-minute documentary, a rare example of additional discs devoted to special features.) This LaserDisc version is still the only way to see the films this way, the story time rearranged to present the 79 years covered by the story straight through, rather than in flashback and alternating sequences (although a VHS version with a somewhat cropped image was also released).

Director's cuts were an interesting development, in that the promise of a release version of a film being the director's cut (also for a film release called "final cut"), has been the holy grail of Hollywood filmmakers for decades. The idea that LaserDiscs could be the place where scenes could be restored and a film put into the form their directors originally wanted was an inviting possibility. Not surprisingly, directors whose films were often badly shortened or recut by studios (that is, great and often adventurous directors) had a particular chance to present something closer to what they originally wanted. Orson Welles, Terry Gilliam, Michael Cimino, Francis Ford Coppola (as mentioned), Sergio Leone, and Sam Peckinpah all had opportunities to see extended versions of their studio-cut films. These could also be described as directors interested in making longer films and/or films with ambitious time structures. In 1991, the director Peter Bogdanovich both added seven minutes and re-edited other sequences for the LaserDisc release of *The Last Picture Show* (1971). Rather than simply including deleted scenes as a special feature and placing them separately from the films themselves, these new or extended versions offered the promise of restoring what had never been seen before, in the way the filmmakers had intended. This practice also anticipates our digital era, when no media work need be considered finished, which has led, for example, to six different versions of Ridley Scott's *Blade Runner* (1982) getting video releases. Scott was also amusingly involved in a director's cut of his film *Alien* (1979) in 2003 which was actually a minute shorter, as Scott has offered in interviews that he is quite happy with the original version, but that the studio requested this recutting so that there was an alternate video version. This feels similar to reports of "extra" scenes now being filmed for new movies with the specific intention of including them as "deleted" scenes. LaserDiscs set in motion all these possibilities; restoring deletions, recutting by directors, and tacking on abandoned material.

Director (and Other) Commentaries

Another feature that began on LaserDiscs took advantage of the technology's capability to offer alternative soundtracks. One opportunity this afforded was for directors to add their own voice-over commentary to their films. This was done extensively, so much so that many of these tracks are still only available on LaserDiscs and were not carried over later to DVDs of the same films.[10] These include commentaries by directors such as Louis Malle, Terry Gilliam, John Schlesinger, and Sydney Pollack. (Pollack, for his film *They Shoot Horses, Don't They?* (1969) when it was released on LaserDisc as a "Limited Autographed Edition", personally signed 2,500 copies, each given an individual number. The director Robert Wise did the same in 1995 for the LaserDisc release of *The Day the Earth Stood Still* (1951).) A number of LaserDisc titles, particularly from the Criterion Collection, had commentary tracks from noted film scholars, such as Robert Carringer, Howard Suber, Maurice Yacowar, and Jeanine Basinger. This use of scholars has sometimes carried over to DVD, but far less frequently, and none of the LaserDisc scholar commentaries were carried over. It should be remembered, though, that the very first audio commentary was on the Voyager Company's 1991 release of *King Kong* (1933), by film historian Ron Haver, so scholars got there ahead of directors.

LaserDiscs usually had one supplementary track, which has expanded on DVDs to allow for multiple additional audio tracks. For a brief time, roughly around 1997, there were simultaneous releases of some titles on both LaserDisc and DVD. When the film *Contact* (1997) was released, the LaserDisc included a commentary by its director Robert Zemeckis and the film's producer, which was included also on the DVD version, which also had another audio track of Jodie Foster and others not on the LaserDisc. Some might feel that additional tracks are a useful technological advance. Once again, though, it was LaserDiscs which began the practice and where there was more experimentation.

It is worth noting that once LaserDiscs allowed for director commentaries, it also allowed for directors to refuse to do director commentaries. Stephen Spielberg has never done one, saying that films should only be shown with their original soundtracks, so as not to distract the audience. Similarly, Woody Allen has refused to do commentaries to any of his films.

Making-Of Documentaries

There has been an extensive number of fictional and documentary films about the making of other movies. What has become a form of its own, perhaps too rigid a form, are the approximately 20-minute featurettes which have become a staple of "special features", a practice begun with a vengeance on LaserDiscs. You know what they look like: some on-the-set footage, interviews with a director and a star or two, perhaps some background information on the film's subject. Like other special features, these have greatly expanded on DVDs and Blu-ray discs, to become sometimes-hours-long, as for example, The Beast Within: The Making of 'Alien' (2003), included with the 2003 DVD box set, *The Alien Quadrilogy*. It runs 171 minutes long, a good deal longer than any of the *Alien* films themselves. Special DVD and Blu-ray releases of Hitchcock films have also had hour-long documentaries (at least) about the making of each film.

While sometimes well done and informative, going back to the days of LaserDiscs, these documentaries are primarily indicators of that desire mentioned elsewhere to provide added features, in order to entice fans and film scholars who might already have copies of the films on VHS, which rarely ever had these extras. Starting with LaserDiscs too, making-of documentaries were especially of interest on science fiction titles and other effects-dependent features, where there was always interest in how effects were accomplished, particularly as we were then moving into the age of CGI.

Other Special Features

LaserDiscs were an experimental playground for seeing what could be done interactively. Maps, "slide shows" of historical material related to the films, complete shooting scripts, storyboards, filmographies, and visual essays were just a few of the efforts. Buying the LaserDisc of *Fellini Satyricon* (1969) also got you some pages on screen from the original Plutarch. If you owned the LaserDisc of the William Wyler western *The Big Country* (1958), you also got the entire Jerome Moross score on a separate track, so that you could watch the film with music alone. To give just one only slightly extreme example, included on the 1992 Voyager LaserDisc of *The Player* (1992) were these listed features (in addition to the new widescreen transfer):

- Audio commentary by director Robert Altman, screenwriter Michael Tolkin, and director of photography Jean Lepine
- Video interviews with Altman, Tolkin, Lepine, and stars Tim Robbins, Greta Scacchi, and Whoopi Goldberg
- Interviews with 20 Hollywood screenwriters about their real-life experiences in "the industry"
- Six deleted scenes
- "Map of the stars" charting the film's 65 star cameos
- Annotated photo history of films about Hollywood
- American and Japanese trailers and TV spots

The plethora of offerings were both an undeniable thrill to scholars and fans, and also suggested the era to come, when websites would supplement what a film offered and DVDs could pack so many special offerings that they could not fit on the same disc with the movie, as in the case mentioned of *Alien*. (LaserDiscs did not usually include additional discs for special features, except in rare cases. They were part of the same disc as the movie, in this technology which could at most offer an hour of material per side. So a feature film in the long play CLV format might leave just a little room for any extra stuff.) What is evident too, as was discussed a bit earlier, is that there had been a symbiotic relationship between film schools and disc producers which still continues today with DVDs and Blu-ray discs, where a prime market for LaserDiscs was schools and their professors, and they also prompted the interest in additional materials (and were often contributors to them, as were their students).

Video might have killed the radio star, but LaserDiscs killed 16mm, or to put it another way, Voyager killed Films Inc. It is hard to imagine there was once a time when companies made large amounts of money sending out film titles for brief periods for showing in schools. LaserDiscs, which sold for about 25% of what a single film rental cost, killed that market quickly, especially because discs offered high-quality video which neither the 16mm prints nor VHS could match. Now, you cannot even find a major film rental company like Films Inc. on Wikipedia. One of its competitors, Blackhawk Films, defunct in 1987 and originally with stronger roots in preservation, at least gets an entry.

Voyager and the Criterion Collection

The Voyager Company, started in 1983, put out their first two LaserDisc titles that year in what they called "the Criterion Collection", which were *Citizen Kane* (1941) and *King Kong* (1933). Beginning so auspiciously, the Criterion Collection released about 200 titles over the next 15 years, concluding not so well with *Armageddon* (1998). (Still not on DVD is the LaserDisc's commentary track, where, besides the director Michael Bay, one of the stars of the film, Ben Affleck, also appears; mainly, as I recall nearly 20 years later, to complain a number of times about plot holes in the film, such as whether one can learn to be an astronaut in a week.) In association the next year with Janus Films, the Criterion Collection was more than reason enough to own a LaserDisc player. As their next two discs were Hitchcock's *The 39 Steps* (1935) and *The Lady Vanishes* (1938), a cinephile in those days went broke with the newly available titles. The uncut and beautifully widescreen version of Max Ophuls's final film *Lola Montes* (1955) (their #12 release), was enough to put most of the people I knew into a serious swoon. The Criterion Collection depended in good part on the art house titles Janus Films had previously rented on 16mm, from such stalwarts as Bergman, Fellini, and Antonioni, which itself was a major contribution. They went well beyond that by releasing beautiful versions of classic American films such as Lumet's *12 Angry Men* (1957), Kubrick's *The Killing* (1956), and of course Frank Capra's *It's a Wonderful Life* (1946). (If you were a Capra fan, you could also go for Voyager's discs of *Arsenic and Old Lace* (1944) and the lesser-known *Lady for a Day* (1933).) While sometimes Criterion titles were already available on LaserDisc from other companies, what also made the Criterion Collection so great was their devotion to exhaustive additional materials, well beyond the conventional special features so common to other LaserDiscs.

Titles from the Criterion Collection have already come up frequently in this essay, because their releases have most interesting uses of LaserDisc technology for commercial titles. To speak of the Criterion Collection is to recall what made LaserDiscs innovative and great, both in uses of the technology and in available titles. Experimental special features, close cooperation with directors on approved editions, scholarly commentaries, production stills, storyboards, ambitious "making-of" documentaries are all to be found on their titles. The company has since re-released a good number (but far from all) of their titles on DVD and Blu-ray, and it is a tribute to the LaserDisc versions that the additional materials have often been carried over intact.

Many of the early Criterion Collection titles are now difficult to locate, as are most early LaserDiscs (though I just noticed on eBay a new, sealed copy of Renoir's *The River* (1951) for just $1.99). Another great Voyager experiment was an entirely audio-only disc, in order to cram in as many radio shows as possible, the 1988 disc *Orson Welles and the Mercury Theater of the Air*. This included 15 radio programs and further audio of reminiscences from former Mercury Theater players. To release an audio-only LaserDisc (seemingly an oxymoron) required putting mono sound (fine for old radio shows) on each of the two available audio tracks and the additional digital audio tracks, for a total of about nine hours. It was smart, though, to do this on LaserDisc, where the name Orson Welles already meant quite a lot, even though the LaserDisc had no video.

Criterion still exists as the best of the companies releasing DVD and Blu-ray disc titles, and its commitment both to world cinema and to collections of lesser-known titles by important directors remains unparalleled. Once again, though, LaserDiscs are both where this started and where the major contributions were first made.

Interactivity

The significant role of interactivity in modern digital technologies, especially video games, received a major boost from LaserDiscs, in a variety of forms. LaserDiscs are not only a transitional technology between VHS and DVDs, they mark a significant step forward to active viewer control and engagement. For one, the technology allowed for the first time the notion of accessing databases rather than just watching or skipping around a film. One great example of this is the Voyager Company title, *Vienna: The Spirit of a City* (1991), produced by the filmmaker Titus Leber. The LaserDisc included 15,000 images of art works in Viennese museums, as well as 20 minutes of motion video. Any frame could be accessed by number (an index in booklet form was provided) and then held on the screen for extended viewing. This was possible owing to another significant advance that LaserDisc technology allowed: a still frame was just a laser light over a spinning disc, so no harm was done to the disc when viewed this way. Still frames were virtually impossible with VHS tapes, because freeze frames would cause stretching of the tape. (Some VHS machines, the better ones, even had a feature that automatically disengaged the still image after a few seconds, to protect the tape.) The combination of the ability to store large numbers of images (and moving image sequences) together with this still frame capability was a powerful step forward afforded by LaserDiscs.

Probably less known, but equally important, was the appearance of LaserDisc players with a serial interface, which allowed for computer control of the disc. While there had been Betamax systems able to control movement of the tape by computer as early as 1979, a LaserDisc and a computer was a much more powerful combination, owing not just to the ability to access still frames, but to do so very rapidly, usually in under a second. It was this fast access that also made LaserDiscs suitable for video games, using computer control. *Dragon's Lair* (1983) was the first very popular example, later to have a life on subsequent game systems. (Let us not forget Dirk the Daring making his first appearance on the 1983 LaserDisc arcade version.) What the serial interface (also known in computer circles as RS-232, a still popular communications protocol) also made possible was programming access to specific frames and sections, rather than entering long numbers by hand with a remote control, and presumably having to keep track of each number. (This would be the difference between entering a number like 39856 to get to the Rosebud sled in *Citizen Kane*, or pressing a button on a computer screen to do exactly the same thing.) Pioneer Electronics was the major LaserDisc player manufacturer offering some versions (at a greater price) of LaserDisc players equipped with this serial interface capability. These players were generally advertised as being for educational uses, as home players were, sadly, never equipped with this ability.

This is a good time to discuss the idea of "chapters", itself a LaserDisc innovation. Because of the high numbers involved in accessing individual frames (as well as needing to remember those numbers), LaserDiscs incorporated the notion of chapters, sometimes called "scenes". These could be employed by disc manufacturers so that easy access was afforded to sections of a disc. Chapters generally totaled around 15–20 per disc for most movies, so these numbers were much more easily entered by hand (or reached by clicking "Next"). This segmentation was also commonly used for music discs, in order to easily access specific songs. This was a sort of acknowledgment of the difficulty of entering numbers on a remote control for more specific passages, as the chapter designations were completely arbitrary and a creation of the disc publisher. One such educational application was the "Periodic Table Videodisc" produced in 1995 by the Department of Chemistry at the University of Wisconsin. The LaserDisc came with a computer program to present each element of the periodic table, which were all on the LaserDisc. Such applications today would undoubtedly be entirely done digitally, with LaserDisc segments being replaced by digital video directly within the program, thus eliminating the need for the cumbersome hybrid system of the computer-LaserDisc player. That, however, is one of the pleasures of looking back at now obsolete technologies. Doing things first required genuine invention, not just a more efficient way of doing things. And, of course, these early efforts paved the way for later applications with more powerful newer technologies.

As computer-controlled disc players allowed for creating as many segments as the program-creator chose, these systems also made part of the program the interface to those locations. What the Pioneer system allowed was a set of simple commands to replace all the capabilities of a remote control. Remarkably, the same computer commands (such as sending "P" to play a disc, or "SF" to step forward a single frame), have been kept consistent by Pioneer as the technology has moved on to DVD players and even current Blu-ray models. As before with LaserDiscs, not all DVD and Blu-ray players have a serial interface, but those that do, employ basically the same commands as computer-controlled LaserDisc players from the 1990s. Technologies can become obsolete, but may live on in evolving forms. It is not just a matter of whether a given device has been replaced; those new technologies can often be closely dependent upon concepts created by their predecessors.

The software to control LaserDisc players was essentially the product of two companies, Microsoft and Asymetrix. The latter, a company begun in 1990 by Paul Allen, one of the founders of Microsoft, had a product called "ToolBook", an authoring application not unlike Apple's Hypercard. ToolBook, though, had a robust programming language built in which could be used to send commands to a LaserDisc player, among other devices (including CD players to control music). Sample ToolBook applications were included in the 1990s by IBM in their Advanced Academic Systems, computers that were marketed to universities. Microsoft had already developed its Visual Basic as an easier-to-use and more powerful offshoot of the Basic programming language, and serial control was also available there. Microsoft further developed what it called VBA, or Visual Basic for Applications, which meant that all the programs in the Microsoft Office package, including Excel and Access, could also incorporate this same language. That this was possible in the early 1990s to control LaserDiscs continues to be an amazingly early combination of video material and computer control, which can easily be seen today in everything from YouTube to every HTML web-based application on the Internet. LaserDiscs were where all this interest in incorporating video into digital applications began.

Another project made possible by LaserDisc technology, highlighting these new interactive possibilities, was the "Aspen Movie Map" produced in 1978 at MIT, in what was to become the MIT Media Lab. Using the group's own form of computer control of a LaserDisc, the user could drive streets in the city of Aspen, turning when they wished, and stopping at any building in order to view more closely what went on within. This was certainly an ingenious and innovative accomplishment, and not so far from how video games employed these devices, especially travel, branching, navigating, and employing a deep database of possibilities. So university projects developed innovative uses for LaserDiscs that sometimes paralleled what entertainment and game companies were also exploring.

A Few Final and Amusing LaserDisc Features

To some, LaserDiscs meant karaoke, a connotation is still has today. Pioneer marketed special LaserDisc machines (called, not surprisingly, LaserKaraoke) which had a direct microphone input on the front of the player, so that mixing a voice with the music on the disc was easily accomplished. Pioneer also sold a series of disc sets called "50's Hits" on through to the 1980s. Each decade had six discs, so 40 years' worth of bar sing-along material took up 48 discs. If that was not enough, Pioneer also had a five-disc set of country songs as well as a Christmas set. (I have seen listings elsewhere for over one hundred LaserDisc Karaoke titles.[11] For a last mention of eBay, each multiple disc set is advertised as selling for $12, with the seller (sadly) reporting that he found them at an estate sale.) The same machines could play regular LaserDiscs, as well as music CDs, as they were basically regular players with the extra audio inputs. (All LaserDisc players could also play audio CDs, although in those days if you had a LaserDisc player, you very likely also had a CD player. Now it seems fitting that two obsolete technologies could reside together in one machine, a double obsolescence.) In another odd carryover from the analog to the digital, there are websites that present karaoke-ready songs copied directly from LaserDiscs. They also provide an opportunity, should one wish, to see what a LaserKaraoke was like.[12] LaserDiscs can take the credit (or the blame) for what is now an extensive industry for digital karaoke, everything from portable mp3 versions to wireless bluetooth systems.[13]

As it took a while to appear for HDTVs (and is now in the process of disappearing there), 3-D also had a brief experimental life on LaserDisc. It required the plastic anaglyph red-cyan glasses that both 1950s 3-D movies employed as well as a few 3-D comic books from that time. If you bought a 3-D LaserDisc title, there were usually two to four paper glasses included, which kept the size of your 3-D viewing parties pretty small, unless you found ways to order more. Like a reversion to the early affinities between movies and carnivals, or the first few offerings in the early 1950s of three-screen Cinerama, a prime 3-D LaserDisc from 1995 was *World's Greatest Roller Coaster Thrills*, which was a disc of exactly what it sounds like. (Volume 2 followed a year later.) There were a few 3-D movie titles, such as the inevitable *House of Wax* (1953), as well as *Jaws 3-D* (1983), *Amityville 3-D* (1983), and *Freddy's Dead: The Final Nightmare* (1991). Even music LaserDisc titles sometimes were in 3-D. I will shamefully admit to having owned *The Judds – Love Can Build a Bridge* (1991), which only came with two pairs of glasses. As with karaoke, Pioneer produced some 3-D LaserDiscs of their own.[14]

These offshoots of regular LaserDiscs, besides being amusing to recall, are interesting for again asserting the transitional and hybrid nature of LaserDiscs. Both karaoke and 3-D are now to be found in superior digital versions (if anything about karaoke can be described as superior). Both existed before LaserDiscs, but found homes for a time on a technology well suited to the niche consumers who went for this sort of stuff. While there were VHS karaoke tapes, once again, you had to sit there and skip from title to title to find the song you were ready to sing, and 3-D on VHS, also occasionally attempted, suffered once more from inferior video quality in comparison to LaserDiscs.

Conclusion

I will confess that the more one thinks about LaserDiscs, the more one misses the darned things. There was a quirkiness and experimental quality to the offerings, so that one cannot help seeing the transition to the digital as a reduction of the initial fun. The brief era of LaserDiscs is like early television, a time we have gotten past but cannot help but marvel at, for the possibilities so quickly put on display that were (maybe) lost as the technology developed further. It feels a little like preferring the kid to the adult it becomes, even if we recognize the maturity and added intelligence that follows childhood. LaserDiscs represent, for those who were around to enjoy them, an innocent time, when the joy of viewing high-quality movies with the chance to interact with them eventually led many to careers related to the stuff they were watching.

Notes

1 www.ebay.com/sch/Vintage-Electronics/183077/bn_1643015/i.html
2 https://en.wikipedia.org/wiki/A_Personal_Journey_with_Martin_Scorsese_Through_American_Movies
3 https://en.wikipedia.org/wiki/Blockbuster_LLC
4 See, for example, Alan Abramson, *The History of Television, 1942 to 2000*, North Carolina: McFarland Publishing, 2003, page 190.
5 A brochure is preserved at www.nacinc.com/datasheets/archive/DF-16C.pdf.
6 A brief YouTube demo of which can be seen at www.youtube.com/watch?v=AsUIQdM40pc.
7 www.youtube.com/watch?v=AGigrMXElcs

8 www.youtube.com/watch?v=0LrPe0rwXOU

9 https://en.wikipedia.org/wiki/The_Godfather_(film_series)

10 The website LaserDisc Database, valuable in its own right, at http://forum.lddb.com/viewtopic.php?f=32&t=82 and http://forum.lddb.com/viewtopic.php?f=13&t=81 offers lists of over 100 films whose commentary tracks were still not made available on DVD as of 2010.

11 See, for example, https://discount99.us/laserdisc%20karaoke, where everyone from John Tesh to ZZ Top can be found.

12 https://archive.org/details/karaoke1 is a good offering from that 1950s set.

13 A website listing the Best Karaoke machines as of September 2018 can be found at www.lifewire.com/best-karaoke-machines-4118378. Regrettably, there is not a remaining videodisc version among all the available choices.

14 A bizarre 40-minute non-3-D YouTube of "3D Museum Pioneer LaserActive" can be viewed on YouTube at www.youtube.com/watch?v=bF2Y6v6GXhE.

25

PERFECT SOUND FOREVER?

How the Compact Disc Sowed the Seeds of Its Own Demise

Jason Curtis

Compact Disc Digital Audio (to give the music version of the compact disc its full name) has now been with us for 35 years, an enormous achievement for any consumer music format, and in that time has sold in its billions, making the recording industry massive profits in the process. However, sales of music albums on compact disc reached their peak as long ago as the year 2000 with unit sales in the US of 942.5 million. Since then, sales have fallen steadily to just 99.4 million units in the US in 2016 (RIAA, 2017a). In the UK, compact discs peaked later in 2004 at 162.4 million units, but have since fallen to 47.6 million in 2016 (BPI, 2017). The compact disc music single effectively disappeared around the mid-2000s, having begun to decline as early as 1997 in the US (RIAA, 2017a). While it is too soon to talk of the end of the compact disc, there might come a day soon when it ceases to be a mainstream format in the same way that the vinyl LP record ceased to be by the early 1990s.

Although the vinyl LP never disappeared entirely, and is now enjoying something of a revival, sales in 2016 in the US were still just 17.2 million units, down from a peak in 1977 of 344 million (RIAA, 2017a). While compact disc sales are still significant, they will most likely continue to decline and, unlike the vinyl LP, the compact disc is less likely to enjoy a revival due to feelings of nostalgia. It does not have the same tactile nature or impressive artwork, and it could be argued that it does not enjoy the same associations with a perceived golden age of albums; it is noticeable that many of the recent vinyl releases are re-releases of albums of the 1960s, 1970s, and 1980s.

How did we end up here? Is it the decisions of the record companies that have led to the decline of the compact disc, or is it the nature of the physical format itself? Was it inevitable that eventually music distribution and consumption would move to the virtual world?

As the compact disc ceases to become the dominant music format, what do we lose? For those who grew up with physical formats, the move to on-line music distribution marks a big shift in how we consume music. Physical album formats meant we tended to buy and listen to an album in the way its creator intended (although the compact disc did begin to erode this); with on-line music distribution, it becomes much easier to just buy or listen to the tracks we like. We can now use our music devices to shuffle or randomize tracks, and with streaming services we can ask the software to select tracks we might like from any of the albums available.

Sales from streaming services have now overtaken those of permanent downloads, and we move yet further from a model of music ownership to music access, when much of the recorded music ever produced is available. Does this devalue music? When it was expensive to buy an album, there were difficult choices to be made; opting to buy the latest Michael Jackson album meant you had to wait and save up for the Madonna album (unless of course you got a friend to make a copy on tape for you). When you bought an album, you wanted to listen to it carefully, read the liner notes, and look at the artwork. When there is so little cost (other than your time) involved in listening to a new Adele track on a streaming service, do we still value the music as much?

This essay will investigate some of these issues, as well as looking at the rise of the compact disc and the factors that have led to its decline.

The Rise of the Compact Disc

Philips and Sony had been working on optical disc technology for some years prior to the launch of Compact Disc Digital Audio in 1982. Philips announced its Video Long Play disc as early as 1972 (this eventually became the LaserDisc). The Video Long Play disc stored analog video on a large optical disc similar in size to an audio LP. Philips then decided to work on an Audio Long Play disc which was to have contained analog audio, but quality issues led to the adoption of digital audio. In 1979, they announced the disc to the press, by now called the compact disc in reference to Philips' own Compact Cassette design (more commonly known as cassettes), and began to look for a partner to work with. Sony of Japan agreed to work with them, and had also been working on digital audio; they had already introduced a means of recording digital audio using videotape (onto both its Betamax and U-Matic formats) with a PCM processor. Sony then decided to look at replacing tape with an optical disc, and demonstrated a 30-cm optical disc in 1977. Sony also worked on the error-correction system that became part of the Red Book standard for the compact disc.

The so-called "Red Book" standard for Compact Disc Digital Audio was published jointly by Philips and Sony in June 1980, and other companies were invited to license the technology. In October 1982, Sony unveiled the first compact disc player for sale, the Sony CDP-101, and in November, Philips launched the CD100 player.

During the course of designing the compact disc, a decision was made to set the play length at 74 minutes, allegedly so a single disc could play Beethoven's Ninth Symphony in its entirety. This decision meant that this was the first mainstream consumer music playback format that was shaped by the music, rather than shaping the way music was played back. No longer were there two "sides" to an album, with a cassette or LP needing to be turned over, or an 8-Track cartridge with its four "programs" meaning that some tracks got split over two programs or running orders re-arranged, or even a 45 rpm or 78 rpm record needing to be turned over or changed every three or four minutes. The new compact disc allowed for an incredible 74 minutes of continuous playback.

Of course, the playing time of a compact disc did begin to shape the way music was played back, as artists and record companies were no longer limited to the length of an LP and albums became longer, or additional content such as live tracks, alternate mixes, "B" sides, hidden tracks, and other items were added. It could be argued as to whether this was always a good thing for music, but it meant the compact disc had an advantage over the LP and was another inducement to consumers to replace their existing LPs with compact discs.

The physical size of the compact disc was partially dictated by the need to fit 74 minutes of music on it, and was increased from a suggested size of 11.5 cm by Philips to 12 cm by Sony. At this size, it still meant that in-car compact disc players could fit into standard-size apertures in dashboards, and also meant that later on, CD-ROM drives would fit in the space used by 5¼-inch computer disk drives.

The compact disc had to be marketed to the music industry before it could be marketed to consumers, and initially there was reluctance to embrace the new digital format. The record industry was in downturn in 1982, and some in the record industry felt that another form of music hardware would confuse the consumer, and was just another way for Philips to make money on yet another format after the cassette. There was also a feeling during this time of campaigns such as the UK's "Home Taping is Killing Music" trying to discourage the copying of albums onto cassettes, that putting out pristine copies of albums on compact disc merely meant pristine master copies for pirates to copy (Milner, 2010: page 213).

The Compact Disc Group was formed in 1983 and brought together the hardware manufacturers and the record labels to promote the compact disc. Both sides needed to be sure the other was on board, as makers of players needed to know there would be something to play on their machines, and the record labels needed to know that players for their discs would be plentiful and fall in price (Milner, 2010: page 217). The group grew to represent more than 50 companies, and marketed the compact disc directly to consumers through tours of nightclubs to provide demonstrations, as well as marketing directly to artists by providing copies of their albums re-mastered to compact disc (Knopper, 2009: page 30).

The resistance of the record labels eventually gave way, and in part this was helped by Philips' ownership of PolyGram, and the cooperation between Sony and CBS (Knopper, 2009: page 25). Music retailers also needed to be convinced to carry the new compact discs, given that they had already invested in fixtures and fittings to display both LP records and cassettes; carrying compact discs as well meant they might have the same album in three different formats and this was just a few years since the peak of the 8-Track cartridge that was quickly being phased out.

What helped convince the record labels and music retailers to market the compact disc was the much higher price they could charge for them compared to the LP record, as well as promising early sales figures in Japan for both players and discs. From world sales of 5 million units in its first full year of 1983, unit sales of compact discs finally overtook those of LP records in 1988 with sales of 400 million units (Shepherd *et al.*, 2003: page 508). With greater profits available from the sales of compact discs over those of LP records, there was a move in the late 1980s to push consumers more quickly toward compact discs. The record labels introduced a new returns policy that meant retailers could not return unsold LP records for credit, and this effectively pushed the risk of underselling artists onto retailers. Very quickly, compact discs took over sales space in record shops (McLeod, 2005: page 525).

By 1996, sales of new LP records were down to just 20 million units, and by this time, sales of pre-recorded cassettes were also falling from their worldwide peak in 1992 of 1,552 million units (Shepherd *et al.*, 2003: page 508), the year in which compact disc sales finally overtook cassette sales. Compact discs initially sold for $16.95 in the US, compared to around $8.98 for an LP (and even less for a cassette). Some of this increase could be explained away as increases in the cost of manufacture, which was initially more than for an LP, although within five years the price of making a disc had fallen to around $1.45 (Milner, 2010: page 222). By renegotiating artists' contracts, the record industry could make even more profits on the sales of compact discs. Prices of compact discs were kept artificially high for a number of years, but did finally begin to fall in the 1990s. Then, between around 1995 and 2000, the record companies used a scheme called "Minimum Advertised Price" to prop up the retail price. This scheme, which resulted in an antitrust case against five record companies, discouraged retailers from discounting the prices of compact discs and was intended to end the price wars by large discount retailers by making agreements with retailers that the record companies would pay the advertising costs for stores that sold compact discs above a certain price. One of the effects of this scheme was to protect specialty music shops from undercutting by large retailers, but it was alleged to have added around $2–$5 to the price of each disc (Labaton, 2000). Once the scheme was outlawed in 2000, large discount retailers in the US took a larger share of the market and drove many specialty music retailers out of business.

Although compact discs were expensive, they sounded much better on the average consumer's home music center. LP records played on cheap turntables could sound really bad, and could suffer from poor speed control and noisy motors, as well as the pops and crackles caused by the nature of the vinyl record itself (Milner, 2010: page 219). Both vinyl records and the needles used to play them wore over time, and records could easily be damaged by worn or chipped needles that were unlikely to be replaced. Cassettes offered inferior sound quality, and occasionally the tape itself would get chewed in the mechanism.

Compact disc players did not suffer from these problems, and even cheap players could sound good. There was no needle to wear, and the discs themselves could be played any number of times without wearing out. Although it was advised in the notes accompanying compact discs to handle them carefully, they could stand up to much more abuse than vinyl LP records. The compact disc was an enormous success, and by the year 2000, compact disc albums and singles combined accounted for 90.5% of unit sales of music in the US (RIAA, 2017a). However, this was also the year that unit sales of compact disc albums peaked at 942.5 million, and unit sales have declined almost every year since (RIAA, 2017a).

The Seeds of the Compact Disc's Demise

As early as 1985, the compact disc was established in the market with the Dire Straits album *Brothers in Arms* (1985) selling more copies on compact disc than on LP, and being the first compact disc to surpass the one million sales mark. In the same year, the CD-ROM, the first in a long line of variations on the compact disc design was introduced. CD-ROM (Compact Disc-Read Only Memory) was a read-only data storage medium, read by a CD-ROM drive controlled by a computer. The first such drive was the standalone Philips CM 100, but it was not until 1988 that the so-called "Yellow Book" standard for CD-ROM was published, and it was not until the 1990s that the CD-ROM drive became ubiquitous with the rise of the multimedia PC. CD-ROMs could hold text, images, video, and sound files, and were ideal for applications such as multimedia encyclopedias and games, being able to store 650 MB of data compared to just 1.44 MB on a high-density floppy disk.

A multimedia PC with a set of speakers could play back audio compact discs using its CD-ROM drive, but of course this also meant that audio compact discs could be read by the CD-ROM drive, and interpreted by the computer as if they were another form of data disk. Once software for "ripping" (extracting) the contents of compact discs became available, the computer could treat them as just another type of mass storage medium, and even the file structure of the compact disc made this easy, with each track or song being interpreted as a separate computer file.

Computer storage capacity was not sufficient in the early 1990s for there to be much point in "ripping" the contents of compact discs; you would need to store the contents somewhere and 650 MB of storage was still expensive. However, once the files could be written to another compact disc, or be compressed and passed over the Internet, the ability to read the compact disc on a computer became a problem. No previous recorded music format had given consumers such freedom to manipulate its content. LP records could be copied to cassette, but the content had to be recorded in real time, was still analog, and was inferior to the original. The CD-ROM drive effectively changed all this, and now perfect digital copies could quickly be made.

In 1987, Sony launched a second digital format, the Digital Audio Tape or DAT. This was a magnetic tape format, similar in size to a cassette. What made it truly different to the compact disc when it was launched was that it was a recordable format, and this was a major worry to the recorded music industry, already concerned over home taping onto cassettes. Unlike DAT though, the cassette lost fidelity with each generation, so a copy of an original source could never sound as good as the original, and a copy of the copy would sound even worse, limiting how far copies were likely to be distributed. With DAT, however, each copy could be as good as the original for as many generations as required.

Much of the music industry responded by refusing to put out albums on DAT. The Recording Industry Association of America (RIAA) threatened to sue anyone introducing a consumer DAT recorder in the US, and eventually successfully lobbied to have a Serial Copy Management System installed on recorders, preventing more than one generation of copying. The Serial Copy Management System provisions were part of the US Audio Home Recording Act of 1992, which also imposed levies on recordable media such as blank DAT tape, and later on blank CD-R Audio media for use in consumer audio recorders.

The US computer industry counter-argued that the levies on blank media should not be applied to their products, and this later led to the anomalous situation in the US whereby a blank CD-R for use in a consumer audio recorder was subject to a levy, but a CD-R for use in a computer CD-R burner was not, even though they were physically identical (apart from a Disc Application Flag identifying the disc as suitable for use in an audio recorder). Other countries also imposed levies on blank media, sometimes extending this to cassette tapes and to standard CD-Rs, and later to MiniDiscs and USB sticks.

DAT was a commercial failure as a consumer audio format, but in the early 1990s, Philips and Sony tried again to introduce new recordable formats, the Digital Compact Cassette (DCC) and MiniDisc, respectively. The DCC was an attempt to replace Philips' own aging analog cassette design, which in 1992 was finally overtaken in pre-recorded form by the compact disc in terms of unit sales. When the compact disc was introduced, it was felt that consumers would be happy with a choice of two formats; the compact disc was to replace the LP at the higher end of the market, and the cassette would remain as the cheaper option, but it now needed to be replaced. DCC players were backward-compatible with cassettes, so consumers could still play their existing music collections. Sony introduced their MiniDisc in 1992, consisting of a small magneto-optical disc in a protective caddy, but with the same capacity as the compact disc.

Neither the DCC nor the MiniDisc were very successful in the marketplace though more pre-recorded music was available than on DAT, and by 1996 one of their key selling points—the ability to record—began to be eroded by the introduction of affordable consumer compact disc burners. The specification for the recordable compact disc was published by Philips and Sony in 1988 (the so-called "Orange Book" standard), and the first professional CD-R burners were introduced in 1991, with Pioneer introducing a consumer CD-R burner in 1996.

As prices fell, and CD-R/CD-RW burners were installed in computers, consumers could now make perfect copies of their compact discs, and some chose to recoup their investment in the equipment by selling these copies (Witt, 2016: page 67). What made this easier was the lack of a levy on the cost of blank CD-Rs for use in computer CD-R burners (due to the exemption of computer media), the compatibility of CD-R discs (when burnt as an audio disc) with all the world's existing compact disc players, and no digital rights management built into the compact disc standard. Indeed, buyers of copied compact discs could create further copies if they so wished as there was no Serial Copy Management System to prevent copies of copies being made, as there was with DAT, DCC, and MiniDisc.

So, by the late 1990s, compact discs could be read by a computer, and the tracks "ripped" and copied to a writable CD-R. This was, so far, little different from home taping of music onto cassette tapes. What was new at this time was the rapid spread of the Internet, and the availability of technology to take the large music files from a compact disc and make them many times smaller for distribution over the Internet. This technology was MP3 encoding, and had been developed over many years using the science of psychoacoustics, or audio masking. Using psychoacoustics meant the developers of the encoding software could program it to leave out those elements of music that the human ear would not perceive. A compact disc copy of the *a capella* Suzanne Vega song "Tom's Diner" (1987) was used as a test piece to refine the algorithm, and in 1994 the first MP3 encoder (L3Enc) was released by the Fraunhofer Institute as shareware, followed by other encoders.

An MP3 encoder could take a file ripped from a compact disc and compress it to around one-tenth the size (depending on the encoding bit rate chosen). Suddenly, it became practical to share music over the Internet, and while dial-up connections were still slow (it could still take three hours to download an average album even in MP3 format (Brown *et al.*, 2001: page 190)), this would soon change with the adoption of broadband Internet. In August 1996, what is believed to be the first pirated MP3 track, Metallica's "Until It Sleeps" (1996), appeared on the Internet on the IRC (Internet Relay Chat) protocol, ripped from the album *Load* (1996) on compact disc and compressed in MP3 format using L3Enc (Witt, 2016: page 73).

Suddenly the compact disc was, in a technical sense, obsolete. The means now existed to distribute music on-line and had the record industry chosen to do so, it could have avoided many of the costs of producing, distributing, and retailing music. It chose not to, and it took several more years until it tried. Ironically, one use the compact disc performed very well was as a means of smuggling music out of the pressing plants and into the hands of pirates for uploading to the Internet (Witt, 2016: page 67), and if discs could be smuggled out before an album reached the shops, this made the resulting files even more desirable.

By 1999, Napster was making the on-line sharing of music very easy and free. No longer did consumers need to search for tracks in chat rooms or multiple websites; by simply installing the Napster program on their computer they could search for tracks held on other people's computers, and share their own music ripped from their compact discs through a peer-to-peer network.

While some artists made their music freely available in MP3 format, much of the music available over the Internet was taken from compact discs, and the early fears of the music industry that placing pristine master copies in the hands of consumers would lead to piracy came to fruition on a scale and via means never envisaged when the compact disc was designed.

Faced with the sharing of millions of ripped and compressed compact disc tracks over the Internet, the record industry eventually took notice and did several things. First, in 2001, it tried setting up on-line music stores selling music through a subscription model, such as Pressplay (Sony and Universal) or MusicNet (EMI, Warner and BMG). These were far from as elegant as Napster, and had serious limitations on usage such as the number of tracks you could stream or download each month. In addition, if you wanted to listen to artists from, say, both Sony and EMI, you had to subscribe to both services (Knopper, 2009: page 144). Second, the RIAA prosecuted over 20,000 individuals they accused of sharing pirated MP3s. Finally, the RIAA tried to prevent the sale of early MP3 players such the Rio PMP3000, arguing that they breached the US Audio Home Recording Act of 1992.

The Act, however, was introduced to deal with the perceived threat from Digital Audio Tape, and the computer industry had successfully lobbied to be exempted from its provisions. Since MP3 players connected to a computer rather than to a piece of audio equipment to copy content, the RIAA failed in its legal challenge, paving the way for devices such as the Apple iPod.

Apple launched the iPod music player in 2001, and this used Apple's iTunes software to manage songs on the device and the user's computer, and included the ability to rip music from compact discs. Suddenly, it was possible to store 1,000 songs on a device significantly smaller than a portable compact disc player, and with a much longer battery life. In 2003, the record labels finally agreed to license music to Apple to sell, and the iTunes Store was launched. Now it was possible to legally buy music through an easy-to-use interface, integrate it with the music ripped from your own compact discs, and copy it to a portable player.

Apple's iTunes was built on a more traditional model of paying for individual songs, but newer streaming services such as Spotify (launched in 2008) were subscription services in which a monthly fee is paid, allowing unlimited access to content, or limited access with advertisements for free. Streaming music services accounted for the largest share of music industry revenues in 2016, with 51% of total sales, having seen strong growth over the previous few years, with physical music sales (compact discs and LPs) representing just under 22% (RIAA, 2017b).

Was the Compact Disc Really So Great?

The compact disc certainly provided the average listener with a great improvement in sound quality over previous music formats. However, when compared with an LP record, the compact disc does lack in several areas. One of these is in the packaging and artwork associated with music. A compact disc in a standard-size jewel case has a cover area of just one-fourth that of an LP cover. While compact disc artwork is still important, it does not have the impact of an LP cover, especially one in gatefold sleeve form. Artwork originally designed for LP covers did not always translate well when the album was later released on compact disc cases, and detail, as well as impact, was often lost.

This criticism could also have been made about cassette packaging, perhaps even more so, since the cover of the case had a different aspect ratio, and often the artwork was cropped to fit the inlay card. There were some notable exceptions to compact disc packaging; to give just two examples, in 1993, the Pet Shop Boys released "Very" in an orange jewel case with raised bumps, and in 1997, Spiritualized released "Ladies and Gentlemen We Are Floating in Space" containing a compact disc in a pharmaceutical-style blister pack, complete with dosage instructions. For the majority of releases though, compact disc packaging used the standard jewel case, or the later Digipack cover.

A recent fashion is the Japanese mini-LP, a compact disc with packaging that mimics that of the original 12-inch LP release, often including replicas of the inner sleeves, inserts and even stickers on the cover promoting the contemporary hit single release. While these are attractive in themselves, they also demonstrate some of what has been lost in the transition to the much smaller size.

Aside from the artwork, playing a compact disc was just not as involving as playing an LP record. With a compact disc, you place the disc in a tray (or in the slot) and it gets taken into the player or you close the lid. Changing tracks involves pressing a button instead of moving the needle, and there is no need to turn the disc over as there is no second side. Playing a vinyl LP is much more of a ritual and involves carefully removing the record from its sleeve, placing it carefully on the turntable, and putting the needle in the correct position to start (assuming the turntable is not fully automatic). You can then sit back and enjoy the cover artwork and liner notes as the tonearm gently bobs up and down on its journey to the center of the disc. Looking at a compact disc, you cannot see where each track ends, nor can you see the dynamics of each track as you can on an LP record (some people can recognize an album simply by the pattern of the grooves on an LP's surface); the playing surface of a compact disc is simply a shiny blank.

With an LP record, the center label artwork was still visible when the record was playing. All LP records ended with a locked groove as this stopped the stylus travelling into the center label, but some records used the locked grooves to play an endlessly-repeating loop. Perhaps the most famous example of this was at the end of The Beatles's *Sgt. Pepper's Lonely Hearts Club Band* (1967). A number of records also had written messages inscribed into the run-out grooves. LP records, then, perhaps had more personality than compact discs, even aside from their imperfections—the pops and crackles that characterize the sound.

While the compact disc differed from the LP record in a number of ways, it shared many features with its close relation, the CD-ROM. As the CD-ROM drive became ubiquitous in the 1990s, it simply emphasized the way in which the compact disc was just another data storage format, in effect identical to the CD-ROM, except that it contained files of music data instead of programs, video, images, or text. Even the process of putting a compact disc into the tray or slot of a CD-ROM drive was identical to putting it into an audio compact disc player. So whether the user was installing a new printer driver, loading a game, or putting on some music, the process was pretty much the same. With the rise of the writable compact disc (the CD-R and later CD-RW), the compact disc became a means of creating backups of data, as well as burning audio compact discs, and the distinction became further blurred.

When the compact disc was launched, it was marketed as offering perfect sound. Terms such as "near CD-quality" are still used, as if the sound quality of a compact disc is the standard to aspire to. However, the compact disc does not offer perfect fidelity, and makes some compromises that formats such as MP3 simply extend even further. The Red Book standard for compact discs specifies a sampling bit rate lower than the original professional master recording, to allow 74 minutes of music to be squeezed onto a 12-cm disc (Rothenbuhler, 2012: page 47). The actual sampling bit rate of a compact disc recording is 1411 kbps, as opposed to 320 kbps for an MP3 recorded at high quality. However, the compact disc has still lost some of the quality of the original performance, since at its sampling rate it can only represent frequencies up to 22 kHz. This means that very high frequency sounds will be lost, even though these sounds are at a higher frequency than most human ears can hear.

Whatever the quality arguments over compact disc, the Red Book standard did need to make some compromises to fit sufficient music onto a 12-cm disc and these have never changed, to ensure that all compact discs will work in all compact disc players. Rothenbuhler (2012: page 47) argues that the main issue with the compact disc (or indeed any digital music format) is that it has a predetermined limit on the amount of data available. This limit has remained the same since the Red Book standard was produced in 1980, and regardless of whether this limit is too low, the compact disc was marketed as providing "perfect sound forever", which meant that there was no room to market any further music formats based on their quality alone. Indeed, later formats such as DAT, DCC, and MiniDisc were never marketed as offering better than "CD-quality" sound, but promised improvements in portability, backward compatibility, or the ability to record as well as playback.

As Rothenbuhler (2012: page 46) also points out, since there is a limit to the data stored on the compact disc, and nothing else can be extracted, there is little point in investing in better-quality compact disc players; the very best engineered compact disc player available cannot make a compact disc sound significantly better. Conversely, better turntables can extract more information from the grooves of an LP, and audiophiles can spend more money on equipment knowing that the investment is likely to reap benefits in terms of improved sound quality from LP records.

Most listeners are perfectly happy with the sound of compact discs, and are also willing to accept the compromise of "near CD-quality" to listen to compressed sound files in "lossy" formats such as MP3, AAC, or Ogg Vorbis, due to the convenience they offer of being able to listen to music anywhere, and the vast choice of music that can be offered on a small device without the need to carry separate physical media. Indeed, most listeners could probably not tell the difference between "CD-quality" sound and that of an MP3 file recorded at a high bit rate.

What Do We Lose with the Decline of the Compact Disc? Ownership of Music versus Access

What do we mean by ownership? If someone owns a copy of an album on a compact disc (or LP or cassette), they are granted the right to play that music in private, but if they want to play it in public a license would be required. If they want to resell the compact disc they can do so, and they can lend the compact disc to a friend. However, under UK law, they are not allowed to copy the compact disc, even for the purposes of "format shifting", for example, copying it to iTunes so it can be synched to an iPod (an exemption to copyright law in the UK was granted to allow legal format shifting in 2014, but this exemption was challenged by the music industry and format shifting was again declared illegal in 2015 (Teare, 2015)).

So even if someone owns a compact disc, their ownership of the content has certain legal limitations. However, due to the flawed design of the compact disc, there are no practical limitations to doing many of these illegal activities; it is easy to copy a compact disc into iTunes, or burn a copy onto a CD-R. As sales of the compact disc decline and music moves on-line, consumers are still subject to some of the existing restrictions on what they can do with the music they buy, but are now also faced with further restrictions. They can no longer lend music to a friend (although it is possible to lend to family members with difficulty) (Wood, 2014) and it has been declared illegal in the US for consumers to resell digital music once they have decided they no longer want to listen to it (Campbell, 2013). It seems that even as music is freed from the limitations of physical media, it becomes subject to the additional technological and legal restrictions of on-line music retailing.

Ownership becomes even more of an issue with streaming music services such as Spotify. Now for a monthly fee, consumers can access most of the world's recorded music, but if they choose not to pay for a monthly subscription, they might be limited in what they can listen to, or made to suffer intrusive advertising. Streamed music then simply becomes a utility like water or electricity, normally available on demand but under the control of the supplier.

So, even if someone owns a compact disc, they do not "own" the music contained in it; however, having ownership of the disc itself does give them the ability to do things they cannot do with digital music, like lend it to a friend or sell it on. Furthermore, having music stored on compact disc means it is possible to arrange and re-arrange a music collection as the owner chooses, perhaps by genre or age, or even by the color of the packaging should they so choose to, things that are difficult to do with digital music.

The End of the Music Album?

The music album has been with us since the introduction of the long-play vinyl record in 1948, but the term "album" dates back to the early 20th century and sets of 78 rpm shellac discs packaged together like a photo album, so that a number of shellac discs, each playing for just three or four minutes per side, could be kept together to make up a complete work. Initially, these were usually classical or operatic works, but popular music later became available packaged as an album of discs.

With the introduction of the long-play vinyl record, artists could now combine 10–12 tracks on a single disc to create an album of music, and by the 1960s this became a new art form in popular music, with attention paid not just to the quality of individual songs, but the way they linked together over the two sides of the disc to form a cohesive whole, whether from the mood of the songs or an overall concept. The compact disc's longer overall playing time, with no need to turn the disc over halfway through playing, meant a change in the nature of the music album, but only in as much as there was now no need to think of the album as having two discrete programs; artists did not need to consider how side A finished and how side B should begin again, and there was more freedom to fill a larger space with more, and perhaps longer, tracks (or bonus tracks).

What really changed with the compact disc was the easy programmability of playback. Now it was easy to skip tracks at the press of a button or play the tracks randomly, and it was also possible to program the playback of a whole album, perhaps skipping any tracks that were disliked, or changing the order of tracks. Now, if someone did not like the track "Rocky Racoon" (1968) on *The Beatles* (1968, also known as the "White Album"), they need never listen to it again.

Services such as the iTunes Store took this a logical step further. The tracks of an album are now unbundled as if they were back on separate shellac discs, and if someone likes just one song on an album (perhaps the single), that is all they need to buy, so there is no need to buy a compact disc containing that track bundled with a dozen others they might consider "filler". MP3 players such as the iPod offer the facility to shuffle play the tracks on a device, along with all the other tracks of the albums contained within it, so consumers might never hear a song within the context in which the artist intended it to be heard.

Of course, the 45 rpm "single" has also been around since 1949, and singles were very important to music sales, initially in their own right, but later as a means of promoting sales of albums (Newman, 2014). However, music albums were where the profit lay, especially when compact disc singles became the major means of distributing singles, since the cost of manufacturing a compact disc single was the same as that of a compact disc album; so the record industry pushed consumers toward albums, which in some cases might have just one or two songs that the listener actually wanted. Knopper (2009: page 105) argues that this helped encourage the downloading of individual songs, initially through file-sharing, and later through services such as the iTunes Store. In the first year that download sales of individual songs were recorded by the RIAA in 2004, singles accounted for the most unit sales since 1981, but this does not take into account file-sharing, meaning that the actual figure for downloads of single tracks would be considerably higher.

The single track has once again become the preeminent form of music sale, and Newman (2014) argues that this has implications for music artists, who now need to release songs that each stand on their own commercially; experimentation with tracks that do not have wide appeal is no longer so desirable to the music industry, even if they might have worked well in the context of an album.

Conclusion

What is the future for the compact disc? Having been with us for over 30 years, it could be a few years yet before it disappears, since while sales are falling every year, the decline is slowing; in 2016, there were still 99.4 million compact disc albums sold in the US alone (RIAA, 2017a). While the vinyl LP is enjoying a much-hyped revival, in 2016 compact discs still outsold them in the US by nearly six-to-one. However, compact discs were outsold in 2016 by over eight-to-one by downloaded tracks and albums (even before taking account of streamed music) (RIAA, 2017a).

In 2014, the compact disc still represented 64% of total album sales in the UK, and Ingham (2015) argues that this figure could represent a fairly stable level for some years, having declined relatively slowly over the past few years. The problem with this is that the idea of the music album itself could be under threat from the increasing dominance of single-track downloads and streaming music. Another danger is that the installed base of compact disc players might not be around much longer. Most computers no longer include an optical drive, and as listeners buy MP3 player docks or Bluetooth speakers to make the most of their investment in downloaded music or streaming subscriptions, they might not replace their compact disc players. DVD and Blu-ray video discs are facing competition from streaming video and are suffering declining sales (Sweney, 2017) and it is possible that the players (that are backward-compatible with compact discs) might not be replaced by consumers in future.

The compact disc can be seen as a transitional format in some respects, bridging the gap between physical analog music formats and virtual digital ones. It was the first digital music format for consumers, and allowed listeners more freedom to choose how to listen to an album by making use of the ability to skip tracks or play tracks in a particular order (or in no order at all). The much-reduced size of the artwork on compact disc packaging prefigured the use of album artwork as little more than an icon on an MP3 player. Its portability, like that of the cassette, allowed listeners their choice of music on the move, and this has been taken even further with the MP3 player that allows a vast array of choice without having to carry tapes or discs. No longer is there any need for "perfect sound", just sound that is good enough to listen to over the ambient noise of our listening environments.

For all its imperfections, the compact disc is still a physical format, and in its decline we lose the most obvious feature of all, the ability to hold a work of recorded music in our hands. Perhaps that does not matter as much as we think. While we cannot sit back and read the liner notes or study the artwork on an MP3 file, we can now look up information about the band, or read reviews of the music as we listen, or even interact with the musicians on social media. While we cannot browse our friends' music collections by looking through their compact disc collections, we can instead get on-line suggestions for music we might like based on other people's preferences. While we might no longer give compact discs as gifts, we can instead give iTunes or Spotify vouchers, in essence giving the gift of music rather than a specific example of it.

Did the compact disc really sow the seed of its own decline? It put digital copies of music into listeners' hands, without any digital rights management, and by extending the format into a computer readable version (the CD-ROM), Sony and Philips made it easy to copy the music content onto a computer. From here it was just a small step to compress that content and make it distributable over the Internet. The pricing model for the compact disc and the push toward bundling content onto higher-margin albums could arguably have made the downloading of music more attractive than it would otherwise have been.

The decline of the compact disc would perhaps have happened anyway, following the trend away from individual physical copies in other areas such as photography, video, and data storage, but the nature of the compact disc hastened its decline, and did indeed sow the seeds of its own demise.

References

British Phonographic Industry (2017) BPI Official UK Recorded Music Market Report for 2016 [online] Available from: www.bpi.co.uk/media/1277/bpi_press_release-end_of_year_2016_recorded_music_market_report_3jan2017.docx [Accessed: September 1, 2017]

Brown, B., Sellen, A.J., and Geelhoed, E. (2001). *Music Sharing as a Computer Supported Collaborative Application*. Proceedings of the Seventh European Conference on Computer-Supported Cooperative Work, 16–20 September 2001, Bonn, Germany. Kluwer, pp. 179–198.

Campbell, M. (2013) Judge denies right to resell iTunes songs, digital media still protected under copyright laws. *AppleInsider* [online] April 13, 2013. Available from: http://appleinsider.com/articles/13/04/01/judge-denies-right-to-resell-itunes-songs-digital-media-still-protected-under-copyright-laws [Accessed: July 21, 2017]

Ingham, T. (2015) Why the music business would be mad to let the CD die. *Music Business Worldwide* [online] January 2, 2015. Available from: www.musicbusinessworldwide.com/music-business-mad-let-cd-die/ [Accessed: July 21, 2017]

Knopper, S. (2009) *Appetite for Self-Destruction: The Spectacular Crash of the Record Industry in the Digital Age*. London: Simon & Schuster.

Labaton, S. (2000) 5 music companies settle federal case on cd price-fixing. *The New York Times* [online] May 11, 2000. Available from: www.nytimes.com/2000/05/11/business/5-music-companies-settle-federal-case-on-cd-price-fixing.html [Accessed: July 21, 2017]

McLeod, K. (2005) MP3s Are Killing Home Taping: The Rise of Internet Distribution and Its Challenge to the Major Label Music Monopoly. *Popular Music and Society 28*(4), pp. 521–531.

Milner, G. (2010) *Perfecting Sound Forever: The Story of Recorded Music*. London: Granta.

Newman, K. (2014) The end of an era: the death of the album and its unintended effects. *Gnovis Journal* [online] February 28, 2014. Available from: www.gnovisjournal.org/2014/02/28/the-end-of-an-era-the-death-of-the-album-and-its-unintended-effects/ [Accessed: July 21, 2017]

Philips Research (n.d.) The history of the CD: The CD family [online] Available from: www.philips.com/a-w/research/technologies/cd/cd-family.html [Accessed: July 21, 2017]

Recording Industry Association of America (2017a) *U.S. Sales Database* [online] Available from: www.riaa.com/u-s-sales-database [Accessed: July 21, 2017]

Recording Industry Association of America (2017b) *News and Notes on 2016 RIAA Shipment and Revenue Statistics* [online] Available from: www.riaa.com/wp-content/uploads/2017/03/RIAA-2016-Year-End-News-Notes.pdf [Accessed: July 21, 2017]

Rothenbuhler, E.W. (2012) The Compact Disc and Its Culture: Notes on Melancholia. In: Bolon, G. (ed.) *Cultural Technologies: The Shaping of Culture in Media and Society*. Routledge Research in Cultural and Media Studies. Abingdon: Routledge, pp. 36–50.

Shepherd, J., Horn, D., Laing, D., Oliver, P., and Wicke, P. (eds.) (2003) *Continuum Encyclopedia of Popular Music of the World: Volume 1, Media, Industry and Society*. London: Continuum.

Sweney, M. (2017) Film and TV streaming and downloads overtake DVD sales for first time. *The Guardian* [online] January 5, 2017. Available from: www.theguardian.com/media/2017/jan/05/film-and-tv-streaming-and-downloads-overtake-dvd-sales-for-first-time-netflix-amazon-uk [Accessed: July 21, 2017]

Teare, I. (2015) Do you copy? UK's new format-shifting exception found illegal. *Technology Law Updates* [online] June 26, 2015. Available from: www.technology-law-blog.co.uk/2015/06/do-you-copy-uks-new-format-shifting-exception-found-illegal.html [Accessed: July 21, 2017]

Witt, S. (2016) *How Music Got Free: The Inventor, the Mogul, and the Thief*. London: Vintage.

Wood, M. (2014) Apple and Amazon Take Baby Steps Toward Digital Sharing. *The New York Times* [online] September 14, 2014. Available from: http://bits.blogs.nytimes.com/2014/09/18/apple-and-amazon-take-baby-steps-toward-digital-sharing/ [Accessed: July 21, 2017]

26

HELLO AGAIN

An Untimely Requiem for the Flip Phone

Paul Benzon

Chris Jordan's *Cell Phones* is a 2007 image from his series *Running the Numbers: An American Self-Portrait*. It is a massive, 60-inch by 100-inch photograph depicting 426,000 cell phones, the number discarded (or "retired", to use Jordan's telling term) in the United States every day at the time of the image's production (Jordan, 2009: 45). In this image, and throughout *Running the Numbers* more generally, Jordan seeks to make material those things that are often dangerously abstract, to pay witness to the "staggering complexity" and scale of human consumption and disposal (Jordan, 2009: 21). Each image in the series visualizes a particular quantity of object. For example, just as *Cell Phones* offers testimony to the number of phones retired each day, *Caps Seurat* (2011) reproduces Seurat's *Isle of La Grand Jatte* through 400,000 bottle caps, the amount consumed in the United States every minute. Yet, precisely because of the scale of these images, their promised exactitude collapses under the literal and figurative weight of the objects they depict, and Jordan's work becomes abstract anew in its concreteness. At full view, *Cell Phones* is indecipherable, appearing almost as static, analog snow, a dead channel. Only when we approach the image more closely or see it in detail does it become clear what it actually depicts: a seemingly uncountable accumulation of cell phones, the silvery white sea of their faux-chrome bodies speckled with the blackness of the spaces left between these objects as they lie alongside and on top of one another. The effect resembles the optical trickery of a Magic Eye stereogram, promising a hidden image out of what seems to be visual noise—pattern here is at once both beside the point and precisely the point.

Premature Burial

At the far end of global disposal and obsolescence, quantification twists back toward uncountability and unknowability—how do we perform a collective autopsy of this mass grave? Jordan describes himself as "appalled by these scenes, and yet also drawn into them with awe and fascination", describing a paradoxical sensation that resonates with the long history of the sublime and, in particular, the technological sublime.[1] However, particularly in the case of *Cell Phones*, the contradictory extremities of the technological sublime emerge not (or at least not only) from the awesome capability of the technology to bestow both life and death, but rather from the technology's own liminal position between life and death. Especially in the sweep of the image's full scale, but also even in detail, any given phone seems potentially still functional to the naked eye—how can we know for sure whether it

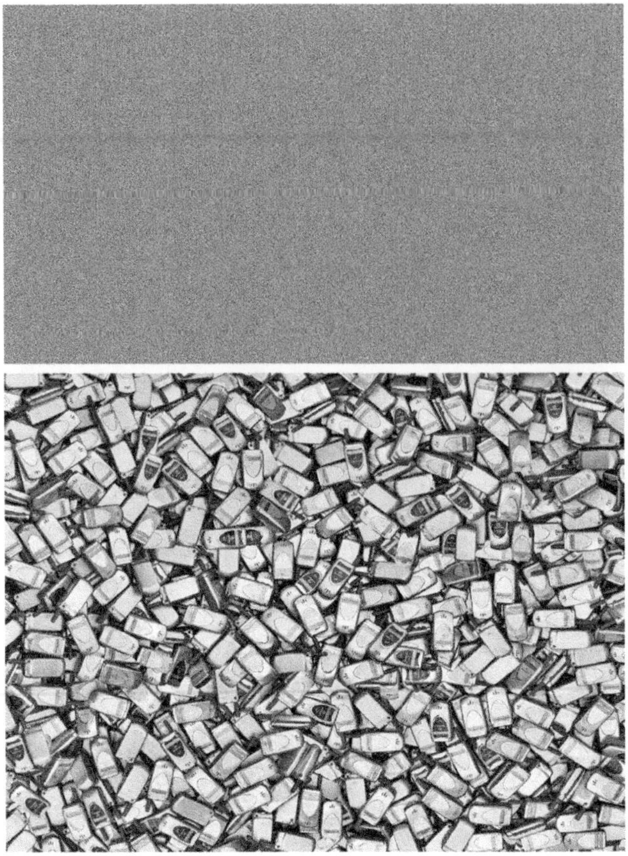

Figure 26.1 Cell Phones, 2007. Full image, 60″ × 100″, and detail. Depicts 426,000 cell phones, equal to the number of cell phones retired in the US every day. © Chris Jordan

works or not, whether it is alive or dead? We might imagine the answer is often neither, and thus Jordan's use of the term "retired" seems particularly pertinent here: in many cases, these devices are still functional but no longer in use, having likely reached the end of their working life but not the end of their life cycle.

Perhaps, then, the question of temporality is paramount in Jordan's image, even more so than the questions of identification or quantification. In comparison to other objects in these series—the car, for example, or the circuit board, or the plastic water bottle—the cell phone occupies a substantively different temporality of consumer use, disuse, and obsolescence.[2] The cell phone has experienced the most rapid adoption of any consumer technology in history (Rainie, 2013). At the same time, thanks to a complex and perpetually accelerating cycle of product innovation and obsolescence, it is bought, used, and discarded perhaps more frequently and more rapidly than any other consumer technology in recent memory. The average smartphone replacement cycle has lengthened in recent years due to the disappearance of the two-year service contract, yet this cycle is still more rapid than that of other consumer electronics (Gryta, 2016). Because of these unusual historical trajectories at both the microscopic and macroscopic levels, the cell phone has a great deal to tell us about the intertwining, recursive vectors of innovation and obsolescence, about how those vectors shape and are shaped by the material particulars of a given device, and about the ideological stakes of those vectors.

The compact history of the cell phone has already seen a number of substantively different form factors, from the earliest "brick" phones of the 1980s to the bar and flip designs of the 1990s and early 2000s to the now seemingly ubiquitous touchscreen slate. These changes in design set in motion corresponding changes in interface, bodily engagement, and other cultural and technological dimensions. Thus, questions of time and quantity also point us to questions of form and materiality. In this essay, I discuss the flip phone as a specific form of cell phone, a subset of the larger category with a particularly compressed and charged history. Even a cursory examination of Jordan's image in close-up detail, along with its 2007 date, suggests that most of the phones shown within it are flip phones, a fitting representation given how pivotal this form factor was in widely popularizing and disseminating the cell phone as a consumer device. In an important sense, the flip phone *was* the cell phone for a number of its crucial early years: its popularity as a specific device helped to accelerate the adoption of the larger category, and it rapidly became all but ubiquitous for a relatively brief window of time, before then virtually disappearing (or at least seemingly so, as we shall see later).

This compression offers a particularly strategic purchase on several questions that are important for a history and theory of media technology and obsolescence—namely, the historical contours of the development and adoption of the cell phone as a quintessential object of digital culture, the social and material stakes of technological emergence and obsolescence, and the methodological question of how we might historicize a particular technology and what it means to do so. Buried in the mass grave of Jordan's photograph, the flip phone is at once both disposed of and preserved. Taking this uneasy image as a point of departure, I trace a media archaeology of the flip phone in the sections that follow. In keeping with the critical strategies of media archaeology, I do not restrict my focus to the flip phone in a strict, linear sense. Indeed, rather than a singular history of the flip phone as a market, technical, or cultural force, I pursue a slant approach to this artifact, tracing the operation of the flip, the hinge, and the fold across different temporalities and technologies, studying the discursive and physical deployment of this mechanism as a marker of complex temporality. After outlining the stakes of a media-archaeological approach to the flip phone, I turn to a prehistory of the flip as a form factor in earlier technologies such as landline telephones and personal computers. I then examine Motorola's MicroTAC and StarTAC models as paradigmatic examples of the high point of the flip phone's market and cultural dominance. Finally, I consider several moments in the recent history of the digital that illustrate an undead return of the flip phone. Through this approach, I show how media historians and theorists might complicate critical conceptions of obsolescence by excavating the unburied of media culture—that which is at once both past and present, forgotten but not gone.

The Hinge of History

My thinking about the flip phone builds on the frameworks that several key media-archaeological thinkers use to imagine new historiographic and political possibilities across time. Jussi Parikka, for example, envisions "time as *pleated*" (2012: 146). Working within this pleated time, Parikka defines media archaeology as

> a way to investigate the new media cultures through insights from past new media, often with an emphasis on the forgotten, the quirky, the non-obvious apparatuses,

> practices and inventions. . . . Media archaeology sees media cultures as sedimented and layered, a fold of time and materiality where the past might be suddenly discovered anew, and the new technologies grow obsolete increasingly fast.
>
> (Parikka, 2012: 2–3)

Lori Emerson imagines a similar figure for the shape of media-archaeological time. Emerson begins her work *Reading Writing Interfaces: From the Digital to the Bookbound* (2014) by focusing on the specific period of the mid-1990s (the moment of the flip phone's widespread emergence, coincidentally) as a temporal hinge opening in multiple directions at once. For Emerson, this hinge "demonstrates how we can wield media archaeology as a conceptual knife that cuts into the present and the near future, not just . . . into the past" (p. 2). Although neither author cites Gilles Deleuze directly in the context of these specific frameworks, we might see in each one an indebtedness, however indirect or mediated, to his thinking about the fold (*le pli*) as a figure of poststructural materiality and time: "A fold is always folded within a fold, like a cavern in a cavern. The unit of matter, the smallest element of the labyrinth, is the fold, not the point . . . The matter-fold is a matter-time" (1992: 6–7).[3]

The pleat, the hinge, the fold, the boomerang: in concretizing temporality through these figures, media-archaeological thinkers have consistently sought to imagine technological change not merely as nonlinear but indeed as recursive and recurrent, torqueing and bending back on itself. These metaphorical historiographic tactics take on an especially loaded meaning in the case of the flip phone for several reasons. The first of these is the relatively compressed timeline of its appearance, disappearance, and reappearance that I have outlined above: to perform an archaeology of the flip phone is not to plumb the deep time of the media, as Siegfried Zielinski describes it, but rather to excavate the recent past as technological change accelerates and bends back upon itself, tracing a tight angle from old to new and back again. More importantly, these figures are important historiographic metaphors for the flip phone because they capture the functionality of the device itself: bending and folding is not only how this artifact circulates across time but also what the device itself does. Because of the centrality of its hinge, the flip phone set in motion a set of material and haptic circumstances for users that was paradigmatically different from that of other cell phone form factors, yet had antecedents in earlier media and communications technologies.

In what follows, then, I take the figure of the hinge as both an object of inquiry and a conceptual framework for excavating the history of flip phone technology. The history of the cell phone as a broader category has most often been told in terms of telecommunications infrastructure (networks, towers, wireless transmissions) and inner technology.[4] The distinguishing feature of the flip phone as a category, however, is the centrality of the hinged folding mechanism to its design, and the resulting ways in which the object interacts with hand and body. It is all too tempting, in the moment of the smartphone's near-ubiquity, to see the flip phone's hinge as a nostalgic throwback or a clumsy curio from a bygone, culturally and technologically "simpler" time. Yet in order to understand the historical and ideological leverage of the flip phone, I argue that we need first to understand its folding mechanism as a historically specific instance of design with its own properties, its own technological genealogy, and its own cultural and technological implications. These properties do not necessarily exist at odds with or pose a direct counterthrust to those of the touch-based smartphone that followed and ultimately eclipsed the flip phone. Yet at the same time, the workings of the flip mechanism stand outside the design ideology of those newer devices. Emerson argues of the haptic politics of smart-device interfaces that

> the shift to the ideology of the user-friendly via the GUI [graphic user interface] is expressed in contemporary multi-touch, gestural, and ubiquitous computing devices, such as the iPad and the iPhone, whose interfaces are touted as utterly invisible and whose inner workings are therefore de facto inaccessible. . . . The iPhone/iPad multitouch interface, which is constantly touted as "magical" or as something that allows us to perform "magic tricks", is invisible in the sense that it constantly seeks to hide its inner workings through glossy attractive packaging.
> (2014: 49, 11)

In the smartphone in particular, the ideological problematics of interface Emerson traces throughout her work converge with the question of design and form factor. As the easy, intangible magic of the GUI becomes inextricable from the polished, impermeable objecthood of the cellphone, vital dimensions of that objecthood conversely become not only metaphorized and skeuomorphic, but also abstracted, seemingly ethereal and invisible. Hermetically sealed, the smartphone seems designed to resist adaptation, hacking, unlocking, and even critical inquiry itself. Moreover, its flat, unbroken face engenders constant engagement, constant imbrication in the networks and economies of information and attention that circulate around the device. This is, of course, far from unique to the smartphone's design; indeed, increased accessibility and engagement is an inevitable effect of mobile technology from its very inception. Yet, design and interface matter here in measurable ways: aside from the more primitive functionality of the flip phone, its closed position is effectively antithetical to the push notifications and text messages that populate the surface of the contemporary smartphone—indeed, while it might be temporarily asleep or (rarely) off, the fixed face of the smartphone is virtually never closed to the flow of information.

I do not intend through these contrasts to suggest that the flip phone was outside of, or impervious to, technocratic power. After all, the inner workings of the flip phone were effectively just as hidden, invisible, and inaccessible as anything encased in the airtight glass of Apple's recent releases, and the stakes of deforming one kind of device rather than the other are only different in abstract terms. Yet the flip phone is paradigmatically different in that the form itself moves. The device's material functionality cannot be assimilated to the hidden magic of the one-piece glass smartphone. Tracing the predecessors and descendants of this functionality—attending to the mechanics of the flip phone's hinge, to when and where it appears and reappears—we find a genealogy of irreducible mechanics, a history of design and form that runs askew of the clean teleology of the iDevice. Collapsing into itself, buried before its time, the flip phone shows us a critical fold in media history.

The History of the Hinge

Motorola introduced the DynaTAC 8000X, the first commercially available mobile phone, in 1984. The DynaTAC was a 28-ounce, one-piece phone, a design colloquially known as a brick phone for its size, weight, and shape, and the first flip phone did not appear until the release of the Motorola MicroTAC in 1989. Yet even prior to either of these early mobile technologies, telephone design has had a long history of existing in tension between two-part and one-part form. Western Electric Model 302, designed by Henry Dreyfuss for the Bell Telephone Company and first introduced in the United States in 1937, is an early example of such a tension. Dreyfuss's design is seminal for its form and shape, but also for its consolidation of the telephone into something resembling a single object. Earlier units had consisted of two pieces, a desktop unit comprised of the handset receiver and rotary dial and a subscriber set box mounted on a nearby wall that connected the unit to the phone network. The Model 302, in contrast, combined these functions into a single object. Yet at the same time, that single object was still irreducibly two parts, a base and a receiver that had to be separated in order for the device to be used. In this sense, Dreyfuss's design folded (pun fully intended) the problem of earlier units into a seemingly single unit. Rather than raising a question of success or failure—whether the design problem in question was "solved" or not—it reveals the complex way in which the telephone has always already been two-parted both within and against its one-partedness, both in spite of and because of innovations that worked to consolidate telephonic technology within a single object. Indeed, the canonization of the Model 302 as a design object is itself a testament to how fully and deeply the bipartite, split mechanics of the folding device run through the dynamics of phone design.

Over forty years after the release of the Model 302, General Telephone and Electric Company (GTE) introduced the first landline flip phone in 1978. The GTE Flip-Phone was small enough to hold in one hand, "the size of a large stapler", and weighed just seven ounces (JS&A National Sales Group, 1978). As a landline designed for compactness and something verging on portability (the unit shipped with a 14-foot cord, "twice as long as a conventional cord" [JS&A National Sales Group, 1978]), it had a hybrid modality that makes it a strangely interstitial antecedent to the mobile flip phone. This hybridity stems from a subtle yet crucial inversion of the mechanics that had characterized phone design since the Model 302. Whereas the 302 purported to condense two pieces into one, marketers and vendors of the GTE Flip-Phone presented it as a one-piece device with the capacity to unfold, becoming two pieces from one. The shrinking size of electronic circuitry—Moore's law at work—allowed designers to place the necessary circuitry for the device inside the handset itself, eliminating altogether the need for the base as a complementary piece. In lieu of this two-part design, the GTE Flip-Phone had a folding panel at the bottom of the unit that covered the keypad, another new feature at the time that was also crucial in facilitating this compressed design. This panel could be laid against a flat surface, keeping the unit in a closed position; when a user picked it up, it would spring open to reveal the keypad.

A 1979 advertisement in the industry journal *Telephony* illustrates this new capability in action: the left side of a two-page spread depicts a sequence of four photographs: a hand reaching for a closed flip phone face down on a surface; the hand in the process of lifting the phone as it unfolds; the hand holding the phone fully open, displaying its keypad; and a close-up shot of a woman holding and speaking on the phone. The accompanying text on the facing page foregrounds the newness of this mechanism as a revolutionary development. Beneath the headline "Our electronic flip-phone telephone will set a lot of tongues wagging", evoking a bodily response that echoes the movement of the device itself, the ad's main copy calls the phone "the most exciting telephone break-through since we invented the dial . . . a novelty that will fascinate young and old alike." This advertisement not only documents the functionality of the flip mechanism, visually explaining this new operation for the vendors in the magazine's audience, but also presents that operation as a transformative change for the telephone, both physically and discursively. Developments in circuitry allowed for consolidation of the device within a single piece, which in turn allowed for portability and foldability to emerge as interdependent features; thus even in the landline, the flip mechanism quickly became synonymous not just with portability but indeed with newness itself.

This confluence of characteristics continued with the introduction of GTE's Flip-Phone II—described in a 1982 advertisement in *U.S. News and World Report* as "like something out of a spy movie . . . one of the most advanced as far as home phones go"—and expanded further with the subsequent introduction of the cordless Flip-Fone 300.[5] Much as the original GTE Flip-Phone increased the mobility of the landline through an extended cord, the Flip-Fone 300 increased the range of the cordless phone from roughly 700 to 1,200 feet without need for an antenna. An optional SuperAntenna tripled this range to 3,600 feet, offering a prefiguration of the range and flexibility of the true mobile phones that were to appear shortly on the consumer market: a 1984 advertisement promises that users could be "blocks away [from the base unit], or in your car, or in a remote building, and still have crystal-clear conversations" (New Horizons, 1984). Viewed with historical hindsight, the ways in which these models anticipate the mobile flip phone in the pressure they put on the limits of landline technology seem logical and obvious almost to the point of tautology. Indeed, their various foldability, compactness, and cordlessness (in the last instance) all incorporate characteristics of the mobile and the flip phone *avant la lettre*. However, two more specific and significant details emerge from this history, the first being the centrality of flip design and functionality to this futuristic trajectory in both practical and discursive terms. The other is the strange absence of these landline flip phones from the dominant history of telephonic innovation, an omission that seems all the more ironic given their close anticipation of the mobile telephony that is now increasingly culturally dominant. In a number of important senses, then, the imagination of a mobile future seems in part dependent upon the erasure of that future's own genealogy—its own technological conditions of possibility.

Early portable computing provides another slant antecedent for the functionality of the mobile flip phone. While, of course, early personal and portable computing served a different practical use than early mobile telephony, it similarly depended on a hinge-based functionality for its defining physical characteristic. In the earliest designs for mobile computers, this hinge functionality emerged as a break from the suitcase-bound portable typewriters that had been available since the early decades of the 20th century and the earlier computers that built on the design of the portable typewriter. In a patent for a "Portable Computer Enclosure" filed in 1979, James D. Murez (1981) writes that earlier designs for portable or tabletop computers

> all relate to a suitcase or typewriter-type package. While, perhaps, suitable for enabling one to carry the computer, their design does not permit what is deemed to be the most efficient use of the enclosure, consonant with opening the computer and using it. The present invention permits use of a single enclosure which satisfies the needs of portability as it is closeable into a suitcase-style cabinet with a tractable carrying handle. Specifically, the keyboard enclosure is hinged to the main frame enclosure in such a manner that it is possible to fold keyboard up against the main frame and to latch the two together.

Much as in the case of Dreyfuss's Model 302, the distinction Murez announces is a subtle, yet crucial matter of objecthood, segmentation, and functionality: whereas earlier portable computers could be placed within a case or package, Murez's design made the case part of the computer itself. Portability was not just incidental or possible through such a design, but quite literally built into that design. The Osborne 1 and the Kaypro II, two of the first widely popular portable computers, followed the design depicted in Murez's patent closely. Released in 1981 and 1982, respectively, these devices consisted of a boxlike rectangular computer unit (each weighing over 25 pounds, and often colloquially referred to as "luggable" rather than portable) with a hinged section on one end; when the computer was placed on a flat surface, a user could fold down this section to reveal a keyboard and a small monitor screen.

Thus the interactive functionality of this object was in many ways the directional opposite of the laptop computers that would follow: whereas those computers include a screen that folds up and away from the keyboard and CPU, the Osborne 1 and the Kaypro II had keyboards that folded down and away from the screen and CPU.

By contrast, the GRiD Compass, designed by William Moggridge in 1979 and first sold in 1982, was one of the first computers to employ this contemporary "upward" hinge (see Figure 26.2). This shift in object functionality, along with the fact that it weighed roughly half what the Osborne and Kaypro did, would seem to suggest that the GRiD was the first truly portable computer in the sense of being a clear and direct predecessor of the contemporary laptop. Indeed, just as the GRiD's hardware was developed for portability, so was its software. Built around a national dialup network called "GRiD Central", the device allowed users to download files, applications, and other data from any telephone connection in a manner that prefigures Google's Chromebook and other cloud-based laptops: an early review remarks that "an executive could toss his computer into a briefcase (and still have room for his personal papers), fly to a meeting, dial up the network, then pull in and send out his files from that remote location" (Somerson, 1983: 324). Taken as a whole, the GRiD was initially received as luxurious and science-fictionally futuristic, a "Bugatti of bits", "so crammed with futuristic electronic devices that it looks like something Luke Skywalker would use to destroy an evil civilization" (p. 322). As this reviewer's hyperbolic analogies suggest, the GRiD's innovations undoubtedly make it tempting to see its upward hinge and networked functionality not only as futuristic in and of themselves but also more specifically as the clearer and more direct antecedent of the flip phone—a warp-speed device that was light years ahead of its competitors, as it were.

However, to divide and distinguish between different modes of portable computer functionality in such a way would be to simplify and rationalize the design history of the folding computer in ways that privilege a linear progression to the present. The hinge mechanism of the flip phone is not the direct descendant of either design, nor quite of both of them, but rather of the tension between them. Contrary to a linear progression, these mirror-image, inverted mechanisms collectively constitute an uneasy first reckoning within the realm of digital technology with what John Naisbitt describes as "high touch" design, an attempt

Figure 26.2 William Moggridge *et al.*, Image from U.S. Patent USD280511 S for "Portable Computer", 1982.

to produce a "counterbalancing human response" to increasing digitization (1982: 39). Indeed, while both design concepts address practical concerns such as compactness and the protection of crucial fragile components, they at the same time serve to conceptualize the device in terms of physical movement and interactivity. While high touch was a widespread concern across a number of design industries in the postmodern period, it was particularly relevant in the context of emergent digital technologies, which were often seen as in need of further humanization and concretization through design in order to counteract their perceived abstraction. Describing this trend, Jeffrey Meikle documents how design theorists of the period

> realized that the designer's responsibility increased greatly as society's dominant technologies became less mechanical, based more on silicon circuitry[.] . . . A device's working elements no longer dictated its size and shape as they had throughout the industrial age. Instead the parameters of shape became arbitrary.
>
> (2005: 202)

Designers of early portable computers, then, imagined these parameters as a way to retain and restore touch through ergonomic interactivity, the haptic demand to move and reshape a device in order to make it function. Prefiguring the design concerns that would characterize the development of the flip phone several years later, these designers laid a structural groundwork in which the digital and the mechanical quite literally folded into one another. Neither one supersedes the other; both instead work, and work around, a hinge moment in the timeline of technological development.

The antecedents of the flip phone, then, are hybrid and interstitial, collectively as well as individually. The flip phone is neither a simple transmutation of the design of the landline, nor a wholesale break from it, nor is it an exact descendant of either the upward- or downward-folding portable computer; its connection to each of these is inexact, uneven, partial, liminal, spectral. Indeed, given that the flip phone emerged from a lineage that included telephonic portability as well as computer portability, perhaps the overlapping contradictions and paradoxes of interface and interactivity I have traced in this section are the most salient characteristics of its genealogy. In all of these contexts the flip mechanism consistently appears as a cutting-edge design attribute, as co-constitutive with a discourse of futurity. Given these prefigurations, there is a rich irony in the fact that the end-result descendant of telephonic and computer portability is paradigmatically not the flip phone, but rather the smartphone, a device that in its single-piece, effectively invisible physical functionality, seems to be the antithesis of the moving parts of the flip design. Both critical voices such as Emerson's and industry mouthpieces alike have long suggested that the smartphone is effectively a handheld computer. A convergence of telephone and computer—or perhaps a subsumption of one by the other—it serves as a phone perhaps last (if at all) among many other functions.[6] In the case of the mainstream industry, at least, these futurist assumptions are deeply bound up in a metanarrative of technocratic progress that turns on seeing the smartphone as a kind of culmination or apotheosis of digital teleology, as perpetually the best that a digital device can be. Such a metanarrative depends in part upon practical developments in technological areas such as battery life, processing power, and wireless data transfer. Yet it also relies upon the abrupt and violent forgetting of an earlier futurism that coalesced around the flip mechanism across a range of technological platforms. This earlier futurism of the flip mechanism was by no means more haptically or socially liberatory than that of the smartphone. In many ways, one discourse contributed to the cultural and economic conditions of possibility for the other, and to suggest that the flip mechanism is somehow wholly outside of such a trajectory would be at best deeply nostalgic. However, tracing the genealogy of this mechanism not only allows us to attend to its being lost, almost willfully forgotten through planned obsolescence, but also draws us to the complex ways in which the presentist futurism of mainstream media discourse works to conceal its past. In the following section, I turn more directly to the design, haptics, and branding of several pivotal flip phone models as a way of tracing the characteristics that were introduced to, and erased from, the dominant narrative of technological development through this device's emergence and obsolescence.

Face Time, Or, Haptic Turns

If the GRiD Compass's design made it, in many senses, the first fully portable computer, the Motorola MicroTAC's flip design made it perhaps the first truly portable phone. The DynaTAC, released in 1984, was the world's first commercially available mobile phone, but its size, weight, and cumbersome brick form factor meant that it was only partly portable, a telephonic counterpart to the luggable computers that preceded the GRiD. Introduced in 1989, the MicroTAC was far more truly portable, a status due both to its reduced weight (at less than half that of its predecessor) and to its introduction of the flip mechanism as a new form factor (see Figure 26.3). The MicroTAC's size and folding design also made it the first mobile phone that could be carried in a belt-clip holster, a feature that would be carried on in its successor, the Motorola StarTAC, in 1996. In fact, the StarTAC 3000, Motorola's first model in the product line, shipped with a holster included, and its manual proclaimed it "the second generation of wearable phones" (with the MicroTAC presumably being the first). In this sense, the flip form factor introduced by Motorola stands not only as a seminal development in the history of portable telephony in general, but also more specifically as a key stage in the genealogy of our current moment of wearable technology. The flip phone was the first device for which manipulation by the body ran parallel to presence on the body. Its hinge made possible an integrated, prosthetic status for the mobile device, a suturing of the body and the network that prefigured the bodily attachments of contemporary devices such as the Fitbit fitness tracker and the Apple Watch.

The MicroTAC and StarTAC set in motion other transformations of the relations between device and body as well. As in the case of the devices I have discussed above, the flip was widely seen as a futuristic innovation, the Star Trek communicator brought into the real, terrestrial world.[7] In addition to transforming the practical portability of the mobile phone, the flip design also crucially altered the haptics of bodily use and telephonic speech. Galit Wellner writes that the StarTAC,[8] the broadly popular successor to the MicroTAC released in 1996, was part of a first wave of mobile phones that was

> optimized for voice communication. . . . [T]he unique clamshell configuration turned out to be well suited for a 'talking face' because it fit nicely between the mouth and the ear. When the cell phone was open, the microphone was close to the mouth—closer than in the candybar or brick layouts.
>
> (2015: 27)

A 1992 British television advertisement for the MicroTAC dramatizes this shift in the mobile industry through an imagined shift in the body itself ("Motorola MicroTAC – Flip 'N' Switch [1992, UK]"). The advertisement begins with a close-up headshot of a man facing forward, yet with the bottom half of his face digitally modified to appear twisted

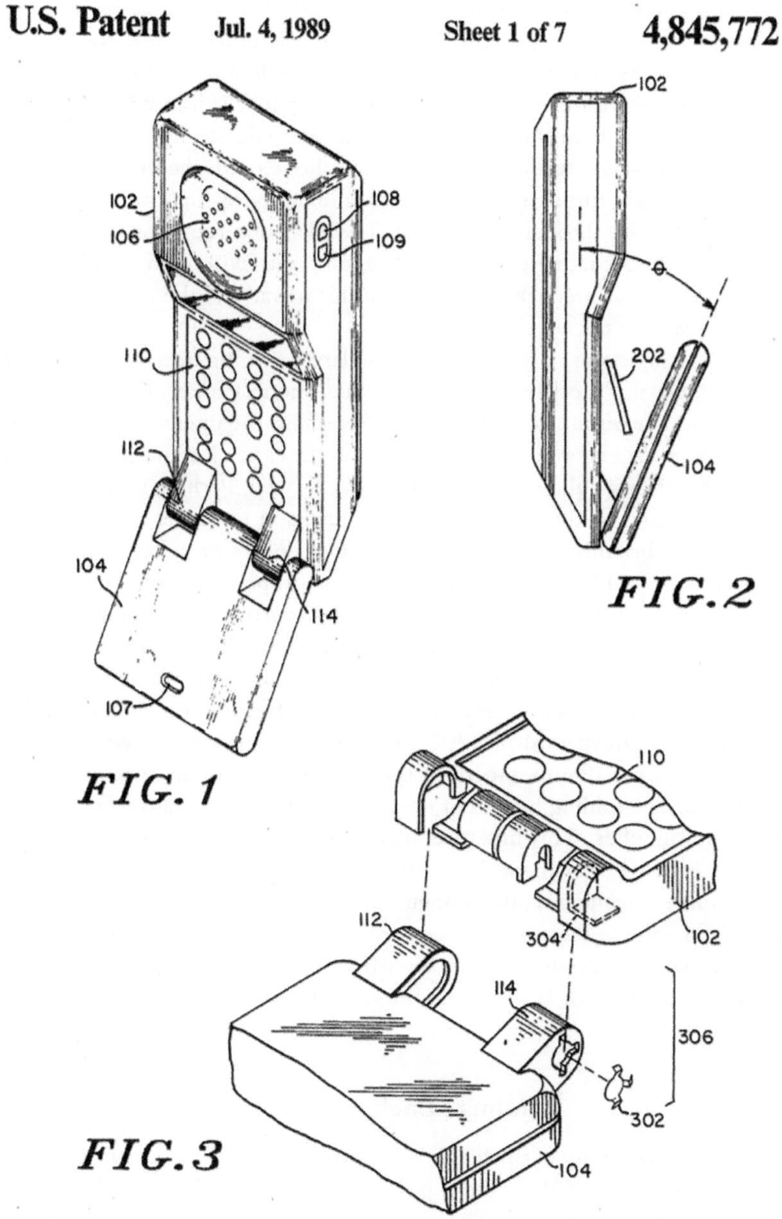

Figure 26.3 Michael P. Metroka *et al.*, Images from U.S. Patent US4845772 A for "Portable radiotelephone with control switch disabling", 1989.

90 degrees to his right. He describes and demonstrates the situation of most "flat" portable phones: "the earpiece goes next to the ear, and the mouthpiece goes next to the mouth", a configuration that seems logical, yet only works because of his distorted head. In the second half of the advertisement, he raises up a MicroTAC, emphatically describing how "Motorola makes a phone like this", a line he punctuates with the opening of the

flip mechanism. Adopting a light, ironic skepticism, he remarks that such a design is "only suited to people who look . . . like . . . this"; with each of these last three words, the bottom half of his face twists back to its proper front-facing position. "Nah, it'll never catch on", he quips, smiling knowingly at the camera, and then closes the flip mechanism with one index finger, ending the advertisement. In this narrative, the cutting-edge technology of digital morphing frames the cutting-edge technology of the flip phone, with the video's posthuman manipulation of the body through technology paradoxically opening up a space for the rehumanization of that body through technology. The ad's humorous twist underscores the stakes of this design decision: rather than the face bending and twisting to the phone—a prospect dramatically invoked as precisely impossible in this narrative—the phone takes on the ability to bend and twist to the face.

This shift in shape and angle was only one part of the larger "haptic turn" set in motion by the introduction of the flip mechanism to the mobile phone market (Goggin, 2009: 242). Gerard Goggin uses the term "haptic turn" in reference to the intensely touch-based technological engagement engendered by the iPhone interface, yet I argue here that the emergence of the flip mechanism allowed for a haptic turn as well. Perhaps most basically, the decreased size and weight of the MicroTAC in comparison to earlier models meant that a mobile phone could be truly carried on the body in a pocket or holster for the first time, providing for an increased bodily engagement. A more important haptic transformation came in the mechanics of the device's form factor: preparing a flip phone for use, whether by answering or placing a call, necessarily entailed unfolding it, transforming it from a closed to an open device. The haptics of this gesture are profoundly interstitial, somewhere in between what came before and what came after, but also somewhere outside of them. The haptics of the Dreyfuss Model 302 inhered in the tension between base and handset, and the haptics of the iPhone inhere in the consolidation of all action within a single digitized interface. The folding haptics of the MicroTAC, by contrast, were at once dual, singular, both, and neither. This unsettled nature pertained to how the mechanics of form factor invited bodily engagement, but also to surface and touch itself. While the Model 302 and its landline descendants offer a singular, fixed location to touch and dial, the iPhone purports to offer infinite locations in one through the "magic" of its touchscreen interface. The flip mechanism introduced with the MicroTAC offers a singular interface, but also—when closed—none at all. It invites high touch interaction precisely by seeming to block against that interaction.

Bending, Back

Following the widespread popularity of the StarTAC, numerous mobile phone manufacturers introduced flip design phones, making the flip the dominant form factor in the United States consumer market for over ten years. Then, as mobile communications began to shift from voice to text and other forms, the popularity of the flip phone waned—first gradually, then suddenly, supplanted by the explosive demand for the iPhone around the turn of the 2010s. In the current cultural moment, when the iPhone has become the fastest-selling, as well as arguably the best-selling, product in history, the flip phone seems almost entirely obsolete.[9] While they are still sold by most major mobile phone vendors, flip phones are largely targeted at what might be delicately called a secondary market, including allegedly technophobic seniors and consumers who want a phone only for "emergency" calls. Such a market seems itself almost past, comprised of old users and old uses, out of touch with the myriad modern applications of the smartphone. Yet the figure of the flip has at the same time re-emerged in a number of contexts in recent years, complicating the simple, teleological narrative of replacement by newer and better technology.

Some of these re-emergences stem from deep nostalgia, a cultural yearning for a time seemingly outside of the complexities of newer technology. Covering the hacking of former Secretary of State Colin Powell's e-mail in the run-up to the 2016 Presidential election, *The New York Times* noted an almost reactionary impulse among other high-ranking politicians. In an article on the event, Michael D. Shear and Nicholas Fandos quoted Senator Richard J. Durbin of Illinois as commenting that "the news of Mr. Powell's hacked emails had him thinking that Senator Chuck Schumer's never-ending use of an old-fashioned flip phone 'makes more sense than ever.'"[10] While there might be some practical legitimacy to the problem and solution Durbin notes, his concerns also more broadly position the flip phone as the technological image of a simple(r) time, devoid of threats to privacy, as well as perhaps devoid of the distraction, isolation, and other cultural effects often associated—however incorrectly—with the ubiquity of smartphone culture.

Pop singer Adele's use of a flip phone in the video for her blockbuster 2015 single "Hello" offers a different cultural figuration of the flip phone as a device of the past. The video follows Adele through a deserted house in the countryside as she reflects on a failed love with a man played by actor Tristan Wilds; memories of their deteriorating relationship, shown in flashback, mix with the present timeline of Adele's singing both within and outside the house. A number of old technologies appear in the video's present timeline, including a typewriter, a corded landline phone (accompanied by a flip-up phone directory), dot-matrix printouts, and a rotary phone booth in the woods, overgrown with weeds and vines. Yet the most notable old technology is the 2000s-era flip phone that Adele uses at the opening of the video, placing a failed call that seems to prefigure and frame the song's telephonic lyrics, epitomized in its chorus "Hello from the other side/ I must have called a thousand times." Indeed, couched among images of the flip phone and its old-media brethren, the song's lyrics seem to be not only about the past but also almost themselves of the past, a belated missive delivered via belated technology. The overarching tone of the video is not one of dysfunction, in which relationships and devices fail to work as planned, but rather of obsolescence, when those things are simply no longer useful. The flip phone serves here as both the medium and the message of nostalgic regret, both a means of articulating desire for the past and the object of that desire itself.[11] Taken together, the Powell scandal and Adele's video situate the flip phone at a complexly undead intersection of technology, time, and culture. Fetishized by Adele, it seems an object of nostalgia—obsolete, no longer functional, figuring an earlier time. And yet at the same time, as Durbin and Schumer show, it also still functions, remaining open to a continued retro usage that re-layers the past onto the present.

And yet it also occupies a third position, a kind of strangely inverted futurity. In November 2016, Apple made public a patent, filed in July 2014, for a "Flexible Display Device", in which it lays out several possible configurations of a folding iPhone, including a single-hinge mechanism similar to the conventional flip form factor, a hinge that would function backwards from its conventional usage, allowing the device to be folded in half with part of its touchscreen facing outwards, and a dual-hinge mechanism that would allow the phone to be folded in three pieces, snake-like (see Figure 26.4). The "Background" section of the patent articulates a striking reversal from Apple's longstanding attachment to the magic of the unbroken surface:

> Liquid crystal displays [in electronic devices] are often mounted under a rigid layer of cover glass. The cover glass protects the liquid crystal display from damage, but the rigid nature of the cover glass and other display layers render the display inflexible.

> Flexible display technologies are available that allow displays to be bent. For example, flexible displays may be formed using flexible organic light-emitting diode (OLED) display technology. It would be desirable to be able to use flexible display technology to provide improved electronic devices.
>
> (Rothkopf *et al.*: 2016)

This proposal radically refigures the iPhone's screen materiality, perhaps its most defining technological, haptic, and ideological characteristic. A digital object that has historically been defined in its purely unitary nature—its pseudo-Platonic ideal of singular form, devoid of visible or palpable mechanics—now takes a potential shape in two or even three parts.[12] This refiguration, in turn, raises the question of how marketing and design forces imagine the historical trajectory of a seemingly obsolete object: suddenly one possible cutting-edge future of the mobile phone is a dramatic return to a superseded, even rejected past. Yet Apple's market dominance and its history of futurist rhetoric suggest that such a design constitutes a play on the history of mobile technology that is more complicated than a simple appeal to nostalgia or retro aesthetics. Indeed, Apple's proposed design imagines a strange kind of corporate media archaeology, like an old steampunk film played backwards. Excavating the very design that it rendered obsolete, the corporation uncannily follows Zielinski's call to "not seek the old in the new, but find something new in the old" (2008: 3).

Here it is the archaeology that produces the archive, reuse that allows for disuse: as Jordan's image shows us, every new product, every early adoption, sends more objects to an unnecessarily early grave. It is history that collapses under the weight of the device, not the other way around. A critical reading of Apple's business ideology makes it tempting to see the flip phone as an exception to this trajectory, as if the corporation's potential resurrection and cannibalization of the hinge means that the flip phone is, or at least previously was, a kind of variantological form of mobile technology. Yet to suggest as much of a device that was so ubiquitous at the height of its popularity and so pivotal to the broader market dissemination of mobile technology would be illogical and self-contradictory at best. Like any technology, the flip mechanism—once new, now old—exerts its weight as both a rule and an exception at different moments in its life cycle. Indeed, Apple's potential excavation of the fold will itself inevitably be replaced, deemed obsolete, and re-consecrated to the landfill. Perhaps, then, the possibility of a folding iPhone makes clear not that the flip phone is "back", nor that it never actually left, but rather that, like any technology, it is once again, and always already was, interstitial: folded, bent, and straightened back and forth across time.

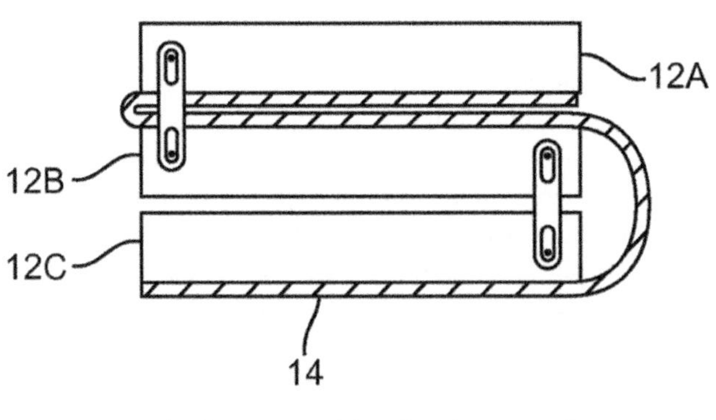

Figure 26.4 Fletcher R. Rothkopf *et al.*, Image from U.S. Patent US9,504,107 B2 for "Flexible Display Devices", 2016.

Notes

I am grateful to Linnea Harris for her assistance with research, to Chris Jordan for allowing reproduction of his artwork, and to the Skidmore College Faculty Development Committee for their support of image licensing for this project.

1 See Nye (1996) for a seminal history of the American technological sublime.
2 In this sense, *Cell Phones* has become more trenchant and layered with the weight of time. This image of old devices now paradoxically feels old itself: some ten years after its release, Jordan's image not only stands in for a discrete timespan of specific disposal circa 2007 but also, in the now-outmoded status of the devices it depicts, invokes a wave of nearly categorical innovation and planned obsolescence between past and present moments.
3 We might add to this group the unnamed narrator of Ralph Ellison's *Invisible Man*. Hiding in a basement filled with 1,369 lights fed by power stolen from the Monopolated Light & Power corporation—a media-archaeological setting if ever there was one—he remarks that the world moves "[n]ot like an arrow, but a boomerang. (Beware of those who speak of the *spiral* of history; they are preparing a boomerang. Keep a steel helmet handy.)" (1952: 6).
4 For examples of these broader histories, see Hanson (2007: chapter 2); Ling (2004: Introduction); and Goggin (2006: chapter 2). For a useful history of critical discourse on mobile telephony, see Goggin (2006: Introduction). Most accounts situate mobile telephony in a media history understood largely in terms of telecommunications technologies such as wireless telegraphy and the two-way radios used by military and police forces in between the World Wars. Paul Levinson's history is a valuable exception to this pattern, focusing on mobility and portability as characteristics connecting a range of devices across time, including the Kodak camera, the transistor radio, and some of the early portable computers I discuss below (see Levinson, 2004: chapter 2, "Information on the Move").
5 The portability of the Flip-Phone II also made it a device of choice for phone hackers of the period: in a December 1986 guide, Phucked Agent 04 of the seminal hacker group Legion of Doom writes that when using telephones to tap into a public line, "[t]he 'all-in-one' handset units without bases are the best (I tend to like QUIK's and GTE Flip Phone II's)."
6 See, for example, Mossberg and Boehret's "The iPhone Is a Breakthrough Handheld Computer", an early review prior to the sale of the first generation of the device in 2007.
7 See Siler (2015) and Wellner (2015) for two instances of this comparison (although see my footnote below on Wellner and the MicroTAC, the StarTAC, and the question of historical accuracy).
8 Wellner describes the StarTAC as "the first handset with a clamshell ('flip') configuration" and thus as representative of the "first historical variation of cell phones" centered on the bodily figure of the "talking head", theorizing the device in bodily terms that are crucial to my argument here (2015: 25). Yet in doing so she strangely ignores the MicroTAC, which preceded the StarTAC and introduced many of the key elements of flip form factor. As such, I discuss these elements as characteristic of the earlier MicroTAC.
9 See Kingsley-Hughes (2016) for a discussion of the iPhone in relation to other product sales histories.
10 Schumer's widely documented use of a flip phone is only one instance of a politician's outdated media practices seemingly becoming strangely, almost retrofuturistically prescient: Shear and Fandos' (2016) article also quotes Lindsay Graham of South Carolina's claim, "I haven't worried about an email being hacked since I've never sent one[.] . . . I'm, like, ahead of my time."
11 Adele's use of the flip phone provided a ripe subject for Internet discussion and memeification following the release of the video. Xavier Dolan, director of the video, commented in an interview that

> I could see the GIFs on Twitter. I'm like, "guys, get over it. It doesn't matter." But the real explanation is that I never like filming modern phones or cars. They're so implanted in our lives that when you see them in movies you're reminded you're in reality. . . . If you see an iPhone or a Toyota in a movie, they're anti-narrative, they take you out of the story.
>
> (Zeitchik, 2015)

Yet the intense reaction to Dolan's directorial decision suggests that the flip phone is anti-narrative as well, and that his comments are, if not somewhat disingenuous, at the very least a strong under-reading of the historical stakes of such an image. If, as he suggests, the flip phone is somehow outside of time, it is so to the extent that it is paradoxically also simply out of time, old, and no longer functional.

12 This pivoting to sell a device with functionality precisely in opposition to the functionality it had previously sold (in this case the unbroken glass screen interface) is a recurring business practice for Apple. For an earlier example, see Levy (2007: 184–186) on the company's shift from emphasizing the control features of the original iPod to emphasizing the randomization and lack of control of the iPod Shuffle.

References

Adele. "Hello." *YouTube*, uploaded by AdeleVEVO, 22 October 2015, www.youtube.com/watch?v=YQHsXMglC9A Accessed 6 December 2016.

Deleuze, Gilles. *The Fold: Leibniz and the Baroque*. Trans. Tom Conley. University of Minnesota Press, 1992.

Ellison, Ralph. *Invisible Man*. 1952. Vintage Books, 1995.

Emerson, Lori. *Reading Writing Interfaces: From the Digital to the Bookbound*. University of Minnesota Press, 2014.

Goggin, Gerard. *Cell Phone Culture: Mobile Technology in Everyday Life*. New edition. Routledge, 2006.

Goggin, Gerard. "Adapting the mobile phone: The iPhone and its consumption." *Continuum*, vol. 23, no. 2, April 2009: 231–244.

Gryta, Thomas. "Americans Keep Their Cellphones Longer." *Wall Street Journal*, 18 April 2016, www.wsj.com/articles/americans-keep-their-cellphones-longer-1461007321. Accessed 27 September 2016.

GTE. Advertisement. *Telephony*, 16 July 1979: 60–61.

GTE. Advertisement. *U.S. News and World Report*, 5 April 1982: 20–21.

Hanson, Jarice. *24/7: How Cell Phones and the Internet Change the Way We Live, Work, and Play*. Praeger, 2007.

Jordan, Chris. *Running the Numbers: An American Self-Portrait*. Museum of Art: Washington State University, 2009.

JS&A National Sales Group. Advertisement. *Popular Science*, November 1978: 25.

Kingsley-Hughes, Adrian. "Billionth iPhone Sold: Does That Make It the Best-Selling Product of All Time?" *ZDNet*, 29 July 2016, www.zdnet.com/article/billionth-iphone-sold-does-that-make-it-the-best-selling-product-of-all-time/. Accessed 17 November 2016.

Kirschenbaum, Matthew G. *Track Changes: A Literary History of Word Processing*. Belknap, 2016.

Levinson, Paul. *Cellphone: The Story of the World's Most Mobile Medium and How It Has Transformed Everything!* Palgrave/St. Martins, 2004.

Levy, Steven. *The Perfect Thing: How the iPod Shuffles Commerce, Culture, and Coolness*. Simon & Schuster, 2007.

Ling, Rich. *The Mobile Connection: The Cell Phone's Impact on Society*. Morgan Kaufmann, 2004.

Meikle, Jeffrey L. *Design in the USA*. Oxford University Press, 2005.

Metroka, Michael P. *et al.* "Portable radiotelephone with control switch disabling." Patent US4845772 A. 4 July 1989.

Moggridge, William G. *et al.* "Portable computer." Patent USD280511 S. 10 September 1985.

Mossberg, Walter S. and Katherine Boehret. "The iPhone Is a Breakthrough Handheld Computer." *AllThingsD*, 26 June 2007, http://allthingsd.com/20070626/the-iphone-is-breakthrough-handheld-computer/. Accessed 27 October 2016.

Motorola Corporation. *StarTAC 3000 User Manual*. 1996.

"Motorola MicroTAC – Flip 'N' Switch (1992, UK)." *YouTube*, uploaded by The Hall of Advertising, 1 June 2012, www.youtube.com/watch?v=18DI8Jghhkg. Accessed 15 November 2016.

Murez, James D. "Portable Computer Enclosure." Patent US 4,294,496. 13 October 1981.

Naisbitt, John. *Megatrends: Ten New Directions Transforming Our Lives*. Warner, 1982.

New Horizons. Advertisement. *Popular Science*, January 1984: 152.

Nye, David E. *American Technological Sublime*. MIT Press, 1996.

Parikka, Jussi. *What Is Media Archaeology? Polity*, 2012.

Phucked Agent 04. "The Legion of Doom! Presents: LOD Reference Guide – Outside Loop Distribution Plant." *Textfiles*, December 1986, http://cd.textfiles.com/group42/PHREAK/OUTLOOP.HTM. Accessed 20 October 2016.

Rainie, Lee. "Cell Phone Ownership Hits 91% of Adults." *Pew Research Center*, 6 June 2013, www.pewresearch.org/fact-tank/2013/06/06/cell-phone-ownership-hits-91-of-adults/. Accessed 13 September 2016.

Rothkopf, Fletcher R. *et al.* "Flexible Display Devices." Patent US20140328041 A1. 6 November 2014.

Shear, Michael D. and Nicholas Fandos. "Concern Over Colin Powell's Hacked Emails Becomes a Fear of Being Next." *The New York Times*, 15 September 2016, www.nytimes.com/2016/09/16/us/politics/email-hacking-colin-powell-congress.html. Accessed 17 November 2016.

Siler, Wes. "Spock and the Motorola StarTac." *Gizmodo*, 27 February 2015, http://gizmodo.com/spock-and-the-motorola-startac-1688496340. Accessed 3 January 2017.

Somerson, Paul. "Boxing the (Grid) Compass." *PC Magazine: The Independent Guide to IBM Personal Computers*, August 1983: 322–332.

Wellner, Galit. *A Postphenomenological Inquiry of Cell Phones: Genealogies, Meanings, and Becoming*. Lexington Books, 2015.

Zeitchik, Steven. "Director of Adele's 'Hello' Video Explains Its Look, Theme and, Oh Yes, Flip Phone." *Los Angeles Times*, 23 October 2015, www.latimes.com/entertainment/movies/moviesnow/la-et-mn-adele-hello-video-song-flip-phone-20151023-story.html. Accessed 6 December 2016.

Zielinski, Siegfried. *Deep Time of the Media: Toward an Archaeology of Hearing and Seeing by Technical Means*. Trans. Gloria Custance. MIT Press, 2008.

27

HD DVD TECHNOLOGIES

John Reid Perkins-Buzo

After the astounding success of standard DVD technologies in the 1990s, the future for optical discs seemed boundless (Dixon, 2005: 58). Along with the continued emergence of high-definition standards for video capture and television broadcast (see Federal Communications Commission website), the need for a high-capacity optical disc format for consumers to enjoy the benefits of these emerging standards also emerged. Any such format would need to store at least four times the data that a standard DVD could hold, and be readable at a much faster rate. It would also need to incorporate features compatible with personal computer operating systems, yet also work with stand-alone players. Standard DVDs lacked these features and the technologies for bringing them to consumers were still under development at that time. What the format would look like was anybody's guess, with the only given being the disc size: 12 cm, that is, the same size as a standard DVD or CD.

Two technologies emerged in the late 1990s to allow high-capacity, interactive optical discs: the blue diode laser and computer chips that could be embedded in consumer level devices and run high-level code. Shuji Nakamura at Nichia Corporation developed the blue diode laser in 1996 as a follow-up to his earlier invention of the blue LED. Standard DVDs use a red laser technology that limits the amount of information recordable on a 12 cm disc due to the length of the wavelength (640 nm). Because of this, each DVD can only contain 4.7 GB per layer of a disc (8.5 GB for a dual-layer disc). With the use of blue lasers, the wavelength for a high-capacity disc device could be reduced to a much lower number, allowing more data to be recorded on the disc. See Figure 27.1 for a comparison of optical disc specifications.

Sony, in collaboration with Pioneer, exhibited an optical disc using a blue laser, DVR Blue, at the Combined Exhibition of Advanced Technologies (CEATEC) in 2000. By February of 2002, Sony rebranded their discs as Blu-ray and began organizing the Blu-ray Disc Founders which consisted of electronics firms Hitachi, LG, Panasonic, Pioneer, Philips, Samsung, Sharp, Sony, and Thomson (Fox, 2002). On May 20, 2002, these companies announced that they would begin licensing the specifications for the disc format with manufacturing of Blu-ray discs to start in June 2002 (Sony, 2002).

Toshiba, serving as chair of the industry standard-setting DVD Forum, together with NEC, proposed a different blue laser disc format named the Advanced Optical Disc in August 2002 (Williams, 2002). The DVD Forum had several members who were also members of the Blu-ray Disc Founders organization (Hitachi, Pioneer, Philips, Sony, and Thomson), and consequently the Forum voted down the proposal twice before an amended ballot rule

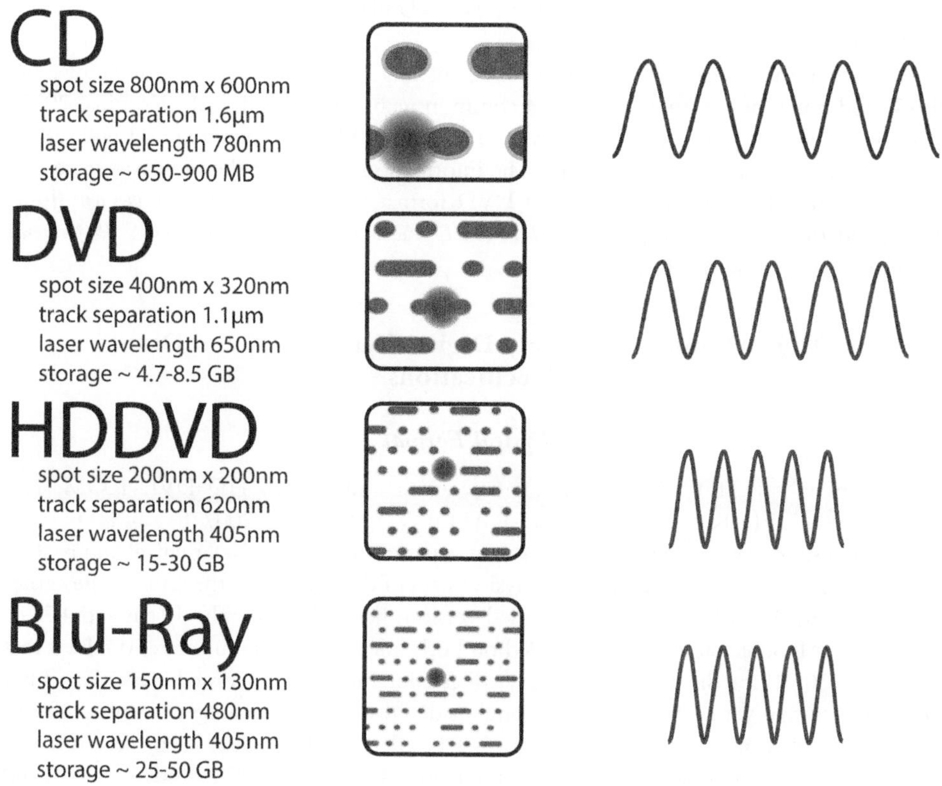

Figure 27.1 Comparison of optical disc specifications.

dismissing abstentions from the vote cleared the way for its passage on November 26, 2003. The format subsequently became known as the HD DVD format after its passage. Still, the HD DVD format failed to acquire a majority of the DVD Forum's votes, which set the scene for a significant optical disc format war (Hara, 2003).

The format war played out within two years, decided by the decisions of major content providers and retail distributors. Prior to the release of any discs, the HD DVD format received exclusive support from Universal Studios, Paramount Pictures, and Warner Bros Pictures. The Blu-ray format had commitments from Columbia Pictures, Walt Disney Pictures, and 20th Century Fox. Paramount would switch to Blu-ray, then back to HD DVD in 2007, while Dreamworks Animation would commit to HD DVD, for a time, also in 2007. Retailers such as Blockbuster, Walmart, and Target found that consumers wanted Blu-ray discs more often than HD DVD discs. Matthew Smith, senior vice president of merchandising at Blockbuster in 2007, reported that although Blockbuster had been renting both Blu-ray and HD DVD titles in 250 stores, consumers were choosing Blu-ray titles more than 70 percent of the time (CNN, 2007). "The consumers are sending us a message. I can't ignore what I'm seeing", he told the Associated Press at the time. The success of Sony's PlayStation 3, which came with a Blu-ray player, made purchasing a separate HD DVD player unnecessary if you owned the console. With the loss of the major content producers and retailers, on February 19, 2008 Toshiba officially announced that it would stop the development of the HD DVD players, conceding the format war to the Blu-ray disc format (Toshiba, 2008).

The two formats differed in several ways, making them physically incompatible, but also shared much in common because of common technologies and goals (Yoshida, 2005). However, since the purpose of this essay is to review HD DVD technologies, I will restrict what follows to its specifics and history. The information comes for the most part from two DVD Forum White Papers on the HD DVD format, *HD DVD Format Overview* (June, 2007), and *Requirements Specification for HD DVD Video Application Functional & Performance Requirements*, Version 1.0 (July, 2005).

High-Definition/Density Digital Versatile/Video Disc Specifications

Physical Format

Four "families" of HD DVD technologies were manufactured: HD DVD-ROM, HD DVD-R, HD DVD-RW, and HD DVD-RAM (DVD FORUM, 2007: 8–12). HD DVD-RAM has a single-layer capacity of 20 GB and was envisioned as a data archival and backup medium, not for consumer video. The physical sizes of the discs were the same as the earlier DVD standard, 12 or 8 cm. The single-sided versions held 15 GB for the 12 cm and 4.7 GB for the 8 cm. Double-sided discs (recorded on both sides of the disc substrate) doubled that amount to 30 GB for the 12 cm. Dual-layer discs containing two recorded media layers per side, doubled that yet again, with 12 cm single-sided discs having dual layers storing 30 GB, and double-sided discs with dual layers storing 60 GB. The 8 cm discs were very uncommon especially in the dual-layer format (they held 9.4 GB for the single-sided version and 18.8 GB for the double-sided). The maximum storage for an HD DVD disc is 60 GB (DVD FORUM, 2007: 5). For compatibility with the original DVD format, the data layer of an HD DVD is 0.6 mm below the surface to protect the data layer from physical damage. The numerical aperture of the optical pick-up head is 0.65, compared with 0.6 for DVD. All HD DVD players are backward-compatible with DVD and CD (DVD FORUM, 2007: 6).

Data Formats

The HD DVD specification includes several file systems, such as ISO 9660 and Universal Disk Format (UDF). However, HD DVD video titles specify UDF version 2.5 for the file system. In this file system, multiplexed audio and video streams, as well as interactive features are stored together in the Enhanced Video Object (EVO) format. This means that advanced contents featuring interactive and network functions were developed and adopted for HD DVD discs so that options such as moving buttons/images and even simple games became available (DVD FORUM, 2005: 5.9).

Audio Formats

The HD DVD audio specification supports up to 24-bit/192 kHz for two channels and up to eight channels of 24-bit/96 kHz encoding (DVD FORUM, 2005: 4.4). HD DVD players are required to decode uncompressed linear PCM, Dolby Digital AC-3, Dolby Digital EX, DTS, Dolby Digital Plus E-AC-3, and Dolby TrueHD (DVD FORUM, 2007: 17). In addition, a secondary sound track may be encoded using one of the standard formats, or in one of the optional audio formats (i.e., DTS-HD High Resolution Audio and DTS-HD Master Audio). The audio is stored in the EVO container along with the video and the advanced interactive content. This makes digital rights and file management less complex than on DVD.

Video Formats

The HD DVD stream may include video encoded with the following formats VC-1, H.264/MPEG-4 AVC, or H.262/MPEG-2 Part 2 (DVD FORUM, 2007: 17). Many resolutions are supported, from low-resolution CIF to standard definition resolutions supported by DVD-Video, and extending to the HDTV formats available by 2006: 720p, 1080i, and 1080p. Many commercial studio-released titles were encoded with VC-1 or H.264/MPEG-4 AVC in 1080i or 1080p format, with supplemental video in 480i or 480p.

Digital Rights Management Security

HD DVD uses the content protection technology AACS (Advanced Access Content System) licensed by AACS LA (Advanced Access Content System Licensing Administrator) for digital rights management security. Unlike DVD, which uses a 40-bit encryption key to encrypt and decrypt its content, AACS has a 128-bit encryption key that is much longer and more difficult to decrypt. The AACS system uses a unique Device Key (different for each HD DVD player) to unlock a second key encrypted within a data block on the content disc. This second Media Key can then be used to decrypt the content. So, even if the disc contents are copied to another device or disc, they cannot be decoded unless the user can extract the key used for encryption. In this way, the recorded contents can be protected from illegal copying. In addition, AACS has a "revocation" feature to exclude the invalid device or disc. For example, AACS "Device Revocation" identifies the player if the encryption is broken and prohibits further playback. It also executes "Content Revocation" that invalidates playback of specified content (DVD FORUM, 2007: 22–25).

Additionally, HD DVDs were released using regional encoding, similar to standard DVDs and Blu-rays. These regions were as follows: Region A/1 covered North America, Central America, South America, Japan, North Korea, South Korea, Taiwan, Hong Kong, and Southeast Asia. Region B/2 held Europe, Greenland, French territories, Middle East, Africa, Australia, and New Zealand. Region C/3 included India, Nepal, Mainland China, Russia, Central and South Asia. As in the case of standard DVDs and Blu-rays, HD DVDs encoded for a given region are not playable in a player not manufactured to play that region's discs.

AACS has been hacked since its introduction in 2006. Hardware HD DVD players have been spoofed and decryption keys have been extracted from PC software players. At this time, computer software that circumvents the original AACS system is widely available. The AACS Licensing Authority introduced AACS2 to strengthen the system in 2012 (AACS), but it might have also been hacked by early 2017 (Silver, 2017; Osborne, 2016).

Interactive Features

HD DVDs use "Advanced Content" instructions and data to allow interactive content to be authored for discs. In particular, HD DVD discs employ a Microsoft Corporation implementation of the Advanced Content specification called the HDi Interactive Format (HDi). HDi features go beyond those of DVD discs to include "Advanced Navigation", "Advanced

Element", and "EVO". Advanced Element is a feature for a highly precise image, character information (text), and sound effects. For example, subtitles using high-resolution fonts are available. Advanced Navigation is a programming feature to enrich interactivity. It consists of "Playlist" and "Markup" data written in XML format and "Scripts" written in ECMA Script. The Playlist can control synchronization between the playback sequence and playback data, and the Markup can control graphics, character information, playback timing of sound effects, and layout. Scripts can control normal playback changes in response to the occurrence of events. By using Advanced Navigation, which allows controlling moving buttons and images through ECMA Script programming, simple games can be developed (DVD FORUM, 2005: 4.2–4.10).

Another aspect of Advanced Content provides for sub-video and sub-audio streams in the EVO file alongside the main video, audio, and sub-pictures (subtitles). This makes it possible for two different types of video and audio to be played at the same time. In addition to the above, Advanced Contents can be received via network in real time. On HD DVD, streaming content from a website while playing the disc contents can be done synchronously (DVD FORUM, 2007: 18).

Future of HD DVD

A decade after their demise, few feel any nostalgia for HD DVD technologies. Why might this be? Recall that they came into being from the competition between Japanese electronics firms to capture market share in the emerging high-definition optical disc market. Toshiba, through its leadership of the DVD Forum, tried to shut out Sony's Blu-Ray Disc Association, but was defeated twice in votes of the Forum's members. After some U.S. government pressure and changes to the voting rules, HD DVD was declared the Forum's official format, but it never received even a majority of votes from the DVD Forum. It was not a popular choice of the electronics industry or the content providers. Some of these were willing to support it at its early stage because of its official status (Universal Studios, Paramount Pictures, and Warner Bros Pictures). However, when consumers made it clear that they preferred Blu-ray discs, the retailers and content providers abandoned HD DVD (Williams 1/5/2008). At that point, "Toshiba said 1 million HD DVD players are currently in the market in North America" (Williams, January 6, 2008). With only a small number of consumers involved, and nearly all the premium content available on Blu-ray, its demise registered little impact on popular culture.

Immediately upon the discontinuance of HD DVD, all major U.S. studios announced support for Blu-ray, and shortly thereafter movies that had been available exclusively on HD DVD were re-released on Blu-ray. In Region 1/A, Paramount released 53 HD DVD titles; Universal, 148 titles; Warner Bros, 126 titles; Weinstein, 11 titles; and smaller content producers (Bandai, BCI, First Look, Magnolia, individual producers), 31 titles. Discovery Channel and other smaller documentary producers released 61 titles, while 45 HD DVD Audio titles were released by several labels. All told, a total of 475 titles were released in Region 1/A on the HD DVD format during the short two years it existed.[1] Of these, 34 were never released in the U.S. on Blu-ray discs, although some are available on HD DVDs released in Japan or Europe.[2] HD DVD titles have attracted some attention as collectables, especially those released exclusively for the format. The number of collectors is small, since the number of players was always small, and grows smaller every year as the machines wear out. However, they still search on web forums, eBay, and Half.com hoping to turn up one of the rare exclusives.

On the geeky end of consumer culture, HD DVD became a source of ridicule, and was almost always paired with the loser of the earlier videotape format war, Sony's Betamax. Jokes combining the two began early in 2008 when Toshiba discontinued the format, and the ThinkGeek specialty on-line store had a 2008 April Fool's advertisement for a "Betamax To HD-DVD Converter" (which is still retrievable on their website). Perhaps, beyond the jokes, the demise of these technologies might make us reflect on the Japanese aesthetic sense of *mono no aware*, that is, the ephemerality and impermanence of all human efforts, which even though they might be serenely beautiful in the moment, yet already are fading and failing.

Finally, HD DVD had a brief rebirth when the People's Republic of China chose it as the basis for its own proprietary high-definition disc system, China Blue High-definition Discs (CBHD; Lawler, 2008). Although CBHDs control the market for consumer high-definition video in China, and are still manufactured (Siglin, 2016), as of fall 2017 their sales are in decline in the face of streaming video services, from both within and without the PRC (Allyn-Feuer, 2009). Whether China will continue to manufacture these discs remains to be seen,[3] and is probably linked to the continued manufacture of Blu-ray, since they are meant to replace Western HD discs on the Chinese market. If CBHDs go out of manufacture, the last continuance of the once prominent HD DVD format will be gone.

Notes

1 Bruce Ames (AVS Forum Special Member) maintains a complete list of U.S. HD DVD titles on the AVS Forum website (last update, June 25, 2012).

2 tkca17 (AVS Forum Member), maintains a complete list of HD DVD exclusives on the AVS Forum website (last update, June 6, 2017).

3 The website for the China High Definition Association (CHDA) Organization as of August 15, 2017, no longer responds to Internet retrieval requests, so the status of the CBHD format cannot be directly ascertained. Their former website (retrieved August 15, 2017) can be seen at the Internet Web Archive.

References

AACS Licensing Authority. "AACS2 for Ultra HD Blu-ray™ is now available for Licensing." Retrieved March 9, 2018 from www.aacsla.com/license/

Allyn-Feuer, Ari. "Homegrown CBHD discs outsell Blu-ray by 3-1 margin in China," *Ars Technica* (August 8, 2009).

CNN. "Blockbuster backs Blu-ray" (June 18, 2007).

Dixon, Douglas. "Next generation DVD authoring," *DV* (February 2005): 58–64.

DVD FORUM. *DVD Forum White Paper – HD DVD Format Overview* (June, 2007).

DVD FORUM. *Requirements Specification for HD DVD Video Application Functional & Performance Requirements, Version 1.0* (July, 2005).

Federal Communications Commission (FCC). *Digital Television*. Retrieved March 9, 2018 from www.fcc.gov/general/digital-television

Fox, Barry. "Replacement for DVD unveiled," *NewScientist* (February 19, 2002).

Hara, Yoshiko. "HD DVD format wins key nod from DVD Forum," *techweb* (November 26, 2003).

Lawler, Richard. "China's Blu-ray competitor CBHD brings HD DVD back from the dead," *Engadget* (July 28, 2008).

Osborne, Charlie. "Pirates dance around AACS 2 encryption to offer UHD Blu-Ray movies online," *ZDNet* (June 13, 2017). www.zdnet.com/article/pirates-dance-around-aacs-2-encryption-to-offer-uhd-blu-ray-movies-online/

Siglin, Tim. "The China problem: government-sponsored formats are incompatible," *StreamingMedia.com* (November/December 2016).

Silver, Curtis. "Massive 4K UHD Blu-Ray torrent file appears online, has AACS 2.0 been cracked?" *Forbes* (May 5, 2017).

Sony Corporation. *Disclosure of Specifications for Large Capacity Optical Disc Recording Format Utilizing Blue-Violet Laser "Blu-ray Disc" Begins* (May 20, 2002). Retrieved March 9, 2018 from www.sony.net/SonyInfo/News/Press_Archive/200205/02-0520E/

Toshiba Corporation. *Toshiba Announces Discontinuation of HD DVD Businesses* (February 19, 2008). Retrieved March 9, 2018 from www.toshiba.co.jp/about/press/2008_02/pr1903.htm.

Williams, Martyn. "CES: Toshiba says HD DVD format isn't dead yet," *Computerworld* (January 6, 2008).

Williams, Martyn. "CES – Warner's Blu-ray Disc move has industry buzzing," *Computerworld* (January 5, 2008).

Williams, Martyn. "Toshiba, NEC share details of blue-laser storage," *PCworld* (August 29, 2002).

Yoshida, Junko. "Sides close to deal on HD disk format," *EETimesAsia* (April 19, 2005).

28

APPENDIX

Timeline of Obsolescence

Mark J. P. Wolf

Unlike "firsts", which do not change except in rare cases where earlier instances are discovered, many "lasts" can change if old technologies are reused after a period of dormancy or supposed death, and many never completely disappear from use. For example, while the production of daguerreotypes largely declined in the 1860s, they have never died out entirely, and there are still artists today who produce them; a 2009 exhibition in Bry Sur Marne, France, featured 182 images by 44 artists, and the 2013 ImageObject exhibition in New York City featured 75 images by 33 artists. Although shellac records stopped being made soon after World War II, 78 rpm records still occasionally reappear in special releases, and although they are usually made of vinyl, there is nothing to stop a company from producing a shellac record again. Just as technologies disappear when demand for them wanes, popular demand can cause an old technology to make a comeback, such as when Kodak Alaris announced on January 5, 2017 that Ektachrome would be returning in both 35mm and Super 8 formats. Thus, obsolescence might not mean death but only dormancy, or simply a fall from dominance that ends mass production and widespread usage. Below, undated entries appear at the beginning of the entry for each year, whereas dated ones are listed in order of occurrence. I would like to thank Jason Curtis for his help in making this list, and more information on some of the technologies mentioned below can be found at Curtis's online Museum of Obsolete Media (www.obsoletemedia.org), an excellent resource with over 525 samples of obsolete media in the museum's collection.

1906 The last commercially recorded Edison brown wax cylinders are produced in Europe. Edison Gold Moulded Records, which are also phonographic cylinders, replace them.

1908 The Russell Hunting Record Company ceases production of Sterling Records, which are black phonographic wax cylinders.

1912 Edison discontinues Gold Moulded Record phonographic cylinders, replacing them with longer-playing Amberol Records, which are also wax cylinders, and Blue Amberol Records, phonographic cylinders made of blue celluloid.

1914 Pathé ceases production of phonographic cylinders.

1922 Indestructible Records, phonographic cylinders made of celluloid, are no longer produced.

1929 **October**: Edison Records closes, and ceases production of all phonographic cylinders and discs.

1930 Orthochromatic nitrate film stocks Cine Negative Film Type E, Type F, and Super Speed Cine negative film are all discontinued.

1932 Pathé vertical-cut records are no longer sold in France.

1933 Kodak ceases production of Beau Brownie Box cameras. **February**: Durium records are no longer produced.

1935 Kodak ceases production of Brownie Box No. 2 cameras.

1941 Kodak Panchromatic Cine Film Type I (1203) is discontinued.

1946 Kodachrome color reversal film 5262 (16mm) is discontinued.

1950 Kodak discontinues producing nitrate film stocks and Kodachrome color reversal film 5265 (16mm).

1951 The United Kingdom ends the production of nitrate film stocks.

1952 Eastman Color Negative film 5247 16D is discontinued.

1956 Eastman film stock Background-X (1230) is discontinued.

1958 Kodachrome color reversal film 5268 (16mm) is discontinued.

1959 Peter Goldmark's Highway Hi-Fi, a record player for cars, ceases production. Eastman Color Negative film 5248 25T is discontinued.

1960 Ten-inch 78 rpm records are discontinued. The last 78 rpm record released in the UK was Elvis Presley's "A Mess of Blues" (1960). Pathescope Ltd. goes into liquidation, ending production of 9.5mm film in England.

1961 RCA ceases production of its record player for cars.

1962 Eastman Color Negative film 5250 50T is discontinued.

1964 RCA Sound Tape Cartridges, which are much larger than cassette tapes, are no longer available.

1968 Eastman Color Negative film 5251 50T is discontinued.

1970 Ampex ceases manufacture of 2-inch helical scan video tape recorders.

1972 IBM ceases production of its Magna-belt dictation system.

1973 **July**: After only a little more than a year on the market, Cartrivision machines are discontinued.

1975 The original Magnavox Odyssey home video game console is discontinued.

1976 After 30 years, Gray Manufacturing Company ceases production of Audograph dictation discs. **July 11**: Keuffel & Esser (K&E) Corporation produced the last slide rule to be manufactured in the United States, and the company presented it to the Smithsonian; the company would later declare bankruptcy in 1982.

1977 **March**: Eastman Color Negative film 5254/7254 100T is discontinued. **September 1**: After only a little more than a year in production, the original Apple I home computer is discontinued.

1978 After making slide rules for over a hundred years, European company ARISTO finally ends slide rule production. Telefunken abandons its Television Electronic Disc (TeD) home video system in favor of VHS videotapes. After only about a year in production, the RCA Studio II home video game console is discontinued. Retail availability of the Coleco Telstar home video game console ends.

1979 Polaroid discontinues the Polavision instant movie film format introduced just two years previously, and ends up writing off $68.5 million in Polavision expenses. **June 1**: The Apple II home computer is discontinued.

1980 Dictaphone ceases production of Dictabelt, a dictation recording format using a plastic belt. Sony abandons the Elcaset home tape recording system, and sells off remaining stock in Finland. The retail availability of Nintendo's Color TV-Game ends. **February 1**: Apple CEO Mike Scott sends out a company-wide memo that decrees that no typewriters should be in use at the company by the end of the year; only word processors would be used.

1981 Milton Bradley's Microvision, the first handheld video game console, is discontinued. **January**: The TRS-80 model I home computer is discontinued. **December 1**: The Apple III home computer is discontinued.

1982 **October 1**: The Apple SilenType computer printer is discontinued. **December 1**: The Apple II Plus, Apple II Euro Plus, and Apple II J-Plus home computers are all discontinued.

1983 The Fairchild Channel F home video game console is discontinued. The Bally Astrocade home video game console is discontinued.

1984 The Emerson Arcadia 2001 home video game console is discontinued. The GCE/Milton Bradley Vectrex home video game console is discontinued. **January 1**: The Apple Lisa home computer is discontinued. **March**: The Texas Instruments TI99/4a home computer is discontinued. **March 20**: The Magnavox Odyssey^2, the last system in the Odyssey line of home video game consoles, is discontinued. **April 24**: The Apple III Plus home computer is discontinued. **May 21**: After less than two years of production, the Atari 5200 home video game console is discontinued. **July**: The SEGA SG-1000 home video game console is discontinued.

1985 The ColecoVision home video game console is discontinued. **January**: The Commodore VIC-20 home computer is discontinued. **January 1**: The Apple Lisa II home computer is discontinued. **March 1**: The Apple IIe home computer is discontinued. **April 1**: The Apple Macintosh XL home computer is discontinued. **October**: The SEGA SG-1000 II home video game console is discontinued. **December 1**: The Apple ImageWriter printer is discontinued.

1986 *The Jewel of the Nile* (1985) is the last feature film to be released on Capacitance Electronic Discs (CEDs), a form of videodisc that recorded composite analog signals on vinyl discs. The Seeburg Background Music System, a background music system using phonograph records, is no longer available. **September 1**: The Apple IIc home computer is discontinued.

1987 **April 2**: The IBM Personal Computer is discontinued.

1988 Kodak no longer produces its disc camera or the Instamatic X line of cameras. Philips and Grundig end distribution of the Video 2000 home video format after losing out to VHS. The 3½-inch microfloppy disk outsells the 5¼-inch minifloppy disk. The Tandy 2000 home computer is discontinued. **November**: Fleetwood Mac's *Greatest Hits* (1988) becomes the last commercial 8-track tape released by a major record label.

1989 **August 1**: The Apple Macintosh SE home computer is discontinued.

1990 Radio Shack stops selling blank 8-track tapes. The Intellivision home video game console is discontinued. **January 15**: The Apple Macintosh II home computer is discontinued. **September 1**: The Apple IIc Plus home computer is discontinued. **October 15**: The Apple Macintosh Plus home computer is discontinued.

1991 Amstrad replaces the 3-inch "Compact Floppy" disk drive in its PCW word-processor product line, the last devices to use this size floppy disk, with a 3½-inch microfloppy disk drive.

1992 Compact Disc sales finally overtake cassette tape sales. The SEGA Master System home video game console is discontinued in North America, and the Atari XEGS home video game console is discontinued. **January 1**: The Atari VCS 2600 and the Atari 7800, both home video game consoles, are discontinued, and the software Apple Writer 1.0 is also discontinued. **September 14**: The Apple Macintosh Classic is discontinued. **December 1**: The Apple IIGS home computer is discontinued.

1993 NEC's PC-Engine home video game console is discontinued in France. **March 15**: Apple's Quadra 700, Macintosh IIsi, and Macintosh LCII are all discontinued. **September 13**: The Apple Macintosh Classic II is discontinued. **November 1**: The Apple IIe Platinum home computer is discontinued.

1994 **April**: Commodore's Commodore 64 home computer and Amiga CD32 are discontinued. **May**: NEC's PC-Engine home video game console, also known as the TurboGrafx-16 in North America, is discontinued in North America. **May 16**: Apple's Macintosh Color Classic is discontinued. **December 16**: NEC's PC-Engine home video game console is discontinued in Japan.

1995 The United States Social Security Administration stops using punched cards for data processing. **January 1**: Apple's QuickTake 100 digital camera is discontinued. **August**: Microsoft launches Windows 95, the first version of its operating system that is not available for retail on 5¼-inch floppy disks (though it was still possible to post a coupon to obtain it on 5¼-inch disks). Fujitsu's FM Towns Marty home video game console is discontinued. **August 14**: The Nintendo Entertainment System (NES) home video game console is discontinued in North America. **November 1**: Apple's Macintosh Color Classic II is discontinued.

1996 Kodak ceases production of Kodachrome 120 film stock. The Pioneer LaserActive, 3DO Interactive Multiplayer, Commodore Amiga 1200, and Atari Jaguar home video game console are all discontinued. **October**: Philips stops production of the Digital Compact Cassette (DCC), an intended successor to the analog cassette tape.

1997 The Apple Bandai Pippin home video game console is discontinued.

1998 Philips ceases publishing software for the CD-i (Compact Disc Interactive) format. QIC, the international trade association formed to promote the use of quarter-inch tape cartridge formats for data storage and backup, becomes inactive. The SEGA Saturn home video game console is discontinued in Europe and North America. **February**: NEC's PC-FX home video game console is discontinued. **February 27**: The Apple Newton handheld computer is discontinued. **August**: Apple launches the iMac computer with no floppy disk drive. **November**: SyQuest, producer of a line of removable rigid-disk storage formats, files for bankruptcy.

1999 The Super Nintendo Entertainment System (SNES) is discontinued in North America.

2000 The SEGA Saturn home video game console is discontinued in Japan.

2001 LG Philips Displays closes a cathode-ray tube plant in Dapon, Taiwan. **March 31**: The SEGA Dreamcast home video game console is discontinued. **September 14**: The Nintendo GameCube is discontinued in Japan. **October 12**: Polaroid Corporation files for Chapter 11 bankruptcy. **November 18**: The Nintendo GameCube is discontinued in North America.

2002 Sony stops producing Betamax video cassette recorders. **April 30**: The Nintendo 64 home video game console is discontinued in Japan. **May 3**: The Nintendo GameCube is discontinued in Europe. **May 17**: The Nintendo GameCube is discontinued in Australia.

2003 VM Labs' Nuon is discontinued. **April 1**: The Super Nintendo Entertainment System (SNES) is discontinued in South Korea. **May 16**: The Nintendo 64 home video game console is discontinued in Europe. **May 23**: LG Philips Displays announces that it is closing its Newport, Wales and Southport, England plants for cathode-ray tube manufacturing. **July 15**: Mitsubishi closes its cathode-ray tube plant in Mexicali, Mexico. **September 22**: Qualstar Corporation, the last manufacturer of 9-track open-reel computer tape drives, announces the final shipment. **September 25**: The NES and the SNES are both discontinued in Japan (where they are also known as the Nintendo Famicom and Super Famicom, respectively). **November 30**: The Nintendo 64 home video game console is discontinued in North America.

2004 Sony ends production of cathode-ray tubes in Japan, and Victor Company of Japan closes its CRT plant in Oyama, Tochigi Prefecture. **January**: Kodak announces that it will stop selling traditional film cameras in Europe and North America, and they cease production of cameras using APS (Advanced Photo System) cartridge film and their entire range of slide projectors. **October 22**: Kodak makes its last carousel slide projector.

2005 Kodak discontinues Kodachrome Super 8 film stock. **March 31**: The Sony PlayStation home video game console is discontinued. **September**: Sony announces that it will be closing down its cathode-ray tube production worldwide. **December**: Sony stops production of Digital Audio Tape (DAT) recorders.

2006 Thomson Consumer Electronics closes its Picture Tube Development Group in Lancaster, Pennsylvania; and Sony closes its CRT centers in Pittsburgh and San Diego. Kodak ceases production of 16mm Kodachrome movie film stock. **January 27**: Western Union sends its last telegram. **June 4**: Microsoft's Xbox home video game console is discontinued in Japan. **June 30**: Microsoft marks Windows 98 for end-of-life.

2007 Around the end of the year, sales of LCD-based TVs finally overtake those of CRT-based TVs. **March 11**: Microsoft's Xbox is discontinued in Europe.

2008 The mass production of piano rolls ends as MIDI files replace them in player pianos. JVC, the company that invented the VHS format, ceases production of standalone VHS video cassette recorders (VCRs). **February 8**: Polaroid announces it will stop producing film for its instant cameras by the end of the year (the Dutch company The Impossible Project is subsequently formed to continue manufacture of some instant films). **February 18**: Analog cellular phone service ceases nationally in the United States. **February 19**: Toshiba announces it will no longer manufacture or market HD DVD players or disc drives, ending the format war with Blu-ray. **March 3**: Sony announces it is ending all production of cathode-ray tubes. **May 27**: Paramount's *Into the Wild* (2007), Warner Bros.' *P.S. I Love You* (2007) and *Twister* (1996) become the last films from major movie studios in the United States to be released on HD DVD.

2009 Manufacture of standard manual office typewriters finally ends when the Indian company Godrej & Boyce announces it will no longer produce them. **January**: Pioneer ceases production of its remaining LaserDisc players. **March**: The PCMCIA (Personal Computer Memory Card International Association), formed to promote standards for and use of PCMCIA memory cards, PC Cards, and the ExpressCard in laptops, is dissolved. **March 2**: Microsoft's Xbox home video game console is discontinued. **June 12**: Analog full-power over-the-air broadcast television ends. **June 22**: After 74 years of production, Kodak announced that it would cease production of Kodachrome color film by the end of the year, due to a decline in sales.

2010 Sony, the last company to commercially manufacture 3½-inch magnetic disks, ceases production of them. The last new car to be factory-equipped with a cassette deck in the dashboard is a 2010 Lexus SC 430. Olympus follows Fuji and switches to Secure Digital (SD) memory cards for its digital cameras, making the xD-Picture Card format, which they jointly developed, obsolete. **September 5**: Anthem Films' *Deadlands: The Rising* (2006), made in a limited run, becomes the last film to be released on HD DVD. **October 22**: Sony ceases selling the Walkman analog cassette player in Japan. **December 30**: Dwayne's Photo, the last store to process Kodachrome, stops accepting rolls of Kodachrome for processing at noon, Parsons, Kansas time.

2011 The term "cassette tape" is removed from the *Oxford English Dictionary*. The Mac Mini no longer has a built-in optical disc drive. **January 18**: The last role of Kodachrome is developed (a roll shot by Dwayne Steinle, owner of Dwayne's Photo, the last store to sell Kodachrome). **February 4**: Kodak announces the discontinuance of Ektachrome 200 on its website. **April 12**: Cisco announces that it is shutting down its Flip Video division.

2012 The last typewriter made in the United Kingdom, an electronic Brother, rolls off the assembly line in Wrexham. **January 19**: Kodak files for Chapter 11 Bankruptcy Protection. **February 9**: Kodak leaves the digital image capture business, phasing out production of digital cameras. **March 1**: Kodak announces the discontinuance of three color Ektachrome films. **August 24**: Kodak announces that it will sell its film, commercial scanner, and kiosk divisions. **September 28**: Kodak announces that it is leaving the inkjet printer business; and Cathode Ray Technology in the Netherlands, the last cathode-ray tube factory in Europe, closes. **December**: Kodak announces that it is discontinuing Ektachrome 100D color reversal movie films 5285 and 7285. **December 20**: Due to bankruptcy, Kodak plans to sell its digital imaging patents for about $525 million. **December 28**: The Sony PlayStation 2 is discontinued in Japan.

2013 Kodak announces it will stop making celluloid acetate for movie film, and *Anchorman 2: The Legend Continues* becomes Paramount's last movie to be released on 35mm film. The Nintendo Wii home video game console is discontinued. **January 4**: The Sony PlayStation 2 is discontinued worldwide. **March**: Sony ends shipments of MiniDisc stereo systems. **July 14**: India's state-run telecommunications company, Bharat Sanchar Nigam Limited (BSNL), the last large-scale telegraph system in the world, sends its last telegram. **October**: The 13-inch MacBook Pro becomes the last Apple product to include a built-in optical disc drive. **December**: By the year's end, all Kodak Ektachrome products are discontinued.

2014 **February**: 92% of the 40,045 theatrical movie screens in the U.S. have already converted to digital. **June 3**: Sony announces that it will discontinue the PlayStation Portable, the only device that uses Sony's own UMD (Universal Media Disc) optical disc that could contain video games, feature-length films, or music. **July 30**: Kodak negotiates with movie studios for an annual movie film order guarantee to preserve the last source of movie film manufacturing in the United States.

2015 Laptops no longer come with VGA connectors. **September 1**: All analog television transmitters are shut down in the United States, by order of the U.S. Government. **September 9**: Apple's iPhone 6 and iPhone 6 Plus are discontinued. **September 29**: The Sony PlayStation 3 is discontinued. **October 1**: OCLC, the Online Computer Library Center, Incorporated, a nonprofit cooperative association that maintains WorldCat, prints its last custom catalog card.

2016 **March**: Sony discontinues its remaining ½-inch professional video tape recorders, including DigiBeta, MPEG IMX, HDCAM, and HDCAM SR formats. **April 20**: Microsoft's Xbox 360 is discontinued. **July**: Japanese company Funai Electronics, the last company to make VCRs, ceases production of them. **October**: The Sony PlayStation 3 is discontinued in North America. **October 27**: Apple discontinues the non-Retina legacy MacBook Pro, the last Apple computer to come with a built-in CD/DVD drive.

2017 **January 5**: Kodak Alaris announces that Ektachrome will return, in 35mm and Super 8 formats, sometime during the fourth quarter of 2017. **January 31**: The Nintendo Wii U is discontinued. **April 17**: Google Spaces is discontinued. **May 29**: The Sony PlayStation 3 is discontinued in Japan. **July 27**: Apple discontinues the iPod Nano and the iPod Shuffle. **October 8**: Work on Windows 10 Mobile ends due to lack of market penetration and interest from app developers. **October 31**: Amazon in the UK ends its LoveFilm by Post DVD and Blu-ray disc rental service citing the decreased demand for disc rental due to streaming. **December 15**: America Online officially discontinues AOL Instant Messenger. Compuserve's Forums are discontinued.

INDEX

Locators in *italics* refer to figures.